Damped Wave Transport and Relaxation

Damped Wave Transport and Relaxation

Professor Kal Renganathan Sharma

School of Chemical and Biotechnology
Shanmugha Arts, Science, Technology & Research Academy
SASTRA Deemed University, Thanjavur, India

2005

ELSEVIER

Amsterdam – Boston – Heidelberg – London – New York – Oxford
Paris – San Diego – San Francisco – Singapore – Sydney – Tokyo

phys
614299781

ELSEVIER B.V.
Radarweg 29
P.O. Box 211, 1000 AE Amsterdam
The Netherlands

ELSEVIER Inc.
525 B Street, Suite 1900
San Diego, CA 92101-4495
USA

ELSEVIER Ltd.
The Boulevard, Langford Lane
Kidlington, Oxford OX5 1GB
UK

ELSEVIER Ltd.
84 Theobalds Road
London WC1X 8RR
UK

First edition 2005

Library of Congress Cataloging in Publication Data
A catalog record is available from the Library of Congress.

British Library Cataloguing in Publication Data
A catalogue record is available from the British Library.

ISBN: 0 444 51943 2

⊗ The paper used in this publication meets the requirements of ANSI/NISO Z39.48-1992 (Permanence of Paper).
Printed in The Netherlands.

Dedication

This book is dedicated to my son Chi. R. Hari Subrahmanyan Sharma (born August 13th 2001)

Preface

The mid 17th century saw the introduction of Newton's law of viscosity and the publication of his *Philosophiae Naturalis Principia Mathematica;*. in 1822 Fourier unveiled his theory, including his law of heat conduction; and in 1855 Fick proposed the first law of diffusion. Since then, numerous developments in fundamental physics have come about and a whole gamut of elucidations has become popular with many physicists and scientists, such as Einstein's theory of relativity with light having the highest velocity, the characterization of Brownian dynamics by Chandrashekar and von Kampen, and Schroedinger's wave equation to describe the movement of molecules in a box. Incorporating this understanding into the universal laws of engineering has, however, yet to happen.

Transient problems in transport phenomena have a variety of applications encompassing drug delivery systems for chemotherapy in bioengineering, heat transfer to surfaces in FBC, fluidized bed combustion boilers in mechanical engineering, simultaneous reaction and diffusion problems in the Zeigler-Natta gas phase, polymerization of polypropylene in chemical engineering, and polyacrylamide gel electrophoresis in bioinformatics. The attention given to transient problems in the leading textbooks currently in vogue represents only a small proportion of the broader heat, mass and momentum transfer discipline - often only 7% of the entire text, and not commensurate with their importance in industry. Often, the problems treated provide a Fourier series solution of a parabolic partial differential equation (PDE) and imply a certain weariness of the student.

In general, the Fourier series does not fully describe all the transient events of significance, for example: the short contact time singularity in heat flux in the widely used surface renewal theories, and the loss of universality when the heat transfer through a small film is considered. In one research problem, i.e., in the heat transfer in fluidized beds to immersed surfaces, for over 50 years, investigators have developed models for overprediction of the surface renewal theory to experiment at short contact times. New resistances were introduced, adding empiricism models that were otherwise derived from first principle models.

They seldom achieve a level of scrutiny higher than the continuum and empirical linear laws that describe forces with flows. Often, the premise of steady state is used. With increased computer resources, more fundamental phenomena can be accounted for and solved. The evaluation of the damped wave equation and the relaxation equation of heat, mass and momentum to represent transient events is one example of such an endeavor. The Euler equation and Navier-Stokes equations can be extended to include the relaxational transport term. At a molecular level, when the heat flux is defined as the energy leaving the surface minus the energy of the molecules entering the surface, it can be modified to include an accumulation of energy term. This may manifest only during transient problems where the accumulation of heat flux or temperature becomes important. In a similar fashion, the mass flux can be modified to include the accumulation of mass, and the shear stress or momentum flux can be modified to include the accumulation of momentum. The happenings in time are as important as those in space, especially in transient problems.

As discussed by Newton, when the apple falls from the tree, the net acceleration decreases to zero at the terminal settling velocity of the apple, due to the changing drag force as a function of velocity. As the velocity increases from zero at rest, the drag force increases proportional to the velocity, changing the resultant force from the difference of gravity and the Archimedes buoyancy force to the resultant gravity minus buoyancy minus drag force. So the rate of acceleration, which can be calculated as the ratio of the resultant force to the mass of the apple, is a pronounced phenomenon during transient events in fluid flow, heat transfer and mass transfer, which is not incorporated in the current theoretical depiction of transient events in the industry.

The myth of a Clausius inequality violation clouded early attempts in the literature to account for the accumulation effects using the equation that came originally from Maxwell and during the mid 20th century from Cattaneo and Vernotte in France. The mere introduction of three terms in the governing equation can lead to the "theoretical possibility" of temperature gradient and heat flux being of the same sign and thus effecting a heat flow from low to high temperature. The second law of thermodynamics is not violated by transient phenomena. Sometimes an improper perspective of the interpretation of model solutions leads to the impression of an inherent flaw in the wave equation with a damping term. At steady state, using Fourier's law and a temperature-dependent heat source, negative temperatures can arise in the solution which need to be interpreted as zero temperature and not a violation of the third law of thermodynamics.

Our primary goal is to encourage the depiction of transient phenomena with a higher level of scrutiny than Fourier's, Fick's and Newton's laws and to seek a connection with molecular phenomena. A case in point is the use of damped wave transport and relaxation equations of heat, mass and momentum. The solution methodologies used to obtain meaningful solutions are: relativistic transformation of coordinates, method of separation of variables, Laplace transforms, and method of complex temperatures. Bounded solutions without any violation of the second law of thermodynamics can be seen. Physical insight is sometimes preferred to mathematical rigor.

The conditions under which subcritical damped oscillations can be found are derived for a finite slab, a cylinder and a sphere and the results depicted in figures. For the evaluation of wave equation effects, the damping term is first removed from the hyperbolic PDE. The solution exhibits symmetry in space and sometimes in time. A zone of zero transfer can be detected in a Zeigler-Natta catalyst during simultaneous reaction and diffusion. The storage coefficient is defined to evaluate the relative contributions of thermal mass and thermal relaxation. New dimensionless groups have been introduced, such as the momentum number, accumulation number, oscillation number, modified Peclet number, heat, mass and momentum, modified Biot number, Fourier modulus, Fick modulus, permeability number, storage number, dimensionless pressure, temperature, concentration, force, stress, heat flux, mass flux and velocity, penetration length, penetration time, velocity of heat, velocity of mass and the velocity of momentum. The thermal lag time associated with realizing a heat disturbance in the interior of a slab, the exterior of a cylinder and a sphere is calculated and expressions provided.

Bessel equations are used extensively in the text and a summary of the relations used, including the generalized Bessel equation, is given as Appendix A. The commonly found inversions of Laplace transforms are available as Appendix B and the reader is referred to numerical inversions of Laplace transforms for expressions not found in Appendix B. Appendix C contains the extended Navier-Stokes equation of motion with the accumulation term included. Experimental evidence for the relaxation time in heat transfer has fallen in the range of a few seconds in the case of dispersed biological materials. This is much higher than the few nanoseconds projected by early investigators of the phenomenon. Chapter 1 provides an introduction to the damped wave transport and relaxation equation. Heat, mass and momentum transfer problems are dealt with separately in Chapters, 2, 3, and 4.

Case studies and applications are discussed in Chapter 5.

Acknowledgements

In early 1987, after the funeral of my paternal grandmother Sankari patti, I met Richard Turton, the last Ph.D student of Emeritus Professor Octave Levenspiel, who introduced me to the problem of overprediction of theory to experiment in fluidized bed to surface heat transfer. I revisited the problem of the incomplete description by Fourier to represent transient heat transfer, with some encouragement from R. Sethuraman (Vice Chancellor of SASTRA Deemed University, Thanjavur), S. Swaminathan (Dean, Students), S. Vaidya Subramaniam (Dean Planning and Development), K.N. Somasekharan (Dean, School of Chemical and Biotechnology).

Special mention is made of Victoria Franchetti Haynes Ph.D, director at Monsanto Plastics. My thoughts and prayers go to Najma Dalal MD. My maternal grandmother, Bagyalakshmi patti, will always remain in my memory. Many thanks to my parents, Sri. S. Kalyanaraman and Mrs. Shyamala Kalyanaraman, for their unfailing support and to Dr. Vidya Krishnan.

Funds for overseas travel to international conferences obtained from several sources are acknowledged, with special mention made of John W. Zondlo, West Virginia University, Morgantown, WV, USA (for travel to Maastricht, the Netherlands to present a paper at the 1987 Conference on Coal Science) and T. R. Rajagopalan Research Cell, SASTRA Deemed University, Thanjavur, India. Funds were received from Nason Pritchard to travel to the World Congress in Chemical Engineering at San Diego, CA in 1996 with some encouragement by Nithiam T. Sivaneri, Associate Chair of Mechanical & Aerospace Engineering at West Virginia University, Morgantown, WV. I was sent to Germany and Italy in 1995 to discuss the construction of the microgravity transport cell from Clarkson University, Pottsdam, NY, USA. Special thanks to Prof. Edward Wegmann and the faculty at the Physics Department at George Mason University for seeing me through 62 conference presentations in 1999, including 17 at the Dallas AIChE meeting. The encouragement of R. Srinivasan and Dr. V. S. Balameenakshi enabled me to present nearly 100 papers at 20 meetings between 1996 and 2000.

Special thanks are due to: Sri. Rangarajan, Chairman and Correspondent, Vel Institutions, Avadi, Chennai; V. Punnusami, the intranet system administrator at the computer laboratory in Vignana Vihar; and Elsevier for working with me on the proofs during the preparation of the manuscript. The blessings of his Holiness Sri. Jayendra Saraswati swamigal and His Holiness Pujyasri Sankara Vijayendra Saraswati swamigal of the Kancheepuram Sankara Mutt through the years are deeply valued.

About the Author

Kal Renganathan Sharma has received three degrees in chemical engineering: a B.Tech from the Indian Institute of Technology, Madras, India, MS and Ph.D from West Virginia University Morgantown, WV, USA. He received unsurpassed marks in mathematics in the X std CBSE examination in 1979. He has held a professional engineer's license from the state of New Hampshire since 1995. His post-doctoral training was with: Monsanto Plastics, R. Shankar Subramanian, former chair and professor at Clarkson University, Potsdam, NY; and Ramanna Reddy, Professor and Director, Concurrent Engineering and Research Center, West Virginia University, Morgantown, WV. He has authored 370 conference papers, eight journal articles, and 54 preprints on the Chemical Preprint Server. This is his third book. Positions he has held include: Adjunct Assistant Professor, Mechanical & Aerospace Engineering, West Virginia University, Morgantown, WV; Visiting Research Assistant Professor, George Mason University, Fairfax, VA; President & Chief Technical Officer, Independent Institute of Technology, Hanover, NH; Principal, Sakthi Engineering College, Chennai; Professor & HOD, Computer Science and then Chemical, Vellore Engineering College, Vellore. Honors received include the Guruvayurappan Award for the most number of papers, "Who's Who" in Science and Engineering, 2004-2005, New York Academy of Sciences, Phi Kappa Phi, Sigma Xi. He is currently a Professor at SASTRA Deemed University, Thanjavur, TN, India.

Table of Contents

1.0 The Damped Wave Conduction and Relaxation Equation

Nomenclature

a	half-width of slab (m)
A	cross-sectional area of the slab (m^2)
C_p	heat capacity (J/kg/K)
d	diameter of molecule (m)
D_{AB}	binary diffusivity (m^2/s)
E	energy coefficient (J/m^2/K)
E'	energy coefficient (J/m/K)
F_j	jth force
G	phonon electron coupling factor
h	heat transfer coefficient (W/m^2/K)
J_k	kth flow
k	thermal conductivity (W/m/K)
L_{kj}	phenomenological coefficients
l	penetration length (m)
m	mass of molecule (kg)
n'	number of molecules per unit volume
P	arbitrary distribution
P_b	ballistic part
P_m	diffusive part
q	heat flux (W/m^2)
t	time (s)
T	temperature (K)
t*	characteristic time
\<u\>	mean molecular speed (m/s)
u	dimensionless temperature for finite medium (T − T$_s$)/(T$_0$ − T$_s$)
u'	dimensionless temperature for semi-infinite medium (T − T$_0$)/(T$_s$ − T)
u"	dimensionless temperature at zero initial temperature (T /T$_s$)
U_y	fluid velocity in the y direction (m/s)
v_h	velocity of heat sqrt(α /τ_r)
x	space (m)

Greek

α	thermal diffusivity (m^2/s)
β	characteristic length on the microscale, (m) (β = E/ρC$_p$)
κ	Boltzmann constant
μ	viscosity (kg/ms)
ρ	density (kg/m^3)
τ	dimensionless time (t/τ_r)
τ_r	relaxation time (s), heat
τ_{mom}	relaxation time (s), momentum

τ_{mr} relaxation time (s), mass

Dimensionless Groups

Be	Storage Number (Area) (ratio of energy mobility to energy storage) $(E/\rho C_p l)$
Be'	Storage Number (Length) (ratio of energy mobility to energy storage) $(E/\rho C_p l^2)$
X	dimensionless distance $x/L\,sqrt(\alpha\tau_r/L^2)$
Fo*	Modified Fourier Number $(\alpha\tau_r/L^2)$
Pe	Peclet number, ratio of convection fluid velocity to heat wave conduction velocity $(U_r/sqrt(\alpha/\tau_r))$
L	dimensionless penetration length $1/\,sqrt(\alpha\,\tau_r)$
q*	dimensionless heat flux $(q/sqrt(k\rho C_p/\tau_r)/(T_s - T_0)$ modified Stanton number $(h/\rho C_p v_h)$
θ	Fourier number $(\alpha t/a^2)$

1.1 Origin

Phenomenological relations that are well established such as Fourier's law of heat conduction, Fick's law of mass diffusion, Newton's law of viscosity, Ohm's law of passage of electrical current are valid at steady state and applicable for common situations. Several technologists, investigators have found that the Fourier series representation does not fully describe all the transient events. Mathematically the infinite series is poorly convergent when applied to certain conditions. The short time limit of the instantaneous surface flux during the transient heat conduction in a semi-infinite medium at an initial temperature at zero subject to a step change in the wall temperature can be seen to be infinite. This has lead to over prediction of theory to experiment when the surface renewal theories have been applied to applications such as fluidized bed heat transfer to surfaces.

Other applications such as in chromatography, CPU overheating, circulating fluidized bed heat transfer to tubes in contact, chemisorption, simultaneous fast reaction and finite speed diffusion in Zeigler Natta catalyst, non-Fick concentration profiles found during silicon doping in VLSI/SLSI, are some examples where the finite speed of wave transmission may become more important than accounted for. When the cross-effects can be neglected the phenomenological laws are linear in nature. They relate the forces to flows. The force for example is the gradient of (1/T) and the flow is the heat flux q during heat transfer. The gradient of temperature not only drives the heat conduction but also has a secondary effect on the mass diffusion and electric phenomenon and so on and so forth. These are the cross-effects. During the Seebeck effect, two dissimilar metal wires are joined and the junctions are maintained at different temperatures. As a result an EMF is generated of the order of 10 μvolts/Kelvin and depends on the sample. In Peltier effect, the current flow drives a heat transfer from one junction to another maintained at the same temperature initially. The Dufour and Soret effects capture the effect of concentration difference on heat flow and thermal gradient on mass diffusion. Phenomenological laws have constants of proportionality such as the thermal conductivity, k in heat conduction, mass diffusion coefficient, D_{AB} in mass diffusion, v the kinematic viscosity in momentum transfer and electrical conductivity in Ohm's Law. A general relation between flows and forces is assumed and is stated as follows;

$$J_k = \sum L_{kj} F_j \qquad (1.1)$$

A unified formalism for flows and forces was provided by Prigogine and Kondepudi, [1998]. They write the entropy production term and for conditions where the linear phenomenological laws are valid the expression takes a quadratic form. A matrix of the phenomenological coefficients was said to be positive definite. The well characterized properties of positive definite matrix were used. The diagonal elements of a positive definite matrix must be positive. A necessary and sufficient condition for matrix L_{ij} to be positive definite is that its determinant and all the determinants of lower dimension obtained by deleting one or more rows and columns must be positive. Thus according to the second law of thermodynamics, the proper coefficients, L_{kk} must be positive. The cross-coefficients can have a plus or minus sign. The elements of the matrix of phenomenological coefficients can be shown to obey the Onsager reciprocal relations. The reciprocal relations [Onsager, 1931] state that,

$$L_{ij} = L_{ji} \qquad (1.2)$$

These relations were noted by Lord Kelvin in the 19th century. But it took Lars Onsager to provide a well-founded theoretical explanation for these relations. His theory is based upon the principle of detailed balance or microscopic reversibility that is valid for systems at equilibrium. He wrote an expression for the entropic change associated with the fluctuations. According to the Einstein's formula and the entropy associated with the fluctuations the Maxwell probability distribution for the fluctuations was derived in terms of the Boltzmann and normalization constants. The fluctuations were assumed to decay according to the linear laws. Using the principle of detailed balance the Onsager reciprocal relations can be obtained. Forces and flows are coupled in general. But they are restricted by the general symmetry principle. Macroscopic causes always have fewer or equal symmetries than the effects they produce.

From a microscopic view, the thermal conductivities of dilute monatomic gases can be predicted by kinetic theory [Bird et al, 1960]. This derivation is revisited here. Consider molecules to be rigid, non-attracting spheres of mass m and diameter, d. The gas is assumed to be at rest and the molecular motion is considered. The following results of kinetic theory for a rigid sphere dilute gas in which temperature, pressure and velocity gradients are small are used;

Mean Molecular Speed $\quad <u> \ = \ sqrt(8\kappa T/\pi m)$ $\qquad$ (1.3)

Wall Collision Frequency $\quad Z \ = \ \frac{1}{4} n' <u>$ $\qquad$ (1.4)
per unit area

Mean free path $\qquad \lambda \ = \ 1/sqrt(2)/(\pi d^2 n')$ $\qquad$ (1.5)

The molecules reaching any plane in the gas have on an average had their last collision at a distance a from the plane, where

$$a \ = \ 2/3 \ \lambda \qquad (1.6)$$

One form of energy that can be transferred upon collisions of molecules is the translational energy. This energy under equilibrium conditions can be written as;

$$\frac{1}{2} m <u^2> \ = \ 3/2 \ \kappa \ T \qquad (1.7)$$

The heat capacity per mole at constant volume, is thus,

$$C_v \ = \ N \ d/dT \ (\frac{1}{2} m <u^2> \ ) = 3/2R \qquad (1.8)$$

Where R is the universal gas constant. This has been found to be satisfactory for monatomic gases up to temperatures of several thousand degrees. In order to determine the thermal conductivity, consider the behaviour of the gas subject to a temperature gradient dT/dy. The heat flux q_y, across any plane of constant y, is *found by summing the kinetic energies of the molecules that cross the plane per unit time in the positive y direction and subtracting the kinetic energies of the equal number that cross in the negative y direction.*

$$q_y \ = \ Z \ [\frac{1}{2} m <u^2>]_{y \cdot a} \ - \ Z \ [\frac{1}{2} m <u^2>]_{y+a} \qquad (1.9)$$

$$= \ 3/2 \ \kappa \ Z \ (\ T|_{y-a} \ - \ T|_{y+a}) \qquad (1.10)$$

It may be assumed that the temperature profile is essentially linear for a distance of several mean free paths. Molecules have velocity representative of their last collision. Accordingly,

$$T|_{y-a} \qquad = \ T|_y \ - \ 2/3 \ \lambda \ dT/dy \qquad (1.11)$$

$$T|_{y+a} \qquad = \ T|_y \ + \ 2/3 \ \lambda \ dT/dy \qquad (1.12)$$

Substituting Eqs. [1.11,1.12] into Eq. [1.10],

$$q_y \qquad = \qquad -1/2n' \ \kappa \ <u> \ \lambda \ dT/dy \qquad (1.13)$$

Equation [1.13] corresponds to the Fourier's Law of heat conduction and the thermal conductivity is then written as;

$$k \qquad = \ \frac{1}{2} n' \kappa <u> \lambda \ = \ 1/3 \ \rho \ C_v <u> \lambda \qquad (1.14)$$

It can be seen that prior to writing Eq. [1.9] the *accumulation of energy* in the incremental volume near the surface considered was neglected. This may be a good assumption at steady state but may not be during short time transient events. Thus considering a time increment t* the heat flux is the energy of molecules out, minus energy of molecules in, minus the accumulation of energy near the surface. The accumulation of energy may be written in terms of the molecular velocity and collision frequency as,

$$q_y \quad = \quad Z \left[\tfrac{1}{2} m <u^2> \right]_{y-a} - Z \left[\tfrac{1}{2} m <u^2> \right]_{y+a}$$
$$- t^* \partial / \partial t \{ Z \left[\tfrac{1}{2} m <u^2> \right]_{y-a} - Z \left[\tfrac{1}{2} m <u^2> \right]_{y+a} \} \qquad (1.15)$$

where t* is some characteristic time constant. To simplify matters Eq. [1.9] is used in Eq. [1.15] to give,

$$q_y = 3/2 \, \kappa \, Z \, (T|_{y-a} - T|_{y+a}) \ - t^* \, \partial q_y / \partial t \qquad (1.16)$$

Assuming a linear temperature gradient the heat flux then becomes;

$$q_y \ = -k \, \partial T / \partial y \ - t^* \, \partial q_y / \partial t \qquad (1.17)$$

Eq. [1.17] is the Cattaneo and Vernotte non-Fourier heat conduction and relaxation equation when t* is given by τ_r, the relaxation time. Morse & Feshbach [1953] hypothesized and Cattaneo [1958] and Vernotte [1958] argued for a more physically reasonable wave speed heat conduction and relaxation equation. This is given by Eq. [1.17]. The problem is severe especially at short contact times and at extreme temperatures where the relaxation time is high such as at cryogenic temperatures for materials with a positive coefficient in the temperature variation with thermal conductivity and hot temperatures for materials with negative coefficient in the temperature variation with thermal conductivity. The thermal wave theory is proposed to remove the paradox in classical conduction model assuming an infinite speed of heat propagation. There have been some discussions about this paradox in the literature among others by Biot [1970]. The thermal relaxation time variation with temperature may vary from some materials to some other materials. Some cases, the relaxation time may increase with decrease in temperature and in other cases the relaxation time may decrease with decrease in temperature.

This equation was originally introduced by Maxwell [1867]. His work on kinetic theory of gases and momentum transfer through the collision of molecules inspires the consideration of finite speed of heat conduction in solids. According to the wave model of heat conduction and relaxation the heat flux equilibrates to the imposed temperature gradient by a relaxation phenomenon of lattice vibrations. The diffusion term dampens the wave propagation. The relaxation time may also be viewed as a weighted effect of the two. For zero relaxation times, the Fourier model remains and at infinite relaxation times the governing equation reverts to the wave equation. It can be shown that when the dissipative or accumulation term exceeds in value compared with exp(τ) dominance of wave phenomenon over diffusion can be seen [Ozisik and Tzou, 1994]. When the temperature increases the lattice vibrations can be expected to increase and thus decreasing the relaxation time.

1.2 Wave Dominant and Fourier Regimes

The Eq. [1.17] is a non-Fourier heat conduction and relaxation equation. Noting the three terms in the equation, some investigators suspected that the heat flux term, q and the thermal gradient term $\partial T/\partial y$ may have the same sign under certain conditions [Bai and Lavine, 1995]. In the Fourier expression both terms will have to have opposite signs for staying within the Clausius inequality. The heat flows only from a higher temperature to a lower temperature. A negative gradient along with a negative sign as a prefix gives a positive heat flux indicating heat flow from higher to lower temperature. If the signs of the heat flux and the temperature gradient are the same then the heat is described as going from lower to higher temperature. This is a second law violation. In order to stay within the second law the heat flux and thermal gradient must have opposite signs in Eq. [1.17]. For other cases, the equation does not depict heat conduction and relaxation but only then it is a mathematical artefact. Same signs for thermal gradient and heat flux will happen only when the flux accumulation is extremely large. In these cases it will later be shown that the heat flux will lag or experience inertia and be nearly zero that the thermodynamics laws will not be violated. For the cases where there is a second law violation it can be shown that the corresponding heat transfer situation is physically unrealistic by another consideration. Eq. [1.17] can be combined with the energy balance in 1 dimension to give the governing equation as follows.

Energy balance Equation: $\quad -\partial q_y/\partial y = (\rho C_p)\partial T/\partial t$ $\qquad$ (1.18)

Differentiating Eq. [1.18] wrt to time and Eq. [1.17] wrt to y and eliminating the cross derivative term $\partial^2 q/\partial t \partial y$ between the two equations the governing equation becomes:

$$k\,\partial^2 T/\partial y^2 = (\rho\,C_p\,\tau_r\,)\,\partial^2 T/\partial t^2 + (\rho C_p)\,\partial T/\partial t \qquad (1.19)$$

$$\alpha\,\partial^2 T/\partial y^2 = \tau_r\,\partial^2 T/\partial t^2 + \partial T/\partial t \qquad (1.20)$$

Eq. [1.20] is also called the telegraph equation. It is the damped wave conduction and relaxation governing equation. It is a hyperbolic partial differential equation and can be referred to as WRENCH, Wave Relaxation Equation for Non-Fourier Conduction of the hyperbolic type. The decay of a RLC circuit can also be given by the above equation. The transverse vibration in a string is another phenomenon that can be depicted by this expression.

Let $\partial T/\partial t \quad = p$ $\qquad$ (1.21)

Then when $\tau_r\,\partial p/\partial t \gg p$ Or when $\ln(\,p\,) \gg t/\tau_r$ or when $p \gg \exp(t/\tau_r)$ or when $\partial T/\partial t \gg \exp(t/\tau_r)$, Eq. [1.20] becomes:

$$\alpha\partial^2 T/\partial y^2 = \tau_r\,\partial^2 T/\partial t^2 \qquad (1.22)$$

Eq. [1.22] can be recognized as the wave equation [Ozisik and Tzou, 1994]. Thus the term $p = \partial T/\partial t$ is the damping term that dampens the wave. Thus when the accumulation term is very large, larger than an exponential rise, the governing equation for heat conduction is given by the wave equation. On the other hand, when the accumulation term is much smaller compared with the exponential rise, the governing equation reverts to the parabolic Fourier transient heat conduction equation and is given by;

$$\alpha\, \partial^2 T/\partial y^2 = \partial T/\partial t \qquad (1.23)$$

Eq. [1.20] is the governing equation for the transient temperature in 1 dimensional heat conduction and relaxation. A similar equation in heat flux can be derived by eliminating the second cross derivative in temperature, i.e., $\partial^2 T/\partial t\partial y$, between Eqs. [1.17, 1.18] to give;

$$\alpha\, \partial^2 q_y/\partial y^2 = \tau_r\, \partial^2 q_y/\partial t^2 + \partial\, q_y/\partial t \qquad (1.24)$$

When $\partial q_y/\partial t \gg \exp(t/\tau_r)$ wave dominated heat conduction and relaxation can be expected to occur. Or when $q_y \gg (\tau_r)\exp(t/\tau_r) + c'$ the wave dominance will prevail. In Figure 1 the instantaneous time rate of surface heat flux from a surface renewal model using the Fourier parabolic transient heat conduction equation, along with the exponential rise in time is shown. Regimes of wave dominance and reversion to Fourier equation is shown. A cross-over point can be identified. For short time scales, the wave equation can describe the time events and for reasonably long times the Fourier's regime is dominant. For intermediate values, the damped wave equation in its entirety need be solved.

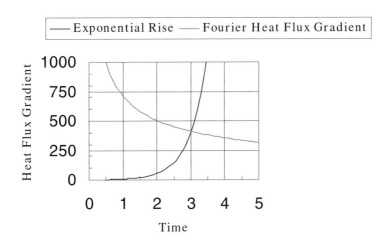

Figure 1.1: **Wave Dominant and Fourier Regimes of Heat Conduction and Relaxation**

8

Worked Example 1.1

The thermal conductivity is given to vary linearly with temperature for some metals as follows;

$$k = k_0 (1 - \beta T) \qquad (1.25)$$

Derive the governing equation in one dimension in terms of temperature as given in Eq. [1.20] and obtain the regimes of wave dominant transfer. The non-Fourier wave heat equation including the temperature variation with thermal conductivity is ;

$$q_y = - k_0 (1 - \beta T) \partial T/\partial y - \tau_r \partial q_y /\partial t \qquad (1.26)$$

β is the coefficient of variation of thermal conductivity with temperature. For some materials the thermal conductivity decrease with increase in temperature. One such example is copper. For some other materials the thermal conductivity increase with increase in temperature such as in Aluminum. The energy balance equation in 1 dimension with no heat source or sink is rewritten from Eq. [1.18] as;

$$- \partial q_y/\partial y = (\rho \, C_p) \, \partial T/\partial t \qquad (1.27)$$

Differentiating the constitutive equation wrt to y and the energy balance equation wrt to time and eliminating the second order cross-derivative of heat flux wrt to time and space between the two equations the governing equation in terms of temperature in 1 dimension becomes:

$$(\rho \, C_p) \, \partial T/\partial t + (\rho \, C_p \tau_r) \partial^2 T/\partial t^2 = k \, \partial^2 T/\partial y^2 - k_0 \, \beta \, (\partial T/\partial y)^2 \qquad (1.28)$$

Obtaining the non-dimensionalized form of the equation by the following substitutions;

$$u = (T - T_s)/(T_0 - T_s) ; \quad \tau = (t/\tau_r) ; \quad Y = y/\mathrm{sqrt}(\alpha \, \tau_r) \qquad (1.29)$$

$$\partial u/\partial \tau + \partial^2 u/\partial \tau^2 = \partial^2 u/\partial Y^2 - \alpha_* \, \beta^* \, (\partial u/\partial Y)^2 \qquad (1.30)$$

When $\partial u/\partial \tau \gg \exp(\tau)$, $\partial^2 u/\partial \tau^2 \gg \partial u/\partial \tau$ and in space when;

$$\partial^2 u/\partial Y^2 \gg -\alpha_* \, \beta^* \, (\partial u/\partial Y)^2 \qquad (1.31)$$

$$\text{Let } (\partial u/\partial Y) = m \qquad (1.32)$$

$$\text{Then; } \partial m/\partial Y \gg -\alpha_* \, \beta^* \, (m)^2 \qquad (1.33)$$

$$\text{Or } m = (\partial u/\partial Y) \gg 1/ (\alpha_* \, \beta Y) \qquad (1.34)$$

The governing equation will revert to the wave equation;

$$\partial^2 u/\partial \tau^2 \quad = \partial^2 u/\partial Y^2 \qquad (1.35)$$

Thus the regime of wave dominance occurs at high rate of heat accumulation when, $\partial u/\partial \tau \gg \exp(\tau)$, and when the thermal gradient in space, $\partial u/\partial Y \gg 1/(\alpha_* \beta * Y)$. Now in the regime of wave dominance it can be inferred that

$$\partial u/\partial \tau = -\alpha_* \beta * (\partial u/\partial Y)^2 \qquad (1.36)$$

Although this PDE is nonlinear, the functions will separate out by the method of separation of variables. This equation can be seen to be valid when the dimensionless temperature varies with the reciprocal of dimensionless time and the square of the space dimensions.

Worked Example 1.2

Consider the convection term in the energy balance equation. Derive the governing equation for temperature and discuss the regimes of wave and Fourier dominance.

The energy balance equation in 1 dimension including the convection term can be written as;

$$- \partial q_y/\partial y - (\rho\, C_p) U_y \partial T/\partial y = (\rho\, C_p)\, \partial T/\partial t \qquad (1.37)$$

where U_y is the constant velocity of the fluid flowing in the y direction. Differentiating the above equation wrt t

$$- \partial^2 q_y/\partial y \partial t - (\rho C_p) U_y \partial^2 T/\partial y \partial t = (\rho\, C_p)\, \partial^2 T/\partial t^2 \qquad (1.38)$$

Differentiating the non-Fourier damped wave expression as given in Eq. [1.17] wrt to y,

$$\partial q_y/\partial y = -k\, \partial^2 T/\partial y^2 - \tau_r\, \partial^2 q_y/\partial y \partial t \qquad (1.39)$$

Eliminating the second order cross derivative of the heat flux wrt to time and space between the differentiated energy balance constitutive law equations;

$$-(\rho C_p)\partial T/\partial t-(\rho C_p)U_y \partial T/\partial y = -k\, \partial^2 T/\partial y^2 + (\rho C_p \tau_r)\partial^2 T/\partial t^2 + (\rho C_p \tau_r)U_y\, \partial^2 T/\partial y \partial t \quad (1.40)$$

Using the dimensionless variables as defined before in Eq. [1.29], the equation becomes;

$$\partial^2 u/\partial Y^2 = \partial u/\partial \tau + \partial^2 u/\partial \tau^2 + Pe \partial^2 u/\partial Y \partial \tau \qquad (1.41)$$

where the ratio of the velocity of the fluid to the velocity of the wave heat transmission can be defined as a Peclet number. Thus $Pe = U/v_h$. For the wave conditions to dominate,

$$\text{Say } \partial u/\partial \tau = p; \text{ Then when } p = -Pe \; \partial p/\partial Y \qquad (1.42)$$

$$\text{Or } p = C'\exp(-Y/Pe)$$

Or When $\partial u/\partial \tau = C''\exp(-Y/Pe)$, the governing equation will revert to a wave equation. Another way of arriving at the wave dominant regime is when the heat accumulation term, $\partial u/\partial \tau \gg \exp(\tau)$, the wave term $\partial^2 u/\partial \tau^2 \gg \partial u/\partial \tau$ and when the thermal gradient, $\partial u/\partial Y \gg Pe \; \partial u/\partial \tau$ the cross derivative $Pe \partial^2 u/\partial Y \partial \tau$ can be neglected compared with the flux balance term $\partial^2 u/\partial Y^2$ the wave dominant equation is reverted to from the governing equation that contains the convection, conduction and wave relaxation terms.

1.3 Boundary Layer in Time and Scaling of the Governing Equation

Let $\delta(x)$ be the time taken for a point inside the medium to realize the presence of the thermal disturbance imposed at the boundary. At $x = 0$, δ is zero and monotonically increase with increase in x. δ has units of time and is a function of x only. At the boundary the thermal disturbance is seen instantaneously and subsequently as x increases there is a growth in the "boundary later in time". $\delta(x)$ [Renganathan, 1990] can be used as a scaling parameter to evaluate the relative magnitudes of different terms in the governing equation. Consider the governing relation given in Eq. [1.20] for temperature in 1 dimension obtained using the non-Fourier heat conduction and relaxation equation without convection.

$$\text{Let } \gamma = t/\delta(x) \qquad (1.43)$$

Eq. [1.20] becomes;

$$1/\delta(x) \, (\partial T/\partial \gamma) + \tau_r /\delta^2 \, (\partial^2 T/\partial \gamma^2) = \alpha \, (\partial^2 T/\partial x^2) \qquad (1.44)$$

Multiplying throughout the equation by δ,

$$(\partial T/\partial \gamma) + \tau_r /\delta \, (\partial^2 T/\partial \gamma^2) = \alpha \delta \, (\partial^2 T/\partial x^2) \qquad (1.45)$$

For regions close to the boundary,

$$(\tau_r /\delta) \, (\partial^2 T/\partial \gamma^2) \gg \alpha \delta \, (\partial^2 T/\partial x^2) \qquad (1.46)$$

This is especially true when the velocity of heat is small, So the governing equation becomes;

$$(\partial T/\partial \gamma) + \tau_r /\delta \, (\partial^2 T/\partial \gamma^2) = 0 \qquad (1.47)$$

$$\text{or } \tau_r \, \partial^2 T/\partial t^2 \quad + \quad \partial T/\partial t \qquad = 0 \qquad (1.48)$$

This is a second order ODE in time and can be solved for. For a semi-infinite medium at an initial temperature T_0 subject to a constant wall temperature for times greater than zero, T_s.

$$(T - T_0)/(T_s - T_0) \ = \ C_1 (\, 1 \, \text{-} \, \exp(t/\tau_r)) \qquad (1.49)$$

When the δ is large the Fourier heat equation of the parabolic type reverts and the error function can be written as;

$$(T - T_0)/(T_s - T_0) \ = \ (\, 1 \, \text{-} \, \text{erf}(x/\text{sqrt}(4\alpha t)) \qquad (1.50)$$

1.4 Poor Convergence of Fourier Series

The Fourier series representation of transient temperature that is currently in use is revisited for a finite slab. The surface flux is found to be poorly convergent. The same kind of problem is also found for the error function depiction of transient temperature for the semi-infinite medium. In this section, tangible examples are shown where the mathematical description of Fourier conduction is unbounded. Later in the text, applications are shown where the unbounded depiction lead to over prediction of theory and incomplete understanding of the physical processes underlying the phenomena. This along with the empirical nature of Fourier's law of heat conduction gives impetus to evaluate non-Fourier expressions to describe heat conduction.

Case 1: Unbounded Surface Flux in a Finite Slab of Thickness 2a - CWT

Consider a finite slab of thickness 2a at an initial temperature of T_0 subject to a constant wall temperature of T_s at either edge for times greater than zero. The Fourier's law of heat conduction in 1 dimension can be written as;

$$q_x \ = \ \text{-}k \, \partial T/\partial x \qquad (1.51)$$

The energy balance, in 1 dimension, can be obtained for a thin slice of cross-section area A and thickness Δx at a distance x from the origin at the center of the slab for a time interval Δt, as follows;

(Energy in) - (Energy out) $\pm$ (Energy Sink/Source) = Energy Accumulated

$$A \, q_x|_x \, \Delta t \quad \text{-} \quad A q_x|_{x+\Delta x} \, \Delta t \qquad = (\rho \, C_p) \, \Delta T \, A \, \Delta x \qquad (1.52)$$

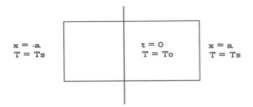

Figure 1.2 Finite Slab of Thickness of 2a Subject to CWT

Dividing throughout Eq. [1.52] by $\Delta x \Delta t$ and in the limit of Δt and Δx and going to zero the energy balance equation is obtained as;

$$-\partial q_x/\partial x = (\rho\, C_p)\, \partial T/\partial t \qquad (1.53)$$

The governing equation, for 1 dimensional heat conduction, in terms of transient temperature as a function of space and time, according to Fourier's law, can be obtained by combining the Fourier's law and the energy balance equation. This is achieved by differentiating Eq. [1.51] with respect to x and substituting the result into Eq. [1.53] to yield;

$$\alpha \partial^2 T/\partial x^2 = \partial T/\partial t \qquad (1.54)$$

The governing relation given by Eq. [1.54] can be non-dimensionalized by the following suppositions. Let, $u = (T - T_s)/(T - T_0)$; $\theta = \alpha t/a^2$; $X = x/a$ and Eq. [1.54] becomes;

$$\partial u/\partial \theta = \partial^2 u/\partial X^2 \qquad (1.55)$$

The Fourier series solution for the parabolic partial differential equation in two variables given by Eq. [1.55] can be obtained by the method of separation of variables. The dimensionless temperature can be represented as a product of two functions of 1 variable each.

$$u = V(\theta)\, \phi(X) \qquad (1.56)$$

V is a function of dimensionless time only and ϕ is a function of dimensionless space only. Eq. [1.31] after separating the terms containing V and the terms containing ϕ on either side of the equation, becomes;

$$V'/V = \phi''/\phi = -\lambda_n^2 \qquad (1.57)$$

where λ_n^2 is a constant. The choice of the negative sign for the constant is to stay consistent with the decaying exponential in time of transient temperature. Solving the two ordinary differential equations shown in Eq. [1.57];

$$\phi = c_1 \, Sin(\lambda_n X) + c_2 \, Cos(\lambda_n X) \qquad (1.58)$$

$$V = c_3 \, exp(-\lambda_n^2 \theta) \qquad (1.59)$$

It can be expected that the transient temperature profile will be symmetrical wrt to the x = 0 axis in the finite slab. Hence at.

$$X = 0, \quad \partial u/\partial X = 0 \qquad (1.60)$$

$$Or \quad c_1 \, \lambda_n \, Cos(\lambda_n X) + c_2 \, \lambda_n \, Sin(\lambda_n X) = 0 \qquad (1.61)$$

In order to obey Eq. [1.36] it can be seen that c_1 need to be set to zero. Thus,

$$\phi = c_2 \, Cos(\lambda_n X) \qquad (1.62)$$

The boundary condition arises from the condition that for times greater than zero the wall temperature of either edge were maintained at T_s.

$$At \, X = 1, \quad u = 0 = c_2 \, VCos(\lambda_n) \qquad (1.63)$$

$$Thus, \, \lambda_n = \pi/2, \, 3\pi/2, \, 5\pi/2, \, 7\pi/2.......= (2n-1)\pi/2 \,, \quad n = 1,2,3,4... \quad (1.64)$$

These are the Eigen values and problems of this type are generally referred to as Eigenvalue problems [Ozisik, 1985]. The transient dimensionless temperature is given by an infinite series,

$$u = \sum_1^\infty c_n \, Cos(\lambda_n X) \, exp(-\lambda_n^2 \theta) \qquad (1.65)$$

The constant c_n can be obtained by using the orthogonality property. The initial condition in time is used (u = 1, θ = 0) and the LHS and RHS of the equation multiplied by $Cos(\lambda_m X)$. Then either sides of the equation are integrated in the limits between 0 and a and the orthogonality property is invoked. If two solutions of the eigenvalue problem, one for the eigenvalue λ_n, and another for λ_m are multiplied and then integrated over the region from X = 0 to 1, the resulting integral vanishes if $\lambda_n \neq \lambda_m$ and equals a constant N if $\lambda_n = \lambda_m$. The normalization integral is evaluated and the constant becomes;

$$C_n = \int_0^1 Cos(\lambda_m X) \, dX \, / \int_0^1 Cos^2(\lambda_m X) \, dX = 4(-1)^{n+1}/(2n-1)\pi \qquad (1.66)$$

The transient temperature is then;

$$u = \sum_1^\infty 4(-1)^{n+1}/(2n-1)\pi \, Cos(\lambda_n X) exp(-\lambda_n^2 \theta) \qquad (1.67)$$

The surface flux at X =1 is obtained as;

$$q' = (aq_x)/(k(T_0 - T_s) = u = \sum_1^\infty 4(-1)^{n+1}/(2n-1)\pi \, (\lambda_n) Sin(\lambda_n) exp(-\lambda_n^2 \theta) \quad (1.68)$$

$$or \, q' = \sum 2(-1)^{n+1} Sin(\lambda_n) exp(-\lambda_n^2 \theta) \qquad (1.69)$$

In the short time limit of $\theta = 0$,

$$q' = 2(1 + 1 + 1 + 1 + \ldots\ldots) = \text{unbounded} \qquad (1.70)$$

Thus the Fourier series representation of transient temperature predicts an infinite heat flux in the zero time limit at the surface.

Case 2: Unbounded Surface Flux in a Semi-Infinite Medium Subject to a Step Change in Temperature using Fourier's Law

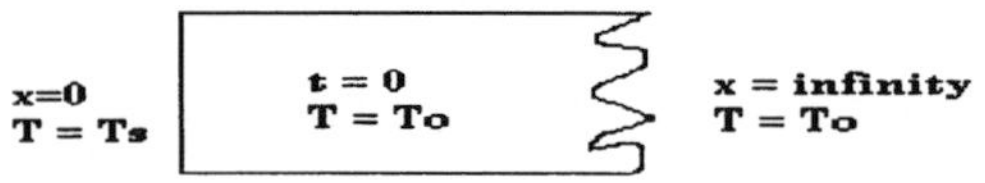

Figure 1.3 Semi-Infinite Medium with Step Change in Temperature at Surface

The governing equation for heat conduction in a semi-infinite medium using Fourier constitutive law is given by Eq. [1.54] and is;

$$\alpha \partial^2 T/\partial x^2 = \partial T/\partial t \qquad (1.71)$$

Let $u' = (T - T_0)/(T_s - T_0)$, then;

$$\alpha \partial^2 u'/\partial x^2 = \partial u'/\partial t \qquad (1.72)$$

The initial and boundary conditions are as shown in Figure 1.3;

$$t = 0, \quad u' = 0 \text{ for all } x \qquad (1.73)$$

$$t > 0, \quad x = 0, \quad u' = 1 \qquad (1.74)$$

$$t > 0, \quad x = \infty, \quad u' = 0 \qquad (1.75)$$

A similarity variable, $\eta = x/\text{sqrt}(4\alpha t)$ is defined and the governing equation transformed into the new variable as follows;

$$\partial u'/\partial t \quad = -1/4 \partial u'/\partial \eta \ (x) \ t^{-3/2}/\alpha^{1/2} = -\eta/2t \partial u'/\partial \eta \qquad (1.76)$$

$$\alpha \partial^2 u'/\partial x^2 \quad = \partial^2 u'/\partial \eta^2 \ (1/4t) \qquad (1.77)$$

Thus Eq. [1.71] becomes,

$$-2\eta \ \partial u'/\partial \eta \ = \ \partial^2 u'/\partial \eta^2 \qquad (1.78)$$

Integrating both sides,

$$\partial u'/\partial \eta = c' \exp(-\eta^2) \tag{1.79}$$

$$\text{or } u' = 1 - \text{erf}(x/\text{sqrt}(4\alpha t)) \tag{1.80}$$

The transient temperature in a semi-infinite medium subject to a step change in temperature at one of the walls using the Fourier's law and the method of similarity transformation is given as a error function solution in Eq. [1.80]. The surface flux is given by;

$$q_x = k(T_s - T_0)/\text{sqrt}(\pi \alpha t) \tag{1.81}$$

In the short time limit, the surface flux given by Eq.[1.81] is unbounded and approaches infinity. This has resulted in a number of phenomena where the critical parameter of interest is the short times such as in the heat transfer coefficient between [Renganathan and Turton, 1998] fluidized beds and immersed surfaces, the theory over predicts the experimental values. Thus in this section, atleast two tangible examples are given where mathematically the Fourier series representation of transient temperature gives singularities.

1.5 Light Speed Barrier and Finite Speed of Heat Propagation

According to the Einstein's theory of relativity the speed of light is the speediest of velocities. When one approaches the speed of light the classical Newton laws need to be rewritten. The Fourier heat conduction equation as given in Eq. [1.71] in one dimension has the property that any perturbation in temperature is propagated with infinite speed. As an illustration, consider the heat conduction in an infinite medium subject to the initial condition:

$$T = T_p \delta(x) \tag{1.82}$$

Where δ is the Kronecker delta function. The transient temperature to this initial value problem can be obtained by Laplace transforms and is given by;

$$T = T_p 2(\pi \alpha t)^{1/2} \exp(-x^2/4\alpha t) \tag{1.83}$$

On examining the solution for the transient temperature it can be seen that for times greater than zero the temperature is finite at all points in the infinite medium except at infinite location [Landau and Lifshitz, 1959]. It can be inferred that the heat pulse has travelled at infinite speed. This is in conflict with the light speed barrier stated by the theory of relativity of Einstein. The fact that any speed of a moving object, including the thermal wave, must be less than the speed of light was examined by Kelly [1968] for diffusion. By using the covariant form of the Boltzmann transport equation, he showed that a finite value of thermal diffusivity cannot co-exist with an infinite value of thermal wave speed based on the same velocity of molecular collisions.

1.6 Hyperbolic, Parabolic and Elliptic PDE

Partial differential equations, PDEs, arise when the number of independent variables is more than one. One important application of PDEs is the Fourier heat equation. It consists of two independent variables in one dimensional problem, i.e., time and space. The space can be represented by three variables for the three space coordinates and the PDE may have upto four independent variables. An *m dimensional, n^{th} order* PDE is written in terms of m independent variables, one dependent variable such as, u, the dimensionless temperature. All of the partial derivatives of the dependent variable, u, are written wrt to the independent variables such as t, x, y, z, etc up to order n. The solution for u is a continuous function with all the nth order derivatives also continuous. When the solution, when substituted back obeys the PDE the solution is said to be the general solution. The general solution contains n arbitrary functions where n is the order of the PDE.

Compared with the ODE where the general solution contains n constants the general solution of PDE contains n functions. These can be determined by using boundary conditions in space and domains of integration using the initial and final time conditions. In addition to the overall order of the PDE, orders of the independent variables can also be noted. Thus for example in Eq. [1.54] the overall order of the PDE, n is two. The order of the space dimension is two and that of the time component is one. In another two examples, in Eqs. [1.20, 1.41] the order of the PDE, n is two and the order in space and time dimensions are both two. The number of conditions at the boundary and domain of integration is equal to the order in space and time domain respectively. Thus in order to completely solve Eq. [1.54], two boundary conditions in space and one condition in time, usually the initial condition, is needed. For Eqs. [1.20, 1.41] two boundary conditions in space and two time conditions are needed. The use of time conditions is discussed in later sections to depict the heat conduction and relaxation process in a physically realistic manner. The use of the final condition at steady state along with the initial condition is used. The PDE can be linear, quasilinear and nonlinear. Let the general nth order PDE be written as;

$$\phi \left(u, t, x,y,z...., \partial u/\partial t, \partial u/\partial x, \partial u/\partial y, \partial u/\partial z, \; \partial^n u/\partial t^k \partial x^{n-k},.. \; \partial^n u/\partial t^n, \partial^n u/\partial x^n,.. \; \partial^n u/\partial z^n \right) = 0$$

When the function, ϕ, is a polynomial of degree one wrt u and to each of its partial derivatives of any order the PDE is said to be *linear*. When ϕ, is a polynomial of degree one wrt the partial derivatives of u of the highest order, the PDE is said to be *quasilinear*. In other cases the PDE is *non-linear* [Varma and Morbidelli, 1997].

Consider a general second order PDE with space and time as the independent variables of the form;

$$A \, \partial^2 u/\partial x^2 + 2B\partial^2 u/\partial x \partial t + C\partial^2 u/\partial t^2 + D\partial u/\partial x + E\partial u/\partial t + F = 0 \qquad (1.84)$$

A determinant Z is defined as;

$$Z = \begin{bmatrix} A & B \\ B & C \end{bmatrix} \qquad (1.85)$$

When the determinant, Z is zero the PDE is said to be *parabolic*. When the determinant Z is less than zero the PDE is said to be *hyperbolic*. When the determinant Z is positive, the PDE is said to be *elliptic*.

For example, when Eq. [1.71] is written in the form shown in Eq. [1.84], A = α, B = C = 0 and E = -1. The determinant Z as defined in Eq. [1.85] is then zero. Hence the Fourier heat equation is said to be parabolic. The Cattaneo and Vernotte non-Fourier damped wave heat conduction and relaxation equation as given in Eq. [1.20] when written in the generalized second order form as in Eq. [1.84] gives the following;

$$A = -\alpha; \quad B = D = F = 0; \quad C = \tau_r, \quad E = 1 \qquad (1.86)$$

The determinant $Z = -(\alpha \tau_r)$. As the thermal diffusivity and relaxation time values are always positive for real materials the Z is negative. Hence Eq. [1.20] is hyperbolic. An example of the elliptic equation is the Laplace equation. The heat conduction at steady state in two dimensions can be written as;

$$\partial^2 u/\partial x^2 + \partial^2 u/\partial y^2 = 0 \qquad (1.87)$$

Comparing Eq. [1.87] with the general form for second order PDE given in Eq. [1.84], A = 1, B = D = E = F = 0 and C = 1. The determinant Z is 1. This is positive and the Eq. [1.87] is said to be elliptic in nature.

1.7 Non-Fourier Equations for Heat Conduction

Impetus to evaluate Non-Fourier expressions to describe heat conduction is given by the realization of the empirical nature of Fourier's law of heat conduction, the occurrence of mathematical singularities of the Fourier series representation of transient heat conduction in important conditions, physically unrealistic implication of infinite speed of propagation of heat and over prediction of theory to experiment in important industrial process such as fluidized bed heat transfer to surfaces, chromatography, CPU overheating, adsorption, gel electrophoresis, restriction mapping, laser heating of semiconductors during manufacture of semi-conductor devices and drug delivery systems. The Cattaneo and Vernotte equation generalizes the heat conduction phenomena and the Fourier's law becomes a special case when the relaxation time becomes zero. Other forms of non-Fourier heat conduction is the ballistic transport equations due to Chen, [2001], dual phase lag model by Tzou [1997] and the EPRT, equation of phonon transport proposed by Majumdar, [1993], microscopic two-step model suggested by Qiu and Tien [1992]. An infinite order PDE was presented by Sharma [2003] using Taylor series expansion to fully describe the transient events. A third order PDE was suggested by Sharma [2003t21].

Fourier's law breaks down also at small scales. The empirical law that relates the temperature gradient to the heat flux through a proportionality constant called the thermal conductivity needs a meaningful definition of temperature. What would be the thermal gradient in small systems where the mean free path of the molecules is larger than the thickness of the medium of transmission ? Temperature at a point is usually defined under a local thermodynamic equilibrium [Bejan, 1988] and realistically the defined location should be greater than the mean free path λ. When the thickness of the medium is much greater than the mean free path of the molecules the thermal gradient is established and the thermal conductivity can be defined. When the thinness of the medium is comparable or smaller in dimension compared with λ, the heat flux across the film can no longer be described by Fourier's law. This is known as the Casimir limit [1938] and the flux is described by an expression similar to the one used in radiation heat transfer [Swartz and Pohl, 1989]. The mechanism is attributed to phonon transport. Phonon is the energy of the waves quantized. The heat transport, for example, in dielectric crystalline materials is believed to be primarily by atomic or crystal vibrations. These vibrations travel as waves and the energy of the waves quantitated is the phonon, [Kittel, 1986]. Mazumdar [1993] developed the EPRT, the equation of phonon radiative transfer. The physics of crystal vibrations was considered. The energy of the crystal vibrations was quantitated in packets of $2\pi h\omega$, where h is the Plack's constant and ω is the frequency of the crystal vibrations. Each quantum of energy was a phonon. The statistics of phonons in thermodynamic equilibrium followed the Bose-Einstein distribution. The Boltzmann equation that can be used to study any transport phenomenon, is given as a function of an arbitrary distribution, P.

$$\partial P/\partial t + v \cdot \nabla P + a \cdot \partial P/\partial v = \partial P/\partial t \big|_{scatt} \qquad (1.88)$$

where, a is the acceleration of the particle or phonon or item of interest, P is the distribution function which depends on the particle position, r, velocity, v, and time t. The speed of phonons is nearly that of sound and does not change significantly over a wide frequency range. As a result, the term, $\cdot \partial P/\partial v$, can be neglected. The $v \cdot \nabla P$ represents the advection or drift of the distribution. The rate of change of P due to scattering is given by $\partial P/\partial t \big|_{scatt}$. The solution of P needs to be obtained. Mazumdar [1993] concluded from his study of EPRT that for steady state heat conduction the EPRT equation reduces to the Fourier's law when the film thickness is much greater than the mean free path of the molecule. For transient heat conduction, the EPRT behaves like the wave equation indicating that a heat pulse is transported as a wave that gets damped within the medium by phonon scattering. In the acoustically thick limit, the Cattaneo and Vernotte hyperbolic heat equation can be derived from the transient EPRT. In the acoustically thin limit of heat conduction across the thin film, where the probability of phonon scattering within the film is very low, the EPRT yields the familiar law of radiative transfer where the difference in temperature raised to the power 4 is needed to estimate the heat flux. Bai and Lavine [1991] applied hyperbolic heat equation to thin film superconductors.

The mechanism of ballistic transport and nonlocal phenomena was attempted to be included in the constitutive law for heat conduction by Chen [2001]. Ballistic transport mechanism is suspected to play a larger role when the thermal gradient is large and/or when the system under consideration is small. They established the ballistic-diffusive equations as an approximation to the transient Boltzmann equations. The Boltzmann equations are considered to be difficult to solve. They recommend numerical tools to solve these equations. They write the distribution function, P as a sum of two components,

$$P = P_m + P_b \qquad (1.89)$$

The P_b, originates from the boundary and serves as the ballistic part and the P_m originates from the internal domain and gives the diffusive part. The temperature also is written as a sum of two components, the ballistic and diffusive parts to the internal energy. They add a ballistic term to the Cattaneo and Vernotte equation [1.20]. They calculate the ballistic and diffusive heat fluxes separately.

The Cattaneo and Vernotte non-Fourier damped wave conduction and relaxation equation and the thermal wave theory was proposed to remove the paradox in the Fourier conduction model assuming an infinite speed of propagation. The equation is given by;

$$\partial^2 T/\partial y^2 = (\tau_r/\alpha)\partial^2 T/\partial t^2 + (1/\alpha)\partial T/\partial t \qquad (1.90)$$

where the velocity of heat conduction is given by;

$$v_h = (\alpha/\tau_r)^{1/2} \qquad (1.91)$$

Eq. [1.69] becomes the Fourier's parabolic heat equation when the velocity of heat is infinite. The Cattaneo and Vernotte equation was found to be admissible within the framework of the second law of thermodynamics [Tzou, 1993]. The relaxation time was suggested to be viewed as the time-lag between the heat flux vector and the temperature gradient when the response time is short. It is also viewed as the physical instant at which the intrinsic length scale in conduction transits to that in wave. The relaxation time is related to macroscopic properties such the wave speed and thermal diffusivity by Eq. [1.70]. Some investigators have related the relaxation time to the electron-phonon collisions and the volumetric heat capacities of the electrons and the metal lattice [Ozisik and Tzou, 1992].

Molecular collision models on the basis of quantum mechanics and statistical mechanics have been applied to the Fourier paradox. The kinetic Boltzmann transport equation is used to extract macroscopic quantities such as temperature from the distribution of molecules. The velocity and frequency of molecular collisions are used to obtain the thermal diffusivity and wave velocity. [Chester, 1963, Cheng, 1989]. Advances in the thermodynamic consideration for the wave behaviour in heat conduction include the extension of the Onsager thermodynamics. [Luikov, 1966]. Weymann [1967] showed that the infinite propagation velocity in heat conduction results from neglecting the atomistic structure of matters. When such effects were accounted for, the modified form of heat equation takes the hyperbolic form of the damped wave conduction and relaxation equation.

The response in time of the relaxation time has been studied by considering 1 dimensional problems in space. Baumeister and Hamill [1969, 1971] studied the temperature wave in a semi-infinite solid subject to a step change in wall temperature. The interface temperature for two suddenly contacting media was studied by Fauske [1973]. The rate dependent response in temperature waves was studied by Chen and Amos [1970]. The effect of time rate of change of thermal wave speed was incorporated by Luikov [1976]. The temperature wave across the thin film media was studied by Letcher [1969] and Kao [1977]. The characteristic lines in thermal wave propagation were studied by Wiggert [1977]. Carey and Tsai [1982] made an attempt for the finite difference formulation. Cylindrical waves were considered by Wilhelm and Choi [1975] in metals.

Joseph and Preziosi [1989, 1990] introduced notions of an effective thermal conductivity and an effective heat capacity to interpret the relaxation behaviour of heat and energy. Ozisik and Vick [1984] predicted the growth and decay of a thermal pulse in one-dimensional solid. The temperature ripple propagating with a finite speed has finite height and width. Frankel et al [1987] studied the reflection of thermal waves by interfaces in composite media. They demonstrated that the flux-formulation in the thermal wave theory is more convenient to use for problems involving flux special boundary conditions. Glass [1986, 1987b, 1987c] studied the effect of thermal conductivity change with temperature on the propagation of thermal waves for a pulsed energy source under different boundary conditions in a semi-infinite medium. They also looked at absorbing and emitting medium. The non-Fourier response of a solid subjected to an oscillatory surface flux was found to correspond to the practical situation of irradiation of a solid by a pulse laser. Non-Fourier effect can be an important event even after a "long time", after the initial transient if the thermal disturbance is oscillatory with the period of oscillation of the same order of magnitude as the thermal relaxation time.

Kim [1990] studied the axisymmetric pulse surface heat source using two-dimensional hyperbolic heat equation. The thermal shock formation around a fast-moving heat source [Tzou, 1989 1990c, 1991a] and a rapidly propagating crack tip [Tzou, 1990a, 1990b] provided similar situations compared with those in high speed aerodynamics. Subsonic, transonic, and supersonic temperature waves were found to exist in transition of the thermal Mach number. They showed that the thermal energy tend to accumulate in a preferential direction around a rapidly moving heat source. Temperature contours surrounding the heat source in the subsonic region was found to be similar to those in diffusion. In transonic and supersonic region, normal and oblique shock waves were found to exist. For Mach number great than or equal to one, the shock surface was found at a angle of $\mathrm{Sin}^{-1}(1/\mathrm{Ma})$ measured from the trailing edge of the heat source. When Ma=2 the thermal shock was found at an angle of 30^{0}.

Thermal shock formation around a rapidly propagating was found to have a similar structure. The thermal shock angle was shown to be $\mathrm{Sin}^{-1}(1/\mathrm{Ma})$ measured from the trailing edge of the crack tip. In transition from the heat-affected zone of the thermally undisturbed zone across the shock surface, a finite jump of temperature was found to result. The singularity of the temperature gradient was shown to vanish at the crack tip. In series of papers Tzou summarized the experimental observations and comparisons of transonic and supersonic temperature waves. Joseph and Preziosi [1989, 1990] provide a review of constitutive equations for thermal wave behavior. Constitutions between the heat flux vector and the temperature gradient including the Jeffreys type and the Guyer and Krumhansl model for second sound propagation in dielectric materials are discussed. Coupling of the thermal relaxation behaviour with the mass and momentum transfer in fluidlike structures have been given by Choi and Wilhelm [1976]. Mass transport was discussed by Roetman [1975] and Sieniutycz [1981].

Taitel [1972] had considered the hyperbolic heat problem in a finite slab, in the framework of hyperbolic heat conduction. He obtained the transient temperature and pointed out that the absolute value of the temperature change may exceed the difference with the wall temperature. This is called the temperature overshoot paradox. This was later analyzed by Barletta and Zanchini [1997] using an entropic balance and the determination of conditions were the Clausius inequality is violated. In both the references, there were four conditions used for time and space boundary conditions. The initial temperature at time zero is held a constant. However the slope with the time domain of the temperature at time zero is replaced with the FINAL Condition for the time domain. i.e., at steady state the transient temperature will decay out to a constant value or to zero in the dimensionless form. This consideration will be shown in a later section to change the nature of the solution considerably to a well bounded expression that is bifurcated. For small values of the slab the transient temperature is subcritical damped oscillatory. For other values the Fourier series representation is augmented by a modification to the exponential time domain portion of the solution. In this study the use of the FINAL condition at steady state as the fourth condition to give a bounded solution in obeyance of Clausius inequality is evaluated for different problems.

Relativistic transformation of coordinates was used to convert the two variables PDE to a single variable Bessel equation in Sharma [2003d]. This is after removing the damping term by dividing the equation by a decaying exponential. Bounded solution [Sharma, 2001t8] in the open interval was found by analytical methods. The conditions where the pulsations in temperature will occur for a finite slab, cylinder and sphere were derived [Sharma, 2004c, 2003t6, 2003t7, 2004t1]. They were shown to be subcritical damped oscillatory. An approximate solution to the hyperbolic heat equation was derived by Renganathan and Turton [1989]. The boundary layer in time is used to scale the PDE into two domains. One of them a function of time only and the other that is valid at longer times reverts to the parabolic Fourier solution. This was used to modify the Mickley and Fairbanks [1955] packet theory to predict fluidized beds to surface heat transfer coefficients. The relaxation time enabled better comparison of experiment to theory especially at short contact times compared with several previous studies. The previous studies found the overperdiction of theory to experiment and sometimes resort had to be made to introduce empirical parameters to obtain fit of experiment to theory. Sharma and Turton [1998a, 2001c] provided a mesoscopic correlation between the heat transfer coefficients in gas solid fluidized beds to immersed surfaces. Renganathan and Turton [1989], Sharma [1999, 1999c] showed that the heat flux can reverse in direction under a periodic boundary condition for certain values of the substrate thermal properties in high speed microprocessors.

The storage coefficient was defined [2003e, 2003t12] and shown to be a critical parameter in the design of substrates. A difference of 95% in fluctuations observed in a thin film flux gauge mounted on different substrates was shown between two different experimental investigators. The case where the amplitude was dampened was where the flux reversal was pronounced. When the surface temperature decreases in magnitude, it can be seen that for small regions near the surface, the temperature will be higher than that of the wall temperature causing a back flux of heat. In the case under consideration the sensors were intended to measure the time varying instantaneous heat transfer coefficient in gas-solid fluidized beds to immersed surfaces. The back flux dampens the fluctuations caused by the phenomena and the measured values underrepresented the actual occurred values. More discussions about this will ensue in a later section.

Sharma [2003a] discussed the different scales of scrutiny, i.e., *microscopic*, *mesoscopic* and *macroscopic*. Macroscopic variables often are used to describe engineering phenomena. It can be realized that these variables are often times repeated time averages of microscopic phenomena. Thus the pressure, temperature, velocity, acceleration variables are macroscopic variables and should be derivable from microscopic variables such as the mean free path of the molecule, mass of the molecule, frequency of collisions, intermolecular interactions etc. The stochastic nature of microscopic phenomena is difficult to treat mathematically and applied at the ease in which macroscopic laws are used in practice. A mesoscopic approach is intermediate in level of scrutiny compared with the microscopic and macroscopic. The time scale of scrutiny can be such that the a few time averages prior to the macroscopic variables the parameters of interest can be studied. This may better account for the observations and lead to a better understanding of how the heat is conducted at short times and small scales. The onset of periodicity when the time events are accounted for was looked at. Mesoscopic approach deals with intermediate level of scrutiny that considers temporal fluctuations which is often averaged out in a macroscopic approach without going into the molecular mean free path or microscopic approach. Transient heat conduction cannot be fully described by Fourier representation. The non-Fourier effects or finite speed of heat propagation effect is accounted for by some investigators using the Cattaneo and Vernotte non-Fourier heat conduction equation. A generalized expression to account for the non-Fourier or thermal inertia effects suggested by Sharma (2003c) as;

$$q = -k\partial T/\partial x - \tau_r \, \partial q/\partial t - \tau_r^2/2! \, \partial^2 q/\partial t^2 - \tau_r^3/3! \, \partial^3 q/\partial t^3 - \ldots \qquad (1.92)$$

This was obtained by a Taylor series expansion in time domain.

Manifestation of higher order terms in the modified Fourier's law as periodicity in the time domain is considered in this section. When a CWT is maintained at one end of a medium of length l where l is the distance from the isothermal wall beyond which there is no appreciable temperature change from the initial condition during the duration of the study the transient temperature profile is obtained by the method of Laplace transforms. The space averaged heat flux is obtained and upon inversion from Laplace domain found to be a constant for the case obeying Fourier's law; $1 - \exp(-\tau)$

using the Cattaneo and Vernotte non-Fourier heat conduction equation, and upon introduction of the second derivative in time of the heat flux the expression becomes, $1 - \exp(-\tau)\,(\mathrm{Sin}(\tau) + \mathrm{Cos}(\tau))$. Thus the periodicity in time domain is lost when the higher order terms in the generalized Fourier expression is neglected.

Consider a long slab with dimensionless length L, at initial temperature T_0, subjected to sudden change in temperature at one of the ends to T_1. The average temperature as a function of time is obtained. The heat propagative velocity is $V_h = \mathrm{sqrt}(\alpha/\tau_r)$. The initial condition

$$t = 0, \ Vx, \ T = T_0 \qquad (1.93)$$

$$t > 0, \ x = 0, \ T = T_1 \qquad (1.94)$$

$$t > 0, \ x \ge L\,(\alpha\tau_r), \ T = T_0 \qquad (1.95)$$

Obtaining the dimensionless variables ;

$$u' = (T - T_0)/(T_1 - T_0); \ \tau = t/\tau_r \ ; \ X = x/\mathrm{sqrt}(\alpha\tau_r) \qquad (1.96)$$

The energy balance on a thin spherical shell at x with thickness Δx is written. The governing equation can be obtained after eliminating q between the energy balance equation and the derivative with respect to x of the flux equation and introducing the dimensionless variables;

$$\partial u'/\partial \tau \ + \ \partial^2 u'/\partial \tau^2 \ = \ \partial^2 u'/\partial X^2 \qquad (1.97)$$

Let the space averaged temperature over dimensionless length L be <u>. Obtaining the Laplace transform of Eq. [1.97] and substituting the boundary conditions letting L tend to infinity;

$$\underline{u} \ = \ 1/s \, \exp(-x\,\mathrm{sqrt}(s(s+1))) \qquad (1.98)$$

The space average temperature in Laplace transform coordinates;

$$L{<}\underline{u}{>} \ = \ 1/s^{3/2}\mathrm{sqrt}(s+1) \qquad (1.99)$$

The dimensionless flux when non-dimensionalized and transformed to Laplace domain is;

$$\underline{q} \ = \ 1/s^{1/2}\mathrm{sqrt}(s+1)\,\exp(-x\,\mathrm{sqrt}(s(s+1))) \qquad (1.100)$$

$$\text{where } q = q''/\mathrm{sqrt}(k\,C_p\,\rho\,/\tau_r\,)/(T_1 - T_0) \qquad (1.101)$$

The space average of the transformed flux expression letting L tend to infinity is;

$$L <\underline{q}> = \ 1/s(s+1) \qquad (1.102)$$

In the time domain it is;

$$L <q> \ = \ 1 - \exp(-\tau) \qquad (1.103)$$

By using the convolution property. If only the Fourier's law is obeyed the transformed expression for heat flux becomes;

$$q \quad = \quad 1/s^{1/2} \exp(-x \sqrt{s})) \qquad (1.104)$$

The space average of the transformed flux expression letting L tend to infinity is

$$L<q> = \quad 1/s \qquad (1.105)$$

In the time domain it is;

$$L<q> = 1 \qquad (1.106)$$

If the generalized Fourier's law is used the transformed expression for heat flux becomes;

$$q \quad = \quad 1/s^{1/2} \exp(-s/2) \exp(-x \sqrt{s}) \exp(s/2) \qquad (1.107)$$

The space average of the transformed flux expression letting L tend to infinity is;

$$L<q> = \quad 1/s \qquad \exp(-s) \qquad (1.108)$$

In the time domain it is;

$$L<q> = S_1(\tau) = 0 \quad \text{for } 0 < \tau < 1 \qquad (1.109)$$

$$= \quad 1, \ \tau > 1$$

In a similar fashion, when,

$$q = -k \partial T/\partial x - \tau_r \ \partial q/\partial t - \tau_r^2/2! \ \partial^2 q/\partial t^2 \qquad (1.110)$$

The transformed expression for heat flux becomes,

$$q \quad = \quad 1/s^{1/2} \sqrt{s} (s + 1 + s^2/2 \ ) \exp(-x \sqrt{s(s+1 + s^2/2} \ )) \qquad (1.111)$$

Where $q = q''/\sqrt{k \ C_p \ \rho \ /\tau_r \ })/(T_1 - T_0)$

The space average of the transformed flux expression letting L tend to infinity is;

$$L<q> = \quad 1/s(s+1 + s^2/2) \quad = \quad 2/s((s+1)^2 + 1 \qquad (1.112)$$

In the time domain it is;

$$L<q> = \quad 1 - \exp(-\tau) \ (Sin(\tau) + Cos(\tau)) \qquad (1.113)$$

By using the convolution property. Thus the space averaged dimensionless heat flux for CWT boundary condition in a medium of dimensionless length L, using the Fourier's law of heat conduction was found to be a constant. The space averaged dimensionless heat flux for CWT boundary condition in a medium of dimensionless length L, using the Cattaneo and Vernotte non-Fourier's law of heat conduction was found to be $1 - \exp(-\tau)$.

The space averaged dimensionless heat flux for CWT boundary condition in a medium of dimensionless length L, using the generalized Fourier's law with infinite partial derivatives of heat flux in time domain of heat conduction was found to be

$$S_1(\tau) = 0 \quad \text{for } 0 < \tau < 1; \qquad (1.114)$$

$$S_1(\tau) = 1 \quad \text{for } \tau > 1$$

This represents a jump in the instantaneous space averaged heat flux. The space averaged dimensionless heat flux for CWT boundary condition in a medium of dimensionless length L, using the generalized Fourier's law with partial derivatives of heat flux terms included upto the second derivative in time domain of heat conduction was found to be;

$$1 - \exp(-\tau)(\mathrm{Sin}(\tau) + \mathrm{Cos}(\tau)) \qquad (1.115)$$

Thus, with the neglect of the higher order flux terms with respect to time the periodicity of heat flux present in the time domain averages out. In a mesoscopic analysis the manifestation of the periodicity is determined upon inclusion of the second derivative in time domain. Earlier reports have indicated the onset of periodicity for CWT when the dimensions of the medium is less than a certain value using the Cattaneo and Vernotte non-Fourier heat conduction equation. The terms in the infinite series after a given term for any given geometry contribute to the periodic fluctuations even when the wall is isothermal.

Sharma [1998b] measured the pressure fluctuations in gas-solid fluidized beds over a wide range of fluidization particle size, superficial fluidizing velocities using fast response pressure probes and recorded the data using a PC hooked up data acquisition analog to digital board. Saddle points were detected in the probability distribution of the pressure fluctuations. This denoted the periodicity found in the fluctuations. A new distribution was suggested by Sharma [2001t10] to capture the periodicity in the probability distribution. The generalized normal distribution can be used to account for the saddle points and in a special case reverts to the Gaussian normal distribution. This was used to characterize the pressure fluctuations. In a later section, the hyperbolic damped wave momentum transfer and relaxation equation is used to model the pressure fluctuations observed in gas-solid fluidized beds [Sharma, 2003t24]. In a similar fashion the intraocular pressure changes in the eyeball was modelled with interesting applications in the disease of glaucoma [Sharma, 1999a].

The heat transfer coefficients during turbulent flow were modelled using the damped wave heat conduction and relaxation equation by Sharma [1998c, 1999f]. The shoulder in the experimentally obtained instantaneous heat transfer coefficients was not clear. As transient events are time varying often times rapid, the occurrence of a horizontal constant line was not well understood. The thermal time constant of a bubble using the parabolic Fourier and hyperbolic non-Fourier equations was derived [Sharma, 1999d, 2001t2, 2003t5]. The gas-solid mass transfer coefficients [Sharma, 1999g], rectilinear migration of a drop in a transient temperature gradient due to thermocapillary force [Sharma, 1999e] were modelled using the non-Fourier hyperbolic equation. The convective and conduction contributions during heat transfer between the circulating fluidized beds and a horizontal tube was obtained [Sharma, 2001a, 2001b]. The contribution of the fluid flowing toward the tube and away from the tube was differentiated. Thus the convective contribution is additive in one case and subtractive in the other. Adding both the solutions and dividing by two gave the conductive component and taking the difference and dividing by two gave the convective component. In later section, the derivation is provided in detail. The pulse decay in infinite medium was studied by the method of Laplace transforms and the solution from the Fourier parabolic equation was compared with [Sharma, 2003aa] the non-Fourier hyperbolic equation.

The critical upper shape limit and lower cycling limit on a nuclear fuel rod during the autocatalytic neutron fission reactions and diffusion was derived by Sharma [2003b, 2003t11]. The infinite order PDE was solved for to obtain a correction in the time domain for the surface renewal theory [2003c, 2003t14]. The relativistic transformation of coordinated was used to convert the hyperbolic PDE into a Bessel special differential equation. The general solution was obtained by solving for the functions using the boundary and time conditions instead of the constants in the case of ODEs [Sharma, 2003d]. This method was used to derive the solution in the cylindrical coordinates [Sharma, 2003t2] and in the spherical coordinates [Sharma, 2003t3]. The effect of a temperature dependent heat source in the governing equation obtained from the non-Fourier damped wave heat conduction and relaxation equation was studied [Sharma, 2003f]. The method of separation of variables was used to obtain an exact solution for the transient temperature in a sphere subject to a step change in the surface [Sharma, 2003g]. The critical thickness of high temperature barrier coating was obtained by Sharma [2003h, 2004b] prior to the origin of subcritical damped oscillations in temperature.

The hyperbolic momentum propagative equation was used to predict the flow of a stationary fluid at zero time, near the wall suddenly subject to motion [Sharma, 2003i] using the method of transformation into relativistic coordinates. The design of adsorber to reduce toxicity in water and molecular sieve design to clean-up army washeries using hyperbolic non-Fourier equation was developed [Sharma, 2003j, 2003k]. Adsorption is a transient event and the finite speed diffusion can play a critical role in it. Restriction fragment mapping using gel acrylamide electrophoresis includes a critical step of diffusion. The effect of finite speed of propagation of mass accounted for by the damped wave hyperbolic diffusion and relaxation equation was explored by Sharma [2003, 2003l]. The diffusion issued in Sanger's method of obtaining the DNA sequences was discussed in Sharma [2003t20]. The confounding effect of charge on the migration of fragments was explored in Sharma [2004t4].

During the analysis of reaction and diffusion problems and exploring the role of finite speed of propagation of mass diffusion using the hyperbolic non-Fick diffusion and relaxation equation a region of zero concentration was found. This happened when the surface of a finite object was subject to a simple first order reaction and the species was allowed to diffuse into the object. Beyond a certain distance from the surface the species never migrated to those locations. This is consistent with the mass inertia associated with the propagation of the concentration disturbance and also the penetration length was shorter than the surface to center distance of the object. The damped wave solution was in terms of a Bessel composite function. At the first zero of the Bessel was the location where the species dropped to a zero concentration. The thermal inertia time and the half-life of the reaction can be compared with each other to quantitate the significance of this finding. For simplification as a first approximation the first order surface reaction rate constant was considered as a reciprocal of the relaxation time of mass wave diffusion and relaxation as shown in Sharma [2004a, 2004d, 2004t3, 2004t6]. The simultaneous fast reaction and finite speed diffusion was studied and exact solutions derived by the method of separation of variables for a slab, cylinder and sphere by Sharma [2003t8, 2003t9, 2003t10].

Transient fixed bed examination using hyperbolic PDE was undertaken [Sharma, 200t2] and studies on classical transient problems such as unsteady evaporation [Sharma, 2001t3], unsteady diffusion with first order reaction [Sharma, 2001t4], gas absorption with rapid chemical reaction [Sharma, 2001t5], diffusion and reaction in laminar flow past a soluble flat plate [Sharma, 2001t6], quenching of a steel billet [Sharma, 2001t7] using the damped wave transmission and relaxation equation was studied using analytical method of solution. During doping of silicon and the manufacture of microprocessor chips, investigators have found that the concentration profiles found could not be explained using Fick's law of mass diffusion. Sharma [2001t11, 2002t2, 2002t9] looked at the use of wave diffusion and relaxation equation to study the theory to experiment discrepancy. The temperature change of thermal conductivity and the asymptotic analysis of the resulting PDEs were explored in Sharma [2001t9]. The measurement of thermal diffusivity used currently and some issues in acquiring the thermal relaxation times was discussed by Sharma [2001t12]. The constant wall flux boundary condition was analyzed in Sharma [2002t1. 2003t15]. It is generally agreed that the Heisler-Grober charts that have been in use for five decades gives poor prediction near the origin, i.e., at short time scales. The equations to derive the lines in the charts were derived from the hyperbolic damped wave conduction and relaxation equation [Sharma, 2002t4]. The chemisorption of organic sulfides onto Palladium for compliance with the EPA mandate for automobile emissions was explored using the non-Fick PDE [Sharma, 2002t8]. During reverse osmosis the deposition of salt on the semi-permeable membrane changes with time, and this concentration polarization layer was studied in light of the finite speed diffusion and overall efficiency by Sharma [2002t8].

The microscopic two-step model [Qiu and Tien, 1992] takes into account the heat transfer through microstructures when the response time is short and becomes comparable to the phonon-electron relaxation time. They consider the excitation of electron gas and the metal-lattice heating by phonon-electron interactions as a two-step process. The two processes were coupled through the phonon-electron coupling factor. The equations are;

$$C_e\, \partial T_e/\partial t \ = \ \nabla.(k\nabla\, T_e) \ - \ G(T_e \ - \ T_l) \qquad (1.116)$$

$$C_l\partial T_l/\partial t \ = G(T_e \ - \ T_l) \qquad (1.117)$$

The coupling factor G, was given by;

$$G = \pi^4 (n_e v_s K)^2/k \qquad \text{with } v_s = K(6 \pi^2 n_a)^{-1/3}T_D/2\pi h \qquad (1.118)$$

where, v_s is the speed of sound, h is the Planck's constant. The temperature of the electron gas was T_e.

The dual phase lag model. DPL, was proposed by Tzou [1995, 1997]. They account for the non-zero times required for the heat flux and temperature gradient to gradually become established in response to the thermal disturbances. The Fourier's law assumes that the thermal gradient and heat flux become established immediately. Microtime heating of metal thin films of the order of less than 1 micron in the order of a few femtoseconds requires an alternate model to Fourier's law to describe the time events. This may also become important in macroscale where the length and time scales are relatively large as in the transient heating of and on the orders of 1 cm and 1 second. The one dimensional DPL, model is given by;

$$q + \tau_q \partial q/\partial t = -k \, \partial T//\partial x - k \, \tau_t \partial^2 T/\partial t \partial x \qquad (1.119)$$

where the thermal lags or delays, τ_q and τ_t are approximately the times needed for the gradual response of heat flux and thermal gradient respectively. $\partial q/\partial t$ and $\partial^2 T/\partial t \partial x$ represent transient behaviour of heat flux and temperature gradient during the response. After the gradual response is complete, the heat flux and thermal gradient achieve the values given by Fourier's law. These are interpreted [Antaki, 1998] as the effects of "thermal inertia" and "microstructural interaction" respectively. τ_q is the delay in establishing heat flux and associated conduction through a solid. τ_t is the delay in establishing the thermal gradient across the solid during which the conduction occurs through its small-structure structures. Values of thermal lags are estimated to be small for continuous solids. For gold the relaxation time of the flux gradient was estimated for gold at 0.7 femtosecond and 89 femto seconds for the relaxation time of the thermal gradient. However, for non-continuous solids the lags were estimated to be relatively large and shown as 8.9 seconds for τ_q and 4.5 seconds for τ_t for one type of sand [Tzou, 1997]. When τ_t is zero the DPL equation reduces to the Cattaneo & Vernotte non-Fourier damped wave conduction and relaxation equation. Further when the relaxation times, τ_q and τ_t are zero the DPL equation reduces to Fourier's Law.

1.7.1 Third Order PDE to describe Non-Fourier Heat Conduction

The alternate expression for Fourier with up to second derivative in time of the flux retained in the Taylor series expansion of the time rate of change of accumulation of heat at the surface can be written as [Sharma, 2003t21]

$$q = - k\ \partial T/\partial x\ -\ \tau_r\ \partial q/\partial t\ -\ \tau_r^2/2!\ \partial^2 q/\partial t^2 \qquad (1.120)$$

Consider a finite slab at initial temperature T_0, subjected to sudden contact with fluid at temperature at to T_1. The transient temperature as a function of space and time is obtained. The heat propagative velocity is V_h = sqrt(α/τ_r). The initial condition

$$t = 0,\ Vx,\ T = T_0 \qquad (1.121)$$

$$t > 0,\ x = 0,\ \partial T/\partial x = 0 \qquad (1.122)$$

$$t > 0,\ x = \pm a,\ -k\partial T/\partial x = - h\ (T_1 - T) \qquad (1.123)$$

Obtaining the dimensionless variables;

$$u = (T - T_1)/(T_0 - T_1);\ \tau = t/\tau_r\ ;\ X = x/\text{sqrt}(\alpha\ \tau_r) \qquad (1.124)$$

The energy balance on a thin spherical shell at x with thickness Δx is written. The governing equation can be obtained after eliminating q between the energy balance equation and the derivative with respect to x of the flux equation and introducing the dimensionless variables;

$$\partial u/\partial \tau\ +\ \partial^2 u/\partial \tau^2\ +\ \tfrac{1}{2}!\partial^3 u/\partial \tau^3 = \partial^2 u/\partial X^2 \qquad (1.125)$$

The solution is obtained by the method of separation of variables.

$$\text{Suppose } u = \exp(-n\tau)\ w(X, \tau) \qquad (1.126)$$

Substituting Eq. [1.136] in Eq.[1.135], and by choosing n = 2/3,

$$-10w/27\ + 1/2\ \partial^3 w/\partial \tau^3 + 1/3\ \partial w/\partial \tau = \partial^2 w/\partial X^2 \qquad (1.127)$$

The method of separation of variables can be used to obtain the solution of Eq. [1.127]. Let $W = V(\tau)\ \phi (X)$. Eq. [1.137] and separating the variables that are a function of X only and τ only;

$$\phi" + \lambda^2 \phi = 0 \qquad (1.128)$$

The solution for Eq. [1.128];

$$\phi = c_1 \text{Sin}(\lambda X) + c_2 \text{Cos} (\lambda X) \qquad (1.129)$$

It can be seen that $c_1 = 0$ as the slope of temperature with respect to X, at X = 0 is 0. Now from the BC at the surface ,

$$\partial u/\partial X = Bi\ u \qquad (1.130)$$

where $Bi = h\ \sqrt{\alpha\tau_r}/k$ \qquad (1.131)

$$- \lambda Sin((\lambda\ X) = Bi\ Cos(\lambda\ X) \qquad (1.132)$$

$$or\ \lambda_n/Bi = Cot(\lambda_n\ a/\sqrt{\alpha}\ \tau_r) \qquad (1.133)$$

for small a,

$$\lambda_n = sqrt(h\ \alpha\tau_r/ak) + n\pi = sqrt(h/S/a) + n\pi \qquad (1.134)$$

where $S = \rho\ C_p/\tau_r$ is the storage coefficient \qquad (1.135)

Now,

$$V'''/2 + V'/3 = V(10/9\ -\lambda^2) \qquad (1.136)$$

Or the auxiliary equation is;

$$r^3/2 + r/3 = -\gamma \qquad (1.137)$$

where $\gamma = 10/27\ -\lambda^2$. Vieta's substitution can be used to solve Eq. [1.148],

$$Let\ r = \omega - 2/9\omega \qquad (1.138)$$

Then Eq. [1.137] becomes,

$$(\omega^3)^2 + 8/27^2 + \gamma\omega^3 = 0 \qquad (1.139)$$

Eq. [1.139] is a quadratic in ω^3. Upon obtaining the solution to Eq. [1.139], it can be used to obtain the 3 roots, $r_1,\ r_2,\ r_3$. The imaginary roots occur when,

$$D = B^3/27 + C^2/4 > 0\ where,\ C,\ B\ are; \qquad (1.140)$$

$$r^3 + 2r/3 + 2\gamma = 0 \qquad (1.141)$$

$$B = 2/3 \qquad (1.142)$$

$$C = 20/27\ - 2\lambda^2 \qquad (1.143)$$

$$i.e.,\ 8/27^2 + (20/27\ -2\lambda^2)^2/4 > 0 \qquad (1.144)$$

For any λ, the contributions from all the terms in the infinite series in time domain will be pulsating. Further, when,

$$q = -k\partial T/\partial x - \tau_r\ \partial q/\partial t - \tau_r^2/2!\ \partial^2\ q/\partial t^2 \qquad (1.145)$$

The Laplace transform of the governing equation for the flux in terms of temperature becomes;

$$q = 1/s^{1/2}sqrt(s + 1 + s^2/2)exp(-xsqrt(s(s+1 + s^2/2))) \qquad (1.146)$$

where $q = q''/sqrt(k\ C_p\ \rho\ /\tau_r)$. The space average of the transformed flux expression letting L tend to infinity is;

$$L < q > = 1/s(s+1 + s^2/2) \qquad (1.147)$$

$$= 2/s((s+1)^2 + 1) \qquad (1.148)$$

In the time domain it is;

$$L < q> = 1 - \exp(-\tau)(\mathrm{Sin}(\tau) + \mathrm{Cos}(\tau)) \qquad (1.149)$$

by using the convolution property. It can be seen from above, that the nature of the solution to the governing equation when upto the second derivative of flux in time is taken in the constitutive equation is periodic in nature. Also the frequency of the periodic component can be seen to be $1/\tau_r$ or $w* = 1$. For purposes of determining the periodic component consider as a semi-infinite medium subjected to CWT boundary condition. Redefining,

$$u = (T - T_0)/(T_1 - T_0) \qquad (1.150)$$

So suppose that $V = f(X)\exp(-i\tau)$, then Eq. [1.127] becomes,

$$-10f/27 + 1/2 -i^3 f + 1/3 -if = \partial^2 f/\partial X^2 \qquad (1.151)$$

$$f'' = f(-10/27 + i(1/2 - 1/3)) = f(E + iF) \qquad (1.152)$$

$$\text{where } E = -10/27; \ F = 1/6$$

or $W = A \exp(-\mathrm{sqrt}(E + iF)X) \exp(-i\tau)$ $\qquad (1.153)$

$$\mathrm{sqrt}(E + iF) = G + iH \qquad (1.154)$$

squaring both sides and equating the real and imaginary parts,

$$E = G^2 - H^2; \qquad F = 2GH \qquad (1.155)$$

$$-10/27 = G^2 - H^2 \qquad (1.156)$$

$$H^4 -10/27 H^2 - 1/144 = 0 \qquad (1.157)$$

$$\text{Or } H^2 = (10/27 \pm \mathrm{sqrt}(100/27^2 + 1/36))/2 \qquad (1.158)$$

$$\text{So, } W = A \exp(-G - iH)X) \exp(-i\tau) \qquad (1.159)$$

$$\text{Or } u = \text{Real Part}(A\exp(-2\tau/3) \exp(-G - iH)X) \exp(-i\tau)) \qquad (1.160)$$

$$= A\exp(-2\tau/3) \exp(-GX) \mathrm{Cos}(HX + \tau)) \qquad (1.161)$$

G and H are given by Eqs. [1.158, 1.155]. From the BC, at $X = 0$

$$\text{or } A = \exp(2\tau/3)/\mathrm{Cos}(\tau) \qquad (1.162)$$

$$\text{or } u = \exp(-GX) \, \text{Cos}(HX + \tau))/ \, \text{Cos}(\tau) \qquad (1.163)$$

1.8 Finite Speed Conduction, Diffusion and Momentum Transfer

The WRENCH equations, wave relaxation equation for non-Fourier conduction of the hyperbolic type for heat conduction, diffusion and momentum transfer are given by;

$$q + \tau_r \partial q/\partial t = -k \, \partial T//\partial x \qquad (1.164)$$

For heat conduction and relaxation the relaxation time is given by τ_r. Research is ongoing on how the relaxation time varies with temperature and how does it vary from one material to another. The velocity of heat conduction and relaxation is given by the square root of the ratio of the thermal diffusivity to the relaxation time of heat.

$$v_h = \text{sqrt}(\alpha/\tau_r) \qquad (1.165)$$

The relaxation time for mass diffusion, τ_m is given in the damped wave diffusion and relaxation equation and is;

$$J + \tau_m \partial J/\partial t = -D \, \partial C/\partial x \qquad (1.166)$$

The velocity of mass diffusion is given by the square root of the binary diffusivity divided by the relaxation time of mass diffusion.

$$V_m = \text{sqrt}(D/\tau_m) \qquad (1.167)$$

This equation reverts at zero relaxation times to the Fick's law which relates the concentration gradient and binary diffusivity with the mass flux in a linear fashion. The damped wave momentum transfer and relaxation is given by;

$$\Im_{xy} + \tau_{mom} \partial \Im /\partial t = -\mu \, \partial v/\partial x \qquad (1.168)$$

τ_{mom} is the relaxation time for momentum transfer. The velocity of momentum transfer is given by;

$$v_{mom} = sqrt(\nu/\tau_{mom}) \qquad (1.169)$$

where the kinematic viscosity is given by ν and the relaxation time of momentum transfer is τ_{mom}. When the relaxation time of momentum becomes zero the constitutive relation reverts to the Newton's law of viscosity.

These constitutive laws can be combined the equation of energy balance, equation of mass balance, and equation of momentum balance to give the governing equations for a given situation and the resulting PDEs solved for.

1.9 Experimental Measurement of Relaxation Times

Estimates were made of the relaxation time of heat conduction and relaxation for real materials. Luikov [1966] suggested that the relaxation times can range from 10^{-3} to 10^{3} seconds depending on the process intensity. Brazhinov et al provide estimates of 20 to 30 seconds for meat products [1975]. Michalowski et al [1982] and Mitura et al [1988] mention that for the falling drying rate period the average value of the relaxation time is of the order of several thousand seconds. For homogeneous substances, the relaxation time values for gases of the order of 10^{-8} to 10^{-10} for gases, 10^{-10} to 10^{-12} for liquids and dielectric solids were quoted by Sieniutycz [1977].

Experimental evidence for hyperbolic wave propagative and relaxation heat equation was provided by Peshkov who measured a velocity of heat of 19 m/s in liquid helium at 1.4 K [1944]. Jackson and Walker [1971] measured the thermal conductivity, second sound and phonon-phonon interactions in Sodium Fluoride at 3.4 K. Experimental confirmation of the phenomena was discussed at low temperature in Bismuth at 3.4 K by Narayanamurthi and Dynes [1972]. Churchill and Brown [1989] measured a second sound along with Chester [11]. Mitra, Kumar, Vedavarz and Moellami [1995] report relaxation time of the order of 20-30 seconds for biological materials at room temperature. Per the experimental evidence provided by Tzou [1992], in 4340 steel at 480 ° C, the thermal wave speed measured was 900 m/s and the relaxation time was estimated of the order of 10^{-11} seconds.

An estimate of the relaxation time of mass can be calculated from the Maxwell probability distribution of the velocity of molecules. This is given by Eq. [1.3]. The binary diffusivity divided by the square of the velocity of molecules gives the relaxation time of mass. This can be a first approximation for the estimate of the relaxation times. The order of magnitude for continuous materials is in the order of a few microseconds according to this approach.

Kumar et al [1995] demonstrated that the macroscopic description of the transient heat conduction processes in biological materials is accurately described by the non-Fourier damped wave conduction and relaxation model. They performed four different experiments for different boundary conditions. The temperature histories embedded in the samples were recorded and the experimental results offer compelling evidence of the wave nature of heat conduction in processed meat. The fact that a finite time occurs before the thermocouples embedded within the media register any temperature deviations. This is consistent with the theoretical predictions of lag as shown in Figure 2.1. They also verified the phenomenon of superposition of waves occurring due to two heat waves approaching each other from two sides. They found that the hyperbolic model was a valid macroscopic description of the microscopic model in biological materials. The corresponding thermal characteristic time was found to be approximately 16 seconds.

Kaminski [1990] stated that the relaxation time represents the interaction of structural elements in inner heat transfer. For homogeneous materials this interaction is at the molecular or crystal lattice level. For non-homogeneous materials the structural interactions may be at a higher level thus explaining the observed relatively "higher values" of relaxation times. The measured values of the relaxation times by Kaminski are given in Table 1.1. The evaluation of relaxation time for nonhomogeneous inner structure materials was carried out based upon the measurement of penetration time, thermal diffusivity and thermal wave speed. Penetration time was found by placing a linear heat source and temperature sensor in the material. The device they used consist of a linear heat source in the form of a needle containing a 0.2 mm resistance wire in 0.1 mm electrical insulation supplied by a power stabilizer. The temperature of the heater was estimated to be from 12.8 ^{0}C to 32.1 ^{0}C higher than the ambient temperature. The thermocouple was placed in the needle and parallel to the heater. The measurement included the time delay of heat wave from the heater to the sensor. The thermocouple response time was quantitated to accurately measure the penetration time. While the parabolic Fourier heat equation estimates gave a value of 5 to 10 seconds to see the disturbance the time taken was after 100 to 200 seconds. A method based upon ac calorimetry for the measurement of thermal diffusivity of a given sample was patented by Kato and Nara (1998). This involves periodically heating a sample with a heat source modulated at an operating frequency; detecting an ac temperature of the sample at a detection point of the sample, the ac temperature being a periodic temperature oscillation; measuring a phase of the detected ac temperature by a phase sensitive detection operation; maintaining the phase constant by controlling the operating frequency of the heat source based on the measured phase to realize a constant wave number condition in which both a wave number and a thermal diffusion length are kept constant and determining the thermal diffusivity of the sample based in the relative change in the operating frequency under the constant wave number condition. Thermal diffusivity is determined by measuring the propagating properties of the thermal waves in the sample periodically heated.

**Table 1.1 Experimentally Measured Values of Relaxation
Time, Kaminski [1990], Mitra et al [1995]**

S. No	Material	Relaxation Time
1.	Sodium Bicarbonate	24.5 seconds
2.	Sand	28.7 seconds
3.	Glass Ballotini	10.9 seconds
4.	Ion Exchanger	53.7 seconds
5.	Processed Meat	15.5 seconds

These calculations were rederived using the HHC wave propagative equation which accounts for the finite speed of heat as opposed to the infinite speed implicit in the Fourier representation [Sharma, 2001t12].

The complex temperature representation is used and the real part is used to obtain the thermal diffusivity and thermal relaxation time of the sample.

1.10 Case of Infinite Relaxation Time and D'Alambert's Solution

Consider a semi-infinite medium with one of the ends at a temperature T_s. Let the slab be at an initial zero temperature. The transient temperature profile in the medium is derived as follows. The initial and boundary conditions are;

$$t = 0, \ \forall x > 0, \quad C = 0 \tag{1.170}$$

$$t > 0, \ x = 0, \quad C = C_s \tag{1.171}$$

The governing equation as derived before by elimination of the second order cross derivative in heat flux between the constitutive law differentiated wrt to time and the energy balance equation differentiated wrt to time is given as;

$$\tau_r \, \partial^2 T / \partial t^2 + \partial T / \partial t + \ = \alpha \, \partial^2 T / \partial x^2 \tag{1.172}$$

At infinite relaxation times, the velocity of heat propagation approaches zero and Eq. [1.172] becomes the wave equation;

$$\tau_r \, \partial^2 T / \partial t^2 \ = \alpha (\ \partial^2 T / \partial x^2 \ ) \tag{1.173}$$

Obtaining the dimensionless variables in the following manner;

$$u" = (T/T_s); \quad \tau = t/\tau_r \quad ; X = x/\text{sqrt}(\alpha\tau_r) \qquad (1.174)$$

The governing equation in the dimensionless form is then;

$$\partial^2 u"/\partial\tau^2 = \partial^2 u"/\partial X^2 \qquad (1.175)$$

$$\text{Let } \chi = X + \tau \qquad (1.176)$$

$$\delta = X - \tau \qquad (1.177)$$

Then Eq. [1.175] becomes;

$$\partial^2 u/\partial\chi^2 + \partial^2 u/\partial\delta^2 - 2\partial^2 u/\partial\delta\partial\chi = \partial^2 u/\partial\chi^2 + \partial^2 u/\partial\delta^2 + 2\partial^2 u/\partial\delta\partial\chi \qquad (1.178)$$

$$\text{or } 4\partial^2 u/\partial\delta\partial\chi = 0 \qquad (1.179)$$

The solution to Eq. [1.179] is the D'Alambert's solution to the wave equation at infinite relaxation times and is given by;

$$u = g(\chi) + V(\delta) \qquad (1.180)$$

Where g and V are any functions. This is the D'Alambert's solution at infinite relaxation time [Sharma, 2003t13].

Summary

The origins of a damped wave non-Fouier heat conduction and relaxation equation was traced back to the phenomenological relations between forces and flows and the Onsager reciprocal relations. From a microscopic viewpoint using the kinetic theory of gases the definition of heat flux was revisited. There is room for an accumulation of energy term in the definition of heat flux as the energy of molecules leaving the surface minus the energy of molecules entering the surface. The accumulation of flux term as proposed originally by Cattaneo and Vernotte independently in 1958 may be good first approximation to the accumulation of energy at the surface in the heat flux term especially to describe transient heat conduction. The relaxation time is an added intrinsic property of the material under consideration. The speed of heat conduction is given by the square root of the ratio of the thermal diffusivity and the relaxation time. The wave dominant transfer regime is shown to occur when the accumulation term in the governing equation becomes greater than $\exp(\tau)$ the ballistic term is much greater than the damping term and the wave and the governing equation reverts to the wave equation. When the accumulation term is much smaller than the $\exp(\tau)$ the ballistic term is negligible and the governing equation becomes the Fourier parabolic PDE. The regimes of wave dominance were derived when the thermal conductivity variation in temperature is accounted for and for the case when convection term is accounted for in the governing equation. A boundary layer in time was shown to obtain an asymptotic solution to the damped wave conduction and relaxation equation. Two instances where the Fourier series representation for transient temperature gives a singularity or unrealistic prediction were shown for the surface flux in a finite slab and a semi-infinite slab. The light speed barrier and the motivation for accounting for the finite speed of heat are discussed. The nature of the PDE such as hyperbolic, parabolic and elliptic are outlined. Compared with the ODE where the general solution contains n constants the PDE contains n functions. These can be determined by using boundary conditions in space and domains of integration using the initial and final time conditions. Other forms of non-Fourier heat conduction is the ballistic transport equations due to Chen, [2001], dual phase lag model by Tzou [1997] and the EPRT, equation of phonon transport proposed by Majumdar, [1993], microscopic two-step model by Qiu and Tien [1992]. An infinite order PDE was presented by Sharma [2003] using Taylor series expansion to fully describe the transient events. A third order PDE was suggested by Sharma [2003t21]. These models are mentioned briefly. A review of the research work done in this area is discussed. The experimental measurement of relaxation time is described. A table of relaxation time values for heterogeneous systems are shown and found to be of the order of 15 seconds. This is orders of magnitude higher than the 10^{-9} seconds originally suspected by early investigators like Luikov and Ozisik. At infinite relaxation times the governing equation reverts to the wave equation and the D'Alamberts solution can be provided for this case.

Exercises

1. *Molecules in Equilibrium*
   What ought to be right relation between the velocity of molecules and temperature in a monatomic gas when the system is not in equilibrium. Use a time varying expression for the Boltzmann relation given in Eq. [1.7] and run down the derivation of the Fourier's law for monatomic gases. Can you calculate the accumulation of energy near the surface from the time varying expression for velocity of molecules. What kind of non-Fourier expression results from such an analysis ?

2. *Non-Linear Temperature Gradient*
   Instead of a linear relation for the temperature use a quadratic expression in Eq. [1.11] and Eq. [1.12]. Compare the results for the derived expression for a relation between heat flux and thermal gradient with the Cattaneo & Vernotte equation and Fourier equation. Repeat the analysis for a generalized polynomial for the Temperature gradient. What does this analysis say about the mechanism of heat conduction and relaxation ?

3. *Accumulation of Energy near the Surface*
   How can the Eq. [1.15] improved upon. What ought to be used for the accumulation of energy and how does the flux term figure into the analysis.

4. *Temperature Dependent Heat Source*
   In the energy balance given in Eq. [1.18] include a heat source term that varies with temperature. Derive the governing equation for the temperature in 1 with dimension using the non-Fourier wave heat conduction and relaxation equation.

5. *Three Dimensional Transient Wave Conduction*
   Extend the wave equation reversion as shown in Eq. [1.22] to three spaces dimensions and time. How does this equation compare with the Schrodinger wave equation. What does this say ?

6. *Wave and Fourier Regimes*
   In Figure 1.1, what does the cross-over point signify ? Derive an expression for the Cross-over point from wave dominant to Fourier dominant transfer in terms of the relaxation time and other parameters ?

7. *Ratio Test*

Using the appropriate test for convergence of infinite series to show that Eq. [1.67] is Convergent. What happens to the parameters of convergence in Eq. [1.70]. How about the expression for heat flux for any arbitrary dimensionless time ?

8. *Pulse Boundary Condition*

Test the boundedness of the temperature at the center of an infinite slab when a pulse of heat is injected at the center of the medium. Assume the Fourier's law of heat conduction. Obtain the expressions for the transient temperature by the method of Laplace transforms and examine singularities if present.

9. *PDE Type*

Consider the convective term in the energy balance equation.

$$- \partial q/\partial y - (\rho \, C_p) U_y \partial T/\partial y \; = \; (\rho \, C_p) \, \partial T/\partial t$$

The governing equation for temperature in 1 dimension using the non-Fourier wave heat conduction and relaxation equation is given by Eq. [1.41]. Is the derived PDE hyperbolic, parabolic or elliptic. What is the term that makes the equation hyperbolic ?

10. *Extended Fourier Equation for Conduction of Heat*

Obtain the equivalent of Eq. [1.97] when three terms in the Taylor infinite series is considered for the non-Fourier equation:

$$q \; = \; -k\partial T/\partial x \; - \; \tau_r \, \partial q/\partial t \; - \; \tau_r^2/2! \; \partial^2 q/\partial t^2 \; - \; \tau_r^3/3! \; \partial^3 q/\partial t^3$$

Derive the governing equation for temperature by combining the above relation with the 1 dimensional energy balance equation.

11. *Heat Capacity Variation with Temperature*

The heat capacity C_p is known to vary with temperature as follows;

$$C_p \; = \; C_{p0}(\; 1 + BT)$$

Use this expression in the energy balance equation and obtain the governing equation in temperature in 1 dimension. Derive the conditions where the wave dominant heat conduction will be significant ?

12. *Boundary Layer in Time*

The governing equation including the convection term using the non-Fourier damped wave conduction and relaxation equation is given by Eq. [1.41].

$$\partial^2 u/\partial Y^2 = \partial u/\partial \tau + \partial^2 u/\partial \tau^2 + Pe\partial^2 u/\partial Y\partial \tau$$

Using the boundary layer time scale the different terms in the equation. When is the cross-derivative important. Present the solutions for the respective domains. Repeat the analysis using the Fourier heat conduction equation. Compare the results from the two analyses.

13. *Variation of Thermal Conductivity with Temperature*
Obtain the exact solution for the governing relation given in Eq. [1.31] where the variation of thermal conductivity with temperature is accounted for and the Non-Fourier damped wave conduction and relaxation equation as the constitutive Law. Assume that the solution u consists of a steady state and transient part.

$$u = u^{ss} + u^t$$

The Eq. [1.31] will then become:

$$\partial u^t /\partial \tau + \partial^2 u^t /\partial \tau^2 = \partial^2 u^t \, \partial Y^2 -\alpha_* \beta* (\partial u^t /\partial Y)^2 + \partial^2 u^{ss} \, \partial Y^2 -\alpha_* \beta* (\partial u^{ss} /\partial Y)^2$$

The above equation will be obeyed when;

$$\partial u^t /\partial \tau + \partial^2 u^t /\partial \tau^2 = \partial^2 u^t \, \partial Y^2 -\alpha_* \beta* (\partial u^t /\partial Y)^2 \quad \text{and}$$

$$\partial^2 u^{ss} \, \partial Y^2 = \alpha_* \beta* (\partial u^{ss} /\partial Y)^2$$

The later equation can be solved for by letting, $(\partial u^{ss} /\partial Y) = m$,

$$\partial m/\partial Y = \alpha_* \beta* (m)^2$$

Integrating this expression;

$$-1/m = \alpha_* \beta*Y + C'$$

$$(\partial u^{ss} /\partial Y) = -1/(\alpha_* \beta*Y + C')$$

or $u^{ss} = -\ln(\alpha_* \beta*Y + C')$

The boundary conditions can be applied.   The transient portion of the solution can be solved by the method of separation of variables and the scaling method shown to evaluate the relative magnitudes of the different terms in the governing equation.

14. *Semi-Infinite Cylindrical Medium*

Examine the problem of heating a semi-infinite medium with constant thermal diffusivity, from a cylindrical surface with a radius R. Use the transformation given in section 1.4. Assume a wall flux of q $* = 1 = (q/sqrt(k\rho C_p/\tau_r)/(T_s - T_0)$, where T_0 is the surrounding fluid temperature at time zero. T_s is a reference surface temperature. Obtain the transient temperature using the Fourier conduction equation.

The governing equation for temperature using the Fourier heat conduction equation can be written as;

$$(\rho C_p)\partial T/\partial t = (k/r)\partial(r\partial T/\partial r)/\partial r$$

Let u' = $(T - T_0)/(T_s - T_0)$, ; q* = $q/sqrt(k\rho C_p/\tau_r)/(T_s - T_0)$, X = $r/sqrt(\alpha\tau_r)$, $\tau = t/\tau_r$; S = $(\rho C_p/\tau_r)$ then;

$$\partial u'/\partial t = (\alpha/r)\partial(r\partial u'/\partial r)/\partial r$$

The initial and boundary conditions are given as;

$$t = 0, \quad u = 0 \text{ for all } r > R$$

$$t > 0, \quad r = R, \quad q* = 1$$

$$t > 0, \quad r = \infty, \quad u = 0$$

A similarity variable, $\eta = r/sqrt(4\alpha t)$ is defined and the governing equation transformed into the new variable as follows;

$$\partial u'/\partial t = -1/4\partial u'/\partial\eta (x) t^{-3/2}/\alpha^{1/2} = -\eta/2t\partial u'/\partial\eta$$

$$-2(\eta\partial u'/\partial\eta) = 1/\eta \,\partial(\eta\partial u'/\partial\eta)/\partial\eta$$

Rearranging;

$$-2\eta \,\partial\eta = \partial(\eta\partial u'/\partial\eta)/(\eta\partial u'/\partial\eta)$$

Integrating both sides,

$$-\eta^2 + c = \ln(\eta\partial u'/\partial\eta)$$

$$\eta\partial u'/\partial\eta = c'exp(-\eta^2)$$

$$\text{or } u' = c'\!\int exp(-\eta^2)/\eta \; d\eta + c"$$

This has to be solved for applying the boundary conditions. The heat flux can be written as;

$$q* = - \partial u'/\partial X = -(1/2)(\partial u'/\partial\eta)/ \tau^{1/2}$$

$$= -(4\alpha t)^{1/2} /r \; c'exp(-r^2/4\alpha t)$$

On examining the expression for heat flux there exists a singularity when $r = 0$. c' can be solved for from the boundary condition at the surface;

$$1 = -(4\alpha t)^{1/2}/R \; c'\exp(-R^2/4\alpha t)$$

c' can be eliminated between the solution equation for the dimensionless heat flux and the equation from the boundary condition. Thus,

$$q^* \quad = \quad (R/r) \exp(-(r^2 - R^2)/4\alpha t)$$

A singularity exists at $r = 0$ in the expression for heat flux. The dimensionless temperature can be obtained as;

$$u' \qquad = c'' \qquad - \eta_R \exp(\eta_R^2)\!\int\!\exp(-\eta^2)/\eta \; d\eta$$

Multiplying the numerator and denominator of the integrand by η, and substituting, $p = \eta^2$,

$$u' \qquad = c'' \qquad - (R/(4\alpha t)^{1/2})\exp(R^2/4\alpha t)\int dp/2\exp(-p)/p$$

$$= c'' \; - (R/(4\alpha t)^{1/2})\exp(R^2/4\alpha t)\int[\exp(-p)/p]dp$$

$$= c'' - \eta_R\exp(\eta_R^2)Ei(p)$$

$Ei(p) = \int[\exp(-p)/p]dp$ is the exponential integral. Values for the exponential integral are tabulated in Jahnke and Emde [1943]. The exponential integral can be used to provide a expression for the dimensionless temperature.. There exists a singularity at short contact times in the Fourier solution of temperature for a infinite cylinder.

15.   *Semi-infinite Spherical Medium*
       Examine the problem of heating a semi-infinite medium with constant thermal diffusivity, from a spherical surface with a radius R. Use the transformation given in section 1.4. Assume a wall flux of $Q^* = 1 = (q_s/\text{sqrt}(k\rho C_p/\tau_r)/(T_s - T_0)$, where T_0 is the surrounding fluid temperature at time zero. T_s is a reference surface temperature. Obtain the transient temperature using the Fourier conduction equation.

       Consider a thin shell of thickness Δr at a distance r from the center of the spherical medium. The governing equation for temperature using the Fourier heat conduction equation can be written as;

$$(\rho C_p)\partial T/\partial t \;=\; (k/r^2)\partial(r^2\partial T/\partial r)/\partial r$$

Let $u' = (T - T_0)/(T_s - T_0)$, ; $q^* = q/\text{sqrt}(k\rho C_p/\tau_r)/(T_s - T_0)$,   then;

$$\partial u'/\partial t \ = \ (\alpha/r^2\ )\partial(\ r^2\ \partial u'/\partial r)/\partial r$$

The initial and boundary conditions are given as;

$$t = 0, \quad u \ = 0 \text{ for all } r > R$$

$$t > 0, \quad r = R, \quad q^* \ = \ 1$$

$$t > 0, \quad r = \infty, \quad u = 0$$

A similarity variable, $\eta \ = \ r/\text{sqrt}(4\alpha t)$ is defined and the governing equation transformed into the new variable as follows;

$$\partial u'/\partial t \quad = -1/4\partial u'/\partial \eta \ (x)\ t^{-3/2}/\alpha^{1/2} \ = \ -\eta/2t\partial u'/\partial \eta$$

$$-2(\eta\partial u'/\partial \eta) \quad = \ 1/\eta^2\ \partial(\ \eta^2\partial u'/\partial \eta)/\partial \eta$$

Rearranging; $\qquad -2\eta\ \partial \eta \ = \ \partial(\ \eta^2\partial u'/\partial \eta)/(\eta^2\partial u'/\partial \eta)$
Integrating both sides,

$$-\eta^2\ + c \ = \ \ln(\eta^2\ \partial u'/\partial \eta)$$
$$\eta^2\ \partial u'/\partial \eta \qquad = \ c'\exp(-\eta^2)$$

$$\text{or u'} \qquad = \ c'\!\int\!\exp(-\eta^2)/\eta^2\ d\eta \ + c''$$

This has to be solved for applying the boundary conditions. The heat flux can be written as;

$$q^* \ = \ - \partial u'/\partial X \ = \ -(1/2)(\partial u'/\partial \eta)/\ \tau^{1/2}$$

$$= \ -c'/\eta/X\exp(-r^2/4\alpha t)$$

On examining the expression for heat flux there exists a singularity when $r = 0$. c' can be solved for from the boundary condition at the surface;

$$1 = -c'/\eta_R/X_R\ \exp(-R^2/4\alpha t)$$

c' can be eliminated between the solution equation for the dimensionless heat flux and the equation from the boundary condition. Thus,

$$q^* \ = \ (R^2/r^2)\ \exp(-(r^2 - R^2)/4\alpha t)$$

A singularity exists at $r = 0$ in the expression for heat flux. The singularity in time does not exist. The dimensionless temperature can be obtained as;

$$\text{u'} \qquad = c'' \qquad -(R^2/\ (2\alpha(t\tau_r)^{1/2})\exp(R^2/4\alpha t)\ \int\exp(-\eta^2)/\eta^2\ d\eta$$

In the short time limit there exists a singularity in the Fourier solution for the infinite sphere.

16. *Accumulation of Energy at the Transfer Surface*

In Eq. [1.15] write the accumulation of energy in terms of the velocity of the molecule at y = 0. What resulting expression do you get for the law of heat conduction ?

$$q_y = Z \, [½ \, m <u^2>]_{y-a} - Z \, [½ \, m <u^2>]_{y+a} - \partial/\partial t \{ Z \, [½ \, m <u^2>]_y \}$$

Assuming a linear temperature gradient the heat flux then becomes;

$$q_y = -k \, \partial T/\partial y - E \, \partial T/\partial t$$

where, E is the energy coefficient with units of (J/m^2/K). The governing equation in 1 dimension will then be;

$$\partial T/\partial t = \alpha \, \partial^2 T/\partial y^2 + \beta \, \partial^2 \, T/\partial t \partial y$$

where α is the thermal diffusivity with units of (m^2/s) and $\beta = E/\rho C_p$, with units of (m) is a characteristic distance in the microscale. In dimensionless form,

$$\partial u/\partial \tau = \partial^2 u/\partial X^2 + Be \, \partial^2 u/\partial \tau \partial X$$

where, Be = $\beta/\text{sqrt}(\alpha \tau_r)$ and τ_r can be taken as (l^2/α) and Be = (E/ρC_pl) where l is a characteristic length in the macroscale.

17. *Accumulation of Energy near the Transfer Surface*

In Eq. [1.15] write the accumulation of energy in terms of the temperature gradient at y = 0. What resulting expression do you get for the law of heat conduction ?

$$q_y = Z \, [½ \, m <u^2>]_{y-a} - Z \, [½ \, m <u^2>]_{y+a}$$
$$- \partial/\partial t \{ Z \, [½ \, m <u^2>]_{y-a} - Z \, [½ \, m <u^2>]_{y+a} \}$$

Assuming a linear temperature gradient the heat flux then becomes;

$$q_y = -k \, \partial T/\partial y - E' \, \partial^2 T/\partial t \partial y$$

where, E' is the energy conductivity coefficient with units of (J/m/K)

The governing equation in 1 dimension will then be;

$$\partial T/\partial t = \alpha\, \partial^2 T/\partial y^2 + \beta'\, \partial^2 T/\partial t \partial y^2$$

where α is the thermal diffusivity with units of (m^2/s) and $\beta' = E'/\rho C_p$, with units of (m^2) is a characteristic area in the microscale. In dimensionless form,

$$\partial u/\partial \tau = \partial^2 u/\partial X^2 + Be'\, \partial^3 u/\partial \tau \partial X^2$$

where, $Be' = \beta'/(\alpha \tau_r)$ and τ_r can be taken as (l^2/α) and $Be' = (E/\rho C_p l^2)$ where l is a characteristic length in the macroscale.

2.0 Transient Heat Conduction and Relaxation

Nomenclature

2a	width of the finite slab (m)
C_n	$4(-1)^{n+1}/(2n-1)\pi$ or $(2(1-(-1)^n)/n\pi$ depending on the problem
C_p	heat capacity (J/Kg/K)
f	Planck's constant
f(x)	function of x only
$g(\tau)$	function of time only
h	heat transfer coefficient (w/m^2 K)
I_0	modified Bessel function of zeroth order and first kind
I_1	modified Bessel function of the first order and first kind
I_2	modified Bessel function of the second order and first kind
I_3	modified Bessel function of the third order and first kind
J_0	Bessel function of the zeroth order and first kind
J_1	Bessel function of the first order and first kind
J_2	Bessel function of the second order and first kind
J_3	Bessel function of the third order and first kind
K_0	modified Bessel function of the zeroth order and second kind
k	thermal conductivity (w/m K)
n	indice of the Fourier series
p	dummy variable
q	heat flux (w/m^2/K)
S	storage coefficient, $S = (\rho C_p/\tau_r)$ w/m^3/K
t	time (s)
T	temperature (K)
T_0	initial Temperature (K)
T_1	amplitude of the temperature (K)
T_s	surface Temperature (K)
u	dimensionless temperature for semi-infinite medium
	$u = (T - T_0)/(T_s - T_0)$
	dimensionless temperature for finite slab,
	$u = (T - T_s)/(T_0 - T_s)$
U'''	heat source (J/m^3/K)
U_y	velocity of suspension (m/s)
u_{xx}	second partial derivative of u with respect to X
v_h	velocity of heat ($v_h = \text{sqrt}(\alpha/\tau_r)$, (ms)
V	function of time only
V'	first derivative of V with respect to time
V''	second derivative of V with respect to time
w	wave dimensionless temperature ($w = u\exp(\tau/2)$)
W	wave dimensionless flux ($W = q*\exp(\tau/2)$)
Wxx	second partial derivate of W with respect to X

x	distance (m)
X	dimensionless distance $X = x/\sqrt{(\alpha \tau_r)}$
X_{pen}	penetration distance in semi-infinite medium cartesian coordinates, $X_{pen} = (23.132 + \tau_i^2)^{1/2}$
	Infinite spherical medium, $X_{pen} = (58.7277 + \tau_i^2)^{1/2}$
Y	dimensionless distance $Y = y/\sqrt{(\alpha \tau_r)}$

Greek

α	thermal diffusivity (m^2/s)
σ	entropy production rate per unit volume $(w/m^3/K)$
κ	wave constant, $1/2\sqrt{(1 + 4\alpha \tau_r)}$
τ	dimensionless time (t/τ_r)
τ_r	relaxation time (s)
τ_{lag}	thermal lag time in semi-infinite medium cartesian coordinates, $(X_p^2 - 23.132)^{1/2}$
	infinite medium, spherical coordinates $(X_p^2 - 58.73)^{1/2}$
ρ	density (kg/m^3)
η	relativistic coordinate $(\tau^2 - X^2)$
ϕ	function of space only
ϕ''	second derivative with respect to space
λ_n	$(2n-1)\pi \sqrt{(\alpha \tau_r)}/2a$,
ω	frequency of the temperature wave at the surface
θ	angle of the cone

Dimensionless Groups

B	Ratio of the thermal lag times, $B = \tau_l/\tau_r$.
Bi	Biot number $(h\sqrt{(\alpha \tau_r)}/k)$
Bi*	Storage Number, h/Sa Ratio of convection to storage
X	Dimensionless distance x/L $(\sqrt{(\alpha \tau_r)/L^2})$
Pe	Peclet number (heat) U_y/v_h
q*	dimensionless heat flux $(q/\sqrt{(k\rho C_p/\tau_r)}/(T_s - T_0))$
$U^{\cdot}$	Source number, ratio of heat source to storage coefficient (U'''/S)
$Fo^{\cdot}$	Fourier modulus $(\alpha \tau_l/L^2)$
ω^*	dimensionless frequency $(\omega \tau_r)$

2.0.1 The Ballistic Term

Waves are ubiquitous in nature [Scales, Snieder, 1999] and they are central to the structure of matter and time and to many physical, biological and chemical phenomena. Wave can be seen as a propagating imbalance. For example, in the spring oscillator the kinetic energy and potential energy are interchanged during the oscillation. Hamilton's principle states that the path taken by the dynamical system is the one that minimizes the time integral of the difference between the kinetic and potential energies. It formalizes the interchange between the two forms of energy. Stable equilibriums can drive the occurrence of waves. For example, consider the gravitational or electromagnetic forces that ties matter together. For small perturbations, about a stable equilibrium point, the forces are approximately linear, a linear restoring force implies harmonic oscillation. Coupled systems of oscillators support both propagating and standing disturbances.

Thus the temperature driving force causing a heat flow when perturbed also may travel as damped waves. The restoring phenomena may be captured by the use of the relaxation time in the non-Fourier damped wave Cattaneo and Vernotte equation. The damping arising from the exponential decay of transience. Linearity also implies superposition. This means that the periodic solutions can be added to obtain finite wave 'packets'. Thus waves, are natural consequence of the stability of harmonic motion for small perturbations about an equilibrium state of coupled or extended systems. Waves are well organized. Scattering may disrupt the organization leading to diffusive behavior. For instance, when a wave breaks on a beach. The advective terms in the equation of motion couple all the different length scales in the wave, and the organization seen in the swell is destroyed. Wave propagation can often be described by a linear differential equation. Nonlinearity, sometimes may destroy the waves. The ripples on the surface of the water in the ocean seen from any beach have a period of 5 − 10 seconds. As these ripples approach the beach, their heights increase upto a point where they can no longer support their weight. They break catastrophically. These phenomena can be accounted for by the nonlinear terms in the equation of motion. In the regimes where nonlinearity becomes critical the wave motion turns into turbulent transport.

It can be interesting to calculate the point where the wave regime changes to the turbulent eddy movements. Sometimes the nonlinearity steepens the wave and provides it its identity. As the Fourier heat equation does not take into account any propagation speed it cannot be a fundamental description of the transport of heat. According to this expression, if a heat source is applied at one end of a rod the temperature far from the surface begins to change instantaneously. Maxwell working from kinetic theory [1867] imported a *ballistic* term into the equations of conduction. He ended up with the telegraph equation. The first derivative of temperature or heat flux with respect to time accounted for the diffusive behavior and the second derivative of the temperature or heat flux with respect to time accounts for the ballistic behavior. In conditions where the rate of heat conduction rapidly establishes itself the ballistic term can be neglected. In section 1.0 the regimes of wave dominance and diffusion dominance were discussed (Figure 1.1).

Ballistic heat pulses were observed experimentally as far back as the 1960s. These observations were made at low temperatures. The concept was that the heat is the manifestation of microscopic motion. The resonant frequencies of atoms or molecules in a lattice is of the of 10^{13} Hz, in the infrared region. When molecules move they give off heat. These lattice vibrations were called phonons. When the lattice is cooled to near absolute zero in one instance, the mean free scattering path of the phonons becomes comparable to the macroscopic size of the sample. At this stage, the lattice vibrations exhibit wave like behavior. Our senses of vision and hearing rely on waves. Neurons work by propagation of electric waves through the axons. Ripples of space-time are called gravitational waves. These propagate at the speed of light. Interferometers are becoming available to measure the gravitational waves. Quantum mechanics is a field of science where the waves have a central role. Einstein used the relation,

$$E = hf$$

where the energy equals Planck's constant times frequency. This connects the wave frequency of light with the energy of light's discrete quanta (photons). De Broglie extended this to electrons and other matter. Dissipation generally dampens the wave motion and ultimately everything seems to come to rest.

2.1 Semi-infinite Medium Subject to Constant Wall Temperature

The semi-infinite medium is considered to study the spatio-temporal patterns that the solution of the non-Fourier damped wave conduction and relaxation equation exhibits. This kind of consideration has been used in the study of Fourier heat conduction. The boundary conditions can be different kinds such as the constant wall temperature, CWT or the constant wall flux, CWF, pulse injection, convective, insulated and exponential decay. The similarity or Boltzmann transformation worked out well in the case of parabolic PDE as shown in the previous section. The conditions at infinite width and zero time are the same. The conditions at zero distance from the surface and at infinite time are the same. A figure of the semi-infinite medium at constant wall temperature is shown in Figure 1.3.

2.1.1   Relativistic Transformation of Coordinates

Baumesiter and Hamill [1969] solved the hyperbolic heat conduction equation in a semi infinite medium subjected to a step change in temperature at one its ends using the method of Laplace transform. The space integrated expression for the temperature in the Laplace domain had the inversion readily available within the tables. This expression was differentiated using the Leibniz's rule and the resulting temperature distribution was given for $\tau > X$ as;

$$u = (T - T_0)/(T_s - T_0) = \exp(-X/2) + X \int_X^\tau \exp(-p/2)\, I_1\, (p^2 - X^2)^{1/2}/(p^2 - X^2)^{1/2}\, dp \qquad (2.1)$$

The method of relativistic transformation of coordinates is evaluated to obtain the exact solution for the transient temperature. Consider a semi-infinite slab at initial temperature T_0, imposed by a constant wall temperature, T_s for times greater than zero at one of the ends. The transient temperature as a function of time and space in one dimension is obtained. Obtaining the dimensionless variables;

$$u = (T - T_0)/(T_s - T_0);\ \tau = t/\tau_r; X = x/\mathrm{sqrt}(\alpha \tau_r); q^* = q/(k\rho C_p/\tau_r)^{1/2}/(T_s - T_0) \qquad (2.2)$$

The energy balance on a thin spherical shell at x with thickness Δx is written in one dimension is written as $-\partial q/\partial x = \rho C_p \partial T/\partial t$. The governing equation can be obtained in terms of the heat flux after eliminating the temperature between the energy balance equation and the non-Fourier expression. This is achieved by differentiating Eq. (1.17) wrt to time and the energy balance equation wrt to x and then eliminating the second cross derivative of the temperature with respect to space and time;

$$\partial q^*/\partial \tau + \partial^2 q^*/\partial \tau^2 = \partial^2 q^* \partial X^2 \qquad (2.3)$$

It can be seen that the governing equation for the dimensionless heat flux is identical in form with that of the dimensionless temperature. The initial condition is;

$$\tau = 0, \quad q^* = 0 \qquad (2.4)$$

The Boundary Conditions are;

$$X = \infty, \quad q^* = 0 \qquad (2.5)$$

$$X = 0, \quad T = T_s\,;\ u = 1 \qquad (2.6)$$

Let us suppose that the solution for q^* is of the form $W\exp(-n\tau)$ for $\tau > 0$ where W is the transient wave flux. Then,

$$\exp(-n\tau)W(-n + n^2) + \exp(-n\tau)\, \partial W/\partial \tau\, (1 - 2n) + \exp(-n\tau)\, \partial^2 W/\partial \tau^2 = \exp(-n\tau)(\partial^2 W/\partial X^2) \qquad (2.7)$$

For $n = \frac{1}{2}$ Eq. [2.3] becomes;

$$\partial^2 W/\partial \tau^2 - W/4 = \partial^2 W/\partial X^2 \qquad (2.8)$$

The solution to Eq. [2.8] can be obtained by the following relativistic transformation of coodinates, for $\tau > X$. Let $\eta = (\tau^2 - X^2)$. Then Eq. [2.8] becomes,

$$\partial^2 W/\partial \tau^2 = 4\tau^2 \partial^2 W/\partial \eta^2 + 2\partial W/\partial \eta \qquad (2.9)$$

$$\partial^2 W/\partial X^2 = 4X^2 \partial^2 W/\partial \eta^2 - 2\,\partial W/\partial \eta \qquad (2.10)$$

Combining Eqs. [2.10, 2.11] into Eq. [2.8],

$$4(\tau^2 - X^2)\, \partial^2 W/\partial \eta^2 \ + 4\partial W/\partial \eta - W/4 \ = 0 \tag{2.11}$$

$$\text{or } \eta^2 \partial^2 W/\partial \eta^2 \ + \eta \partial W/\partial \eta - \eta W/16 = 0 \tag{2.13}$$

Eq. [2.13] can be seen to be a special differential equation in one independent variable. The number of variables in the hyperbolic PDE has thus been reduced from two to one. Comparing Eq. [2.13] with the generalized form of Bessel's Eq. [A.30] it can be seen that a = 1, b = 0, c = 0, s = ½, d = -1/16. The order of the solution is calculated as 0 and the general solution is given by;

$$W = c_1\, I_0(1/2\eta^{1/2}) + c_2\, K_0\, (1/2\eta^{1/2}) \tag{2.14}$$

The wave flux W, is finite when $\eta = 0$ and hence it can be seen that c_2 can be seen to be zero. The c_1 can be solved from the boundary condition given in Eq. [2.6]. The expression for the dimensionless heat flux for times τ, greater than X is thus,

$$q^* = c_1\, \exp(-\tau/2)I_0(1/2(\tau^2 - X^2)^{1/2}) \tag{2.15}$$

For large times, the modified Bessel's function can be given as an exponential and reciprocal in square root of time by asymptotic expansion. Consider the surface flux, i.e., when in Eq. [2.15] X is set as zero.

$$q^* = c_1\, \exp(-\tau/2)\exp(\tau/2)\, /\text{sqrt}(2\pi\tau) = c_1/\, \text{sqrt}(2\pi\tau) \tag{2.16}$$

For times, when $\exp(\tau)$ is much greater than the heat flux it can be seen that the second derivative in time of the dimensionless flux in Eq. [2.3] can be neglected compared with the first derivative. The resulting expression is the familiar expression for surface flux from the Fourier parabolic governing equation for CWT in a semi-infinite medium and is given by;

$$q^* = 1/\, \text{sqrt}(2\pi\tau) \tag{2.17}$$

Comparing Eq. [2.16] and Eq. [2.17] it can be seen that c_1 is 1. Thus the dimensionless heat flux is given by;

$$q^* = \exp(-\tau/2)I_0(1/2(\tau^2 - X^2)^{1/2}) \tag{2.18}$$

The solution for q* needs to be converted to the dimensionless temperature u and then the boundary condition applied. From the energy balance,

$$-\partial q^*/\partial X = \partial u/\partial \tau \tag{2.19}$$

Thus differentiating Eq.[2.18] wrt to X and substituting in Eq. [2.19] and integrating both sides wrt τ. For $\tau > X$,

$$u = \int X \exp(-\tau/2)I_1(1/2(\tau^2 - X^2)^{1/2})/\, (\tau^2 - X^2)^{1/2}\, d\tau \ + C(X) \tag{2.20}$$

It can be left as an indefinite integral and the integration constant can be expected to be a function of space. The $C(X)$ can be solved for by examining what happens at the wave front. At the wave front, $\eta = 0$ and time elapsed equals the time taken for a thermal disturbance to reach the location x given the wave speed sqrt(α/τ_r). The governing equations for the dimensionless heat flux and dimensionless temperature are identical in form. At the wave front, Eq. [2.12] reduces to;

$$\partial W/\partial \eta = W/16 \tag{2.21}$$

or
$$W = c'\exp(\eta/16) = c' \tag{2.22}$$

$$u = c'\exp(-\tau/2) = c'\exp(-X/2) \tag{2.23}$$

Thus $C(X) = c'\exp(-X/2)$. Thus

$$u = \int X\exp(-\tau/2)I_1(1/2(\tau^2 - X^2)^{1/2})/(\tau^2 - X^2)^{1/2}\,d\tau \quad + c'\exp(-X/2) \tag{2.24}$$

From the boundary condition in Eq. [2.6] it can be seen that $c' = 1$. Thus, for $\tau >$ X,

$$u = \int X\exp(-\tau/2)I_1(1/2(\tau^2 - X^2)^{1/2})/(\tau^2 - X^2)^{1/2}\,d\tau \quad + \exp(-X/2) \tag{2.25}$$

It can be seen that the boundary conditions are satisfied by the Eq. [2.25] and describes the transient temperature as a function of space and time that is governed by the hyperbolic wave diffusion and relaxation equation. The flux expression is given by Eq. [2.15]. This can be integrated into fluidized bed to surface heat transfer models as shown in Renganathan [1990] and Sharma and Turton [1998].

2.1.2 Regimes of Heat Flux at a Interior Point inside the Medium

It can be seen that expressions for dimensionless heat flux and dimensionless temperature given by Eq. [2.18] and Eq. [2.25] are valid only in the open interval for τ > X. When τ = X, the wavefront condition results and the dimensionless heat flux and temperature are identical and is;

$$q^* = u = \exp(-X/2) = \exp(-\tau/2) \tag{2.26}$$

When X > τ, Eq. [2.9] can be redefined as; $\eta = X^2 - \tau^2$. Eq. [2.12] becomes,

$$\eta^2\partial^2 W/\partial\eta^2 \quad + \eta\partial W/\partial\eta + \eta W/16 = 0 \tag{2.28}$$

The general solution for this Bessel equation is given by;

$$W = c_1 J_0(1/2\eta^{1/2}) + c_2 Y_0(1/2\eta^{1/2}) \tag{2.29}$$

The wave temperature W, is finite when $\eta = 0$ and hence it can be seen that c_2 can be seen to be zero. The c_1 can be solved from the boundary condition given in Eq. [2.6]. The expression in the open interval or the dimensionless heat flux for times τ, smaller than X is thus,

$$q^* = c_1 \exp(-\tau/2) J_0(1/2(X^2 - \tau^2)^{1/2}) \tag{2.30}$$

On examining the Bessel function in Eq. [2.29] it can be seen that the first zero of the Bessel occurs when the argument becomes 2.4048. Beyond that point the Bessel function will take on negative values indicating a reversal of heat flux. There is no good reason for the heat to reverse in direction at short times. Hence the Eq. [2.30] is valid from the wave front down to where the first zero of the Bessel function occurs.

By using the expression at the wave front for the dimensionless heat flux c_1 in Eq. [2.30] can be solved for and found to be 1. The Eq. [2.30] can also be obtained directly from Eq. [2.18] by using $I_0(\eta) = J_0(i\eta)$. The expression for temperature in a similar vein for the open interval $\tau < X$ is thus;

$$u = \int X \exp(-\tau/2) J_1(1/2(\tau^2 - X^2)^{1/2})/ (\tau^2 - X^2)^{1/2} d\tau \quad + \exp(-X/2) \tag{2.31}$$

Consider a point X_p in the semi-infinite medium. Three regimes can be identified during the heating of this point from the surface as a function of time. This is illustrated in Figure 2.1. The series expansion of the modified Bessel composite function of the first kind and zeroth order was used using a Microsoft Excel spreadsheet on a Pentium IV desktop microcomputer. The three regimes and the heat flux at the wave front are summarized as follows;

1) The first regime is a thermal inertia regime when there is no transfer.

2) The second regime is given by expression Eq. [2.30] for the heat flux and

$$q^* = \exp(-\tau/2) J_0(1/2(X^2 - \tau^2)^{1/2}) \tag{2.32}$$

The first zero of the zeroth order Bessel function of the first kind occurs at 2.4048. This is when,

$$2.4048 = 1/2(X^2 - \tau^2)^{1/4} \quad \text{or} \quad \tau_{\text{lag}} = \text{sqrt}(X^2 - 23.132) \tag{2.33}$$

Thus τ_{lag} is the thermal lag that will ensue before the heat flux is realized at an interior point in the semi-infinite medium at a dimensionless distance X from the surface. In Figure 2.1 one value of X is used, i.e., 5. Thus for points closer to the surface the time lag may be zero. Only for dimensionless distances greater than 4.8096, the time lag is finite. For distances *closer then 4.8096sqrt($\alpha\tau_r$)* the thermal lag experienced *will be zero*. For distances,

$$x > 4.8096\text{sqrt}(\alpha\tau_r) \tag{2.34}$$

The time lag experienced is given by Eq. [2.33] and is sqrt$[X^2 - 4\beta_1^2]$ where, β_1 is the first zero of the Bessel function of the first kind and zeroth order and is 2.4048. In a similar fashion, the penetration distance of the disturbance for a considered instant in time, beyond which the change in initial temperature is zero can be calculated as;

$$X_{pen} = (23.132 + \tau_i^2)^{1/2}$$

3) The third regime starts at the wavefront and described by Eq. (15). This is shown in Figure 2.0 for a point in the interior of the semi-infinite medium at a dimensionless distance 5 from the surface.

$$q^* = c_1 \exp(-\tau/2)I_0(1/2(\tau^2 - X^2)^{1/2}) \qquad (2.35)$$

4) At the wave front, $q^* = u = \exp(-X/2) = \exp(-\tau/2)$

2.1.3 Approximate Solution

The expressions for transient temperature derived in the above section needs integration prior to use. More easily usable expressions can be developed by making suitable approximations. Realizing that a for PDE a set of functions instead of constants as in the case of ODE needs to be solved from the boundary conditions the c in Eq. [2.16] is allowed to vary with time. This results in an expression for transient temperature that is more readily available for direct use of the practitioner. Extensions to three dimensions in space are also straight forward in this method.

In this section, the exact solution for the CWT problem in semi-infinite medium in 1 dimension is revisited since the discussion by the method of Laplace transforms by Baumeister & Hamill. An expression that does not need further integration is attempted to be derived in this section. Consider a semi-infinite slab at initial temperature T_0, subjected to sudden change in temperature at one of the ends to T_1. The heat propagative velocity is Vh = sqrt (α/τ_r). The initial condition

$$t = 0, Vx, \quad T = T0 \qquad (2.36)$$

$$T > 0, \quad x = 0, \quad T = T1 \qquad (2.37)$$

$$T > 0, x = \infty \quad T = T_0 \qquad (2.38)$$

Obtaining the dimensionless variables;

$$u = (T - T_0)/(T_1 - T_0) \quad ; \tau = t/\tau_r \quad ; X = x/sqrt(\alpha \tau_r) \qquad (2.39)$$

The energy balance on a thin spherical shell at x with thickness Δx is written. The governing equation cab be obtained after eliminating q between the energy balance equation and the derivative with respect to x of the flux equation and introducing the dimensionless variables;

$$\partial u/\partial \tau \; + \; \partial^2 u/\partial \tau^2 \quad = \; \partial^2 u/\partial X^2 \qquad (2.40)$$

Suppose $u = \exp(-n\tau)\, w\,(X, \tau)$. By choosing $n = \frac{1}{2}$, the damping component of the equation is removed. Thus for $n = \frac{1}{2}$, Eq. [2.40] becomes,

$$-w/4 \; + \; \partial^2 w/\partial \tau^2 \; = \; \partial^2 w/\partial X^2 \qquad (2.41)$$

By inspecting the flux solution obtained by method of Laplace transforms, [Sharma, 2003], consider the transformation variable η as;

$$\eta \; = \; \tau^2 - X^2, \; \text{for } \tau > X \qquad (2.42)$$

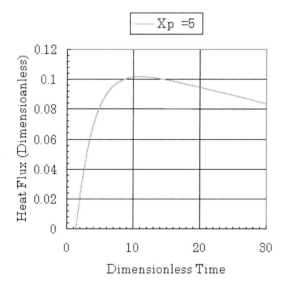

Figure 2.1   Three Regimes of Heat Flux in an Interior Point in a Semi-Infinite Medium

$$\partial w/\partial \tau \; = \; (\partial w/\partial \eta\,)2\tau \qquad (2.43)$$

$$\partial^2 w/\partial \tau^2 \; = \; (\partial^2 w/\partial \eta^2\,)4\tau^2 \; + \; 2(\partial w/\partial \eta\,) \qquad (2.44)$$

In a similar fashion,

$$\partial^2 w/\partial X^2 = (\partial^2 w/\partial\eta^2)4X^2 + 2(\partial w/\partial\eta) \qquad (2.45)$$

Substituting Eqs. [2.44 – 2.46] into Eq. [2.42]

$$(\partial^2 w/\partial\eta^2)4(\tau^2 - X^2) + 4(\partial w/\partial\eta) \quad -w/4 = 0 \qquad (2.46)$$

$$\eta^2\partial^2 w/\partial\eta^2 + \eta\partial w/\partial\eta - \eta w/16 = 0 \qquad (2.47)$$

Eq. [2.47] can be recognized as the modified Bessel equation.

$$w = c_1 I_0(sqrt(\eta)/2) + c_2 K_0(sqrt(\eta)/2) \qquad (2.48)$$

For $X = 0$, u is 1 or finite and hence it can be seen that w is also finite and hence $C_2 = 0$. Writing the expression for u,

$$\text{So } u = c_1 exp(-\tau/2)(I_0(sqrt(\eta)/2)) \qquad (2.49)$$

From BC, Eq. [2.49] becomes,

$$1 = c_1 exp(-\tau/2)I_0(\tau/2) \qquad (2.50)$$

c_1 can be eliminated by dividing Eq. [2.50] by Eq. [2.49] to yield;

$$u = (I_0(sqrt(\tau^2 - X^2))/2))/(I_0(\tau/2)) \qquad (2.51)$$

for $\tau > X$. for $\tau < X$ or $t < x/v_h$

$$u = J_0(sqrt(X^2 - \tau^2))/2))/(I_0(\tau/2)) \qquad (2.52)$$

Eq. [2.52-2.51] is shown in Figure 2.1 for dimensionless heat flux in a semi-infinite slab maintained at constant wall temperature at one of its edges. It can be inferred that an expression in time is used for for c_1. A domain restricted solution for short and long times may be in order. For long times, $I_0(\tau/2)$, approximates as $exp(\tau/2)/sqrt(2\pi\tau)$.

$$1 = c_1 exp(-\tau/2)(I_0(\tau/2)) = c_1 exp(-\tau/2) exp(\tau/2)/sqrt(2\pi\tau) \qquad (2.53)$$

$$c_1 = sqrt(2\pi\tau) \qquad (2.54)$$

$$\text{or } u = sqrt(2\pi\tau) exp(-\tau/2)(I_0(sqrt(\tau^2 - X^2))/2) \qquad (2.55)$$

Thus the temperature solution in a semi-infinite medium when subject to a step change in temperature at one of the ends is obtained as an open interval solution. The dimensionless flux as earlier derived is;

$$q = q''/sqrt(k C_p \rho/\tau_r)(T_1 - T_0) = exp(-\tau/2)(I_0(1/2sqrt(\tau^2 - X^2)) \qquad (2.56)$$

Further the instantaneous surface flux is when $X = 0$;

$$q''/\text{sqrt}(k\ C_p\ \rho\ /\tau_r\ )\ (T_1 - T_0\ )\ =\ \exp\ (-\tau/2)I_0\ (\tau/2\ ) \tag{2.57}$$

2.1.4 Extension to 3-Dimensions

Eq. [2.41] in 3 dimensions can be written as;

$$\partial u/\partial\tau\ +\ \partial^2\ u/\partial\tau^2\quad =\ \partial^2\ u/\partial X^2\ +\ \partial^2\ u/\partial Y^2\ +\ \partial^2\ u/\partial Z^2 \tag{2.58}$$

After dividing by $\exp(-\tau/2)$, the equation becomes,

$$-w/4\ +\ \partial^2\ w/\partial\tau^2\ =\ \partial^2\ w/\partial X^2 +\ \partial^2\ w/\partial Y^2\ +\ \partial^2\ w/\partial Z^2 \tag{2.59}$$

Let $$\eta\ =\ \tau^2 - X^2\ -\ Y^2\ -\ Z^2 \tag{2.60}$$

$$\partial^2\ w/\partial\eta^2\ )4(\ \tau^2 - X^2 -\ Y^2\ -\ Z^2\ )\ +\ 8(\partial\ w/\partial\eta\ )\qquad -w/4 = 0 \tag{2.61}$$

$$\eta^2\partial^2\ w/\partial\eta^2\ +2\eta\partial\ w/\partial\eta\ -\ \eta\,w\ /16\ \ = 0 \tag{2.62}$$

Comparing Eq. [2.63] with the generalized Bessel equation (Eq. [2.104]),

$b = 0;\ a = 2;\ c = 0;\ s = \frac{1}{2};\ d = -1/16\ ;\ p = 2.\ \text{sqrt}(1/4) = 1\ (\text{order})$

$$w\ =\ c_1\eta^{1/2}\ I_1(\text{sqrt}(\eta)/2)\ + c_2\,\eta^{1/2}\ K_1(\text{sqrt}(\eta)/2)\ ) \tag{2.63}$$

It can be deduced that c_2 is zero as w is finite when $\eta = 0$. The boundary condition as a point temperature at the origin gives rise to for $\tau > \text{sqrt}(X^2 + Y^2 + Z^2)$;

$$u\ =\ [\tau\ /\ \text{sqrt}(\tau^2\ - X^2 - Y^2 - Z^2\ )]\ I_1\ (\text{sqrt}(\tau^2\ - X^2 - Y^2 - Z^2\ )/2)\ /\ I_1(\ \tau/2) \tag{2.64}$$

For $\text{sqrt}(X^2 + Y^2 + Z^2) > \tau$,

$$u\ =\ [\tau\ /\ \text{sqrt}(X^2 + Y^2 + Z^2 - \tau^2\ )]\ J_1\ (\text{sqrt}(\ X^2 + Y^2 + Z^2 - \tau^2)\ /2)\ /\ I_1(\tau/2) \tag{2.64}$$

Worked Example 2.1

It can be seen that $w = c\, I_0\, (\mathrm{sqrt}(\tau^2 - X^2)\,)/2)$ is a solution to the damped wave conduction and relaxation equation with the damping term removed;

$$\partial^2 w/\partial\tau^2 \ - \ w/4 \ = \ \partial^2 w/\partial X^2 \tag{2.65}$$

Under what conditions can c be a function of time? Let ψ be a solution to Eq. [2.65]. Then;

$$-\psi/4 \ + \ \partial^2 \psi/\partial\tau^2 \ = \ \partial^2 \psi/\partial X^2 \tag{2.66}$$

Substituting $c\,(\tau)\psi$ in Eq. [2.64] we find when $c\psi$ will also be a solution.

$$\partial^2 w/\partial\tau^2 \ = \ \partial^2 c/\partial\tau^2\ \psi \ + \ c\ \partial^2\psi/\partial\tau^2 \ + 2\partial c/\partial\tau\ \partial\psi/\partial\tau \tag{2.67}$$

$$\partial^2 w/\partial X^2 \ = \ c\ \partial^2\psi/\partial X^2 \tag{2.68}$$

Substituting Eqs. [2.67, 2.68] into Eq. [2.66];

$$\partial^2 c/\partial\tau^2\ \psi \ + \ c\ \partial^2\psi/\partial\tau^2 \ + 2\partial c/\partial\tau\ \partial\psi/\partial\tau \ - c\psi/4 \ = \ c\ \partial^2 \psi/\partial X^2 \tag{2.69}$$

Eq. [2.69] can be written as two equations;

$$c\ \partial^2\psi/\partial\tau^2 \ - \ c\psi/4 \ = \ c\ \partial^2 \psi/\partial X^2 \quad\quad \text{or} \quad\quad -\psi/4 \ + \ \partial^2 \psi/\partial\tau^2 \ = \ \partial^2 \psi/\partial X^2 \tag{2.70}$$

$$\partial^2 c/\partial\tau^2\ \psi \ + 2\partial c/\partial\tau\ \partial\psi/\partial\tau \ = 0 \tag{2.71}$$

Eq. [2.70] is obeyed when $c \neq 0$.   When c is chosen in a fashion that Eq. [2.71] is obeyed then, $c\psi$ will also be a solution to the damped wave conduction and relaxation equation. When,

$$\partial p/\partial\tau \ = \ -2c'p, \text{ where, } \ p \ = \partial c/\partial\tau,\ c' \text{ is a constant} \tag{2.72}$$

or when, $\ p \ = c''\exp(-2c'\tau) \ $ or when $c = -c''/2c'\exp(-2c'\tau) \ + c'''$ (2.73)

the $c\psi$ will also be a solution to the wave equation where ψ is the solution. c can be a exponential in time as shown in Eq. [2.73].

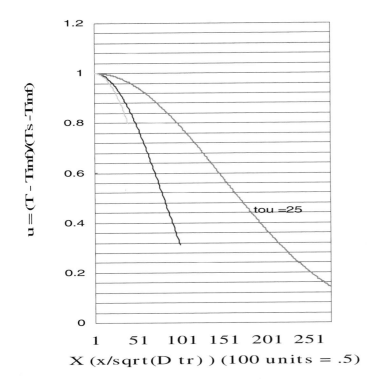

Figure 2.2 Transient Wave Conduction and Relaxation in a Semi-infinite Slab Subject to Constant Wall Temperature

Worked Example 2.2

Verify $w = c\, I_0\, (\sqrt{\tau^2 - X^2}\,)/2)$ is a solution to the Eq. [2.41].

$$-w/4 \;+\; \partial^2 w/\partial\tau^2 \;=\; \partial^2 w/\partial X^2 \tag{2.74}$$

$$w = c \sum_{n=0}^{n=\infty} (\tau^2 - X^2)^n/4^{2n}/n!^2 \tag{2.75}$$

$$\partial w/\partial\tau = 2c\tau \sum_{n=0}^{n=\infty} n(\tau^2 - X^2)^{n-1}/4^{2n}/n!^2 \tag{2.76}$$

$$n=\infty \qquad\qquad\qquad n=\infty$$

$$\partial^2 w/\partial\tau^2 \;=\; 2c \sum_{n=0} \; n(\tau^2 - X^2)^{n-1}/4^{2n}/n!^2 \;+\; 4c\tau^2 \sum_{n=0} \; n(n-1)(\tau^2 - X^2)^{n-2}/4^{2n}/n!^2 \tag{2.77}$$

$$\partial^2 w/\partial X^2 \;=\; -2c \sum_{n=0}^{n=\infty} \; n(\tau^2 - X^2)^{n-1}/4^{2n}/n!^2 \;+\; 4cX^2 \sum_{n=0}^{n=\infty} \; n(n-1)(\tau^2 - X^2)^{n-2}/4^{2n}/n!^2 \tag{2.78}$$

Now, when Eq. [2.77] – Eq. [2.78] = Eq. [2.75] /4 the PDE in Eq. [2.74] is obeyed. Eq. [2.77] - Eq. [2.78] is;

$$4c/16 \sum_{n=0}^{n=\infty} \; ((\tau^2 - X^2)^{1/2}/4)^{2n-2}/n!/(n-1)! \;+\; c16/64 \sum_{n=0}^{n=\infty} \; (\tau^2 - X^2)/4)^{2n-2}/(n-2)!/n! \tag{2.79}$$

$$\text{or } c/2 I_{-1}(\text{sqrt}(\tau^2 - X^2)/2)/\text{sqrt}(\tau^2 - X^2)) \;+\; c/4 \, (I_{-2}(\text{sqrt}(\tau^2 - X^2))/2) \tag{2.80}$$

Using the recurrence relation;

$$I_0(\beta x) \;=\; I_2(\beta x) \;+\; 2 \, I_1(\beta x) / (\beta x) \tag{2.81}$$

$$\text{and I-p}(\beta x) \;=\; I_p(\beta x) \tag{2.82}$$

Eq. [2.74] is thus verified by using the recurrence relations and rearranging Eq. [2.80] according to Eq. [2.81, 2.82].

$$c/4 \, (I_2(\text{sqrt}(\tau^2 - X^2))/2) \;+\; c/4(2I_1(\text{sqrt}(\tau^2 - X^2)/2)/ \text{sqrt}(\tau^2 - X^2))) \;=\; cI_0(\text{sqrt}(\tau^2 - X^2)/2)/4$$

Worked Example 2.3

Verify that $w = c\text{Sin}(\kappa X - \omega\tau) + d\text{Cos}(\kappa X - \omega\tau)$ is a solution to the wave equation,

$$-w/4 \;+\; \partial^2 w/\partial\tau^2 \;=\; \partial^2 w/\partial X^2 \tag{2.83}$$

$$\partial^2 w/\partial X^2 \;=\; -\kappa^2 \, (c\text{Sin}(\kappa X - \omega\tau) + d\text{Cos}(\kappa X - \omega\tau)) \tag{2.84}$$

$$\partial^2 w/\partial\tau^2 \;=\; -\omega^2(c\text{Sin}(\kappa X - \omega\tau) + d\text{Cos}(\kappa X - \omega\tau)) \tag{2.85}$$

Now inorder to obey Eq. [2.83]

$$\text{Eq.}[2.85] – \text{Eq.}[2.84] \;=\; w/4$$

Thus,

$$(\kappa^2 - \omega^2) \, (c\text{Sin}(\kappa X - \omega\tau) + d\text{Cos}(\kappa X - \omega\tau)) \;=\; c\text{Sin}(\kappa X - \omega\tau) + d\text{Cos}(\kappa X - \omega\tau))/4 \tag{2.86}$$

or $\qquad (\kappa^2 - \omega^2) = \frac{1}{4}$ $\qquad\qquad$ (2.87)

or $\qquad \kappa^2 = \omega^2 + \frac{1}{4}$ $\qquad\qquad$ (2.88)

When $\omega\tau = v_h t$, $\omega = \text{sqrt}(\alpha\tau_r)$, where v_h is the velocity of heat propagation $= \text{sqrt}(\alpha/\tau_r)$.

Then, $\kappa = 1/2\text{sqrt}(1 + 4\alpha\tau_r)$ $\qquad\qquad$ (2.89)

The implications are that the dimensionless temperature,

$u = \exp(-\tau/2)\,(c\text{Sin}(\,x\text{sqrt}(1 + 1/4\alpha\tau_r\,) - v_h t) + d\text{Cos}(\,x\text{sqrt}(1 + 1/4\alpha\tau_r\,) - v_h t)\,)$

will obey the damped wave governing equation;

$$\partial u/\partial\tau + \partial^2 u/\partial\tau^2 = \partial^2 u/\partial X^2$$

This is a solution to the damped wave conduction and relaxation equation. The general solution of the equation is however a different matter. This was discussed in the main section for the semi-infinite medium. The dimensionless heat flux was given after applying the boundary conditions to the general solution. The dimensionless temperature can be obtained from the energy balance and by integrating the flux expression appropriately.

The approximate solution to the dimensionless temperature can be written in a more useful form by realizing that the general solution of the PDE consists of n arbitrary functions for a nth order PDE as compared with a ODE of nth order with n arbitrary constants. The wave function that obeys the governing PDE and can be used for certain applications. The wave velocity and lag terms are captured in the composite variable of $(\kappa X - \omega\tau)$.

Worked Example 2.4

Discuss the transformation, $\eta = \text{sqrt}(\tau^2 - X^2)$, in the open interval, for $\tau > X$, in the transformation of the PDE given by Eq. [2.83] into a Bessel equation.

$$\partial w/\partial\tau = \partial w/\partial\eta\;\tau/\text{sqrt}((\tau^2 - X^2)) = \partial w/\partial\eta\;\tau/\eta \qquad (2.90)$$

$$\partial^2 w/\partial\tau^2 = (\tau^2/\eta^2)\,\partial^2 w/\partial\eta^2 - \partial w/\partial\eta\;\tau^2/\eta^3 + \partial w/\partial\eta/\eta \qquad (2.91)$$

$$\partial^2 w/\partial X^2 = (X^2/\eta^2)\,\partial^2 w/\partial\eta^2 - \partial w/\partial\eta\;X^2/\eta^3 - \partial w/\partial\eta/\eta \qquad (2.92)$$

The governing equation then becomes;

$$(\tau^2 \partial^2 w/\partial\eta^2\ )\ \partial^2 w/\partial\eta^2\ -\ \partial w/\partial\eta\ \tau^2/\eta^3\ -\ w/4 +\ 2\ \partial w/\partial\eta/\eta\ =(X^2/\eta^2)\ \partial^2 w/\partial\eta^2\ -\ \partial w/\partial\eta\ X^2/\eta^3 \tag{2.93}$$

or
$$\eta^2/\eta^2\ \partial^2 w/\partial\eta^2\ +(2-1)/\eta\ \partial w/\partial\eta\ -w/4 = 0 \tag{2.94}$$

$$\eta^2 \partial^2 w/\partial\eta^2\ +\ \eta\ \partial w/\partial\eta\ -\eta^2 w/4 = 0 \tag{2.95}$$

Compared with the generalized Bessel equation [A.30] ;

$$b=0;\ a = 1;\ c = 0;\ d = -1/4;\ s = 1$$

The order p of the Bessel solution is then $p = 1/s\ \mathrm{sqrt}((1-a)^2/4\ -\ c)) = 0$.
and $\mathrm{sqrt}(|\,d\,|/s\ ) = \tfrac{1}{2}$

$$\text{and}\ \ w\ =\ c_1 I_0(\mathrm{sqrt}(\eta)/2)\ \ + c_2\ K_0(\mathrm{sqrt}(\eta)/2)$$

Thus the substitution , $\eta = \mathrm{sqrt}(\tau^2 - X^2)$, results in the same Bessel equation as the substitution , $\eta = (\tau^2 - X^2)$.

Worked Example 2.5

The c_1 was eliminated between the Bessel solution and the boundary condition at the surface. Backing out an expression for c_1 it can be seen to be a mild function of time. In Worked Example 2.1 the $\dot{c}_1$ being a function of time upto exponential was tolerated, still obeying the hyperbolic damped wave conduction and relaxation equation. Assume a $g(\tau)$ such that,

$$g(\tau)\ I_0(\tau/2)\ = \exp(\tau/2) \tag{2.97}$$

Examine $g(\tau)$. Differentiating Eq. [2.97] wrt time,

$$g'I_0(\tau/2)\ + g/2 I_1(\tau/2)\ =\ \tfrac{1}{2}\ \exp(\tau/2) \tag{2.98}$$

Invoking Eq. [2.108] for g,

$$g'\ I_0(\tau/2)\ + I_1(\tau/2)\ \exp(\tau/2)/\ I_0(\tau/2)/2 =\ \tfrac{1}{2}\ \exp(\tau/2) \tag{2.99}$$

$$g'I_0(\tau/2)\ =\ 1/2\exp(\tau/2)(\ 1-\ \ I_1(\tau/2)/\ I_0(\tau/2)\ ) \tag{2.100}$$

Invoking Eq. [2.108] for $I_0(\tau/2)$

$$dg/g\ \ \ \ \ =\ d\tau/2(\ 1-\ \ I_1(\tau/2)/\ I_0(\tau/2)\ ) \tag{2.101}$$

Integrating both sides,

$$\ln(g)\ \ \ \ =\ \int_0^{\infty} d\tau/2(\ \ (\ 1-\ \ I_1(\tau/2)/\ I_0(\tau/2)\ ) \tag{2.102}$$

$$g\ \ \ \ \ \ \ =\ \exp(\int d\tau/2((\ 1-\ \ I_1(\tau/2)/\ I_0(\tau/2)\ ) \tag{2.103}$$

In the limit of infinite dimensionless time g becomes, 1. In the limit of zero time, from Eq. [2.103] g becomes 1. Thus 1 is a good supposition for c_1. The results presented by eliminating c_1 can yield a more usable form of the solution.

Worked Example 2.6

Obtain the solution of the following equation for the dimensionless temperature in a semi-infinite medium subject to CWT at one of its surfaces after the damping term is removed by the method of Laplace transforms.

$$-w/4 \ + \ \partial^2 w/\partial \tau^2 \ = \ \partial^2 w/\partial X^2 \tag{2.104}$$

Obtaining the Laplace transform of the damped wave conduction and relaxation equation with the damping term removed from the equation.

$$-\underline{w}/4 + s^2\underline{w} \quad = d^2\underline{w}/dX^2 \tag{2.105}$$

At t= 0, u = 0 as given in the time and boundary conditions in Eq. [2.36-2.38]. w = $u\exp(\tau/2)$ and is also zero. So is $\partial w/\partial \tau \ = \ \partial u/\partial \tau \exp(\tau/2) + u/2\exp(\tau/2) = 0$ at time = 0. Thus solving the second order ODE in the Laplace domain;

$$\underline{w} \quad = \quad a_1 \exp(+X\mathrm{sqrt}(s^2 - \tfrac{1}{4}) \ ) + \ a_2 \exp(-X\mathrm{sqrt}(s^2 - \tfrac{1}{4}) \tag{2.106}$$

At X = ∞, u = 0 and $\underline{w}$ = 0. Hence a_1 can be seen to be zero.

$$\underline{w} \quad = \quad + a_2 \exp(-X\mathrm{sqrt}(s^2 - \tfrac{1}{4}) \tag{2.107}$$

From the boundary condition at X = 0,

$$u = 1 \quad \text{and} \ w = \exp(\tau/2) \tag{2.108}$$

The Laplace transform is;

$$\underline{w} \ = \ 1/(s-1/2) \tag{2.109}$$

Thus, a_2 can be seen to be 1/(s-1/2)

$$\underline{w} \quad = \quad 1/(s-1/2)\exp(-X\mathrm{sqrt}(s^2 - \tfrac{1}{4}) \tag{2.110}$$

From the Laplace inversion tables [Mickley, Sherwood and Reed, 1957],

$$L^{-1} \ (1/(s^2-1/4)^{1/2}\exp(-X\mathrm{sqrt}(s^2 - \tfrac{1}{4}) \ = \ I_0 1/2\mathrm{sqrt}(\tau^2 - X^2) \tag{2.111}$$

Eq. [2.122] can be rewritten as;

$$\underline{w} = (1/(s^2-1/4)^{1/2}\exp(-X\mathrm{sqrt}(s^2 - \tfrac{1}{4}) \ [(s+1/2)/(s -1/2)]^{1/2} \ = \ h(s)g(s) \tag{2.112}$$

64

$$h(s) = (1/(s^2-1/4)^{1/2}\exp(-X\mathrm{sqrt}(s^2 - \tfrac{1}{4}) \; ; h(\tau) = I_0 1/2\mathrm{sqrt}(\tau^2 - X^2) \tag{2.113}$$

$$g(s) = [(s+1/2)/(s -1/2)]^{1/2} \tag{2.114}$$

$g(\tau)$ can be obtained by multiplying the numerator and denominator by sqrt(s+1/2) and expressing the result as a sum of two terms;

$$g(s) \quad = s/\mathrm{sqrt}(s^2 - \tfrac{1}{4}) + 1/2/\mathrm{sqrt}(s^2 - \tfrac{1}{4}) \tag{2.115}$$

Obtaining the Laplace inverse of Eq. [2.115]

$$g(\tau) \quad = I_0(\tau/2) + 1/2[\mathrm{sqrt}(\pi)/\mathrm{gamma}(1/2)\, I_0(\tau/2) \tag{2.116}$$

An approximation to Eq. [2.116] is c exp(τ/2) at short and long times as shown in Worked Example 2.5, where,

$$c = 1 + 1/2\mathrm{sqrt}(\pi)/\mathrm{gamma}(1/2) \tag{2.117}$$

By using the convolution property of Laplace inversion;

$$w = c\int_0^\tau \exp((\tau-p)/2)\, I_0(1/2\mathrm{sqrt}(p^2 - X^2))dp \tag{2.118}$$

2.2 Relativistic Transformation of Cylindrical Coordinates in Infinite Medium

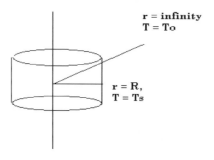

r = infinity
T = To

r = R,
T = Ts

Figure 2.3 Semi-Infinite Medium in Cylindrical Coordinates Heated from a Cylindrical Surface

Consider a fluid at a initial temperature T_0. The surface of the cylinder is maintained at a constant temperature T_s for times greater than zero. The heat propagative velocity is given as the square root of the ratio of the thermal diffusivity and relaxation time. V_h = sqrt(α/τ_r). The two time conditions, initial and final and the two boundary conditions are;

$$t = 0, \ r > R, \ T = T_0 \tag{2.119}$$

$$t > 0, \ r = R, \ T = T_s \tag{2.120}$$

$$r = \infty, \ t > 0, \ T = T_0 \tag{2.121}$$

The governing equation in temperature is obtained by eliminating the second cross-derivative of heat flux wrt to r and t between the non-Fourier damped wave heat conduction and relaxation equation and the energy balance equation in cylindrical coordinates (Figure 2.4). Considering a cylindrical shell of thickness Δr,

$$\Delta t(\ 2\pi r L \ q_r \ - \ 2\pi(r + \Delta r)L \ q_{r+\Delta r} \ ) =(\ (\rho C_p) \ 2\pi L r \Delta r \ \Delta T) \tag{2.122}$$

In the limit of Δr, Δt going to zero, the energy balance equation in cylindrical coordinates becomes;

$$-1/r \partial(r q_r)/\partial r \ = \ (\rho C_p)\partial T/\partial t \tag{2.123}$$

The non-Fourier damped wave equation is;

$$q_r \ = \ -k \ \partial T/\partial r \ - \ \tau_r \ \partial q_r/\partial t \tag{2.124}$$

Multiplying Eq. [2.124] by r and differentiating wrt to r and then dividing by r,

$$(1/r)\partial(r q_r \ )/\partial r \ = \ (-k/r)\partial(\ r\partial T/\partial r)/\partial r \ - \ (\tau/r) \ \partial 2(r q r)/\partial t\partial r \tag{2.125}$$

Differentiating Eq. [2.123] wrt t.

$$-1/r \partial^2(r q_r)/\partial t \partial r \ = \ (\rho C_p)\partial^2 \ T/\partial t^2 \tag{2.126}$$

Substituting Eq. [2.126] and Eq. [2.125] into Eq. [2.123] the governing equation in temperature is obtained as;

$$(\rho C_p \tau_r)\partial^2 \ T/\partial t^2 \ + \ (\rho C_p) \ \partial T/\partial t \ = \ (k/r)\partial(\ r\partial T/\partial r)/\partial r \tag{2.127}$$

Obtaining the dimensionless variables,

$$u = (\ T - T_0 \ )/(T_s - T_0); \ \tau = (t/\tau_r); \ X = r/sqrt(\alpha \tau_r) \tag{2.128}$$

The governing equation in the dimensionless form can be written as;

$$\partial u/\partial \tau \ + \ \partial^2 \ u/\partial \tau^2 \ = \ \partial^2 u/\partial X^2 \ + 1/X \ \partial u/\partial X \tag{2.129}$$

The damping term is removed from the governing equation. This is done realizing that the transient temperature decays with time in a exponential fashion. The other reason for this manoeuvre is to study the wave equation without the damping term. Let $u = w\exp(-\tau/2)$ and the damping component of the equation is removed to yield;

$$-w/4 \; + \; \partial^2 w/\partial\tau^2 \; = \; \partial^2 w/\partial X^2 \qquad + 1/X \; \partial w/\partial X \qquad (2.130)$$

Eq. [2.130] can be solved by using the method of relativistic transformation of coordinates. Consider the transformation variable η as;, $\eta = \tau^2 - X^2$, for $\tau > X$. The governing equation becomes as shown in the above sections,

$$(\partial^2 w/\partial\eta^2)4(\tau^2 - X^2) \; + \; 6(\partial w/\partial\eta) \; -w/4 = 0 \qquad (2.131)$$

$$4\eta^2\partial^2 w/\partial\eta^2 \; + 6\eta\partial w/\partial\eta \; - \; \eta w /4 \; = 0 \qquad (2.132)$$

or
$$\eta^2\partial^2 w/\partial\eta^2 \; + 3/2\eta\partial w/\partial\eta \; - \; \eta w/16 \; = 0 \qquad (2.134)$$

Comparing Eq. [2.134] with the generalized Bessel equation as given in Eq. [2.96], the solution is;

$$a = 3/2; \; b = 0; \; c = 0; \; d = -1/16; \; s = \tfrac{1}{2}$$

The order p of the solution is then $p = 2 \; \mathrm{sqrt}(1/16) = \tfrac{1}{2}$

$$W = c_1 I_{1/2} (1/2 \; \mathrm{sqrt}(\tau^2 - X^2)/(\tau^2 - X^2)^{1/4} \; + \; c_2 I_{-1/2} (1/2 \; \mathrm{sqrt}(\tau^2 - X^2)/(\tau^2 - X^2)^{1/4} \qquad (2.135)$$

c_2 can be seen to be zero as W is finite and not infinitely large at $\eta = 0$. An approximate solution can be obtaining by eliminating c_1 between Eq. [2.131] and the above equation. It can be noted that this is a mild function of time, however. As the general solution of PDE consists of n arbitrary functions when the order of the PDE is n compared with n arbitrary constants for ODE. From the boundary condition at, $X = X_R$,

$$1 = \exp(-\tau/2) \; c_1 I_{1/2} (1/2 \; \mathrm{sqrt}(\tau^2 - X_R^2)/(\tau^2 - X_R^2)^{1/4} \qquad (2.136)$$

$$u = [(\tau^2 - X_R^2)^{1/4}/(\tau^2 - X^2)^{1/4}][I_{1/2} (1/2 \; \mathrm{sqrt}(\tau^2 - X^2) / I_{1/2} (1/2 \; \mathrm{sqrt}(\tau^2 - X_R^2)] \qquad (2.137)$$

In terms of elementary functions, Eq. [2.139] can be written as;

$$\text{or} \quad u = [(\tau^2 - X_R^2)^{1/2}/(\tau^2 - X^2)^{1/2}][\mathrm{Sinh}(1/2 \; \mathrm{sqrt}(\tau^2 - X^2) /\mathrm{Sinh}(1/2 \; \mathrm{sqrt}(\tau^2 - X_R^2)] \qquad (2.138)$$

In the limit of X_R going to zero, the expression becomes;

$$\text{or} \quad u = [\tau/(\tau^2 - X^2)^{1/2}][\mathrm{Sinh}(1/2 \; \mathrm{sqrt}(\tau^2 - X^2) / \mathrm{Sinh}(\tau/2)] \qquad \text{for } \tau > X, \qquad (2.139)$$

for $X > \tau$,

$$u = [(X_R^2 - \tau^2)^{1/4}/(X^2 - \tau^2)^{1/4}][J_{1/2}(1/2 \text{ sqrt}(X^2 - \tau^2)/I_{1/2}(1/2 \text{ sqrt}(\tau^2 - X^2)] \qquad (2.140)$$

Eq. [2.142] can be written in terms of trigonometric functions as;

$$u = [(X_R^2 - \tau^2)^{1/2}/(X^2 - \tau^2)^{1/2}][\text{Sin}(1/2 \text{ sqrt}(X^2 - \tau^2)/\text{Sinh}(1/2 \text{ sqrt}(\tau^2 - X^2)] \qquad (2.141)$$

In the limit of X_R going to zero, the expression becomes;

$$\text{or } u = [\tau/(X^2 - \tau^2)^{1/2}][\text{Sin}(1/2 \text{ sqrt}(X^2 - \tau^2)/\text{Sinh}(\tau/2)] \qquad (2.142)$$

The dimensionless temperature at a point in the medium at $X = 7$ for example is considered and shown in Figure 2.4. Three different regimes can be seen. The first regime is that of the thermal lag and consists of no change from the initial temperature. The second regime is when

$$\tau_{lag}^2 = X^2 - 4\pi^2 \quad \text{or } \tau_{lag} = \text{sqrt}(X_p^2 - 4\pi^2) = 3.09 \qquad \text{when } X_p = 7 \qquad (2.143)$$

For times greater than the time lag and less than X_p the dimensionless temperature is given by Eq. [2.142]. For dimensionless times greater than 7, the dimensionless temperature is given by Eq. [2.139]. For distances *closer to the surface compared with* 2π *the time lag will be zero.*

Worked Example 2.7

Obtain the governing equation for heat flux in the radial direction. Solve for the heat flux using the method relativistic transformation of coordinates. From the solution of the radial heat flux obtain the exact solution for the temperature for a semi-infinite medium heated from a cylindrical surface at constant wall temperature T_s at $r = R$.

Eliminating the second order cross-derivative of temperature wrt to time and radial position the governing equation terms of the radial flux is given by;

$$\partial q_r/\partial t + \tau_r \partial^2 q_r/\partial t^2 = (\alpha/r)\partial/\partial r(r(\partial q_r/\partial r)) \qquad (2.144)$$

Obtaining the equation in the dimensionless form by using the following change of variables;

$$u = (T - T_0)/(T_s - T_0); \tau = t/\tau_r; X = x/\text{sqrt}(\alpha \tau_r);$$

$$q^* = q/(k\rho C_p/\tau_r)^{1/2}/(T_s - T_0) \qquad (2.145)$$

$$\partial q_r/\partial \tau + \partial^2 q_r/\partial \tau^2 = 1/X\partial/\partial X(X\partial q^*/\partial X) \qquad (2.146)$$

$$\text{or } \partial q_*/\partial \tau + \partial^2 q_*/\partial \tau^2 = (1/X)\partial q^*/\partial X + \partial^2 q_*/\partial X^2 \qquad (2.147)$$

Let us suppose that the solution for q* is of the form $W\exp(-\tau/2)$ for $\tau > 0$ where W is the transient wave flux. Then,

$$\partial^2 W/\partial \tau^2 - W/4 = \partial^2 W/\partial X^2 + (1/X)\partial W/\partial X \qquad (2.148)$$

Consider the transformation, $\eta = \tau^2 - X^2$. Eq. [2.148] becomes;

$$4\eta^2 \partial^2 w/\partial \eta^2 + 6\eta \partial w/\partial \eta - \eta w /4 = 0 \qquad (2.149)$$

$$\text{or } \eta^2 \partial^2 w/\partial \eta^2 + 3/2\eta \partial w/\partial \eta - \eta w /16 = 0 \qquad (2.151)$$

In Eq. [2.150] the 6 comes from a two from the ballistic term transformation to the relativistic coordinate domain, 2 from the diffusion term, 2 from the thermal gradient

term and 2 from the term $(1/X)\partial W/\partial X$. Comparing the Eq. [2.151] with the generalized Bessel equation given in Eq. [2.96], $a = 3/2$; $b = 0$; $c = 0$; $s = \frac{1}{2}$; $d = -1/16$

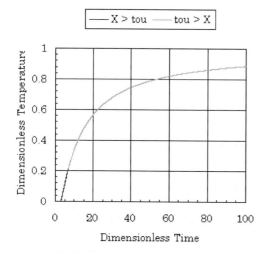

Figure 2.4: Dimensionless Temperature in an Infinite Cylinder at a Point Xp 7

The order of the Bessel solution would be; $p = 1/2$; sqrt($|d|/s$) = ½. Hence the solution to Eq. [2.142] can be written as ;

$$W = c_1 I_{1/2} (1/2 \text{ sqrt}(\tau^2 - X^2)/(\tau^2 - X^2)^{1/4} + c_2 I_{-1/2} (1/2 \text{ sqrt}(\tau^2 - X^2)/(\tau^2 - X^2)^{1/4} \qquad (2.152)$$

c_2 can be seen to be zero as W is finite and not infinitely large at $\eta = 0$.

$$\text{and} \quad q^* = c_1 \exp(-\tau/2) I_{1/2}(1/2 \text{ sqrt}(\tau^2 - X^2)/(\tau^2 - X^2)^{1/4} \qquad (2.153)$$

From Eq. [2.123], in the dimensionless form;

$$-1/X \partial/\partial X(X q_*) = \partial u/\partial \tau$$

$$u = -\int 1/X \, \partial/\partial X(X \, c_1 \exp(-\tau/2) \, I_{1/2}(1/2 \text{ sqrt}(\tau^2 - X^2)/(\tau^2 - X^2)^{1/4})d\tau + C(X) \qquad (2.154)$$

C(X) need be solved from the conditions at the wavefront. At the wavefront, by substituting $\eta = 0$ in Eq. [2.150],

$$\partial w/\partial \eta = w/24$$

$$w = c' \exp(\eta/24) \qquad (2.155)$$

At $\eta = 0$, $u = c' \exp(-X/2)$

From the boundary condition at $X = X_R$

$$C(X) = \exp(X_R - X)/2 \qquad (2.156)$$

2.3 Relativistic Transformation of Spherical Coordinates in a Infinite Medium

Consider a fluid at an initial temperature T_0. The surface of a solid sphere is maintained at a constant temperature T_s for times greater than zero (Figure 2.5). The heat propagative velocity is given as the square root of the ratio of the thermal diffusivity and relaxation time , $v_h = \text{sqrt}(\alpha/\tau_r)$.

The two time conditions, initial and final and the two boundary conditions are;

$$t = 0, \ r > R, \ T = T_0 \qquad (2.157)$$

$$t = \infty, \ T = T_s \text{ for all R}$$

70

$$t > 0, \quad r = R, \quad T = T_s \tag{2.158}$$

$$r = \infty, \quad t > 0, \quad T = T_0 \tag{2.159}$$

The governing equation in temperature is obtained by eliminating the second cross-derivative of heat flux wrt to r and t between the non-Fourier damped wave heat conduction and relaxation equation and the energy balance equation in spherical coordinates. Considering a shell of thickness Δr at a distance r from the center of the solid sphere,

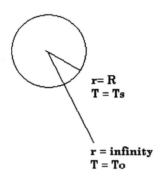

r= R
T = Ts

r = infinity
T = To

Figure 2.5 Semi-Infinite Medium Heated from a Solid Spherical Surface

$$\Delta t(4\pi r^2 \, q_r - 4\pi(r + \Delta r)^2 \, q_{r+\Delta r}) = ((\rho C_p) \, 4\pi r^2 \Delta r \, \Delta T) \tag{2.160}$$

Dividing the Eq. [2.158] throughout with $\Delta r \Delta t$, and in the limit of Δr, Δt going to zero, the energy balance equation in cylindrical coordinates become;

$$-1/r^2 \, \partial(r^2 q_r)/\partial r = (\rho C_p)\partial T/\partial t \tag{2.161}$$

The non-Fourier damped wave equation is;

$$q_r = -k \, \partial T/\partial r - \tau_r \, \partial q/\partial t \tag{2.162}$$

Multiplying Eq. [2.160] by r^2 and differentiating wrt to r and then dividing by r^2,

$$(1/r^2)\partial(r^2 q_r)/\partial r = (-k/r^2)\partial(r^2\partial T/\partial r)/\partial r - (\tau/r^2) \, \partial^2(r^2 q_r)/\partial t \partial r \tag{2.163}$$

Differentiating Eq. [2.161] wrt t.

$$-(\tau / r^2)\, \partial^2 (r^2 q_r)/\partial t \partial r = (\rho C_p)\partial^2\, T/\partial t^2 \tag{2.164}$$

Substituting Eq. [2.164] and Eq. [2.163] into Eq. [2.160] the governing equation in temperature is obtained as;

$$(\rho C_p \tau_r)\partial^2\, T/\partial t^2 + (\rho C_p)\partial T/\partial t = (k/r^2)\partial(\,r^2\partial T/\partial r)/\partial r \tag{2.165}$$

Obtaining the dimensionless form of the governing equation,

$$\partial u/\partial \tau + \partial^2\, u/\partial \tau^2 = \partial^2 u/\partial X^2 + 2/X\, \partial u/\partial X \tag{2.166}$$

where,
$$u = (\,T - T_0\,)/(T_s - T_0); \quad \tau = (t/\tau_r); \quad X = r/\mathrm{sqrt}(\alpha \tau_r)$$

The damping term is removed from the governing equation. This is done realizing that the transient temperature decays with time in an exponential fashion. The other reason for this manoeuvre is to study the wave equation without the damping term. Let $u = w\exp(-\tau/2)$.

$$-w/4 + \partial^2\, w/\partial \tau^2 = \partial^2\, w/\partial X^2 + 2/X\, \partial w/\partial X \tag{2.167}$$

Eq. [2.167] can be solved by using the method of relativistic transformation of coordinates. Consider the transformation variable η as;
for $\tau > X$, $\eta = \tau^2 - X^2$

$$(\partial^2\, w/\partial \eta^2\,)4(\,\tau^2 - X^2\,) + 8(\partial\, w/\partial \eta\,) - w/4 = 0 \tag{2.168}$$

$$4\eta^2 \partial^2\, w/\partial \eta^2 + 8\eta\partial\, w/\partial \eta - \eta w /4 = 0 \tag{2.169}$$

or
$$\eta^2 \partial^2\, w/\partial \eta^2 + 2\eta\partial\, w/\partial \eta - \eta w/16 = 0 \tag{2.171}$$

Comparing Eq. [2.175] with the generalized Bessel equation as given in Eq. [2.96], the solution is; $a = 2$; $b = 0$; $c = 0$; $s = \tfrac{1}{2}$; $d = -1/16$. The order of the Bessel solution would be; $p = 2\,\mathrm{sqrt}(\tfrac{1}{4}) = 1$; $\mathrm{sqrt}(|\,d\,|/s) = \tfrac{1}{2}$. Hence the solution to Eq. [2.171] can be written as ;

$$w = c_1 I_1(1/2\,\mathrm{sqrt}(\tau^2 - X^2)/(\tau^2 - X^2)^{1/2} + c_2 K_1\,(1/2\,\mathrm{sqrt}(\tau^2 - X^2)/(\tau^2 - X^2)^{1/2} \tag{2.172}$$

c_2 can be seen to be zero as W is finite and not infinitely large at $\eta = 0$. The solution is in terms of a composite modified Bessel function of the first order and first kind. Therefore the heat flux can be written as;

$$u = c_1 \exp(-\tau/2)I_1\,(1/2\,\mathrm{sqrt}(\tau^2 - X^2)/(\tau^2 - X^2)^{1/2} \tag{2.173}$$

From the boundary condition at the solid surface;

$$1 = c_1 \exp(-\tau/2)I_1\,(1/2\,\mathrm{sqrt}(\tau^2 - X_R^2)/(\tau^2 - X_R^2)^{1/2} \tag{2.174}$$

Dividing Eq. [2.179] by Eq. [2.180] the solution for u can be given in a more usable form for $\tau > X$;

$$u = [(\tau^2 - X_R^2)^{1/2}/(\tau^2 - X^2)^{1/2}] \, [I_1 \, (1/2 \, \text{sqrt}(\tau^2 - X^2)/ \, I_1 \, (1/2 \, \text{sqrt}(\tau^2 - X_R^2)] \tag{2.175}$$

$$u = [(\tau^2 - X_R^2)^{1/2}/(X^2 - \tau^2)^{1/2}] \, [J_1(1/2 \, \text{sqrt}(X^2 - \tau^2)/ \, I_1 \, (1/2 \, \text{sqrt}(\tau^2 - X_R^2)] \tag{2.176}$$

Eq. [2.176] can be written for $X > \tau$ by using the relation of $I_p(ix) = i^p J_p(x)$. For $X = \tau$, the solution at the wave front result. This can be obtained by solving Eq. [2.175] at $\eta = 0$. In the limit of X_R going to zero,

$$\text{for } \tau > X, \quad u = [\tau/(X^2 - \tau^2)^{1/2}] \, [I_1 \, (1/2 \, \text{sqrt}(\tau^2 - X^2)/ \, I_1 \, (\tau/2)] \tag{2.177}$$

$$\text{For } X > \tau; \quad u = [\tau/(X^2 - \tau^2)^{1/2}] \, [J_1 \, (1/2 \, \text{sqrt}(X^2 - \tau^2)/ \, I_1(\tau/2)] \tag{2.178}$$

Seventeen terms were taken in the series expansion of the modified Bessel composite function of the first kind and first order and the Bessel composite function of the first kind and first order respectively and the results plotted in Figure 2.6 for a given $Xp = 9$ using a Microsoft Excel spreadsheet on a Pentium IV desktop microcomputer. Three regimes can be identified. The first regime is that of the thermal lag and consists of no change from the initial temperature. The second regime is when

$$\tau_{1,m}^2 = X^2 - (7.6634)^{\wedge}2 \quad \text{or } \tau_{1,m} = \text{sqrt}(X^2 - 7.6634^2) = 4.72 \text{ when } X_p = 9 \tag{2.179}$$

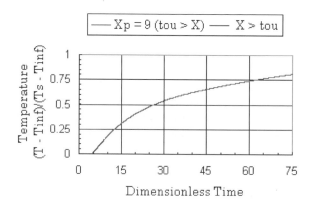

Figure 2.6 Dimensionless Temperature in the Exterior Point in a Infinite Sphere at $X_p = 9$

The first zero of $J_1(x)$ occurs at $x = 3.8317$. The 7.6634 is twice the first root of the Bessel function of the first order and first kind. For times greater than the time lag and less than X_p the dimensionless temperature is given by Eq. [2.196a]. For dimensionless times greater than 7, the dimensionless temperature is given by Eq. [2.196b]. For distances closer to the surface compared with $7.6634\text{sqrt}(\alpha\tau_r)$ the thermal lag time will be zero. The ballistic term manifests as a thermal lag at a given point in the medium.

Worked Example 2.8

Consider a semi-infinite cone with the step change in temperature imposed on the apex of the cone. Using the relativistic transformation of coordinates obtain the transient temperature distribution in the cone. The initial temperature is T_0 and the apex temperature for times greater than zero is T_s.

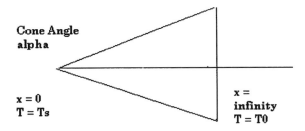

Figure 2.7 Semi-infinite Cone with Constant Apex Temperature

Consider a thin slice of thickness Δx at a distance x from the origin. Th ehalf angle of the cone is θ. The energy balance in the thin slice without any reaction for a time interval Δt can be written as;

$$[\pi x^2 \tan^2\theta \; q_x - \pi (x + \Delta x)^2 \tan^2\theta \; q_{x+\Delta_x}] \, \Delta t = (\rho C_p) \, \Delta T \, (1/3\pi(x + \Delta x)^3 \tan^2\theta - 1/3\pi x^3 \tan^2\theta) \qquad (2.180)$$

Dividing throughout the above equation by $\Delta x \Delta t \pi \tan^2\theta$ and taking the limit as Δt and Δx goes to zero;

$$-(1/x^2)\partial(x^2 q_x)/\partial x = (\rho C_p) \, \partial T/\partial t \qquad (2.181)$$

Eq. [2.198] can be combined with the damped wave conduction and relaxation equation as follows;

$$q_r = -k \, \partial T/\partial x - \tau_r \, \partial q_r/\partial t \qquad (2.182)$$

Multiplying the above equation by x^2 and then differentiating with respect to x and then dividing by x^2;

$$1/x^2 \, \partial(x^2 q_x)/\partial x = -k/x^2 \, \partial(x^2 \partial T/\partial x)/\partial x - \tau_r /x^2 \partial^2(x^2 q_r)/\partial x \partial t \qquad (2.183)$$

Differentiating Eq. [2.181] wrt to time;

$$-(1/x^2)\partial^2 (x^2 q_x)/\partial x \partial t = (\rho C_p) \, \partial^2 T/\partial t^2 \qquad (2.184)$$

Combining Eqs. [2.185], [2.184] and [2.182],

$$\alpha/x^2 \, \partial(x^2 \partial T/\partial x)/\partial x = \partial T/\partial t + \tau_r \, \partial^2 T/\partial t^2 \qquad (2.185)$$

Eq. [2.185] can be converted to the dimensionless form. Obtaining the dimensionless variables,

$$u = (T - T_0)/(T_s - T_0); \quad \tau = (t/\tau_r); \quad X = r/\mathrm{sqrt}(\alpha \tau_r) \qquad (2.186)$$

The governing equation in the dimensionless form can be written as;

$$\partial u/\partial \tau + \partial^2 u/\partial \tau^2 = \partial^2 u/\partial X^2 + 2/X \, \partial u/\partial X \qquad (2.187)$$

The governing equation obtained for the cone is identical to that obtained for the sphere. The damping term is removed from the governing equation. This is done realizing that the transient temperature decays with time in a exponential fashion. The other reason for this manoeuvre is to study the wave equation without the damping term. Let $u = w\exp(-\tau/2)$. Then,

$$-w/4 + \partial^2 w/\partial \tau^2 = \partial^2 w/\partial X^2 + 2/X \, \partial w/\partial X \qquad (2.188)$$

Eq. [2.188] can be solved by using the method of relativistic transformation of coordinates. Consider the transformation variable η, for $\tau > X$ as; $\qquad \eta = \tau^2 - X^2$. Eq. [2.189] becomes;

$$(\partial^2 w/\partial \eta^2)4(\tau^2 - X^2) + 8(\partial w/\partial \eta) - w/4 = 0 \qquad (2.189)$$

$$4\eta^2 \partial^2 w/\partial \eta^2 + 8\eta \partial w/\partial \eta - \eta w /4 = 0 \qquad (2.190)$$

or $\qquad\qquad \eta^2 \partial^2 w/\partial \eta^2 + 2\eta \partial w/\partial \eta - \eta w/16 = 0 \qquad (2.192)$

Comparing Eq. [2.192] with the generalized Bessel equation as given in Eq. [2.96], the solution is;

$$a = 2; \quad b = 0; \quad c = 0; \quad s = \tfrac{1}{2}; \quad d = -1/16$$

The order of the Bessel solution would be $p = 2 \, \mathrm{sqrt}(\tfrac{1}{4}) = 1;$ $\mathrm{sqrt}(|d|/s) = \tfrac{1}{2}$
Hence the solution to Eq. [2.192] can be written as ;

$$w = c_1 I_1(1/2 \text{ sqrt}(\tau^2 - X^2)/(\tau^2 - X^2)^{1/2} + c_2 K_1 (1/2 \text{ sqrt}(\tau^2 - X^2)/(\tau^2 - X^2)^{1/2} \qquad (2.193)$$

c_2 can be seen to be zero as W is finite and not infinitely large at $\eta = 0$. The solution is in terms of a composite modified Bessel function of the first order and first kind. Therefore the heat flux can be written as;

$$u = c_1 \exp(-\tau/2) I_1 (1/2 \text{ sqrt}(\tau^2 - X^2)/(\tau^2 - X^2)^{1/2} \qquad (2.194)$$

From the boundary condition at the solid surface;

$$1 = c_1 \exp(-\tau/2) I_1(\tau/2)/(\tau) \qquad (2.195)$$

Dividing Eq. [2.195] by Eq. [2.194] the solution for u can be given in a more usable form as for $\tau > X$.

$$u = [\tau/(\tau^2 - X^2)^{1/2}] [I_1 (1/2 \text{ sqrt}(\tau^2 - X^2)/ I_1 (\tau/2)] \qquad (2.196)$$

$$u = [\tau/(X^2 - \tau^2)^{1/2}] [J_1 (1/2 \text{ sqrt}(X^2 - \tau^2)/ I_1 (\tau/2)] \qquad (2.197)$$

For $X > \tau$, Eq. [2.197] may be written. For $X = \tau$, the solution at the wave front result. This can be obtained by solving Eq. [2.190] at $\eta = 0$. At the wave front,

$$u = \exp(-X/2) \qquad (2.198)$$

Worked Example 2.9

Obtain the governing equation for heat flux in the radial direction for the sphere as shown in Figure . Solve for the heat flux using the method relativistic transformation of spherical coordinates. From the solution of the radial heat flux obtain the exact solution for the temperature for a semi-infinite medium heated from a spherical surface at constant wall temperature T_s at $r = R$.

Eliminating the second order cross-derivative of temperature wrt to time between the energy balance equation and the damped wave conduction and relaxation equation and radial position the governing equation terms of the radial flux is given by;

$$\partial q/\partial t + \tau_r \partial^2 q/\partial t^2 = (\alpha/ r^2)\partial/\partial r(r^2(\partial q/\partial r)) \qquad (2.199)$$

Obtaining the equation in the dimensionless form by using the following change of variables;

$$u = (T - T_0)/(T_s - T_0); \tau = t/\tau_r; X = x/\text{sqrt}(\alpha \tau_r); q^* = q/(k\rho C_p/\tau_r)^{1/2}/(T_s - T_0) \qquad (2.200)$$

$$\partial q_r/\partial \tau + \partial^2 q_r/\partial \tau^2 = 1/X^2 \partial/\partial X(X^2(\partial q^*/\partial X)) \qquad (2.201)$$

$$\text{or } \partial q_r/\partial \tau + \partial^2 q_r/\partial \tau^2 = (2/X)\partial q^*/\partial X + \partial^2 q_r/\partial X^2 \qquad (2.202)$$

Let us suppose that the solution for q^* is of the form $W\exp(-n\tau)$ for $\tau > 0$ where W is the transient wave flux. For $n = \frac{1}{2}$ as shown in the above sections Eq. [2.202] becomes;

$$\partial^2 W/\partial \tau^2 - W/4 = \partial^2 W/\partial X^2 + (2/X)\partial W/\partial X \qquad (2.203)$$

Consider the transformation, $\eta = \tau^2 - X^2$
Eq. [2.203] becomes;

$$4\eta^2 \partial^2 w/\partial \eta^2 + 8\eta \partial w/\partial \eta - \eta w /4 = 0 \qquad (2.205)$$

$$\text{or } \eta^2 \partial^2 w/\partial \eta^2 + 2\eta \partial w/\partial \eta - \eta w /16 = 0 \qquad (2.206)$$

In Eq. [2.205] the 8 comes from a two from the ballistic term transformation to the relativistic coordinate domain, 2 from the diffusion term, 2 from the thermal gradient term and 4 from the term $(2/X)\partial W/\partial X$. Comparing the Eq. [2.206] with the generalized Bessel equation given in Eq. [2.96], $a = 2$; $b = 0$; $c = 0$; $s = \frac{1}{2}$; $d = -1/16$. The order of the Bessel solution would be $p = 2$ sqrt($\frac{1}{4}$) $= 1$; sqrt($|d|$ $/s$) $= \frac{1}{2}$. Hence the solution to Eq. [2.203] can be written as;

$$W = c_1 I_1(1/2 \text{ sqrt}(\tau^2 - X^2)/(\tau^2 - X^2)^{1/2} + c_2 K_1 (1/2 \text{ sqrt}(\tau^2 - X^2)/(\tau^2 - X^2)^{1/2} \qquad (2.207)$$

c_2 can be seen to be zero as W is finite and not infinitely large at $\eta = 0$. The solution is in terms of a composite modified Bessel function of the first order and first kind. Therefore the heat flux can be written as;

$$q^* = c_1 \exp(-\tau/2)I_1 (1/2 \text{ sqrt}(\tau^2 - X^2)/(\tau^2 - X^2)^{1/2} \qquad (2.208)$$

The temperature solution can be obtained by substituting Eq. [2.208] into Eq. [2.161] in the dimensionless form and integrating the same.

2.4 Finite Slab at Constant Wall Temperature

Consider a finite slab of width 2a with an initial temperature at T_0. The sides of the slab are maintained at constant temperature of T_s. The governing equation in the dimensionless form is then;

$$\partial u/\partial \tau \; + \; \partial^2 u/\partial \tau^2 \; = \; \partial^2 u/\partial X^2 \qquad (2.209)$$

Where,

$$u \; = \; (T - T_s)/(T_0 - T_s); \; \tau \; = \; t/\tau_r \quad ; \; X \; = \; x/sqrt(\alpha \tau_r) \qquad (2.210)$$

The initial condition is given as follows

$$t = 0, \; Vx, \; T = T_0 \; ; \; u = 1 \qquad (2.211)$$

Boundary Conditions in Space

$$t > 0, \; x = 0, \; \partial T/\partial x = 0 \quad ; \quad \partial u/\partial x = 0 \qquad (2.212)$$

$$t > 0, \quad x = \pm a, \quad T = T_s \; ; \quad u = 0 \qquad (2.213)$$

The fourth and final condition in time,

$$t = \infty, \; Vx, \; T = T_s \qquad ; \quad u = 0 \qquad (2.214)$$

The governing equation was obtained by a 1 dimensional energy balance (in – out + reaction = accumulation). This is achieved by eliminating q_x between the damped wave conduction and relaxation equation and the equation from energy balance ($-\partial q/\partial x = (\rho C_p)\partial T/\partial t$). This is achieved by differentiating the constitutive equation with respect to x and the energy equation with respect to t and eliminating the second cross derivative of q with respect to r and time. This equation is then non-dimensionalized. The solution is obtained by the method of separation of variables.

$$\text{Let } u = V(\tau) \; \phi(X) \qquad (2.215)$$

Eq.[2.209] becomes,

$$\phi^{\bullet}(X)/\phi(X) \; = \; (V'(\tau) + V^{\bullet}(\tau))/V(\tau) \; = -\lambda_n^2 \qquad (2.216)$$

$$\phi(X) \; = \; c_1 Sin(\lambda_n X) + c_2 Cos(\lambda_n X) \qquad (2.217)$$

From the boundary conditions,

$$\text{At } X = 0, \; \partial \phi/\partial X = 0, \qquad \text{So, } c_1 = 0 \qquad (2.218)$$

$$\phi(X) \; = \; c_1 Cos(\lambda_n X) \qquad (2.219)$$

$$0 \; = \; c_1 Cos(\lambda_n X_a) \qquad (2.220)$$

$$(2n-1)\pi/2 = \lambda_n X_a \qquad (2.221)$$

$$\lambda_n \; = \; (2n-1)\pi \; sqrt(\alpha \; \tau_r)/2a \; , \quad n = 1,2,3... \qquad (2.222)$$

The time domain solution would be,

$$V = \exp(-\tau/2)\,(c_3\exp(\text{sqrt}(1/4 - \lambda_n^2)\,\tau) + c_4\exp(-\text{sqrt}(1/4 - \lambda_n^2)\tau)) \qquad (2.223)$$

$$\text{or } V\exp(\tau/2) = (c_3\exp(\text{sqrt}(1/4 - \lambda_n^2)\,\tau) + c_4\exp(-\text{sqrt}(1/4 - \lambda_n^2)\tau)) \qquad (2.224)$$

From the final condition $u = 0$ at infinite time. So is $V\phi\exp(\tau/2) = W$, the wave temperature at infinite time. The wave temperature is that portion of the solution that remains after dividing the damping component either from the solution or the governing equation. For any non-zero ϕ, it can be seen that at infinite time the LHS of Eq. [2.236] is a product of zero and infinity and a function of x and is zero. Hence the RHS of Eq. [2.236] is also zero and hence in Eq. [2.224] c_3 need be set to zero. Hence,

$$u = \sum_1^\infty c_n\exp(-\tau/2)\,\exp(-\text{sqrt}(1/4 - \lambda_n^2)\,\tau)\,\text{Cos}(\lambda_n X) \qquad (2.225)$$

Where λ_n is described by Eq.[2.222] C_n can be shown using the orthogonality property to be $4(-1)^{n+1}/(2n-1)\pi$. It can be seen that Eq. [2.225] is bifurcated. As the value of the thickness of the slab changes the characteristic nature of the solution changes from monotonic exponential decay to subcritical damped oscillatory. For $a < \pi\,\text{sqrt}(\alpha\tau_r)$, even for $n = 1$, $\lambda_n > \tfrac{1}{2}$. This is when the argument within the square root sign in the exponentiated time domain expression becomes negative and the result becomes imaginary. Using Demovrie's theorem and taking real part for small width of the slab,

$$u = \sum_1^\infty c_n\exp(-\tau/2)\,\text{Cos}(\text{sqrt}(\lambda_n^2 - 1/4)\,\tau)\,\text{Cos}(\lambda_n X) \qquad (2.226)$$

Eqs [2.237, 2.238] can be seen to be well bounded. The solution obtained by Taitel [19], for the centerline temperature of the finite slab is given below. They considered a constant wall temperature and the initial time conditions included a $\partial T/\partial t = 0$ term in addition to the initial temperature condition. The solution they presented is as follows:

$$u = \sum_0^\infty b_n\exp(-\tau/2)\exp(-\tau/2\,\text{sqrt}(1 - 4(2n+1)^2\pi^2\alpha\tau/a^2)) + \sum_0^\infty c_n\exp(-\tau/2)\,\exp(+\tau/2\,\text{sqrt}(1 - 4(2n+1)^2\pi^2\alpha\tau/a^2))$$

$$(2.227)$$

Multiplying both sides of the equation by $\exp(\tau/2)$,

$$u\exp(\tau/2) = W = \sum_0^\infty b_n\exp(-\tau/2\,\text{sqrt}(1 - 4(2n+1)^2\pi^2\alpha\tau/a^2)) + \sum_0^\infty c_n\exp(\exp(+\tau/2\,\text{sqrt}(1 - 4(2n+1)^2\pi^2\alpha\tau/a^2)))$$

$$(2.228)$$

At infinite times, the LHS of Eq. [2.228] is 0 times ∞ and is zero. The RHS does not vanish. Thus the expression given by Taitel [1992] and later discussed as a temperature overshoot, may be as a result of the growing exponential term in the above expression. Compared with this, the above solutions Eqs. [2.229, 2.230] are bounded. Eqs. [2.229, 2.230] becomes zero after some time. This would be time taken to reach steady state. Thus, for $a \geq \pi\,\text{sqrt}(\alpha\,\tau_r)$

$$u = \sum_1^\infty c_n\exp(-\tau/2)\,\exp(-\text{sqrt}(1/4 - \lambda_n^2)\,\tau)\,\text{Cos}(\lambda_n X) \qquad (2.229)$$

Where $c_n = 4(-1)^{n+1}/(2n-1)\pi$ and $\lambda_n = (2n-1)\pi\,\text{sqrt}(\alpha\,\tau_r)/2a$

The centerline temperature for a particular example is shown in Figure 2.6. Eight terms in the infinite series given in Eq. [2.229] were taken and the values calculated on a 1.9 GHz Pentium IV desktop personal computer. The number of terms was decided on the incremental change or improvement obtained by doubling the number of terms. The number of terms was arrived at a 4% change in the dimensionless temperature.

The Taitel paradox is obviated by examining the final steady state condition and expressing the state in mathematical terms. The W term which is the dimensionless temperature upon removal of the damping term needs to go to zero at infinite time. This resulted in our solution being different from previous reports [Taitel, 1972, Barletta and Zanchini 1997] and is well bounded. The use of FINAL condition is may be what is needed for this problem to be used extensively in engineering analysis without being branded as violating second law of thermodynamics. The conditions were the touted violations of second law are not physically realistic. A bifurcated solution results. For small width of the slab, a $<$ π sqrt(α τ_r), the transient temperature is subcritical damped oscillatory. The centerline temperature is shown in Figure 2.7.

$$u = \sum_0^\infty c_n \exp(-\tau/2) \; \text{Cos}(\text{sqrt}(\lambda_n^2 - 1/4) \; \tau) \; \text{Cos}(\lambda_n X) \qquad (2.230)$$

An exact well bounded solution that is bifurcated depending on the width of the slab is provided. The transient solution to the damped wave Cattaneo and Vernotte non-Fourier hyperbolic wave propagative and relaxation equation is obtained by the method of separation of variables. A well bounded infinite series expression is provided. The temperature overshoot identified by Taitel [1972] is obviated by examining the final steady state condition and expressing the state in mathematical terms. A bifurcated solution results. For small width of the slab, a $<$ π sqrt(α τ_r), the transient temperature is subcritical damped oscillatory. Taitel [1972] had considered the problem in finite slab, in the framework of hyperbolic heat conduction. He obtained the transient temperature and pointed out that the absolute value of the temperature change may exceed the difference with the wall temperature. This has been discussed as the temperature overshoot problem in the literature. This was later analyzed by Barletta and Zanchini [1997] using an entropic balance and the determination of conditions were the Clausius inequality is violated. Relativistic transformation of coordinates was used to convert the two variables PDE to a single variable Bessel equation in Sharma [2003]. This is after removing the damping term by dividing the equation by a decaying exponential. The conditions where the pulsations in temperature will occur for a finite slab, cylinder and sphere were derived [2003]. They were shown to be subcritical damped oscillatory.

80

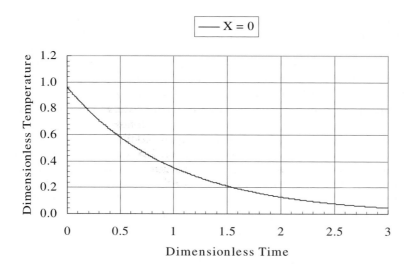

Figure 2.8 Centerline Temperature in a Finite Slab at Constant Wall Temperature (large a) (a = 0.86m, $\alpha = 10^{-5}$ m^2/s, τ_r = 15 s)

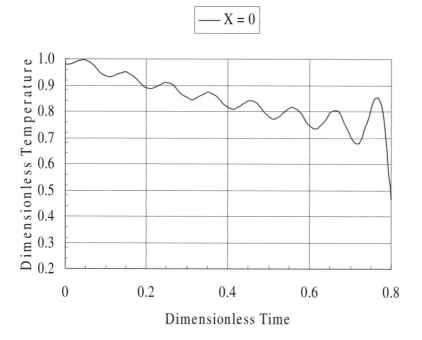

Figure 2.9 Centerline Temperature in a Finite Slab at CWT (Small a) (a = 0.001m, $\alpha = 10^{-5}$ m^2/s, τ_r = 15 s)

In both [Taitel 1972 and Barletta and Zanchini , 1997] , there were four conditions used for initial and boundary constraints. The two in space domain are retained here. The initial temperature at time zero is also retained. However the slope with the time domain of the temperature at time zero is replaced with the FINAL Condition for the time domain. i.e., at steady state the transient temperature will decay out to a constant value or to zero in the dimensionless form. This consideration is shown to change the nature of the solution considerably to a well bounded expression that is bifurcated. For small values of the slab the transient temperature is subcritical damped oscillatory. For other values the Fourier series representation is augmented by a modification to the exponential time domain portion of the solution. In this section, the use of the FINAL condition at steady state as the fourth condition to give a bounded solution in obeyance of Clausius inequality was achieved.

2.5 On the Use of Storage Coefficient

The storage coefficient, S, with the units of W/m^3 /K, and given by $S = (\rho C_p /\tau_r)$ can be a critical parameter in the design of substrates of high speed processors in a similar vein to the thermal conductivity k (W/m/K) and heat transfer coefficient $(W/m^2/K)$. The ratio of the thermal mass to the relaxation time of the material may be an indicator of the heat stored in the material. Especially during periodic phenomena this may become an important consideration. With the increase in speed of the microprocessors the periodic heating of the surface of a substrate becomes of interest. Given the time scales of the period of the microprocessor and some of the short time scale anomalies of Fourier heat conduction it is of interest to evaluate other expressions other than Fourier such the Cattaneo & Vernotte non-Fourier heat conduction and relaxation Equation, Dual Phase Lag Model subject to a periodic boundary condition.

2.5.1 Hyperbolic PDE as Governing Equation

Consider a semi-infinite slab at initial temperature T_0, imposed by a periodic temperature at one of the ends by $T_0 + T_1 Cos[wt]$. The transient temperature as a function of time and space in one dimension is obtained. Obtaining the dimensionless variables;

$$u = (T - T_0)/(T_1); \quad \tau = t/\tau_r; \quad X = x/sqrt(\alpha \, \tau_r) \quad (2.231)$$

The energy balance on a thin shell at x with thickness Δx is written. The governing equation is obtained after eliminating q between the energy balance Equation and the derivative with respect to x of the flux Equation and introducing the dimensionless variables;

$$\partial u/\partial \tau \; + \; \partial^2 u/\partial \tau^2 \quad = \; \partial^2 u/\partial X^2 \tag{2.232}$$

The initial condition is;

$$t = 0, \; T = T_0 \; ; \; u = 0 \tag{2.233}$$

The Boundary Conditions are;

$$X = \infty, \; \; T = T_0 \; ; \; u = 0 \tag{2.234}$$

$$X = 0, \; \; T = T_0 + \; T_1 \, Cos(\omega t). \; ; \; u \; = Cos(\omega^* \tau) \tag{2.235}$$

Let us suppose that the solution for u is of the form $f(x)exp(-i\omega^*\tau)$ for $\tau > 0$. Where, ω is the frequency of the temperature wave imposed on the surface and the T_1 is the amplitude of the wave. Eq. [2.232] becomes;

$$(-i\omega^*) \, f \, exp(-i\omega^*\tau) \; + (i^2\omega^{*2} \,)f \, exp(-i\omega\tau) \; = \; f'' \, exp(-i\omega^*\tau) \tag{2.236}$$

$$i^2 \; f(\omega^{*2} + i\omega^*) \quad = \quad f'' \tag{2.237}$$

$$f(X) \; = \quad c \; exp(-iX\omega^* sqrt(\omega^* + i)) \tag{2.238}$$

d can be seen to be zero as at $X = \infty$, $u = 0$.

$$u \; = c \; exp(-iX\omega^* sqrt(\omega^* + i) \,)exp(-i\omega^*\tau) \tag{2.239}$$

From the boundary condition at $X = 0$,

$$Cos(\omega^*\tau) = \; Real \; Part \; (cexp(-i\omega^*\tau) \,) \; or \; c = 1 \tag{2.240}$$

$$Or \; u \; = \; exp(-X\omega^*(A + iB)exp(-i\omega^*\tau) \; = \; exp(-A\omega^* X)exp(-i(BX\omega^* + \omega^*\tau) \tag{2.241}$$

$$Where \; A + iB = isqrt(\omega^* + i) \tag{2.242}$$

Squaring both sides

$$A^2 - B^2 \; + 2ABi \; = i^2 \, (\omega^* + i) = -\omega^* \; -i \tag{2.243}$$

$$A^2 - B^2 \; = -\omega^* \; ; \; 2AB = -1 \; or \; B = -1/2A \tag{2.244}$$

$$Or \; A^2 \; - 1/4A^2 \; = -\omega^* \tag{2.245}$$

$$A^2 = (-\omega^* \pm \text{sqrt}(\omega^{*2}+1))/2 ; \quad B = -1/2A \tag{2.246}$$

Obtaining the real part,

$$(T - T_0)/(T_1) = u = \exp(-A\omega^*X)\text{Cos}(\omega^*(BX + \tau)) \tag{2.247}$$

The time lag in the propagation of the periodic disturbance at the surface is captured by the above relation Thus the boundary conditions can be seen to be satisfied by Eq. [2.247]. In a similar vein to the supposition of $f(x)\exp(-i\omega^*\tau)$ the heat flux q, can be supposed to be of the form, $q^* = g(x)\exp(-i\omega^*\tau)$. From Eq. [2.261] and the suppositions of u and q we have;

$$g = f/(1 - i\omega^*) \tag{2.248}$$

Combining the f from Eq. [2.250] into Eq. [2.260]

$$q^* = -\omega^*(A + iB) \exp(-X\omega^*(A + iB)\exp(-i\omega^*\tau) \tag{2.249}$$

$$= -\omega^*(A + iB)\exp(-A\omega^*X)\exp(-i(BX\omega^* + \omega^*\tau) \tag{2.250}$$

$$= -\omega^*(A + iB)\exp(-A\omega^*X)(\text{Cos}(BX\omega^* + \omega^*\tau) + i\text{Sin}(BX\omega^* + \omega^*\tau)) \tag{2.251}$$

Obtaining the real part

$$q = \text{sqrt}(kS) \, \omega^* \exp(-A\omega^*X)(B \, \text{Sin}(\omega^*(BX + \tau)) - A\text{Cos}(\omega^*(BX + \tau))) \tag{2.252}$$

Where $q^* = q/\text{sqrt}(k \, S)$, k the thermal conductivity and S the storage coefficient $(\rho C_p/\tau_r)$. Thus the sustained part of the solution is periodic with a time lag from the periodic boundary condition imposed on the surface. The heat flux is determined by the negative temperature gradient and the accumulation term in the Cattaneo and Vernotte Equation. For certain values it can be seen that the heat flux can reverse in direction. The flux reversal can be an issue in substrate design of high speed processors.

2.5.2 Parabolic PDE as Governing Equation

The above analysis is repeated for the parabolic PDE as governing Equation.

$$\partial u/\partial \tau = \partial^2 u/\partial X^2 \tag{2.253}$$

where instead of τ_r a characteristic time is used, l^2/α, where l is some penetration length in the semi-infinite medium and α the thermal diffusivity of the material. The solution given in Eq. [2.271] becomes;

$$f = c \exp(-X\text{sqrt}(i\omega^*)) \tag{2.254}$$

84

or u = c exp(-Xsqrt(iω*)) exp(-iω*τ)   (2.255)

or u = c exp(-X(C + iD) exp(-iω*τ)   (2.256)

C + iD = sqrt(iω*) ; Squaring both sides, and equating the real and imaginary
  parts (C = D).

$$C = D = (\omega^*/2)^{1/2}$$ (2.257)

From the boundary condition given c can be seen to be 1.

Or $u = \exp(-(\omega^*/2)^{1/2}X) \exp(-i(\ \omega^*\tau + X(\omega^*/2)^{1/2}$ (2.258)

Obtaining the real parts,

$u = \exp(-(\omega^*/2)^{1/2}X) \cos(\ \omega^*\tau + X(\omega^*/2)^{1/2})$ (2.259)

Thus the storage coefficient in addition to the thermal conductivity of the
substrate and frequency of oscillation is an important parameter in the design of
substrates in high speed processors as heat sink devices.

2.5.3 The Dual Phase Lag Model

The non-Fourier dual phase lag model is given by Eq. [2.274] by Tzou [1997];

$$q = -k\partial T/\partial x\ -\ \tau_r\ \partial q/\partial t\ -\ k\tau_t\partial^2 T/\partial t\partial x$$ (2.260)

The time lags τ_r and τ_t are the times taken for gradual response of heat flux and
temperature gradient respectively. The transient behavior of the heat flux is captured
by $\partial q/\partial t$ and $\partial^2 T/\partial t\partial x$ represents the transient nature of the temperature gradient. After
the gradual response is complete the heat flux and temperature gradient achieve the
values given by Fourier's law. In general, the time lags are non-zero times accounting for
the effects of thermal inertia and microstructural interactions. The time delay in
establishing heat flux and associated conduction through a solid may tend to induce
thermal waves with sharp wave fronts separating heated and unheated zones in the
solid. The sharp wave fronts are smoothed by the time lag to the temperature gradient.
Values of thermal lags for discontinuous solids can be of the order of 4 – 9 seconds. It
remains small for continuous solids such as gold. It can be seen that Eq. [2.260] becomes
the Fourier's law of heat conduction when both the time lags are zero and when the time
lag to temperature gradient is zero the Equation becomes the Cattaneo and Vernotte
Equation.
        The periodic component analysis as conducted for the Cattaneo & Vernotte
Equation and the Fourier Equation is repeated for the DPL Equation as follows. The
governing Equation in 1 dimension is given by;

$$\partial u/\partial \tau \quad + \quad \partial^2 u/\partial \tau^2 \qquad = \partial^2 u/\partial X^2 \quad + B \, \partial^3 u/\partial \tau \partial X^2 \tag{2.261}$$

Where, $B = \tau_r/\tau_r$. Suppose $u = f(x) \exp(-i\omega^*\tau)$. When $B = 0$, the governing Equation becomes Eq. [2.232] from Cattaneo & Vernotte and when $B = 1$, the Equation becomes Eq. [2.253] from Fourier's law. Plugging the supposition for u into Eq. [2.261],

$$f'' = f \omega^* /(1 + B^2 \, \omega^{*2})((B\text{-}1) \, \omega^* - i \, (1 + B\omega^{*2}) \tag{2.262}$$

$$u = c \exp(\text{-}X \text{sqrt}(\omega^* /(1 + B^2 \, \omega^{*2})((B\text{-}1) \, \omega^* - i \, (1 + B\omega^{*2})) \exp(\text{-}i\omega^*\tau) \tag{2.263}$$

From the boundary condition at $X = 0$, c can be seen to be 1.

Or $u = c\exp(\text{-}X \, (E + i \, F)) \exp(\text{-}i\omega^*\tau)$ \hfill (2.264)

Where $E + iF = \text{sqrt}(\omega^* /(1 + B^2 \, \omega^{*2})((B\text{-}1) \, \omega^* - i \, (1 + B\omega^{*2}))$ \hfill (2.265)

Squaring both sides

$$E^2 - F^2 + 2iEF = (\omega^* /(1 + B^2 \, \omega^{*2}))((B\text{-}1) \, \omega^* - i \, (1 + B\omega^{*2})) \tag{2.266}$$

$$E^2 - F^2 = \omega^{*2} (B\text{-}1)/(1 + B^2 \, \omega^{*2}); \quad F = \text{-} (1 + B\omega^{*2})/2E/(1 + B^2\omega^{*2}) \tag{2.267}$$

$$\text{or } E^2 \text{-}(1 + B\omega^{*2})^2 /(1 + B^2 \, \omega^{*2})^2 / 4E^2 = \omega^{*2} (B\text{-}1)/(1 + B^2 \, \omega^{*2}) \tag{2.268}$$

$$\text{or } E^4 - \omega^{*2} (B\text{-}1)/(1 + B^2 \, \omega^{*2}) E^2 - (1 + B\omega^{*2})^2 /(1 + B^2 \, \omega^{*2})^2 / 4 = 0 \tag{2.269}$$

$$E^2 = (\omega^{*2} (B\text{-}1) \pm \text{sqrt}(\omega^{*4} (B\text{-}1)^2 + \omega^{*2} (1 + B\omega^{*2})^2))/(2 /(1 + B^2 \, \omega^{*2})) \tag{2.270}$$

Eq. [2.284] reduces to the result for the case of Cattaneo and Vernotte as would be expected when $B = 0$ as in the case of the Cattaneo and Vernotte non-Fourier heat conduction and relaxation Equation. When $B = 1$ Eq. [2.263] reduces to the results that would be expected in the case of the Fourier's law.

$$u = \exp(\text{-}EX)\text{Cos}(FX + \omega^*\tau) \tag{2.271}$$

where E and F are given by Eq. [2.270] and [2.267] respectively.

Appropriate choice of the substrate material can keep the dampening of the amplitude of fluctuations in check. The design of the substrate will depend on the following parameters;

1) Thermal Conductivity of the substrate (k), with units of W/m/K.
2) Storage coefficient of the substrate $S = C_p \, \rho/\tau_r$ with units of $W/m^3/K$.
3) ω the frequency of the oscillation of the temperature wave at the surface

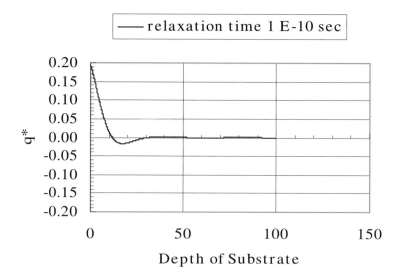

Figure 2.10 Dimensionless Heat Flux with Depth of the Substrate $\tau = 10$;
$\tau_r = 10^{-10}$ **sec; Damped Wave Conduction and Relaxation**

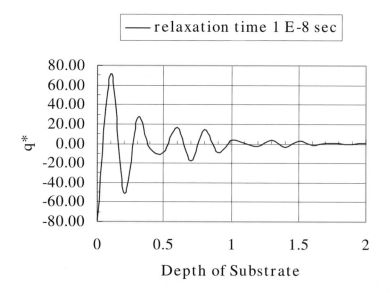

Figure 2.11 Dimensionless Heat Flux with Depth of the Substrate
$\tau = 10$; $\tau_r = 10^{-8}$ **sec; Damped Wave Conduction**

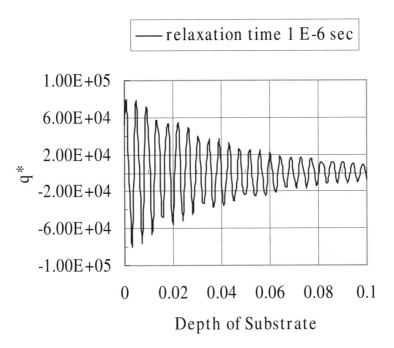

Figure 2.12 Dimensionless Heat Flux with Depth of the Substrate
$\tau = 10$; $\tau_r = 10^{-6}$ sec;   Damped Wave Conduction

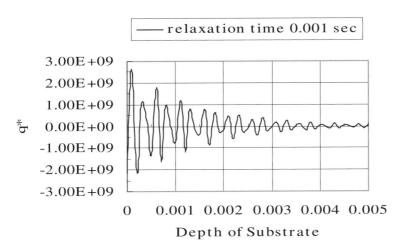

Figure 2.13 Dimensionless Heat Flux with Depth of the Substrate
$\tau = 10$; $\tau_r = 10^{3}$ sec; Damped Wave Conduction

The reversal of heat flux can be observed. The extent of reversal and attenuation of the periodic temperature at the surface driving the problem can be seen to vary with the storage coefficient of the substrate. The location where the heat flux becomes zero can be seen for low values of relaxation time. The amplitude of the disturbance dampening and the time lag of response can be seen in the predictions of the Cattaneo & Vernotte model and the dual phase lag model. In Figures 2.10-2.13 is shown the dimensionless heat flux as a function of distance from the surface for a considered time instant $\tau = 10$ and for relaxation time of 10^{-10}, 10^{-8} and 10^{-6} and 0.001 seconds. The attenuation is subcritical damped oscillatory. The flux reversal can be observed. The storage coefficient is lumped into the dimensionless heat flux term and can be seen to be a major causative factor in the attenuation of the heat pulse. The frequency of the microprocessor considered was 2 GHz. The over damped systems can be seen.

2.6 Transient Temperature in a Sphere at CWT

Consider a sphere at initial temperature T_0. The surface of the sphere is maintained at a constant temperature T_s for times greater than zero. The heat propagative velocity is given as the square root of the ratio of thermal diffusivity and relaxation time, $V_h = \text{sqrt}(\alpha/\tau_r)$.

The initial and boundary conditions are;

$$t = 0, \quad Vr, \quad T = T_0 \tag{2.272}$$

$$t > 0, \quad r = 0, \quad \partial T/\partial r = 0 \tag{2.273}$$

$$t > 0, \quad r = R \quad T = T_s \tag{2.274}$$

The governing equation can be obtained by eliminating q_r between equations (1) and the equation from energy balance of in − out − accumulation. This is achieved by differentiating the constitutive equation with respect to r and the energy equation with respect to t and eliminating the second cross derivative of q with respect to r and time. Thus,

$$\tau_r \, \partial^2 T/\partial t^2 + \partial T/\partial t = \alpha \, \partial^2 T/\partial r^2 + 2\alpha/r \, \partial T/\partial r \tag{2.275}$$

Obtaining the dimensionless variables ;

$$u = (T - T_s)/(T_0 - T_s); \tau = t/\tau_r; X = r/\text{sqrt}(\alpha \tau_r) \tag{2.276}$$

The governing equation in the dimensionless form is then;

$$\partial u/\partial \tau + \partial^2 u/\partial \tau^2 = \partial^2 u/\partial X^2 + 2/X \ \partial u/\partial X \qquad (2.277)$$

The solution is obtained by the method of separation of variables. First the damping term is removed by the substitution, $u = \exp{-\tau/2} \ w \ (X, \tau)$.

$$-w/4 + \partial^2 w/\partial \tau^2 = \partial^2 w/\partial X^2 + 2/X \ \partial w/\partial X \qquad (2.278)$$

The method of separation of variables can be used to obtain the solution of Eq. [2.292].

$$\text{Let } W = V(\tau) \ \phi \ (X) \qquad (2.279)$$

Plugging Eq. (2.293) into Eq. (2.292), and separating the variables that are a function of X only and τ only;

$$\phi" + 2\phi'/X + \lambda^2 \phi = 0 \qquad (2.280)$$

$$V" = V(1/4 \ -\lambda^2 \ ) = 0 \qquad (2.281)$$

The solution for Eq. (2.294) is the Bessel function of 1/2th order and first kind;

$$\phi = c_1 J_{1/2} (\lambda \ X) + c_2 J_{-1/2} (\lambda \ X) \qquad (2.282)$$

It can be seen that $c_2 = 0$ as the concentration is finite at $X = 0$. Now from the BC at the surface,

$$\phi = c_1 J_{1/2} (\lambda \ R/\sqrt{\alpha \tau_r}) = 0 \qquad (2.283)$$

$$\text{or } \lambda_n \ R/(\alpha \tau_r)^{1/2} = (n-1)\pi, \text{ for } n = 2,3.4... \qquad (2.284)$$

The solution for Eq. [2.295] is the sum of two exponentials. The term containing the positive exponential power exponent will drop out as with increasing time the system may be assumed to reach steady state and that the points within the sphere will always have temperature values less than that at the boundary.

Thus, $\qquad\qquad V = c_4 \exp(-\tau \ (\ 1/4 \ -\lambda_n^2)^{1/2}) \qquad (2.285)$

Or, $\qquad\qquad u = \sum_0^\infty c_n J_{1/2} (\lambda_n \ X) \ \exp(-\tau \ /2 - \tau(\ 1/4 \ -\lambda_n^2)^{1/2}) \qquad (2.286)$

The c_n can be solved for from the initial condition by using the principle of orthogonality for Bessel functions. At time zero the LHS And RHS are multiplied by $J_{1/2}(\lambda_m \ X)$. Integration between the limits of 0 and R is performed. When n is not m the integral is zero from the principle of orthogonality. Thus when n = m, Thus

$$c_n = -\int_0^R J_{1/2} (\lambda_n \ X) / \int_0^R J_{1/2}^2 (c\lambda_n \ X) \qquad (2.287)$$

It can be noted from Eq. [2.300] that when

$$1/4 < \lambda_n^2 \qquad (2.288)$$

the solution will be periodic with respect to time domain. This can be obtained by using De Movries theorem and obtaining the real part to $\exp(-i\tau(-1/4 + \lambda_n^2)^{1/2})$. Also it can be shown that for terms in the infinite series for greater than 2 and for,

$$R < 2\pi(\alpha\tau_r)^{1/2} \tag{2.289}$$

the contribution to the solution will be periodic. Thus a bifurcated solution is obtained. Also from Eq. [2.300] it can be seen that all terms in the infinite series will be periodic. i.e., even for n =2 when Eq. [2.303] is valid,

$$u = \sum_2^\infty c_n J_{1/2}(\lambda_n X) \exp(-\tau/2 \cos(\tau(\lambda_n^2 - 1/4)^{1/2}) \tag{2.290}$$

Thus the transient temperature profile in a sphere is obtained for a step change in temperature at the surface of the sphere using the modified Fourier's heat conduction law. A lower limit on the radius is obtained to avoid cycling of temperature in the time domain during transience. A lower limit on the radius of the sphere subject to CWT at the surface exist and found to be $2\pi(\alpha\tau_r)^{1/2}$ in order to avoid pulsations in temperature with respect to time. The exact solution for transient temperature profile using finite speed heat conduction is derived by the method of separation of variables. It is a bifurcated solution. For certain values of lamda the time portion of the solution is cosinous and damped and for others it is a infinite series of Bessel function of the first kind and 1/2th order and decaying exponential in time. Also it can be shown that for terms in the infinite series with n greater than 2, the contribution to the solution will be periodic for small R. The exact solution is bifurcated.

2.7 Pulse Injection in an Infinite Medium

Consider an infinite medium at an initial temperature T_0. The center of the medium is subjected to a pulse injection of heat, Q (J/m^2) at time zero. Mathematically this is given by the dirac delta function $Q \delta(\tau)$ at x = 0. The heat propagative velocity is given as the square root of the ratio of thermal diffusivity and relaxation time.

$$V_h = \text{sqrt}(\alpha/\tau_r) \tag{2.291}$$

The initial condition

$$t = 0, Vx, T = T_0 \tag{2.292}$$

$$t > 0, x = 0, q = Q \delta(\tau) = 0 \tag{2.293}$$

$$t > 0, x = \infty, T = T_0 \tag{2.294}$$

The governing equation can be obtained by eliminating q_x between the damped wave conduction and relaxation equation and the equation from energy balance of in − out = accumulation $(-\partial q/\partial x = (1/\alpha)\partial T/\partial t)$. This is achieved by differentiating equation

(3) with respect to x and the energy equation with respect to t and eliminating the second cross derivative of q with respect to r and time. Thus,

$$\tau_r \, \partial^2 T/\partial t^2 + \partial T/\partial t = \alpha \, \partial^2 T/\partial x^2 \qquad (2.295)$$

Obtaining the dimensionless variables ;

$$u = (T - T_0)/(T_0) ; \quad \tau = t/\tau_r; \quad X = x/sqrt(\alpha \, \tau_r) \qquad (2.296)$$

The governing equation in the dimensionless form is then;

$$\partial u/\partial \tau + \partial^2 u/\partial \tau^2 = \partial^2 u/\partial X^2 \qquad (2.297)$$

The solution is obtained by the method of Laplace transforms.

$$\bar{u} = A \exp(-sqrt(s(s+1))X) \qquad (2.298)$$

$$\bar{q}^* = -d \bar{u}/dX \qquad (2.299)$$

(at X = 0, q is a pulse injection; at time 0 it is infinity and later it is zero)

$$\text{where } q^* = q/T_0 \, sqrt(kC_p \, \rho/\tau_r) \qquad (2.300)$$

$$\bar{q}^* = A \, sqrt(s(s+1)) \, \exp(-sqrt(s(s+1))X) \qquad (2.301)$$

$$\text{At } X = 0, \, \bar{q}^* = Q^* \qquad (2.302)$$

$$\text{Or } A = Q^*/ \, sqrt(s(s+1)) \qquad (2.303)$$

Upon inversion from the Laplace domain.

$$u = Q^* \exp(-\tau/2) \, I_0 \, (1/2 \, sqrt(\tau^2 - X^2)) \qquad (2.304)$$

At X = 0,

$$(T - T_0) = Q/sqrt(kC_p \, \rho/\tau_r) \, \exp(-\tau/2) \, I_0 \, (1/2 \, sqrt(\tau^2 - X^2)) \qquad (2.305)$$

This is in the open interval of $\tau > X$. At large times, $I_0 \, (1/2 \, sqrt(\tau^2 - X^2))$ approximates to $\exp(1/2 \, sqrt(\tau^2 - X^2))/sqrt(\pi \, sqrt(\tau^2 - X^2)$

At X = 0, $\qquad (T - T_0) = Q/sqrt(kC_p \, \rho/\tau_r) \, /sqrt(\pi\tau) \qquad (2.306)$

For the parabolic case,

$$u = A \exp(-sqrt(s)X \qquad) \qquad (2.307)$$

$$A = Q^*/ \, sqrt(s) \qquad (2.308)$$

Then, $\qquad (T - T_0) = Q/sqrt(kC_p \, \rho/\tau_r) \, 1/sqrt(\pi\tau) \exp(-X^2/4\tau) \qquad (2.309)$

At X = 0, $(T - T_0)$ = $Q/\text{sqrt}(kC_p \, \rho/\tau_r) \, 1/\text{sqrt}(\pi\tau)$ (2.310)

Eq. [2.310] can be seen to be the same as Eq. [2.306]. The method of Laplace transforms was used to obtain the transient temperature profile when a pulse of heat is injected in the center of an infinite medium. For the parabolic case, the solution is a Gaussian curve with the composite function as argument. For the hyperbolic case the solution is the generalized Gaussian curve with saddle points with the composite function as argument. At X = 0, and at large times the hyperbolic solution was found to be identical to that obtained by the parabolic solution.

2.8 Temperature Overshoot

Several studies have been undertaken to explore the thermodynamic feasibility of the use of the non-Fourier damped wave conduction and relaxation equation. Gurtin and Pipkin [1968] proposed a theory of extended irreversible thermodynamics. This is based on the assumption that the specific Helmholtz free energy ψ, the specific entropy, s, and the heat flux vector, q, depend on the temperature, T, on the temperature history and on the ∇T history. Coleman et al [1982] presented a treatment of extended irreversible thermodynamics. They attempted to show that the Cattaneo and Vernotte equation is compatible with the second law of thermodynamics. This is only if the specific internal energy u does not depend only on T but also on q. An expression for the entropy production rate σ based upon local equilibrium hypothesis was determined by Barletta and Zanchini [1997] . They provide a proof that the entropy production per unit volume must be non-negative to stay within the bounds of the Clausius inequality. In addition to the energy balance equation and the constitutive, non-Fourier damped wave and relaxation equation they also write a local entropy balance equation. This is stated as follows;

$$\nabla \cdot (q/T) + \rho \, \partial s/\partial t = \sigma \qquad (2.311)$$

Combining the entropy balance equation with the energy balance equation at constant density, ρ, and using the relation du = Tds hold locally, the following expression for the entropy production rate was derived;

$$\sigma = -1/T^2 \, q. \nabla T \qquad (2.312)$$

Combining the WRENCH and entropy balance equation;

$$\sigma = 1/(kT^2)(q.q + \tau_r q.\partial q/\partial t) \qquad (2.313)$$

Whenever the heat flux density q, has a constant direction and decreases at some point so steeply that $|\partial q/\partial t| > |q| \, \tau$ the RHS of Eq. [2.302] becomes negative and the local equilibrium scheme cannot be applied. Both Barletta and Zanchini [1997] and Taitel

[1972] consider the hyperbolic conduction equation in a finite slab at constant wall temperature. They used the time condition of $\partial u/\partial \tau = 0$ in addition to the initial condition. The cases where the interior temperature was found to be greater than the boundary temperature, Barletta and Zanchini [1997] has shown a negative entropy production rate according to Eq. [2.302]. In section 2. the final condition was used to derive a bounded solution for the transient temperature of a finite slab by Sharma [2003].

2.9 Convective Boundary Condition

In this section, the temperature in the finite slab is solved for using the convective boundary condition. Consider a finite slab at initial temperature T_0, subjected to sudden contact with fluid at temperature at to T_1. The transient temperature as a function of space and time is obtained. The heat propagative velocity is $V_h = \text{sqrt}(\alpha/\tau_r)$. The initial condition

$$t = 0, \quad Vx, \quad T = T_0 \qquad (2.314)$$

$$t > 0, \quad x = 0, \quad \partial T/\partial x = 0 \qquad (2.315)$$

$$t > 0, \quad x = \pm a, \quad -k\partial T/\partial x = h\,(T_1 - T) \qquad (2.316)$$

Obtaining the dimensionless variables ;

$$u = (T - T_1)/(T_0 - T_1) \,; \quad \tau = t/\tau_r; \quad X = x/\text{sqrt}(\alpha\,\tau_r) \qquad (2.317)$$

The energy balance on a thin spherical shell at x with thickness Δx is written. The governing equation is obtained after eliminating q between the energy balance equation and the derivative with respect to x of the flux equation and introducing the dimensionless variables;

$$\partial u/\partial \tau + \partial^2 u/\partial \tau^2 = \partial^2 u/\partial X^2 \qquad (2.318)$$

The solution is obtained by the method of separation of variables. First the damping term is removed for purposed of simplifying the mathematics and improving the understanding of the underlying phenomena, by the substitution, $u = \exp(-\tau/2)\, w\,(X, \tau)$. Eq. [2.332] becomes,

$$-w/4 + \partial^2 w/\partial \tau^2 = \partial^2 w/\partial X^2 \qquad (2.319)$$

The method of separation of variables can be used to obtain the solution of Eq. [2.333].

$$\text{Let } W = V(\tau)\, \phi\,(X) \qquad (2.320)$$

Plugging Eq. [2.334] into Eq. [2.333], and separating the variables that are a function of X only and τ only;

$$\phi'' + \lambda^2 \phi = 0 \tag{2.321}$$

$$V'' = V(1/4 - \lambda^2) \tag{2.322}$$

The solution for Eq. [2.335] is the Bessel function of 1/2th order and first kind;

$$\phi = c_1 \sin(\lambda X) + c_2 \cos(\lambda X) \tag{2.323}$$

It can be seen that $c_1 = 0$ as the slope of temperature with respect to X, at X = 0 is 0. Now from the BC at the surface,

$$\partial u / \partial X = -Bi\, u \tag{2.323}$$

where $Bi = h \sqrt{\alpha \tau_r} / k$

$$-\lambda \sin((\lambda X) = -Bi \cos(\lambda X)$$

$$\text{or } \lambda_n / Bi = \cot(\lambda_n a / \sqrt{\alpha \tau_r}) \tag{2.324}$$

for small a,
$$\lambda_n = \text{sqrt}(h\alpha\tau_r/ak) + n\pi = \text{sqrt}(h/S/a) + n\pi \tag{2.325}$$

where $S = \rho C_p / \tau_r$ is the storage coefficient and approximate expression for $\cot\theta$ was used. Now, when

$$\text{sqrt}(h \alpha \tau_r / ak) > \tfrac{1}{2}$$

$$\text{or } a < 4 h/S \tag{2.326}$$

the pulsations in time domain will be seen. Then the transient temperature profile will be given by;

$$u = \Sigma_0^\infty c_n \exp(-\tau/2)\cos(\text{sqrt}(\lambda_n^2 - 1/4)\tau) \cos(\lambda_n X) \tag{2.327}$$

λ_n is given by Eq. [2.342]. c_n can be obtained from the initial condition and the orthogonal property and is found to be $4(-1)^{n+1}/(2n-1) \pi$.

Eq. [2.327] is shown in Figure 2.14 for problem 15 is shown in Figure 2.14. The h/Sa or Biot modulus calculated for the brick wall with convection was 0.36. This is greater than $\tfrac{1}{4}$. Hence the temperature can be expected to undergo subcritical damped oscillations. The centerline temperature is shown for a 10 cm brick subject to convective boundary condition. The time to steady state can be estimated from the x intercept from the Figure 2.14.

For large slabs, the dimensionless temperature can be given by;

$$u = \Sigma_0^\infty c_n \exp(-\tau/2)\exp(\text{sqrt}(1/4 - \lambda_n^2)\tau) \cos(\lambda_n X)$$

This can be extended to two or three dimensions. For the two dimensional problem a double infinite series sum will result and the initial condition can be used to obtain the orthogonality constant. For the three dimensional case the triple sum of infinite series will result.

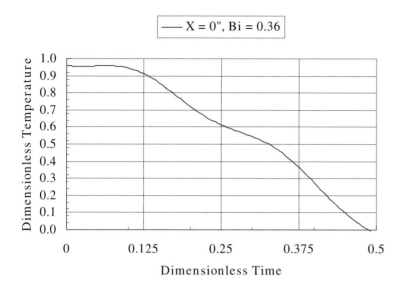

Figure 2.14 Centerline Temperature with Convective Boudary Condition for Small Slab

2.10 Plane of Null Transfer

Consider a semi-infinite slab with an initial temperature at T_0. It is assumed that at the interface of one of the faces of the slab is heated to a higher temperature and then allowed to cool by an exponential decay. The initial condition is given as follows

$$t = 0, \quad Vx, \quad T = T_0 \tag{2.328}$$

Boundary Conditions in Space

$$t > 0, \quad x = 0, \quad T = T_0 + (T_s - T_0) \exp(-t/2\tau_r) \tag{2.329}$$

$$t > 0, \quad x = \infty, \quad T = T_0 \tag{2.330}$$

The fourth and final condition in time,

$$t = \infty, \quad Vx, \quad T = T_0 \tag{2.331}$$

The governing equation can be obtained by a 1 dimensional energy balance as shown in earlier sections,

$$\tau_r \partial^2 T/\partial t^2 + \partial T/\partial t = \alpha \partial^2 T/\partial x^2 \tag{2.332}$$

The term which is a first partial derivative with respect to time of the temperature is the accumulation term and the term with the second partial derivative in time is the relaxation time and the heat conduction term if the one on the RHS with the second derivative in space. Obtaining the dimensionless variables;

$$u = (T - T_0)/(T_s - T_0); \tau = t/\tau_r \quad ; X = x/\text{sqrt}(\alpha \tau_r) \tag{2.333}$$

The governing equation in the dimensionless form is then;

$$\partial u/\partial \tau + \partial^2 u/\partial \tau^2 = \partial^2 u/\partial X^2 \tag{2.334}$$

The solution is obtained by the method of transformation to relativistic variables. Initially the damping term is eliminated using a $u = W \exp(-\tau/2)$ substitution. Eq. [2.353] then becomes,

$$W_{xx} = -W/4 + W_{\tau\tau} \tag{2.335}$$

Eq. [2.335] can be solved by the following transformations to the wave coordinates or canonical form; Let the transformation be as follows;

$$\eta = \tau + X \tag{2.336}$$

$$\xi = \tau - X \tag{2.337}$$

Eq. [2.335] becomes;

$$\partial^2 W/\partial \tau^2 = \partial^2 W/\partial \eta^2 + \partial^2 W/\partial \xi^2 + 2\partial^2 W/\partial \xi \partial \eta \tag{2.338}$$

$$\partial^2 W/\partial X^2 = \partial^2 W/\partial \eta^2 + \partial^2 W/\partial \xi^2 - 2\partial^2 W/\partial \xi \partial \eta \tag{2.339}$$

$$4 \partial^2 W/\partial \xi \partial \eta = W/4 \tag{2.340}$$

At this stage Laplace transforms is obtained in the ξ domain;

$$d\underline{W}/d\eta = \underline{W}/16/s \quad \text{or the solution is } \underline{W} = C' \exp \eta/16s \tag{2.341}$$

At, $\xi = 0$, $\tau = X$. This is at the wave front. The time taken at a point x in the medium to see the disturbance at the boundary traveling with a speed of sqrt(α/τ_r). Only for times greater than this, the problem exists. For times equal to and less than this time, the initial concentration remain and thus $W = u = 0$. Using the boundary condition $W = \exp(-\tau/2)\exp(\tau/2) = 1$ at $\eta = \tau$.

$1/s = C' \exp(\tau/16s)$ or $C' = 1/s \exp(-(\tau/16s))$

Thus, $\underline{W} = 1/s \, exp(\eta - \tau)/16s \qquad = 1/s \, exp(X)/16s$ \hfill (2.342)

The inversion of Eq. [2.362] is within the tables available [Mickley, Sherwod and Reid, 1975]. So,

$u = exp(-\tau/2) \, J_0 \, (\frac{1}{2} \, sqrt(X\tau - X^2))$ \hfill (2.343)

It can be seen that at infinite time the expression becomes zero as specified. The expression is valid only for times greater than the penetration time $t > x/sqrt(D/\tau_r)$. So at the wave front $u = exp(-\tau/2)$. For $X = 0$, the imposed boundary condition $u = exp(-\tau/2)$. The Bessel composite function of the first kind and zeroth order has many zeros and can be seen to be damped oscillatory. In order to amplify the wavy nature of the solution the exponential decay boundary condition was considered in this study. The plane of zero heat transfer can be identified and is when, $X\tau - X^2 = 24.8,$

Thus $X = (\tau \pm sqrt(\tau^2 - 99.2))/2$ \hfill (2.344)

There are two roots to the quadratic equation. The root for the negative contribution for significant times, $(\tau > 9.96)$ occurs before the wave front. This is an interesting result. This is similar to the Krogh solution of the anorexic region in tissue at steady state with Michael Menton kinetics and cylindrical geometry with two layers. Eq. [2.363] is shown in Figure 2.15. The dependence of dimensionless temperature with dimensionless distance changes from concave to convex as the time of scrutiny is increased as shown in the Figure. The plane of zero transfer can be read of the graph for a given time instant under consideration. For a time instant under consideration, after a said distance the dimensionless temperature is zero. There is no further transmission of heat. The decay of the disturbance at the surface is faster compared with the speed of heat propagation to the location of null transfer. So when the time taken for the heat disturbance is much greater compared with the decay time at the surface the null transfer region can be expected. The meaning of the two roots of the equation need further study.

98

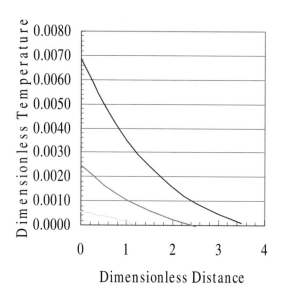

| — Tou = 9.96 — Tou = 12 — Tou = 15 |

Figure 2.15 Plane of Null Transfer in a Semi-infinite Slab

2.11 Temperature Dependent Heat Source

Consider a rod of length l, with one end maintained at temperature T_s. The other end is at the zero temperature (O K). This is the lowest temperature achievable according to the third law of thermodynamics. The entropy is zero at 0 K. At time t = 0, the rod is at 0 K. For times greater than zero, the temperature dependent heat source is allowed to heat the rod. It is of interest to study the temperature distribution in the rod using the non-Fourier damped wave conduction and relaxation equation. A temperature dependent heat source with the source strength U''' w/m³/K is present in the rod. The energy balance on a thin section with thickness Δx is considered at a distance x from the origin for an incremental time Δt. Thus in 1 dimension,

$$(q\,A\big|_x - q\,A\big|_{x+\Delta x} + U'''\,TA\,\Delta x)\,\Delta t \quad = A\,\Delta x\,(\rho C_p)\,\Delta T \qquad (2.345)$$

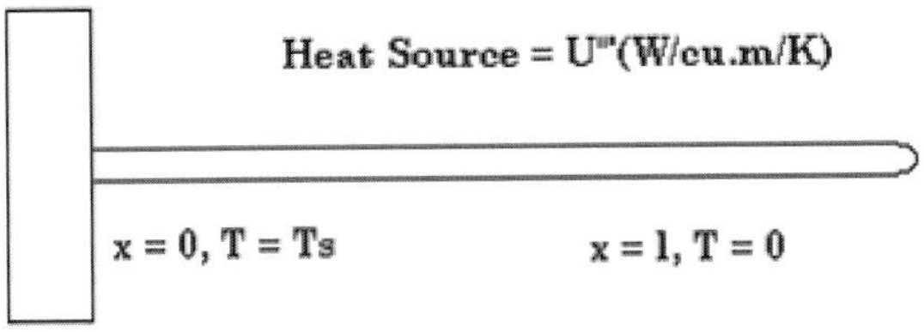

Figure 2.16 Temperature Dependent Heat Source in a Rod

Dividing throughout the equation wrt x and t and taking the limits as Δx, Δt goes to zero and at constant cross-sectional area, the energy balance equation becomes;

$$-\partial q/\partial x + U'''T = ((\rho C_p)\ \partial T/\partial t \qquad (2.346)$$

The non-Fourier damped wave heat conduction and relaxation equation can be written as;

$$q = -k\partial T/\partial x - \tau_r\ \partial q/\partial t \qquad (2.347)$$

The governing equation for the temperature can be obtained by eliminating the heat flux between the energy balance equation and the constitutive law for heat conduction. Thus differentiating the energy balance equation wrt to time and the constitutive equation wrt to x and eliminating the second cross-derivative of flux wrt to time and space;

$$U'''T + k\partial^2 T/\partial x^2 = ((\rho C_p \tau_r)\ \partial^2 T/\partial t^2 + (\rho C_p - U'''\tau_r)\partial T/\partial t \qquad (2.348)$$

Using the dimensionless variables;

$$u = T/T_s\ ; \qquad \tau = t/\tau_r;\ X = x/sqrt(\alpha\tau_r) \qquad (2.349)$$

The governing equation in temperature becomes;

$$U'''u + S\partial^2 u/\partial X^2 = S\partial^2 u/\partial \tau^2 + (S - U''')\partial u/\partial \tau \qquad (2.350)$$

Let $U^* = U'''/S$, Then the dimensionless governing equation can be written as;

$$U^*u + \partial^2u/\partial X^2 = \partial^2u/\partial\tau^2 + (1 - U^*)\partial u/\partial\tau \qquad (2.351)$$

Where, $S = (\rho C_p/\tau_r)$ is the storage coefficient. It has units of w/m³/K. It can be seen from the governing equation that when $S = U'''$, the damped wave conduction and relaxation equation simplifies to a wave equation. The equation reverts to the governing equation seen for the finite slab at constant wall temperature when $U''' = 0$. Thus when $U'''/S = 1$,

$$U^*u + \partial^2u/\partial X^2 = \partial^2u/\partial\tau^2 \qquad (2.352)$$

Let $\eta = X^2 - \tau^2$

For $X > \tau$, the governing equation will transform to ;

$$U^*u + 4\eta\,\partial^2u/\partial\eta^2 + 4\partial u/\partial\eta = 0 \qquad (2.353)$$

$$\eta^2\,\partial^2u/\partial\eta^2 + \eta\partial u/\partial\eta + U^*\eta u/4 = 0 \qquad (2.354)$$

Comparing the above equation with the generalized Bessel equation; $a = 1; b = 0; s = \frac{1}{2}; d = U^*/4; c = 0.\ p = 2.\ \mathrm{sqrt}(0) = 0$; $\mathrm{sqrt}(d)/s = U^{*1/2}$.

$$u = c_1 J_0 (U^{*1/2}(X^2 - \tau^2)^{1/2} + c_2 Y_0 (U^{*1/2}(X^2 - \tau^2)^{1/2} \qquad (2.355)$$

c_2 can be seen to be zero as u is finite at zero η.

Thus, $$u = c_1 J_0 (U^{*1/2}(X^2 - \tau^2)^{1/2} \qquad (2.356)$$

This function exhibits damped wave behavior. This is valid till the first zero. Thus at the first zero of the Bessel function,

$$5.7831 = U^*(X^2 - \tau^2)$$

Thus for, $$X \geq \mathrm{sqrt}(5.7831/U^* + \tau^2) \qquad (2.357)$$

The temperature u will be zero. For short times there will be a good portion of the rod that will not have any temperature. This is even at infinite heat source strength. This is a clear manifestation of the finite speed propagation of the heat. For $\tau > X$,

$$u = c_1 I_0 (U^{*1/2}(\tau^2 - X^2)^{1/2} \qquad (2.358)$$

From the boundary condition at $X = 0$,

$$1 = c_1 I_0 (U^{*1/2}(\tau) \qquad (2.359)$$

Eliminating c_1 between the two equations an approximate solution for u can be written as;

$$u = I_0 (U^{*1/2}(\tau^2 - X^2)^{1/2}/ I_0 (U^{*1/2}(\tau) \qquad (2.360)$$

The general solution for the temperature in the rod with a temperature dependent heat source can be obtained as follows. Let the solution be expressed as a sum of steady state and transient state components of the dimensionless temperature.

Let,
$$u = u^{ss} + u^{\tau} \tag{2.361}$$

Then Eq. [2.378] can be written as;

$$U*u^{ss} + \partial^2 u^{ss}/\partial X^2 = \partial^2 u^{\tau}/\partial \tau^2 + (1 - U*)\partial u^{\tau}/\partial \tau - U*u^{\tau} + \partial^2 u^{\tau}/\partial X^2 \tag{2.362}$$

The steady state component will obey the equation;

$$U*u^{ss} + \partial^2 u^{ss}/\partial X^2 = 0 \text{ with the boundary conditions;} \tag{2.363}$$

$$X = 0, \ u^{ss} = 1 \tag{2.364}$$

$$X = X_1 \ u^{ss} = 0 \tag{2.365}$$

The solution to the second order ODE will then be;

$$u^{ss} = c'Sin(U*^{1/2}X) + c''Cos(U*^{1/2}X) \tag{2.366}$$

From the boundary condition given in Eq. [2.364], c'' can be seen to be 1. From the boundary condition given in Eq. [2.365],

$$c' = -Cot(U*^{1/2}X_1) \tag{2.367}$$

Thus the steady state solution to the temperature is given by;

$$u^{ss} = Cos(U*^{1/2}X) - Cot(U*^{1/2}X_1) Sin(U*^{1/2}X) \tag{2.368}$$

$$\text{or } u^{ss} = Cos(U*^{1/2}X_1) (Cos(U*^{1/2}X)/ Cos(U*^{1/2}X_1) - Sin(U*^{1/2}X)/ Sin(U*^{1/2}X_1)) \tag{2.369}$$

2.11.1 Critical Point of Zero Transfer

It can be seen that at steady state the temperature is periodic with respect to position. This is an interesting result. It can also be noted that the mathematical expression given in Eq. [2.393] can take on negative values. Negative temperature cannot exist as according to the third law of thermodynamics the lowest temperature attainable is 0 K. At 0 K the entropy of any system would be zero. The interpretation of the solution given in Eq. [2.403] in terms of the wave conduction and relaxation is that after a certain location in the rod the temperature will be zero. This can be referred to as the critical point of zero heat transfer. This is shown in Figure 2.17. This was generated using a Microsoft excel for windows software package on a 1.9 GHz pentium IV personal computer. In the Figure 2.6, the heat source $U*$ = U'''/S was taken as 0.5, length of the rod 10 cm, thermal diffusivity 10^5 m^2/s, relaxation time, τ_r as 15 seconds. For $X \geq 3.75$ the temperature comes to the end temperature, of) K imposed on the right end of the rod. Beyond this region there is no heat transfer. It can also be noted that the temperature within the rod exceeds the surface temperature. In this case, this is caused by the temperature dependant heat source. The maximum in temperature occurs at, X = 1.5. Further since the temperature is higher within the rod, the heat flux direction will be in the reverse direction of what it was to begin with neat the zero time. Thus the heat flow will be from the maximum location at $X = 1.5$ toward the $X = 0$ location as well as from the $X = 1.5$ location to the $X = X_1$ location. To begin with the problem was one where the initial temperature was at 0 K. The surface temperature at $X = 0$, was T_s. At short times the heat flux would be from the $X = 0$ location toward the right hand side of the surface. This has been reversed by the time the system reaches steady state. The heat source term is the contributor to the energy.

The heat flux expression at steady state can be written as;

$$q* = -\partial u^{ss}/\partial X = U*^{1/2}\mathrm{Cos}(U*^{1/2}X_1)\ (\ \mathrm{Sin}(U*^{1/2}X)/\ \mathrm{Cos}(U*^{1/2}X_1)\ + \mathrm{Cos}(U*^{1/2}X)/ \mathrm{Sin}(U*^{1/2}X_1))$$

(2.370)

where, $q* = q/\mathrm{sqrt}(k\rho C_p/\tau_r)/(\ T_s)$

It can be seen from Figure 2.18 that several things happen when the steady state heat flux is plotted as a function of the distance in the rod , according to the solution given in Eq. [2.394]. The following distinctions can be recognized from this illustration; There exists a location of maximum heat flux. In Figure 2.18, for the assumed values of the length of the rod, the relaxation time, the ratio of heat source strength with the storage coefficient, the location where the maximum heat flux occurs is at the dimensionless distance, $X = 1.5$.

There is a found a critical location in the rod beyond which there is no heat transfer. This is found to occur at $X = 3.75$ in Figure 2.17, Figure 2.18. There is found two locations in the rod where the heat flux changes in direction. The cross-over locations occur at $X = 0.5$ and again at $X = 2.55$. These are locations of minimum heat flux. The transient portion of the solution will then be;

$$\partial^2 u^\tau/\partial \tau^2 + (1 - U*)\partial u^\tau/\partial \tau = U*u^\tau + \partial^2 u^\tau/\partial X^2$$

(2.371)

The boundary conditions are;

$$X = 0, u^\tau = 0$$

(2.372)

$$X = X_1, u^\tau = 0$$

(2.373)

Eq. [2.397] can be solved for by the method of separation of variables. First the damping term is removed by the substitution, $u^\tau = w\exp(-\tau/2)$. Eq. [2.395] becomes,

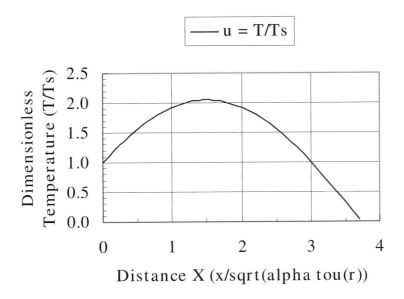

Figure 2.17 Dimensionless Temperature along a Heated Rod

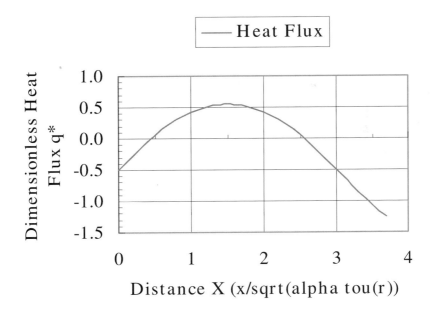

Figure 2.18 Dimensionless Heat Flux along a Heated Rod

$$(1 - U^*)(\partial w/\partial\tau - nw) + (\partial^2 w/\partial\tau^2 + n^2w - 2n\partial w/\partial\tau) = U^*w + \partial^2 w/\partial X^2$$

$$\partial^2 w/\partial\tau^2 + \partial w/\partial\tau(1 - U^* - 2n) + w(-n(1 - U^*) + n^2 - U^*) = \partial^2 w/\partial X^2$$

Letting $n = (1 - U^*)/2$, Eq. [2.371] becomes;

$$\partial^2 w/\partial\tau^2 - w/4(1 + U^*)^2 = \partial^2 w/\partial X^2 \tag{2.374}$$

Eq. [2.400] can be solved by the method of separation of variables.

$$\text{Let } w = V(\tau)\phi(X) \tag{2.375}$$

Eq. [2.400] becomes;

$$V''/V - 1/4(1 + U^*)^2 = \phi''/\phi = -\lambda_n^2 \tag{2.376}$$

The solution for the second order ODEs can be written as follows;

$$\phi = c_1 Sin(\lambda_n X) + c_2 Cos(\lambda_n X) \tag{2.377}$$

From the boundary condition given in Eq. [2.396], $c_2 = 0$. From the boundary condition given in Eq. [2.373],

$$c_1 Sin(\lambda_n X_1) = 0 \tag{2.378}$$

$$(\lambda_n l/sqrt(\alpha\tau_r)) = n\pi, \quad n = 0,1,2\ldots\ldots \tag{2.379}$$

$$\lambda_n = n\pi \, sqrt(\alpha\tau_r)/l \tag{2.380}$$

The time portion of the solution can be written as;

$$V = c_1 exp\tau/2 sqrt((1 + U^*)^2 - 4\lambda_n^2) + c_2 \, exp-\tau/2 sqrt((1 + U^*)^2 - 4\lambda_n^2) \tag{2.381}$$

The transient portion of the solution will decay out to leave the steady state portion of the solution alone. The zero temperature at the x = 1, does the job of removing heat as it is generated in the rod. Thus at infinite time $w = u^\tau exp(\tau/2) = 0$ times infinity = 0. Thus $w = V\phi$ at infinite time = 0. Therefore V = 0 at steady state. Hence the constant c_1 is zero in Eq. [2.381].

$$\text{Thus, } V = c_2 exp-\tau/2((1 + U^*)^2 - 4\lambda_n^2) \tag{2.382}$$

The general solution can be written as;

$$u^\tau = \Sigma_0^\infty c_n exp(-\tau(1 - U^*)/2) exp(-\tau/2((1 + U^*)^2 - 4\lambda_n^2) Sin(\lambda_n X) \tag{2.383}$$

where, λ_n is given by Eq. [2.406]. The c_n can be solved for from the initial condition and is found to be;

$$c_n = 2[1 - (-1)^n]/n\pi \qquad (2.384)$$

Thus the general solution can be written as;

$$u = Cos(U*^{1/2}X_1) (Cos(U*^{1/2}X)/ Cos(U*^{1/2}X_1) - Sin(U*^{1/2}X)/ Sin(U*^{1/2}X_1))$$
$$+ \sum_0^\infty [1 - (-1)^n] 2/n\pi exp(-\tau(1-U*)/2) exp(-\tau/2((1+ U*)^2 - 4\lambda_n^2)Sin(\lambda_n X) \qquad (2.385)$$

where, λ_n is given by Eq. [2.408]. It can be seen from the general solution that even for n = 1 in the infinite series, when;

$$\lambda_n > (1+U*)/2 \quad \text{or when} \quad 1 < 2\pi sqrt(\alpha\tau_r)/(1 + U*) \qquad (2.386)$$

the temperature will undergo subcritical damped oscillations. This is for the cases when $U* < 1$. It can be seen that for $U* > 1$, the "damping term" will begin to grow in amplitude with time and become a runaway.

Worked Example 2.10

Consider the convective heating of a thin rod of length 2l as shown in Figure 2.15. The initial temperature of the rod is maintained at T_0. The heat transfer coefficient for heat transfer between the rod and the surrounding air is given by h (w/m^2/K). The temperature of the surrounding air is maintained at T_s. Obtain the governing equation for transient temperature in the rod using the Non-Fourier damped wave heat conduction and relaxation equation.

The energy balance on a thin section with thickness Δx is considered at a distance x from the origin for a incremental time Δt. Thus in 1 dimension,

$$(q \pi d^2/4|_x - q \pi d^2/4|_{x+\Delta x} + \pi dh(T_s - T)\Delta x) \Delta t = \pi d^2/4\Delta x (\rho C_p) \Delta T \qquad (2.387)$$

where d is the cross-sectional diameter of the thin rod. Dividing throughout the equation by $\Delta x\Delta t$ and taking the limits as Δx, Δt goes to zero and at constant cross-sectional area, the energy balance equation becomes;

$$-\partial q/\partial x + 4h(T_s - T)/d = ((\rho C_p) \partial T/\partial t \qquad (2.388)$$

Using the dimensionless variables;

$$u = (T - T_s) /(T_0 - T_s) ; \tau = t/\tau_r; X = x/sqrt(\alpha\tau_r); q* = q/sqrt(k\rho C_p/\tau_r)/(T_s - T_0) \qquad (2.389)$$

Eq. [2.387] becomes;

$$-\partial q*/\partial X - 4\beta u = \partial u/\partial\tau \qquad (2.390)$$

where, $\beta = h/Sd$; S is the storage coefficient, $S = (\rho C_p/\tau_r)$. The dimensionless group β represents the ratio of the convective heating to the energy stored in the material.

The non-Fourier damped wave heat conduction and relaxation equation in the dimensionless form can be written as;

$$q^* = -\partial u/\partial X - \partial q^*/\partial \tau \tag{2.391}$$

The governing equation for the temperature can be obtained by eliminating the heat flux between the energy balance equation and the constitutive law for heat conduction. Thus differentiating the energy balance equation wrt to time and the constitutive equation wrt to x and eliminating the second cross-derivative of flux wrt to time and space;

$$4\beta u + \partial^2 u/\partial \tau^2 + (1+4\beta)\partial u/\partial \tau = + \partial^2 u/\partial X^2 \tag{2.392}$$

Thus the governing equation in temperature in the dimensionless form is given by Eq. [2.418]. Eq. [2.418] can be solved for by the method of separation of variables. First the damping term is removed by the substitution, $u^\tau = w\exp(-n\tau)$, Eq. [2.392] becomes;

$$\partial^2 w/\partial \tau^2 + \partial w/\partial \tau (1+4\beta - 2n) + w(-n(1+4\beta) + n^2 + 4\beta) = \partial^2 w /\partial X^2 \tag{2.393}$$

Letting $n = (1 +4\beta)/2$, Eq. [2.392] becomes;

$$\partial^2 w/\partial \tau^2 - w/4[(1- 4\beta^*)^2] = \partial^2 w /\partial X^2 \tag{2.394}$$

Eq. [2.420] can be solved by the method of separation of variables.

$$\text{Let } w = V(\tau)\phi(X)$$

Eq. [2.394] becomes;

$$V''/V - 1/4(1 - 4\beta^*)^2 = \phi''/\phi = -\lambda_n^2 \tag{2.395}$$

The solution for the second order ODEs can be written as follows;

$$\phi = c_1 Sin(\lambda_n X) + c_2 Cos(\lambda_n X) \tag{2.396}$$

From the boundary condition at the center of the rod,. i.e.,

$$\partial u/\partial X = 0 = \partial w/\partial X = 0 \tag{2.397}$$

$$\partial\phi/\partial X = c_1(\lambda_n) Cos(\lambda_n X) - c_2(\lambda_n)Sin(\lambda_n X) = 0 \tag{2.398}$$

From Eq. [2.398] it can be seen that c_1 is zero. Hence,

$$\phi = c_2 Cos(\lambda_n X) \tag{2.399}$$

From the boundary condition at $x = \pm 1$, $T = T_s$, i.e.,

$$X = l/\sqrt{(\alpha \tau_r)} = X_1, , \quad u^{\tau} = w\exp(-(1 + 4\beta)/2\tau) = 0 \qquad (2.400)$$

Or
$$0 = c_1 Cos(\lambda_n X_1) \qquad (2.401)$$

$$(2n-1)\pi/2 = \lambda_n X_1 \qquad (2.402)$$

$$\lambda_n = (2n-1)\pi \sqrt{(\alpha \tau_r)}/2l , \quad n = 1,2,3... \qquad (2.403)$$

The time domain solution would be,

$$V = (c_3 \exp(\sqrt{1/4(1 - 4\beta^*)^2 - \lambda_n^2})\tau)+ c_4 \exp(-\sqrt{1/4(1 - 4\beta^*)^2 - \lambda_n^2})\tau)) \qquad (2.404)$$

From the final condition u = 0 at infinite time. So is $V\phi = w = u\exp(1 + 4\beta)\tau/2)$, the wave temperature at infinite time. The wave temperature is that portion of the solution that remains after dividing the damping component either from the solution or the governing equation. For any non-zero ϕ, it can be seen that at infinite time V is zero. Hence c_3 can be set to zero.

$$u = \Sigma_1^\infty c_n \exp(-(1 + 4\beta)\tau/2) \exp(-\sqrt{1/4 - \lambda_n^2})\ \tau) Cos(\lambda_n X) \qquad (2.405)$$

Where λ_n is described by Eq.(2.403). C_n can be shown using the orthogonality property to be $4(-1)^{n+1}/(2n-1)\pi$. It can be seen that the solution given in Eq.(2.443) is

bifurcated. As the value of the thickness of the slab changes the characteristic nature of the solution changes from monotonic exponential decay to subcritical damped oscillatory. For a < $\pi \sqrt{(\alpha \tau_r)}$, even for n =1, $\lambda_n > \frac{1}{2}$. This is when the argument within the square root sign in the exponentiated time domain expression becomes negative and the result becomes imaginary. Using Demovrie's theorem and taking real part for small width of the slab,

$$u = \Sigma_1^\infty c_n \exp(-(1 + 4\beta)/2\tau/2) Cos(\sqrt{\lambda_n^2 - 1/4})\ \tau) Cos(\lambda_n X) \qquad (2.406)$$

2.12 Method of Laplace Transforms

The method of Laplace transform reduces the solution of a linear PDE to an algebraic procedure. It can be used in transient heat conduction and relaxation problems. Relevant boundary conditions are introduced earlier in the analysis and the constants of integration is evaluated automatically. A source for referring this method is Spiegel [180]. The Laplace transform is defined as;

$$L (f(t)) = \int_0^\infty \exp(-st)f(t)\ dt = f(s) \qquad (2.407)$$

0

The Laplace transform of a function f(t) is defined for positive values of t as a function of the new variable s by the integral given in Eq. [2.445]. The transform exists if f(t) meets the following criteria;

1) f(t) is continuous or piecewise continuous in any interval, $t_1 \leq t \leq t_2$ where $t_1 > 0$.

2) $t^n |f(t)|$ is bounded near $t = 0$ when approached from positive values of t for some number n, where $n < 1$.

3) $\exp(-s_0 t)|f(t)|$ is bounded for large values of t for some number s_0.

In most practical problems, if L(f(t)) exists, the Laplace transform of the derivatives exist although this is not necessarily the case. The linear property of the transform is;

$$L\,(a\,f(t) + b\,g(t)) = a\,\underline{f(s)} + b\underline{g(s)} \qquad (2.408)$$

The transform of the derivatives of f(t) are as follows;

$$L\,(\,d^n f(t)/dt^n) = s^n\,\underline{f(s)} - (s^{n-1}(f(0+) + s^{n-2}\,df(0+)/dt + \ldots\ldots + d^{n-1}f(0+)/dt^{n-1}) \qquad (2.409)$$

Eq. [2.447] relates the nth derivative of f(t) to the transform of f(t) itself and the numerical values approached by the lower order derivatives of f(t) as t tends to zero from positive values. The translation property of the transform can be written as;

$$L\,(\exp(at)f(t)) = \underline{f(s-a)} \qquad (2.410)$$

Thus the transform of the product exp(at) f(t) is obtained by replacing s by s-a in the transform of f(t). If,

$$
\begin{aligned}
f(t) \quad &= 0 \qquad && t < a \\
&= g(s-a) \qquad && t \geq a
\end{aligned}
$$

then, $\qquad\qquad \underline{f(s)} \quad = \quad \exp(-as)\,\underline{g(s)} \qquad\qquad (2.411)$

Suppose the function f(t) is such that it is zero for all values of t less than some positive number a and of the form g(t −a) for $t \geq a$. The transform of this function is then found as a product of exp(-as) and the transform of g(t). The inverse transform is obtained by reading from tables. A complete set of tables was given by Campbell and Foster [1948]. The convolution property can be used for transforms whose inversions are not found in the table. The transform can be broken down into a product of two transforms whose inversion exist in the tables such that;

$$\underline{f(s)} \quad = \quad \underline{g(s)}\,\underline{h(s)} \qquad (2.412)$$

where $g(s)$ is the Laplace transform of the function $g(t)$ and $h(s)$ is the Laplace transform of the function $h(t)$. The convolution integral then enables $f(t)$ to be determined:

$$f(t) \quad = L^{-1}(f(s)) \quad = \int_0^t g(t-p)\,h(p)\,dp = g*h \qquad (2.413)$$

2.12.1 Problem of Semi-Infinite Medium in Cartesian Coordinates given a Step Change in Temperature at One of The Walls

Consider the problem of 1 dimensional heat conduction and relaxation in a semi-infinite medium subject to a step change in temperature at one of the walls (Figure 1.3). Obtaining the dimensionless variables;

$$u = (T - T_0)/(T_s - T_0); \qquad \tau = t/\tau_r; \qquad X = x/sqrt(\alpha\,\tau_r) \qquad (2.414)$$

The energy balance on a thin shell at x with thickness Δx is written. The governing Equation can be obtained after eliminating q between the energy balance Equation and the derivative with respect to x of the flux equation and introducing the dimensionless variables;

$$\partial u/\partial\tau + \partial^2 u/\partial\tau^2 = \partial^2 u/\partial X^2 \qquad (2.415)$$

The initial condition is;

$$t = 0, \ T = T_0 \,; \ u = 0 \qquad (2.416)$$

$$t = \infty, \ T = T_s \qquad (2.417)$$

The Boundary Conditions are;

$$X = \infty, \ T = T_0 \,; \ u = 0 \qquad (2.418)$$

$$X = 0, \ T = T_s, \ u = 1 \qquad (2.419)$$

Obtaining the Laplace transform of Eq. [2.415] ;

$$s\,\underline{u} - 0 + s^2\,\underline{u} - 0 - 0 = d^2\underline{u}/dX^2 \qquad (2.420)$$

Solving for the second order ordinary differential equation;

$$\underline{u} = a\exp(+X\sqrt{s(s+1)})) + b\exp(-X\sqrt{s(s+1)})) \tag{2.421}$$

From the boundary condition given by Eq. [2.418] a can be seen to be zero and from the boundary condition given by Eq. [2.419] b can be seen to be 1/s. Thus,

$$\underline{u} = 1/s\exp(-X\sqrt{s(s+1)})) \tag{2.422}$$

Bausemeister and Hamill [1969] defined a another variable H such that,

$$H = \int u \, dx \tag{2.423}$$

The function H in the Laplace domain is thus;

$$\underline{H} = -1/s/\sqrt{s(s+1)}\exp(-X\sqrt{s(s+1)})) + c' \tag{2.424}$$

C' can be seen to be zero from the boundary condition given by Eq. [2.419]. The inversion of the $\underline{H}$ can be obtained by using the convolution property.

$$\underline{H} = g(s)h(s) \tag{2.425}$$

$$g(s) = 1/s \quad ; \quad h(s) = -1/\sqrt{s(s+1)} \exp(-X\sqrt{s(s+1)})) \tag{2.426}$$

$$g(\tau) = 1 \tag{2.427}$$

$$h(\tau) = -\exp(-\tau/2) I_0(1/2 \sqrt{\tau^2 - X^2}) \tag{2.428}$$

For, $\tau > X$, Using the convolution property,

$$H(\tau) = - \int_0^\tau \exp(-p/2) I_0(1/2 \sqrt{p^2 - X^2}) dp \tag{2.429}$$

Using the Leibniz's formula;

$$u = \partial H/\partial X = X \int_0^\tau \exp(-p/2) I_1(1/2 \sqrt{p^2 - X^2}) /\sqrt{p^2 - X^2} \, dp$$

$$+ \exp(-X/2) \tag{2.430}$$

$$u = (T - T_0)/(T_s - T_0) = \exp(-X/2) + X \int_X^\tau \exp(-p/2) I_1 (p^2 - X^2)^{1/2}/(p^2 - X^2)^{1/2} dp \tag{2.431}$$

The dimensionless heat flux is given by;

$$q^* + \partial q^*/\partial \tau = -\partial u/\partial X \tag{2.432}$$

The Laplace transform of Eq. [2.459] is given by;

$$q^* = -1/(s+1)\partial u/\partial X \qquad (2.433)$$

Differentiating Eq. [2.431] wrt X and substituting the result into Eq. [2.433];

$$q^* = 1/(sqrt(s(s+1))exp(-Xsqrt(s(s+1)))) \qquad (2.434)$$

The inversion is found readily in the tables; For $\tau > X$,

$$q^* = exp(-\tau/2) \; I_0(1/2sqrt(\tau^2 - X^2)) \qquad (2.435)$$

For $X > \tau$, $\qquad q^* = exp(-\tau/2) \; J_0(1/2sqrt(X^2 - \tau^2)) \qquad (2.436)$

At the wave front, $X = \tau$,

$$q^* = exp(-\tau/2) \qquad (2.437)$$

The three regimes of heat flux at a interior point in the medium is shown in Figure 2.1.

2.12.2 Constant Wall Flux in a Semi-infinite Medium

Consider the problem of 1 dimensional heat conduction and relaxation in a semi-infinite medium subject to a constant wall flux at one of the walls. Obtaining the dimensionless variables;

$$u = (T - T_0)/(T_0); \quad \tau = t/\tau_r; \qquad X = x/sqrt(\alpha \tau_r) \qquad (2.438)$$

The energy balance on a thin shell at x with thickness Δx is written. The governing Equation can be obtained after eliminating q between the energy balance Equation and the derivative with respect to x of the flux equation and introducing the dimensionless variables;

$$\partial u/\partial \tau + \partial^2 u/\partial \tau^2 = \partial^2 u/\partial X^2 \qquad (2.439)$$
The initial condition is;

$$t = 0, \; T = T_0 \; ; \; u = 0 \qquad (2.440)$$

The Boundary Conditions are;

$$X = \infty, \; T = T_0 \; ; \; u = 0 \qquad (2.441)$$

$$X = 0, \qquad q = Q \quad ; q^* = Q^* \tag{2.442}$$

Obtaining the Laplace transform of Eq. [2.466] ;

$$s\underline{u} - 0 + s^2 \underline{u} - 0 - 0 \qquad = d^2\underline{u}/dX^2 \tag{2.443}$$

Solving for the second order ordinary differential equation;

$$\underline{u} = a\exp(+X\sqrt{s(s+1)}) + b \exp(-X\sqrt{s(s+1)}) \tag{2.444}$$

From the boundary condition given by Eq. [2.468] a can be seen to be zero and from the boundary condition given by Eq. [2.469];

$$Q^*/s = b\sqrt{s(s+1)}/(s+1) \text{ or } b = Q^*/s^{3/2}\sqrt{s+1} \tag{2.445}$$

$$\underline{u} = Q^*/s^{3/2}\sqrt{s+1} \exp(-X\sqrt{s(s+1)}) \tag{2.446}$$

Multiplying the numerator and denominator by sqrt(s+1) and using the linear property of the Laplace transforms;

$$u/Q^* = \exp(-\tau/2) I_0(1/2\sqrt{\tau^2 - X^2}) + \int_0^\tau \exp(-p/2) I_0(1/2 \sqrt{p^2 - X^2})dp \tag{2.447}$$

Thus the temperature profile in a semi-infinite medium subject to a constant wall flux is obtained. The surface temperature is obtained by inversion of;

$$\underline{u} = Q^*/s^{3/2}\sqrt{s+1} \tag{2.448}$$

$$\text{or } u/Q^* = \exp(-\tau/2) I_0(\tau/2) + \tau\exp(-\tau/2) (I_0(\tau/2) + I_1(\tau/2)) \tag{2.449}$$

The flux expression in the Laplace domain can be seen to be;

$$\underline{q^*} = Q^* 1/s\exp(-X\sqrt{s(s+1)}) \tag{2.450}$$

the inversion of this expression can be obtained as derived before and is;

$$q/Q = - \int_0^\tau \exp(-p/2) I_0(1/2 \sqrt{p^2 - X^2})x/\sqrt{p^2 - X^2} dp$$
$$+ \exp(-X/2) \tag{2.451}$$

Worked Example 2.11

Obtain the transient temperature distribution in one dimension in an ice cube during phase change. The surface flux can be taken as a product of melting rate and the latent heat of fusion and bulk density. The melt front can be posed as a boundary condition using the method of Laplace transforms.

The rate of heat added during the melting at the surface $= A\lambda \, dl/dt \, \rho_s = -q_x = +m*\lambda$
Obtaining the dimensionless variables;

$$u = (T - T_{ice})/(T_{ice}); \quad \tau = t/\tau_r; \quad X = x/\sqrt{\alpha \, \tau_r} \qquad (2.452)$$

The energy balance on a thin shell at x with thickness Δx is written. The governing Equation can be obtained after eliminating q between the energy balance Equation and the derivative with respect to x of the flux equation and introducing the dimensionless variables;

$$\partial u/\partial \tau + \partial^2 u/\partial \tau^2 = \partial^2 u/\partial X^2 \qquad (2.453)$$

The initial condition is;

$$t = 0, \quad T = T_{ice}; \quad u = 0 \qquad (2.454)$$

The Boundary Conditions are;

$$X = l^*, \qquad q = \lambda \, dl/dt \, \rho_s \qquad (2.455)$$

$$X = \infty, \quad T = T_{ice} \qquad (2.456)$$

Let $dl/dt = m^*$
$$X = l^*, \quad q/(\lambda \rho_s v_h) = dl^*/d\tau = q^* \qquad (2.457)$$

Obtaining the Laplace transform of Eq. [2.453] ;

$$s\underline{u} - 0 + s^2\underline{u} - 0 - 0 = d^2\underline{u}/dX^2 \qquad (2.458)$$

Solving for the second order ordinary differential equation;

$$\underline{u} = a\exp(+X\sqrt{s(s+1)}) + b\exp(-X\sqrt{s(s+1)}) \qquad (2.459)$$

From the boundary condition given by Eq. [2.445] a can be seen to be zero and from the boundary condition given by Eq. [2.446];

$$q^*/s = b\sqrt{s(s+1)}/(s+1)\exp(-l^*\sqrt{s(s+1)}) \qquad (2.460)$$

$$\text{or } b = q^*/s^{3/2}\sqrt{s+1} \qquad \exp(+l^*\sqrt{s(s+1)}) \qquad (2.461)$$

or $\underline{u}$ = $q*/s^{3/2}$sqrt(s+1)   exp(+l*sqrt(s(s+1))) exp(-Xsqrt(s(s+1)))

$$= q*(s+1)/s^{3/2} \text{ sqrt(s+1) exp(-(X- l*)sqrt(s(s+1)))} \qquad (2.462)$$

Inverting the Laplace transformed expression by using the linear property of transformation;

u   = $q* \exp(-\tau/2) I_0(1/2 \text{ sqrt}(\tau^2 - (X - l*)^2)) +$

$$\int_0^{\tau} q* \exp(-p/2) I_0 \text{ ½ sqrt}(p^2 - (X - l*)^2) \qquad (2.463)$$

Eq. [2.499] is valid, when $\tau > (X - l*)$. When $(X - l*) > \tau$,

u   = $q* \exp(-\tau/2) J_0(1/2 \text{ sqrt}((X - l*)^2 - \tau^2)) +$

$$\int_0^{\tau} q* \exp(-p/2) I_0 \text{ ½ sqrt}((X - l*)^2 - p^2) \qquad (2.464)$$

2.12.3 Convective Heating in a Semi-infinite Medium

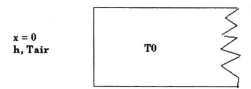

x = 0
h, Tair   T0

Figure 2.19 Convective Heating in a Semi-Infinite Medium

Consider a semi-infinite medium at a initial temperature of T_0 heated by air at a temperature of T_{air} and heat transfer coefficient, h. It is assumed that the heat transfer coefficient is constant with respect to time and the ambient temperature is held constant and at an elevated temperature compared with the initial temperature of the slab. In one dimension the energy balance equation on a thin slice of thickness Δx can be written as;

$$-\partial q/\partial x \quad = (\rho\, C_p)\, \partial T/\partial t \qquad (2.465)$$

Combining Eq. [2.501] with the Cattaneo and Vernotte damped wave heat conduction and relaxation equation the governing equation can be written as;

$$\partial u/\partial \tau \quad + \quad \partial^2 u/\partial \tau^2 \quad = \partial^2 u/\partial X^2 \qquad (2.466)$$

$$\text{where, } u = (T - T_0)/(T_{air} - T_0);\ \tau = t/\tau_r;\ X = x/sqrt(\alpha\tau_r) \qquad (2.467)$$

Obtaining the Laplace transform of the PDE given in Eq. [2.466] and solving for the second order ODE the Laplace transformed dimensionless temperature in the slab is given by;

$$\underline{u} = \quad aexp(+Xsqrt(s(s+1))) \quad + b\ exp(-Xsqrt(s(s+1))) \qquad (2.468)$$

The boundary conditions are given by;

$$X = \infty,\ u = 0 \qquad (2.469)$$

$$\tau = 0,\ u = 0 \qquad (2.470)$$

$$X = 0,\ h(T_{air} - T) = -k\partial T/\partial x \qquad (2.471)$$

$$\text{Or}\quad hsqrt(\alpha\tau_r)/k\ (1-u) = -\partial u/\partial X$$

Let $Bi = hsqrt(\alpha\tau_r)/k$, where Bi = Biot number,

$$Bi(1-u) \quad = -\partial u/\partial X \qquad (2.472)$$

The Laplace transform of the boundary condition given in Eq. [2.471] is given by;

$$Bi/s - Bi\underline{u} = -d\underline{u}/dX \qquad (2.473)$$

From the boundary condition given by Eq. [2.469] a can be seen to be zero and from the boundary condition given by Eq. [2.471] at $X = 0$;

$$Bi/s - Bi\ b = bsqrt(s(s+1))$$

$$\text{Or } b = Bi/s/(Bi + sqrt(s(s+1))) \qquad (2.474)$$
Thus,

$$\underline{u} = Bi/(s(Bi + sqrt(s(s+1)))) exp(-Xsqrt(s(s+1))) \qquad (2.475)$$

Eq. [2.475] can be inverted using the convolution theorem. Let the transformed expression be represented as a product of two functions;

$$\underline{H} = g(s)h(s) \tag{2.476}$$

$$g(s) = \mathrm{sqrt}(s+1)/\mathrm{sqrt}(s)/(Bi + \mathrm{sqrt}(s(s+1))) \tag{2.477}$$

$$h(s) = -Bi/\mathrm{sqrt}(s(s+1)) \ \exp(-X\mathrm{sqrt}(s(s+1))) \tag{2.478}$$

$$h(\tau) = \exp(-\tau/2) \ I_0(1/2 \ \mathrm{sqrt}(\tau^2 - X^2) \tag{2.479}$$

For, $\tau > X$

$h(\tau)$ can be found by inverting the Laplace transformed expression given in Eq. [2.478]. Multiplying the numerator and denominator by $\mathrm{sqrt}(s)(Bi - \mathrm{sqrt}(s(s+1)))$,

$$\mathrm{sqrt}(s(s+1)) \ (Bi - \mathrm{sqrt}(s(s+1))) \ /(s)/(Bi^2 - s(s+1)) \tag{2.480}$$

Eq. [2.511] is in the form of $P(s)/Q(s)$ and the method of residues can be used to obtain the inversion. The poles of $Q(s)$ are 0, r_1 and r_2.

$$r_1, r_2 = -1/2 \pm \ \mathrm{sqrt}(1 + 4Bi^2)/2 \tag{2.481}$$

$$g(\tau) = A \exp(-\tau/2)\exp(+\tau\mathrm{sqrt}(1 + 4Bi^2)/2 \ + B \exp(-\tau/2)\exp(+\tau\mathrm{sqrt}(1 + 4Bi^2)/2 \tag{2.482}$$

$$\text{where, } A = (Bi \ \mathrm{sqrt}(r_1(r_1 + 1)) - r_1(r_1 + 1) \ )/(Bi^2 - 3 \ r_1^2 - 2 \ r_1) \tag{2.483}$$

$$B = (Bi \ \mathrm{sqrt}(r_2(r_2 + 1)) - r_2(r_2 + 1) \ )/(Bi^2 - 3 \ r_2^2 - 2 \ r_2) \tag{2.484}$$

Using the convolution property,

$$u \qquad = \int_0^\tau g(\tau - p)\exp(-p/2) \ I_0(1/2 \ \mathrm{sqrt}(p^2 - X^2)dp \tag{2.485}$$

At zero Biot number the problem reverts to that of the CWT and at infinite Biot number the problem becomes the CWF case. When h varies with time, the boundary condition may have to include the ballistic term of the heat flux also.

2.13　Average Temperature in a Finite Slab with Convective Boundary Condition

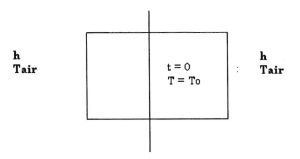

Figure 2.20　Average Temperature in a Finite Slab during Convective Heating

Consider a finite slab at an initial temperature of T_0 heated by air at a temperature of T_{air} and heat transfer coefficient, h. It is assumed that the heat transfer coefficient is constant with respect to time and the ambient temperature is held constant and at an elevated temperature compared with the initial temperature of the slab. In one dimension the energy balance equation on a thin slice of thickness Δx can be written as;

$$-\partial q/\partial x \quad = (\rho\, C_p)\, \partial T/\partial t \tag{2.486}$$

Combining Eq. [2.517] with the Cattaneo and Vernotte damped wave heat conduction and relaxation equation the governing equation can be written as;

$$\partial u/\partial \tau \;+\; \partial^2 u/\partial \tau^2 \quad = \partial^2 u/\partial X^2 \tag{2.487}$$

where, $u = (T - T_{air})/(T_0 - T_{air})$; $\tau = t/\tau_r$; $X = x/\sqrt{\alpha\tau_r}$ (2.488)

The initial, final and boundary conditions are;

$$t = 0, \quad u = 1 \tag{2.489}$$

$$t = \infty, \quad u = 0 \tag{2.490}$$

$$X = 0, \quad \partial u/\partial X = 0 \tag{2.491}$$

$$X = a/\sqrt{\alpha\tau_r}, \quad -h(T_{air} - \langle T \rangle) = -k\partial T/\partial x \tag{2.492}$$

The heat transfer coefficient is defined wrt to the average temperature in the slab for convenience of later illustration.

$$\text{Or, } -Bi <u> = \partial u/\partial X \tag{2.493}$$

Where,. $Bi = h\sqrt{(\alpha\tau_r)}/k$

Eq. [2.515] is integrated wrt to X between 0 and X*, where $X^* = a/\sqrt{(\alpha\tau_r)}$ to give;

$$\partial <u>/\partial\tau \quad + \quad \partial^2 <u>/\partial\tau^2 \quad = \quad -Bi/X^*<u> \tag{2.494}$$

$$\text{Let } Bi^* = h\alpha\tau/ak = = h/Sa \tag{2.495}$$

Where S = storage coefficient, $S = (\rho C_p/\tau_r)$. Eq. [2.494] is a second order ODE with constant coefficients. The solution to the ODE is;

$$<u> \quad = \quad \exp(-\tau/2)(a\exp(+\tau/2\sqrt{(1 + 4Bi^*)}) + b\exp(-\tau/2\sqrt{(1 - 4Bi^*)})) \tag{2.496}$$

or

$$<u> \exp(\tau/2) \quad = \quad a\exp(+\tau/2\sqrt{(1 + 4Bi^*)}) + b\exp(-\tau/2\sqrt{(1 - 4Bi^*)}) \tag{2.497}$$

From the final condition in time as given in Eq. [2.490], at infinite time, the LHS of Eq. [2.497] becomes zero multiplied by infinity and becomes zero. Hence a can be seen to be zero. From the initial condition given by Eq. [2.489] the average initial temperature is also 1. Hence b = 1. Thus the average temperature in the finite slab is given by;

$$<u> \quad = \quad \exp(-\tau/2) \exp(-\tau/2\sqrt{(1 - 4Bi^*)}) \tag{2.498}$$

On examination of Eq. [2.529], when $Bi^* > 1/4$, the average temperature becomes subcritical damped oscillatory. The argument within the square root becomes negative. The square rot of -1 is i. Using De Movrie's theorem, $\exp(-i\theta) = \cos(\theta) - i\sin(\theta)$. The real part can be taken and. For $Bi^* > 1/4$; $h\alpha\tau/ak > 1/4$;

$$<u> \quad = \quad \exp(-\tau/2) \cos(\tau/2\sqrt{(4Bi^* - 1)}) \tag{2.499}$$

The dimensionless frequency of the oscillations are, $\sqrt{(4Bi^* - 1)}$. The frequency of the oscillations are $\sqrt{(4Bi^* - 1)}/\tau_r$. As the Biot number is large the frequency becomes large. As the relaxation time increases the frequency becomes smaller. Eq. [2.536] can be applied for $Bi < 1/4$ and Eq. [2.537] for $Bi > 1/4$. For $Bi^* = 1/4$ the average temperature decays exponentially and is given by $\exp(-\tau/2)$. The thermal time constant of the slab can be defined as the time taken to attain, say $2/3^{rd}$ the final value. The thermal time constants for this definition for different Biot numbers are shown in Table 2.13.1. The thermal time constant for small Biot numbers are given by;

$$\ln(<u>) \quad = \quad -\tau/2 (1 + \sqrt{(1 - 4Bi^*)}) \tag{2.500}$$

$$\text{or } \tau_c = 2\ln(3)/(1 + \sqrt{(1 - 4Bi^*)}) \tag{2.501}$$

where τ_c is thermal time constant as defined. As Biot number is decreased the thermal time constant decreases in this regime. This is for Biot numbers less than 0.25. Given the relaxation time the time constant in seconds can be obtained by multiplying the τ_c by the relaxation time. Thus for Biot number equal to 0.1 and the relaxation time equal to 15 seconds, the thermal time constant can be seen to be 18.6 seconds. The Biot number is given by the ratio of the heat transfer coefficient to the storage coefficient multiplied by the half-width of the slab. As the half-width is increased with the other parameters remaining the same the Biot number decreases and the thermal time constant is found to decrease! For large Biot numbers as the half-width is increased the thermal time constant increases.

Table 2.13.1 Thermal Time Constant for a Finite Slab
with Convective Boundary Condition

S. No.	Biot Number	Thermal Time Constant (t/τ_r)
1.	0.01	1.1
2.	0.1	1.24
3.	0.25	2.2
4.	1	1.116
5.	2	0.795
6.	4	0.575

It can be seen from the table that three different expressions were used to calculate the thermal time constant. The first regime is when the Biot number is small, which is for large slabs, small heat transfer coefficient, large storage coefficients. Large storage coefficient translates to small relaxation times and high thermal masses. The second regime is when the Biot number equals 0.25 when the exponential decay in time is the solution for the transient temperature in dimensionless form. The third regime is when the Biot number is greater than 0.25. This is for small slabs, small storage coefficient. Small storage coefficient of the medium translates to large relaxation times and small thermal mass. In such cases subcritical damped oscillations can be found in the temperature in transience. Within the constraints of not having negative temperature in the expression which would be a violation of the third law of thermodynamics the time averaged temperature for a finite slab is shown for large and small Biot numbers in Figures 2.21 and 2.22 respectively. In the regime of small Biot numbers, the thermal time constant increases with increasing Biot number whereas in the regime of large Biot numbers the thermal time constant decreases with increase in Biot number. The thermal time constant as a function of the Biot number is shown in

120

Figure 2.23. The time taken to steady state can be read from the x intercept.

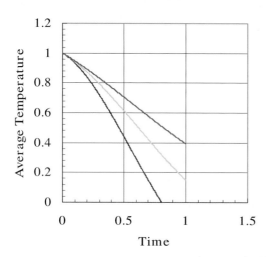

Figure 2.21 **Average Temperature in a Finite Slab with**   Convective
Boundary Condition, Bi > 1/4

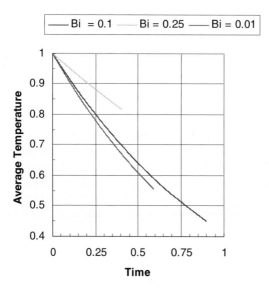

Figure 2.22 **Average Temperature in a Finite Slab with Convective**
Boundary Condition with Bi ≤ 1/4

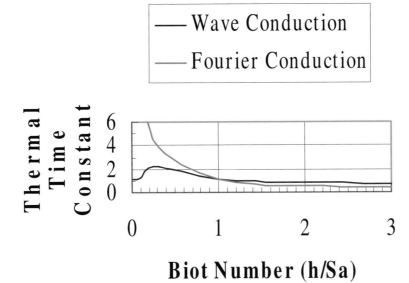

Figure 2.23 Thermal Time Constant Predictions as a Function of Biot Number

Worked Example 2.12

The alternate expression for Fourier with up to second derivative in time of the flux retained in the Taylor series expansion of the time rate of change of accumulation of heat at the surface can be written as [Sharma, 2003t21]

$$q = -k \, \partial T/\partial x - \tau_r \, \partial q/\partial t - \tau_r^2/2! \, \partial^2 q/\partial t^2 \qquad (2.502)$$

Consider a finite slab at initial temperature T_0, subjected to sudden contact with fluid at temperature at to T_1. The transient temperature as a function of space and time need to be obtained. The heat propagative velocity is $V_h = \sqrt{\alpha/\tau_r}$. The initial condition

$$t = 0, \ Vx, \ T = T_0$$

$$t > 0, \ x = 0, \ \partial T/\partial x = 0 \qquad (2.503)$$

$$t > 0, \ x = \pm a, \ -k\partial T/\partial x = -h\,(T_1 - T) \qquad (2.504)$$

Obtaining the dimensionless variables ;

$$u = (T - T_1)/(T_0 - T_1); \quad \tau = t/\tau_r \quad ; \quad X = x/sqrt(\alpha \, \tau_r) \qquad (2.505)$$

The boundary condition given in Eq. [2.535] when h is defined wrt to the average temperature of the slab can be written as;

$$-Bi \langle u \rangle = \partial u/\partial X \qquad (2.506)$$

The energy balance on a thin spherical shell at x with thickness Δx is written. The governing equation can be obtained after eliminating q between the energy balance equation and the derivative with respect to x of the flux equation and introducing the dimensionless variables;

$$\partial u/\partial \tau + \partial^2 u/\partial \tau^2 + \tfrac{1}{2}!\partial^3 u/\partial \tau^3 = \partial^2 u/\partial X^2 \qquad (2.507)$$

Obtain the average temperature in the slab. Eq. [2.538] is integrated wrt to X between 0 and X^*, where $X^* = a/sqrt(\alpha\tau_r)$ to give;

$$\partial \langle u \rangle/\partial \tau + \partial^2 \langle u \rangle/\partial \tau^2 + \tfrac{1}{2}!\partial^3 \langle u \rangle/\partial \tau^3 = -Bi/X^* \langle u \rangle \qquad (2.508)$$

$$\text{Let } Bi^* = h\alpha\tau/ak = = h/Sa \qquad (2.509)$$

Where S = storage coefficient, $S = (\rho C_p/\tau_r)$. Eq. [2.539] is a third order ODE with constant coefficients. The solution to the ODE can be obtained by solving for;

$$r^3/2 + r^2 + r + Bi^* = 0 \qquad (2.510)$$

$$\text{or } r^3 + 2r^2 + 2r + 2Bi^* = 0$$

$$\text{Let } r = (y - 2/3) \text{ Then} \qquad y^3 + py + q = 0$$
$$(2.511)$$

$$p = -20/9; \quad q = 16/27 + 2Bi^*$$

The Vieta's substitution is;
$$y = z - p/3z \qquad (2.512)$$
Then,

$$z^6 + f z^3 - e^3/27 = 0 \qquad (2.513)$$

where $f = q$; $\quad e = p$

$$\text{With } z^3 = u$$

$$u^2 + f u - e^3/27 = 0 \qquad (2.514)$$

The solution to the quadratic is;

$$(-q \pm sqrt(q^2 + 4p^3/27))/2 \qquad (2.515)$$

The root becomes imaginary when,

$$(16/27 + 2Bi*)^2 - 32000/729/27 < 0$$

$$or \quad Bi* < 0.35$$

The roots of the equation are r_1, r_2 and r_3. Then the solution is;

$$<u> = a \, exp(-r_1\tau) + b \, exp(-r_2\tau) + c \, exp(-r_3\tau)$$

$$(2.516)$$

2.13.1 Parabolic Solution

The average temperature for the same problem described above can be solved for using the Fourier's law of heat conduction for purposes of comparison. Let the relaxation time be equal to l^2/α. Then the governing equation can be written as;

$$\partial u/\partial \tau \qquad = \partial^2 u/\partial X^2 \qquad (2.517)$$

where, $u = (T - T_{air})/(T_0 - T_{air})$; $\tau = t/\tau_r$; $X = x/sqrt(\alpha\tau_r)$
$X = a/sqrt(\alpha\tau_r)$, $-Bi <u> = \partial u/\partial X$ $\qquad (2.518)$

Where,. $Bi = hsqrt(\alpha\tau_r)/k$. Eq. [2.549] is integrated wrt to X between 0 and X*, where X* = $a/sqrt(\alpha\tau_r)$ to give;

$$\partial <u>/\partial \tau \qquad = -Bi/X*<u> \qquad (2.519)$$

Let $Bi* = h\alpha\tau_r/ak = = h/Sa$. Where S = storage coefficient, $S = (\rho C_p/\tau_r)$. Eq. [2.563] is a first order ODE with constant coefficients. The solution to the ODE is;

$$<u> = exp(-Bi* \tau) \qquad (2.520)$$

The thermal time constant of the slab can be defined as the time taken to attain, say $2/3^{rd}$ the final value. The thermal time constants for this definition for different Biot numbers are shown in Table 2.13.2. This is also plotted in Figure 2.17. The thermal time constant is given by;

$$\tau_c = ln(3)/Bi*$$

where τ_c is thermal time constant as defined. As Biot number is decreased the thermal time constant increases in this regime.

As can be seen from the Table 2.13.2 the thermal time constant predicted using the damped wave heat conduction and relaxation equation is higher than that predicted using the Fourier's law of heat conduction for the large Biot number regimes.

This is attributable to the thermal inertia effects and the additional ballistic transport mechanism that needed to be accounted for. For the small Biot number regimes the thermal time constant predicted by the Fourier's law is higher. For small Biot number regime Eq. [2.533] can be reduced to an equation which is independent of the Biot number. The average temperature would be a decaying exponential in time. Further for very small Biot number Eq. [2.572] may not be valid as there is singularity at Biot number going to zero. So in reality a plateau can be expected as the Biot number is decreased. Thus for large half-sizes, the zone of null transfer may exist near the center of the slab and the zone that is heated remains the same beyond a threshold size of the slab.

Table 2.13.2 Thermal Time Constant for a Finite Slab with Convective Boundary Condition – Comparison Between Fourier and Damped Wave Conduction and Relaxation

S. No.	Biot Number	Thermal Time Constant	Fourier Thermal Time Constant
1.	0.01	1.1	109.9
2.	0.1	1.24	10.99
3.	0.25	2.2	4.4
4.	1.0	1.116	1.1
5.	2.0	0.795	0.55
6.	4.0	0.575	0.275

Worked Example 2.13

Use the dual phase lag model mentioned in Chapter 1 and finds the conditions where subcritical damped oscillations can be found in a finite slab subject to constant wall temperature boundary condition. (Figure 2.23)

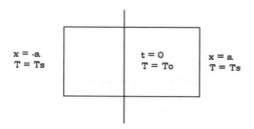

$x = -a$
$T = Ts$

$t = 0$
$T = To$

$x = a$
$T = Ts$

Figure 2.24 Finite Slab of Width 2a Subject to CWT

Consider a finite slab of width 2a with an initial temperature at T_0. The sides of the slab are maintained at constant temperature of T_s. The energy balance on a thin slice of thickness Δx at a distance x in one dimension with no heat generation s given by;

$$\partial q/\partial x = (\rho C_p)\, \partial T/\partial t$$

The non-Fourier equation from the dual phase lag model is given by;

$$q + \tau_q \partial q/\partial t = -k\, \partial T//\partial x - k\, \tau_t \partial^2 T/\partial t \partial x \qquad (2.521)$$

The governing equation in the dimensionless form can be obtained by differentiating the energy balance equation wrt t and Eq. [2.552] wrt x and then eliminating the second cross-derivative of q wrt to t and x and non-dimensionalizing the variables;

$$\partial u/\partial \tau + \partial^2 u/\partial \tau^2 = \partial^2 u/\partial X^2 + \beta \partial^3 u/\partial X^2 \partial \tau \qquad (2.522)$$

Where,

$$u = (T - T_s)/(T_0 - T_s) \quad ; \tau = t/\tau_r \quad ; X = x/\mathrm{sqrt}(\alpha \tau_r) \qquad (2.523)$$
$$\beta = \tau_T/\tau_q$$

The initial condition is given as follows

$$t = 0, \; Vx, \; T = T_0 \; ; \; u = 1 \qquad (2.524)$$

Boundary Conditions in Space

$$t > 0, \; x = 0, \; \partial T/\partial x = 0 \quad ; \quad \partial u/\partial x = 0 \qquad (2.525)$$

$$t > 0, \quad x = \pm a, \quad T = T_s \; ; \quad u = 0 \qquad (2.526)$$

The fourth and final condition in time,

$$t = \infty, \; Vx, \; T = T_s \quad ; \quad u = 0 \qquad (2.527)$$

The solution is obtained by the method of separation of variables. First the damping term is removed by supposing $u = w\exp(-\tau/2)$. Eq. [2.553] becomes,

$$\partial^2 w/\partial X^2 \, (1 - \beta/2) + \beta \partial^3 w/\partial X^2 \partial \tau = \partial^2 w/\partial \tau^2 - w/4 \qquad (2.528)$$

$$\text{Let } u = V(\tau)\, \phi(X) \qquad (2.529)$$

Eq.[2.559] becomes,

$$\phi^{\cdot}(X)/\phi(X)\,(1 - \beta/2 + \beta V'/V) = (V''(\tau)/V(\tau) - 1/4) = -\lambda_n^2 \qquad (2.530)$$

or

$$\phi^{\cdot}(X)/\phi(X) = = (V''(\tau)/V(\tau) - 1/4)\,/\,(1 - \beta/2 + \beta V'/V) = -\lambda_n^2 \qquad (2.531)$$

$$\phi(X) = c_1 Sin(\lambda_n X) + c_2 Cos(\lambda_n X) \tag{2.532}$$

From the boundary conditions,

$$At\ X = 0,\ \partial\phi/\partial X = 0 \tag{2.533}$$

$$So,\ c_1 = 0$$

$$\phi(X) = c_1 Cos(\lambda_n X)$$

$$0 = c_1 Cos(\lambda_n X_a)$$

$$\lambda_n = (2n-1)\pi\ sqrt(\alpha\ \tau_r)/2a\ ,\quad n = 1,2,3... \tag{2.534}$$

The time domain solution would be the solution to the second order ODE;

$$V'' + V'(\beta(2n-1)^2\pi^2\ (\alpha\ \tau_r)/4a^2 + V((2n-1)^2\pi^2\ (\alpha\ \tau_r)/4a^2(\ 1 - \beta/2)\ -1/4) = 0 \tag{2.535}$$

Evan for n = 1, when $b^2 - 4ac$ is negative the system will exhibit subcritical damped oscillations. This is when,

$$\beta^2\pi^4\ (\alpha\ \tau_r)^2/16a^4\ <\ \pi^2\ (\alpha\ \tau_r)/a^2(1 - \beta/2) - 1 \tag{2.536}$$

In terms of the half-width a,

$$16a^4\ -\ 16a^2\ \pi^2\ (\alpha\ \tau_r)(1 - \beta/2)\ +\ \beta^2\pi^4\ (\alpha\ \tau_r)^2 < 0 \tag{2.537}$$

$$or\ (a^2 - r_1)(a^2 - r_2) < 0 \tag{2.538}$$

The roots are;

$$1/2\pi^2\ (\alpha\ \tau_r)(1 - \beta/2)\ \pm\ 1/2\pi^2(\alpha\tau_r)sqrt((1 - \beta) \tag{2.539}$$

Subcritical damped oscillations will occur when, a^2 lies between the two roots, r_1 and r_2. i.e.,

$$\pi^2\ (\alpha\ \tau_r)/2[1 - \beta/2 - (1 - \beta)^{1/2}]\ <\ a^2\ <\ \pi^2\ (\alpha\ \tau_r)/2[\ 1 - \beta/2 + (1 - \beta)^{1/2}\ ] \tag{2.540}$$

For the special case when $\beta = 0$,

$$0 < a < \pi(\alpha\ \tau_r)^{1/2}/2\ [\ 1 + 1\} = \pi(\alpha\ \tau_r)^{1/2} \tag{2.541}$$

This is the result obtained for a finite slab at constant wall temperature using the damped wave heat conduction and relaxation equation. When $\beta = 1$ the subcritical damped oscillations will not occur and this is the Fourier solution. The exact solution can be obtained by solving for the time domain portion of the solution;

$$V'' + V'(\beta\lambda_n^2)\ +\ V(\lambda_n^2\ (\ 1 - \beta/2)\ -1/4))\ = 0 \tag{2.542}$$

The roots of the auxiliary equation are;

$$p_1 \cdot p_2 = -(\beta\lambda_n^2)/2 \pm \tfrac{1}{2}\,sqrt((\beta^2\lambda_n^4 - 2\lambda_n^2\,(2 - \beta) - 1)) \qquad (2.543)$$

$$V = exp(-\tau(\beta\lambda_n^2)/2)\,(c_3 exp(\tau/2 sqrt((\beta^2\lambda_n^4 - 2\lambda_n^2\,(2 - \beta) - 1)))) +$$
$$c_4 exp(-\tau/2 sqrt((\beta^2\lambda_n^4 - 2\lambda_n^2\,(2 - \beta) - 1)))) \qquad (2.544)$$

$$\text{or } Vexp(\tau\beta\lambda_n^2/2) = (c_3 exp(\tau/2 sqrt((\beta^2\lambda_n^4 - 2\lambda_n^2\,(2 - \beta) - 1)))) +$$
$$c_4 exp(-\tau/2 sqrt((\beta^2\lambda_n^4 - 2\lambda_n^2\,(2 - \beta) - 1)))) \qquad (2.545)$$

From the final condition, $u = 0$ at infinite time. So is $Vexp(\tau\beta\lambda_n^2/2)$. W, the wave temperature at infinite time is also zero. The wave temperature is that portion of the solution that remains after dividing the damping component either from the solution or the governing equation. For any non-zero ϕ, it can be seen that at infinite time the LHS of Eq. [2.595] is a product of zero and infinity and a function of x and is zero. Hence the RHS of Eq. [2.595] is also zero and hence in Eq. [2.595] c_3 need be set to zero. Hence,

$$u = \sum_1^\infty c_n exp(-\tau(\beta\lambda_n^2)/2)\,exp(-\tau/2 sqrt(\beta^2\lambda_n^4 - 2\lambda_n^2\,(2 - \beta) - 1))\,Cos(\lambda_n X) \qquad (2.546)$$

Where λ_n is described by Eq.[2.565]. C_n can be shown using the orthogonality property to be $4(-1)^{n+1}/(2n-1)\pi$. It can be seen that Eq.[2.577] is bifurcated. As the value of the thickness of the slab changes the characteristic nature of the solution changes from monotonic exponential decay to subcritical damped oscillatory.

Worked Example 2.14

Consider a naval warhead that is protected using the thermal barrier coating such as the Magnesium Oxychloride Sorrel cement. Develop the temperature profile in the film and obtain the criteria for subcritical damped oscillations to occur in the film when a fire breaks out.

The damped wave heat conduction and relaxation equation is used to describe the transient heat processes. The governing equation is given by;

$$\partial u/\partial\tau + \partial^2 u/\partial\tau^2 = \partial^2 u/\partial X^2 \qquad (2.547)$$

$$\text{where, } u = (T - T_{fire})/(T_n - T_{fire}); \quad \tau = t/\tau_r; \quad X = x/sqrt(\alpha\tau_r) \qquad (2.548)$$

The initial, final and boundary conditions are;

$$t = 0, \quad u = 1$$

$$t = \infty, u = 0$$

$$X = 0, \quad u = 0$$

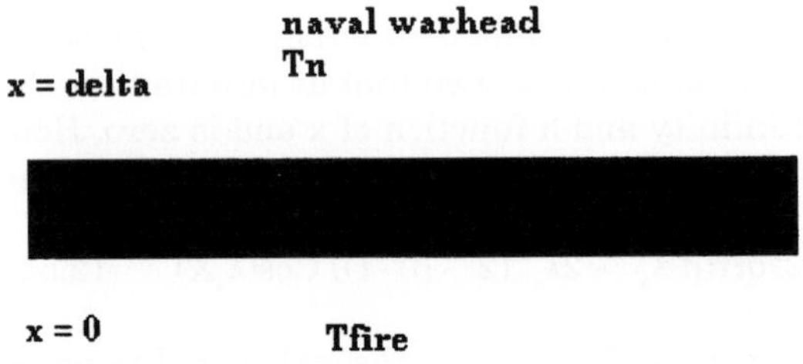

naval warhead
Tn

x = delta

x = 0   Tfire

Figure 2.25 Thermal Barrier Coating

$$X = \delta/\sqrt{(\alpha \tau_r)}, \quad -\partial u/\partial X = h^* (u - 1) \tag{2.549}$$

$$\text{Where,} \quad h^* = h/\sqrt{(k\rho C_p/\tau_r)} \tag{2.550}$$

The damping term can be removed from Eq. [2.578] by a $u = w \exp(-\tau/2)$ substitution. After removal of the damping term,

$$\partial^2 w/\partial \tau^2 - w/4 \quad = \partial^2 w/\partial X^2 \tag{2.551}$$

The non-homogeneity in the Boundary condition can be removed by supposing that the temperature consists of two terms that can be superposed, one that is at steady state and the other that is transient.

$$\text{Let,} \quad u = u^{ss} + u^t \tag{2.552}$$

$$w = w^{ss} + w^t$$

Thus,

$$\partial^2 u^{ss}/\partial X^2 = 0$$

$$u^{ss} = c_1 X + c_2 \tag{2.553}$$

From the boundary condition it can be seen that $c_2 = 0$. From the boundary condition given,

$$\partial u^{ss}/\partial X = c_1 = h^* (c_1 X_\delta - 1) \tag{2.554}$$

$$c_1 = h*/(1 + h*X_\delta)$$

or,

$$u^{ss} = (h*X)/(1 + h*X_\delta) \tag{2.555}$$

The transient portion of the solution will then obey;

$$\partial u^t/\partial \tau \ + \ \partial^2 u^t/\partial \tau^2 \qquad = \partial^2 u^t/\partial X^2$$

$$X = 0, \ u^t = 0$$

$$X = X_\delta, \ -\partial u/\partial X = h* \ u^t \tag{2.556}$$

$$Or - \partial w^t/\partial X = h* \ w^t$$

The transient wave temperature can be solved for by the method of separation of variables.

$$\partial^2 \ w^t/\partial \tau^2 - w/4 \qquad = \partial^2 \ w^t/\partial X^2$$

$$Let, \ w^t = V(\tau)\phi(X)$$

Then,

$$\phi^{\cdot}(X)/ \ \phi(X) = (V'(\tau) + V^{\cdot}(\tau))/ V(\tau) \qquad = -\lambda_n^2$$

$$V = \exp(-\tau/2) \ (c_3 \exp(sqrt(1/4 -\lambda_n^2) \ \tau)+ \ c_4 \exp(-sqrt(1/4 -\lambda_n^2)\tau))$$

$$or \ V\exp(\tau/2) = (c_3 \exp(sqrt(1/4 -\lambda_n^2) \ \tau)+ \ c_4 \exp(-sqrt(1/4 -\lambda_n^2)\tau)) \tag{2.557}$$

From the final condition u = 0 at infinite time. So is $V\phi\exp(\tau/2) = W$, the wave temperature at infinite time. The wave temperature is that portion of the solution that remains after dividing the damping component either from the solution or the governing equation. For any non-zero ϕ, it can be seen that at infinite time the LHS of Eq. [2.588] is a product of zero and infinity and a function of x and is zero. Hence the RHS of Eq. [2.588] is also zero and hence in Eq. [2.588] c_3 need be set to zero.

$$\phi(X) \quad = c_1 Sin(\lambda_n X) + c_2 Cos(\lambda_n X) \tag{2.558}$$

From the boundary condition given by Eq. [2.620], $c_2 = 0$.

$$At \ X = X_\delta, \ \partial\phi/\partial X = h* \ \phi \tag{2.559}$$

$$c_1 Sin(\lambda_n X) = \lambda_n \ c_1 Cos(\lambda_n X)$$

$$\lambda_n = h*tan(\lambda_n X_\delta) \tag{2.560}$$

Taylor series expansion of tanx for small x is given by;

$$\tan(x) = x + 2x^3/3! = x + x^3/3 \tag{2.561}$$

for small values,

$$\tan(\lambda_n X_\delta) = (\lambda_n \delta/\text{sqrt}(\alpha\tau_r)) + (\lambda_n \delta/\text{sqrt}(\alpha\tau_r))^3/3 \tag{2.562}$$

Combining Eqs. [2.593] and Eq. [2.591],

$$\lambda_n = \text{sqrt}(3(\alpha\tau_r)^{3/2}/\delta^3 h^* (1 - h^*\delta/\text{sqrt}(\alpha\tau_r))) + 2n\pi \tag{2.563}$$

$$n = 0, 1, 2, 3, \ldots$$

The temperature solution can be written as;

$$u^t = \Sigma_0^\infty c_n \exp(-\tau/2) \exp(-\tau \, \text{sqrt}(1/4 - \lambda_n^2)) \, \text{Sin} (\lambda_n X) \tag{2.564}$$

The critical thickness of the coating to avoid subcritical damped oscillations in temperature can be calculated by;

$$\text{sqrt}(3(\alpha\tau_r)^{3/2}/\delta^3 h^* (1 - h^*\delta/\text{sqrt}(\alpha\tau_r))) > \tfrac{1}{4} \tag{2.565}$$

for small h^*,

$$\text{for } \delta < (12/h^*)^{1/3} \, \text{sqrt}((\alpha\tau_r)) \text{ oscillations will occur.} \tag{2.566}$$

$$\text{where, } c_n = 2(1 - \text{Cos}(\lambda_n X_\delta))/(\lambda_n X_\delta \cdot 2\text{Cos}(\lambda_n X_\delta)\text{Sin}(\lambda_n X_\delta))$$

Summary

Damped wave propagation can be described by the telegraph equation. As the Fourier's law implies an unrealistic speed of heat propagation, it cannot be a fundamental depiction of transfer of heat. Maxwell developed the telegraph equation by importing a ballistic term that accounts for the accumulation of energy near the transfer surface. The first derivative of temperature or heat flux with respect to time accounted for the diffusive behavior and the second derivative of the temperature or heat flux with respect to time accounts for the ballistic behavior. In conditions where the rate of heat conduction rapidly establishes itself the ballistic term can be neglected. As discussed in section 1.0, the regime of wave dominance is when the accumulation of heat is greater than $\exp(\tau/2)$ and diffusion dominance is when the accumulation of heat is much less than $\exp(\tau/2)$.

Molecular motion gives out heat as phonons. Lattice vibrations may be a source for wave behavior. Dissipation leads to cessation of wave motion. Wave motion transitions to diffusive transport after the accumulation of energy ceases and the linear temperature gradient is established. Baumesiter and Hamill [1969] solved the hyperbolic heat conduction equation in a semi infinite medium subjected to a step change in temperature at one its ends using the method of Laplace transform. The space integrated expression for the temperature in the Laplace domain had the inversion readily available within the tables. This expression was differentiated using the Leibniz's rule and the resulting temperature distribution was given for $\tau > X$ as an integral equation. The method of relativistic transformation of coordinates was used to solve for the hyperbolic PDE for the heat flux in one dimension. The damping term was removed by a $u = W\exp(\tau/2)$ substitution. After obtaining an order reduction to a Bessel differential equation the wave flux solution was found to be a modified Bessel composite function of the first kind and zeroth order.

The transformation $\eta = (\tau^2 - X^2)$ was found suitable for the order reduction, The composite function, $\eta^{1/2} = \mathrm{sqrt}(\tau^2 - X^2)$, can be seen to exhibit spatio-temporal symmetry. The expressions at $+\tau$ and $-\tau$ are the same indicating some memory effects. The boundary condition at the asymptotic limit of large time of the parabolic solution c_1 can be seen to be 1. The derived expression for dimensionless heat flux is integrated from the energy balance equation to yield the dimensionless temperature at CWT in a semi-infinite medium. Three regimes of heat flux was identified in the dimensionless heat flux; the first the thermal inertia or zero transfer regime. The time lag for a interior point in a semi-infinite medium at CWT is $\tau_{lag} = \mathrm{sqrt}(X^2 - 23.132)$. The second regime is a regime of increasing value with time and given by a Bessel composite function of the zeroth order and first kind. The third regime is one of asymptotic exponential decay and is given by a composite Bessel function of the first kind and zeroth order.

By eliminating c_1 between the superposed solution of wave temperature and exponential decay and the equation from the boundary condition a more readily usable expression for the dimensionless temperature was obtained for the CWT in semi-infinite medium problem. The solution to the damped wave hyperbolic PDE $w = c I_0 \eta^{1/2}$ was substituted back in the PDE to see the tolerance for c as a function of time. The equation tolerates an $\exp(\tau/2)$. A more readily usable expression was obtained by solving for the dimensionless temperature in three dimensions in cartesian coordinates.

$w = \mathrm{Sin}(\kappa X - \omega\tau) + d\mathrm{Cos}(\kappa X - \omega\tau)$ was verified to obey the wave equation less the damping term. The transformation, $\eta = (\tau^2 - X^2)^{1/2}$ was also shown to be a suitable transformation for converting the hyperbolic PDE into a Bessel differential equation. $I_0(\tau/2)$ and $\exp(\tau/2)$ is of the same asymptotic order. The governing equation for the wave temperature was solved for by the method of Laplace transforms. A usable form of the solution was obtained for the dimensionless temperature in an infinite medium heated by an infinite cylinder. The solution was found to be for $\tau > X$ a modified Bessel composite function of the 1/2th order and first kind. For $X > \tau$, the solution is found to be a Bessel composite function of the first kind and zeroth order. The thermal time lag to see any change in temperature at an exterior point in the infinite medium, was found to be, τ_{lag} = sqrt(X_p^2 - $4\pi^2$). The three regime of the dimensionless temperature at an exterior point in the infinite medium in 1 dimensional cylindrical coordinates is shown in Figure 2.4. The solution for the dimensionless heat flux is obtained from the governing equation for heat flux.

A usable form of the solution was obtained for the dimensionless temperature in an infinite medium heated by an infinite sphere. The solution was found to be for $\tau > X$ a modified Bessel composite function of the 1st order and first kind. For X > τ, the solution is found to be a Bessel composite function of the first kind and first order. The thermal time lag to see any change in temperature at an exterior point in the infinite medium, was found to be, τ_{lag} = sqrt(X_p^2 − 7.6634^2). The three regime of the dimensionless temperature at an exterior point in the infinite medium in 1 dimensional spherical coordinates is shown in Figure 2.6. The dimensionless temperature is solved for for a right triangular cone with the apex at constant temperature. The heat flux is solved for in the radial direction for the sphere from the governing equation for the heat flux.

The general solution for the temperature distribution in a finite slab subject to constant wall temperature was obtained by the method of separation of variables. The final condition was used a solution that is well bounded and differs from the solution provided by early investigators. The exact solution is bifurcated. For slab width less than πsqrt($\alpha\tau_r$) the temperature is subcritical damped oscillatory. The solution can be broken up into two terms of Cos ($\kappa X + \omega\tau$) and Cos ($\kappa X - \omega\tau$). For larger slabs a product solution similar to the Fourier series solution was obtained and the time domain was augmented on account of the ballistic term. Storage coefficient was defined, (S = $\rho C_p/\tau_r$) with units of W/m^3/K. The temperature response to the periodic boundary condition using the parabolic Fourier model, hyperbolic damped wave conduction and relaxation equation and the dual phase lag model was obtained by the method of complex temperature. The hyperbolic PDE captures the damped wave propagation with a lag. Heat flux reversal can result under certain conditions of the storage coefficient.

The plane of null transfer is identified by imposing a decaying exponential boundary condition in a semi-infinite medium and solving the damped wave transport equation and relaxation by double transformation. The first is the conversion to wave coordinates and the second is the Laplace transform. The quadratic equation is solved for when the first root of the Bessel composite function of the first order and first kind occurs at 2.4048 and the solution found to be X = ($\tau \pm$ sqrt (τ^2 - 99.2))/2. A temperature dependent heat source is imposed on a rod held at O K on one end and constant temperature on the other. A Bessel composite function solution results and the critical length above which the temperature will remain unchanged is derived when $U^* = 1$. This is when, X $\geq$ sqrt($5.7831/U^* + \tau^2$). A steady state temperature profile is derived. The heat flux is obtained for the rod heated by a temperature dependent heat source. A heat flux reversal was found after a said distance in the rod. A critical point of zero transfer was found. The method of separation of variables was used to obtain the general solution for the temperature profile. A bifurcated solution is found. For length smaller than, 1 < 2πsqrt($\alpha\tau_r$)/(1 + U^*), subcritical oscillations in temperature was found. The problem of convective heating of the rod was considered and solution obtained by the method of separation of variables. The method of Laplace transforms was used to solve for the dimensionless temperature in a semi-infinite medium at CWT at one of its ends. The solution for the CWF for the semi-infinite medium in one dimension was obtained by the method of Laplace transforms.

The transient temperature in an ice cube during phase change was obtained by the method of Laplace transforms. The solution for the convective heating of a semi-infinite medium was solved. At zero Biot number the problem reverts to that of the CWT and at infinite Biot number the problem becomes the CWF case. The average temperature in a finite slab subject to convective heating was obtained by solving for the resulting second order ODE is solved for. For Biot numbers, $h\alpha\tau/ak > \frac{1}{4}$, $<u> = \exp(-\tau/2) \cos(\tau/2\sqrt{(4Bi^* - 1)})$ which is subcritical damped oscillatory. The notion of thermal time constant in view of damped wave conduction and relaxation is reviewed. The thermal time constant undergoes a maxima with increase in Biot number. For the region of increase in thermal time constant with increase in half-width the slab may become larger than the critical molecular distances causing a smoother transition to steady state.

The modified Fourier equation with upto the second derivative of heat flux with respect to time was used to obtain the solution for the transient temperature in a finite slab subject to constant wall temperature, CWT. For $Bi^* < 0.35$, the roots of the cubic equation that comes from the auxiliary equation can be found to be imaginary. This implies an oscillatory solution in time domain. When combined with the decaying exponential the solution becomes subcritical damped oscillatory. The parabolic Fourier model solution was obtained and the thermal time constant as a function of the Biot number was obtained. Two limits for the width of the slab were derived where the subcritical damped oscillations can occur. The conditions were the subcritical damped oscillations in a thin film coating in cartesian coordinates are derived. For $\delta < (12/h^*)^{1/3}$ sqrt($(\alpha \tau_r)$ subcritical oscillations can be found. A composite infinite slab is considered and the dimensionless temperature in two semi-infinite slabs was solved for.

Exercises

1.   *Wavefront*
        In deriving Eq. [2.51], c_1 was eliminated between the solution equation and the relation from the boundary condition at $X = 0$. Consider the wave temperature, at the wave front.

$$w = c_1 I_0 (\eta^{1/2})$$

At the wave front it can be seen that $u = \exp(-X/2) = \exp(-\tau/2)$.

$$(\partial^2 w/\partial \eta^2)4(\eta) + 4(\partial w/\partial \eta) - w/4 = 0$$

$$w = c'\exp(-\eta/16)$$

At $\eta = 0$, $w = c'$ and $u = c'\exp(-\tau/2)$. C' can be seen to be one from the boundary condition of $u = 1$ at $X = 0$. Impose the condition at the wave front and solve for c_1. Show that $c_1 = 1$.

$$u = \exp(-\tau/2)I_0(1/2 \; \text{sqrt}(\tau^2 - X^2))$$

Compare the above result with Eq. [2.51]. Show that the asymptotic orders of $\exp(-\tau/2)$ and $1/I_0(\tau/2)$ are the same. Obtain the limits of the ratio of the two functions at infinite time.

$$\underset{\tau \to \infty}{\text{Lt}} \quad \exp(\tau/2)/I_0(\tau/2)$$

$$= (1 + \tau + \tau^2/2! + ...)/(1 + (\tau/4)^2/1!^2 + \; ...)$$

Divide the numerator and denominator by the term with the highest power exponent in τ. For example take three terms in the numerator and two terms in the denominator and divide numerator and denominator by τ^2. Show that as $\tau \to \infty$, the ratio is a constant and hence the two functions $\exp(\tau/2)$ and $I_0(\tau/2)$ are of the same asymptotic order.

2. *Relativistic Transformation for $X > \tau$*
In solving Eq. [2.47] the transformation of, $\eta = \tau^2 - X^2$, was used to obtain a order reduction from hyperbolic PDE to a Bessel differential equation in 1 variable. Use the transformation,

$$\eta = X^2 - \tau^2$$

and obtain the order reduction. For $X > \tau$ obtain a solution that exhibits symmetry in space and time. Discuss the implications of time symmetry and the importance of memory in the process. Show that Eq. [2.52] can be obtained directly from the Bessel differential equation.

3. *Cylindrical Coordinates in Three Dimensions*
In section 2.2, the relativistic transformation of coordinates, was used to solve for the temperature as a function of spatiotemporal variables in 1 dimension for a cylinder at constant wall temperature. Consider the three dimensions in the cylindrical coordinates. Solve for the temperature from the following governing equation given in Eq. [C.15]:

$$\tau_r(\partial^2 T/\partial t^2) + \partial T/\partial t = \alpha(1/r \; \partial/\partial r \; (r\partial T/\partial r) + 1/r^2 \; \partial^2 T/\partial\theta^2 + \partial^2 T/\partial z^2)$$

In the dimensionless form, the governing equation can be written as;

$$\partial^2 u/\partial \tau^2 + \partial u/\partial \tau = (1/X\, \partial u/\partial X) + \partial^2 u/\partial X^2 + \partial^2 u/\partial \zeta^2 + \partial^2 u/\partial Z^2$$

where,

$$u = (T - T_0)/(T_s - T_0); \quad \tau = t/t_r; \quad X = r/sqrt(\alpha t_r); \quad Z = z/\,sqrt(\alpha t_r)$$
$$\zeta = r\theta/\,sqrt(\alpha t_r)$$

After removing the damping term, by a $u = w\exp(-\tau/2)$ substitution, convert the resulting wave equation using the transformation;

$$\eta = \tau^2 - X^2 - \zeta^2 - Z^2$$

Obtain the Bessel solution for the dimensionless temperature as a function of the lumped coordinate and use the constant wall temperature boundary condition.

$$\eta \partial^2 w/\partial \eta^2 + 5/2 \partial w/\partial \eta \quad -w/16 = 0$$

Show that the solution for the dimensionless temperature can be approximated as;

$$u = [(\tau^{3/2})/(\tau^2 - X^2 - \zeta^2 - Z^2)^{3/4}]\, I_{3/2}(1/2\, sqrt(\tau^2 - X^2 - \zeta^2 - Z^2)/\, I_{3/2}(\tau/2)$$

This is for $\tau > sqrt(X^2 + \zeta^2 + Z^2)$ at the limit of zero dimensions of the cylinder,

and for $X^2 + \zeta^2 + Z^2 > \tau^2$

$$u = [(\tau^{3/2})/(X^2 + \zeta^2 + Z^2 - \tau^2)^{3/4}]\, J_{3/2}(1/2\, sqrt(X^2 + \zeta^2 + Z^2 - \tau^2)/\, I_{3/2}(\tau/2)$$

4. *Inverse Problem*
    In the problem of obtaining the transient temperature using the damped wave and relaxation equation in a one dimensional semi-infinite slab, say the temperature in the interior of the medium is given as a function of time.

    $$u\,(X = X_p) = g_0 + g(t)$$

    Obtain the complete general solution for the transient temperature. What is the temperature at the surface at $X = 0$?

5. *Thermal Lag Time, Penetration Distance*
    Find the thermal lag time associated with a interior point at a distance 20

cm the surface in a semi-infinite medium subject to a CWT in one dimension. Find the distance beyond which the temperature will be undisturbed for a time instant of 18 seconds. Repeat the analysis for two dimensional medium   a) $(x,y) = (16,18)$ b) $(15, 15)$ c) 25 seconds. d) 40 seconds. The thermal diffusivity of the medium is 10^5 m^2/s and the relaxation time is 12 seconds. The initial temperature of the medium is 25 $^\circ$ C and the hot surface is held at 60 $^\circ$C.

6.   *Cylindrical Coordinates in Three Dimensions*
   Obtain the governing equation for heat flux in three dimensions in cylindrical coordinates. Obtain the solution to the damped wave conduction and relaxation equation for the case of constant wall temperature using the energy balance equation and relativistic transformation of coordinates.

7.   *Waveform Solution*
   Using the Worked Example 2.3, obtain the transient temperature in a finite slab subject to constant wall flux.

8.   *Right Circular Cone*
   Obtain the solution for the transient temperature in the cone as shown in Worked Example 2.8 in two space directions, r and z in addition to the time domain. Consider the apex temperature as T_s for times greater than higher than the initial uniform temperature of T_0.

9.   *Non-Homogeneity in Wall Temperature*
   Obtain the transient temperature for a finite slab of length L, subject to the following conditions using the damped wave conduction and relaxation equation;

$$x = 0, \quad T = T_1$$

$$x = L, \quad T = T_2$$

$$t = 0, \quad T = T_0$$

10.   *Transient Temperature in a Annulus*
    Obtain the transient temperature for a cylindrical annulus subject to the

following space and time conditions.

$$t = 0, \quad T = T_0$$

$$r = R_i, \quad T = T_i$$

$$r = R_o, \quad T = T_o$$

11. *Thermal Lag Time, Penetration Distance*
Find the thermal lag time in a spherical infinite medium at a distance of 15 cm from the hot surface. Obtain the region beyond which the medium temperature will be undisturbed after a time instant of 35 secs. The thermal diffusivity of the medium is 10^5 m^2/s and the relaxation time is 12 seconds. The initial temperature of the medium is 25 $^\circ$ C and the hot surface is held at 60 $^\circ$C. Discuss the space and time symmetry of the nature of the solution.

12. *Centerline Temperature of the Bar and Subcritical Damped Oscillations*
A rectangular iron bar 5 cm by 4 cm with a thermal conductivity of 55 W/m K, thermal diffusivity of 2.101 E-05 m^2/s, relaxation time of 12 secs is initially at a uniform temperature at 225 $^\circ$C. Suddenly the surfaces of the bar are subjected to a constant wall temperatuer of 225 $^\circ$C from an initial temperature 25 $^\circ$ C. Calculate the center temperature of the bar after 18 secs after the start of cooling using the damped wave conduction and relaxation equation. Obtain the critical dimensions of the iron bar below which the transient temperature can exhibit subcritical damped oscillations. Repeat the analysis for dimensions of the iron bar of 0.3 x 0.2 mm after 9.6 secs.

13. *Periodic Convective Boundary Condition*
Consider a finite slab subject to the convective boundary condition. Using a a space averaged expression for the temperature obtain the governing equation for the transient temperature for the slab using the damped wave conduction and relaxation equation. The heat transfer coefficient is periodic in time and e expressed as;
$$h = h_o + h_A \cos(\omega t)$$

Obtain the transient temperature for the entire slab. Comment on the critical size of the slab. Discuss the attenuation and phase lag.

14. *Attenuation and Phase Lag*
    The transient response to a periodic disturbance for the parabolic, hyperbolic
    and   dual phase lag models as a function of time was studied in section 2.5.
    Obtain a plot as a function of distance for a time instant of $\tau = $ 3,   the   transient
    response to the periodic disturbance. Discuss the attenuation and phase lag.

15. *Brick Wall with Convection*
    A 0.10m thick brick wall with thermal diffusivity of 1 E -05 m^2/s and thermal
    relaxation time of 9 seconds, density 2300 Kg/m^3 and thermal conductivity of 0.5
    W/m K is initially at a temperature of 240 ^{0}C. The wall is suddenly exposed to
    a convective environment at T_∞ is 29 ^{0}C with a heat transfer coefficient of 70
    W/m^2 K. Determine the center temperature at 3/4, 3 and 6 h after exposure to
    the cooler ambient. Obtain the surface temperature at 3 and 6 h. What is the
    time taken to reach steady state ?

16. *Cooling Time of a Orange*
    An orange of diameter of 5 cm is initially at a uniform temperature of 26 ^{0}C. It is
    placed in a refrigerator in which the air temperature is 3.583 ^{0}C. If the transfer
    coefficient between the air and the surface of the orange is h = 61 W/m^2 K,
    determine the time taken for the center of the orange to reach 5^0 C. The thermal
    diffusivity and relaxation time of the orange can be taken as 1.4 E -05 m^2/s and
    14 seconds respectively. The thermal conductivity can be assumed that of water
    at 0.59 W/m K.

17. *Heating a Potato*
    A 3.58 cm diameter potato initially at a uniform temperature of 22 ^{0}C, is
    suddenly dropped into boiling water at 94 ^{0}C. The heat transfer coefficient
    between the water and the surface is h = 40 W/m^2 K. The thermophysical
    properties of the potato can be taken as α = 1.6 E-04 m^2/s, τ_r = 16 seconds,
    and k = 4 W/m/K.   Determine the time required for the centerline
    temperature of the potato to reach 90 ^{0} C. What is the smallest size of the
    potato prior to the onset of subcritical damped oscillations ? What is the time
    taken to reach steady state ?

18. *Convective Boudary Condition in a Finite Slab*
    Obtain the transient temperature for a finite slab of length L, subject to the
    following conditions using the damped wave conduction and relaxation

equation;

$$x = 0, \quad T = T_1$$

$$x = L, \quad -k\partial T/\partial x = h(T - T_a)$$

$$t = 0, \quad T = T_0$$

What are the conditions for the dimensions of the slab when the temperature will undergo subcritical damped oscillations.

19. **Heating of the Earth's Crust**

Consider the earth's crust heated by the sun. The initial temperature of the earth is at T_0, imposed by a periodic temperature at the crust by $T_0 + T_s Cos(wt)$. Obtain the transient temperature as a function of time and space in one dimension within the sphere.

$$u = (T - T_0)/(T_s); \quad \tau = t/\tau_r; X = x/sqrt(\alpha \tau_r)$$

The governing Equation can be obtained in dimensionless form as;

$$\partial u/\partial \tau + \partial^2 u/\partial \tau^2 = 2/X \partial u/\partial X + \partial^2 u/\partial X^2$$

The initial condition is;

$$t = 0, \quad T = T_0 ; \quad u = 0$$

The Boundary Conditions are;

$$X = 0, \quad \partial u/\partial X = 0$$

$$X = X_s, \quad T = T_0 + T_s Cos(\omega t). ; \quad u = Cos(\omega^* \tau)$$

Suppose that the solution for u is of the form $f(x)exp(-i\omega^* \tau)$ for $\tau > 0$. Where, ω is the frequency of the temperature wave imposed on the surface and the T_s is the amplitude of the wave. Show that the solution for the transient temperature in the medium can be given by;

$$u = (X_s/X)^{1/2} Cos((\omega^* \tau)) Real (J_{1/2}(A + iB)X/J_{1/2}(A + iB)X_s$$

where, $A + iB = (\omega^{*2} + i\omega^*)^{1/2}$

20. Obtain the transient temperature using the damped wave equation in spherical

coordinates in the atmosphere in problem 19. The time and space conditions are;

$$X = X_R, \quad u = Cos(\omega^*\tau)$$

$$X = \infty, \quad u = 0$$

$$\tau = 0, \quad u = 0$$

Show that the solution for transient temperature can be obtained by the method of relativistic coordinates and is;

$$u = Cos(\omega^*\tau) \, [(\tau^2 - X_R^2)^{1/2}/(\tau^2 - X^2)^{1/2}] \, [I_1 \, (1/2 \; sqrt(\tau^2 - X^2)/ \, I_1 \, (1/2 \; sqrt(\tau^2 - X_R^2)]$$

This is valid for $\tau > X$. For $X > \tau$;

$$u = Cos(\omega^*\tau) \, [(X_R^2 - \tau^2)^{1/2}/(X^2 - \tau^2)^{1/2}] \, [J_1 \, (1/2 \; sqrt(X^2 - \tau^2)/ \, I_1 \, (1/2 \; sqrt(\; \tau^2 - X_R^2 \,)$$

21. *Periodic Surface Temperature in Cylindrical Coordinates*
Consider Figure 2.3 in cylindrical coordinates. The initial temperature of the cylinder is at T_0, imposed by a periodic temperature at the surface by $T_0 + T_s$ Cos(wt). Obtain the transient temperature as a function of time and space in one dimension within the cylinder.

$$u = (T - T_0)/(T_s); \qquad \tau = t/\tau_r; X = x/sqrt(\alpha \; \tau_r)$$

The governing Equation can be obtained in dimensionless form as;

$$\partial u/\partial \tau \; + \; \partial^2 u/\partial \tau^2 \qquad = 1/X \partial u/\partial X \; + \; \partial^2 u/\partial X^2$$

The initial condition is;

$$t = 0, \quad T = T_0 \, ; \quad u = 0$$

The Boundary Conditions are;

$$X = 0, \quad \partial u/\partial X = 0$$

$$X = X_s, \quad T = T_0 + T_s Cos(\omega t). \, ; \quad u = Cos(\omega^*\tau)$$

Suppose that the solution for u is of the form f(x)exp(-i$\omega^*\tau$) for $\tau > 0$. Where, ω is the frequency of the temperature wave imposed on the surface and the T_s is the amplitude of the wave. Show that the solution for the transient temperature in the medium can be given by;

$$u = \text{Cos}((\omega^*\tau)) \ \text{Real} \ (J_0(A + iB)X/J_{1/2}(A + iB)X_s$$

where, $A + iB = (\omega^{-2} + i\omega^*)^{1/2}$

22. Obtain the transient temperature using the damped wave equation in cylindrical coordinates in the semi-infinite medium in problem 21. The time and space conditions are;

$$X = X_R, \quad u = \text{Cos}(\omega^*\tau)$$

$$X = \infty, \quad u = 0$$

$$\tau = 0, \ u = 0$$

Show that the solution for transient temperature can be obtained by the method of relativistic coordinates and is;

$$u = \text{Cos}(\omega^*\tau) \ [(\tau^2 - X_R^2)^{1/4}/(\tau^2 - X^2)^{1/4}] \ [I_{1/2} \ (1/2 \ \text{sqrt}(\tau^2 - X^2)/ \ I_{1/2} \ (1/2 \ \text{sqrt}(\tau^2 - X_R^2)]$$

This is valid for $\tau > X$. For $X > \tau$;

$$u = \text{Cos}(\omega^*\tau) \ [(X_R^2 - \tau^2)^{1/4}/(X^2 - \tau^2)^{1/4}] \ [J_{1/2} \ (1/2 \ \text{sqrt}(X^2 - \tau^2)/I_{1/2} \ (1/2 \ \text{sqrt}(\tau^2 - X_R^2)]$$

23. *Slab, Cylinder, and Sphere*
Consider a slab of thickness of 11 mm, a cylinder of diameter 11 mm and a sphere of 11 mm diameter each made of aluminum with thermal diffusivity of 10^5 m²/s and thermal relaxation time of 16 secs and thermal conductivity of 240 W/m K initially at a uniform temperature $T_i = 300$ °C. Suddenly they are all immersed into a well stirred bath at $T_\infty = 50$ °C. The heat transfer coefficient between the surface and the fluid is 4000 W/m² K. Calculate the time required for the centers of the slab, sphere and cylinder to cool to 80 °C.

24. Consider a slab of thickness 2 mm, cylinder 2 mm, a sphere of diameter 2mm each made of glass ballontini with thermal diffusivity of 8.8 E -04 m²/s, and thermal relaxation time of 100 milliseconds and a thermal conductivity of 0.1 w/m initially at a uniform temperature T_i of 370 °C. Suddenly they are all immersed into a well stirred bath at T_∞ at 54 ° C. The heat transfer coefficient

between the surface and the fluid is 80 W/m² K. Calculate the time required for the centers of the slab, sphere and cylinder to cool to 80 °C.

25. *Desalination by Ion Exchange*
The ion-exchange is used for   desalination processes to make potable drinking water from sea water.   A series of anion and cation exchange resins are used to remove the sodium and chloride ions. The ion exchanger columns are regenerated by thermal means. Consider a long cylindrical   ion-exchange column with a thermal relaxation time of 53 seconds, thermal conductivity of 8 W/m K and thermal diffusivity of 1   E -05 m²/s. Calculate the minimum height of the ion-exchange column below which the temperature will exhibit subcritical oscillations. Obtain the time taken for the ion-exchange column to be heated from 14 °C to 150 °C. Obtain the transient temperature profile in a cylindrical column using the damped wave conduction and relaxation equation. The heat transfer coefficient may be taken as 370 w/m² K. Obtain the centerline temperature of the ion-exchange column, average temperature of the column.

26. *Thermal Lag and Penetration Distance in Concrete*
A thick concrete slab   with thermal diffusivity of 7   E-07 m²/s, thermal relaxation time of 28.0 seconds is initially at a uniform temperature of 14 ° C. One of its surface temperature is suddenly heated to 62 °C. By describing the heat transfer process as one by damped wave conduction and relaxation, determine the temperature at depth of 3.5 cm from the surface. What is the time lag associated with the point. When does the regime change happen ? What is the   heat   flux at the surface 8 min after the surface temperature is raised if the thermal conductivity of concrete is 0.4 W/m K.

27. *Point of Ignition in Wood*
A thick wood wall with thermal diffusivity of   7.2 E-07 m²/s and thermal relaxation time of   116 seconds and thermal conductivity of 0.21 w/m K is initially at a uniform temperature of 23 °C. The wood may ignite at 437 °C.   If the surface   is exposed to hot gases at T∞,   and the heat transfer coefficient between the gas and the surface is 53 w/m² K. How long will it take for the surface of the wood to reach 437 °C.   How long will it take for a interior point 5 nm from the surface to reach the ignition point ?

28. *Periodic Boundary Condition and Method of Retavistic Transformation*
Consider a semi-infinite slab in 1 dimension subject to a periodic boundary condition

$$T = T_0 + T_s \, Cos(\omega t)$$

The initial temperature of the slab is at T_0 and the temperature at $X = \infty$, is also T_0. The governing equation for the transient temperature described by the damped wave conduction and relaxation equation can be written as;

$$\partial u/\partial \tau + \partial^2 u/\partial \tau^2 = \partial^2 u/\partial X^2$$

This is after obtaining the dimensionless variables as in problem 19. The damping component to the solution can be removed by a $u = w exp(-\tau/2)$ substitution. The resulting equation is;

$$\partial^2 w/\partial X^2 = \partial^2 w/\partial \tau^2 - w/4$$

Using a relativistic transformation of coordinates, $\eta = \tau^2 - X^2$ for $\tau > X$, show that the solution to the transient temperature is;

$$u = Cos\,(\omega^* \tau)\, I_0(1/2\,(\tau^2 - X^2)^{1/2})/I_0(\tau/2)$$

For $X > \tau$,

$$u = Cos\,(\omega^* \tau)\, J_0(1/2\,(X^2 - \tau^2)^{1/2})/I_0(\tau/2)$$

or for large η

$$2u = exp(-\tau/2)[Cos(\omega^*\tau + 1/2\,(X^2 - \tau^2)^{1/2} - \pi/4) + Cos(\omega^*\tau - 1/2\,(X^2 - \tau^2)^{1/2} - \pi/4)]$$

29. *Periodic Boundary Condition and Method of Laplace Transforms*
Tang and Araki (1996) considered a 1 dimensional transient heat conduction and relaxation in an insulated finite medium with a thickness of L, constant thermal properties, and initial temperature distribution $T(x,0) = 0$. From time 0 the external surface at $x = 0$ is exposed to a periodic heat flux with amplitude q_0 and frequency ω. The governing equation and space and time conditions are;

$$\alpha \partial^2 T/\partial x^2 = \partial T/\partial t + \tau_r \, \partial^2 T/\partial t^2$$

$$x = 0, \quad q = q_0 Cos(\omega t)$$

$$x = L, \quad \partial T/\partial x = 0$$

$$t = 0, \quad q = 0, \quad \partial T/\partial t = 0$$

Obtaining the Laplace transform,

$$\alpha d^2 \underline{T}/dx^2 = \underline{T}\,(s + s^2\tau_r)$$

$$\underline{T} = c_1 \exp(-x(s + s^2\tau_r)^{1/2}/\alpha^{1/2}) + c_2 \exp(+x(s + s^2\tau_r)^{1/2}/\alpha^{1/2})$$

$$d\underline{T}/dx = r(c_1 \exp(-rx) - c_2 \exp(+rx))$$

The integration constants can be solved for from the boundary conditions as;

$$c_2 = -\exp(-2rL)rq_o/(s^2 + \omega^2)/(1 - \exp(-2rL))$$

$$c_1 = -rq_o/(s^2 + \omega^2)/((1 - \exp(-2rL)))$$

where, $r = (s^2\tau_r + s)^{1/2}$

Use a series expansion for $(1 - \exp(-2rL))^{-1} = \sum\limits_{0}^{\infty} \exp(-2rnL)$ obtain from the Laplace inversion tables the term by term inversion for the temperature distribution. Use the convolution theorem when necessary.

30. *Thermal Oscillations and Resonance* (Xu and Wang, 2002)
The governing equation using the dual phase lag model in 1 dimension can be written as;

$$1/\alpha(\partial T/\partial t + \tau_q \partial^2 T/\partial t^2) = \partial^2 T/\partial x^2 + \tau_T \partial^3 T/\partial t\partial x^2 + 1/k(U''' + \tau_q \partial U'''/\partial T)$$

where U''' is the volumetric heat source and the time and space conditions are;

$$x = 0,\ L,\ T = 0$$

$$t = 0,\ T = \phi(x),$$

$$t = 0,\ \partial T/\partial t = \psi(x)$$

Let $T = \sum\limits_{1}^{\infty} V(t) \operatorname{Sin}(\lambda_n x)$

where $\lambda_n = n\pi/L$

The Fourier sine series to express ϕ and ψ is as follows;

$$\phi(x) = \sum\limits_{1}^{\infty} \phi_n (\operatorname{Sin}\lambda_n x), \text{ where } \phi_n = 2/L\int\limits_{0}^{L} \phi \operatorname{Sin}(\lambda_n \xi)d\xi$$

$$\psi(x) \quad = \sum_1^\infty \psi_n \, Sin(\lambda_n x), \; where \quad \psi_n \; = 2/L\!\int_0^L \psi Sin(\lambda_n \xi)d\xi$$

Substitute the expression for T and its dependence on the space domain into the governing equation when $U''' = 0$;

$$(1 + \alpha\lambda_n^2\tau_T)(\partial V/\partial t) + \tau_q\,(\partial^2 V/\partial t^2) + \; + \alpha\lambda_n^2\tau_m V = 0$$

$$V(0) = \phi_n \; and \; V_t\,(0) = \psi_n$$

Introduce the damping coefficient, $f_n = 1/\tau_q + \tau_T\,\omega_n^2$ and the natural frequency,

$$\omega_n^2 = \alpha\,\lambda_n^2/\tau_q$$

Thus the equation for the temperature reduces to;

$$\partial^2 V/\partial t^2 + f_n\partial V/\partial t + \omega_n^2 V = 0$$

Show that several different scenarios are possible;

$$f_n^2 - 4\omega_n^2 < 1 \; \text{- overdamped oscillations}$$

$$f_n^2 - 4\omega_n^2 > 1 \; \text{- underdamped oscillations}$$

$$f_n^2 - 4\omega_n^2 = 1 \; \text{- critically damped oscillations}$$

Derive the conditions of resonance by comparing the frequency of the thermal oscillations with that of the frequency of the heat source imposed on the system.

3.0   Transient Mass Diffusion and Relaxation

Nomenclature

a	width (m)
C	concentration (mole/m^3)
D	diffusivity (m^2/s)
h	height of the nuclear fuel rod (m)
$I_0(x)$	modified Bessel function of the zeroth order and first kind
$I_1(x)$	modified Bessel function of the first order and first kind
$I_2(x)$	modified Bessel function of the second order and first kind
J	mass flux (mole/sq.m/s)
$J_0(x)$	Bessel function of the zeroth order and first kind
$J_1(x)$	Bessel function of the first order and first kind
$J_2(x)$	Bessel function of the second order and first kind
$J_3(x)$	Bessel function of the third order and first kind
$K_0(x)$	modified Bessel function of the zeroth order and second kind
$K_1(x)$	modified Bessel function of the first order and second kind
$K_2(x)$	modified Bessel function of the second order and second kind
$K_3(x)$	modified Bessel function of the third order and second kind
k'''	autocatalytic reaction rate (s^{-1})
R	radius of the fuel rod (m)
t	time (s)
x	distance (m)
X	dimensionless distance x/sqrt(Dτ_r)
u	dimensionless concentration (C – C$_s$ /C$_s$)
V	function of X only
V_m	velocity of mass diffusion sqrt(D/τ_r)
W	wave concentration (mol/m^3)
$Y_0(x)$	Bessel function of the zeroth order and second kind
$Y_1(x)$	Bessel function of the first order and second kind
$Y_2(x)$	Bessel function of the second order and second kind
$Y_3(x)$	Bessel function of the third order and second kind

Subscript

0	initial
i	integration constant i, i = 1,2,3,4.
k	indice
r	relaxation time
s	surface

Superscript

ss	steady state
t	transient state

Greek

α	thermal diffusivity (m^2/s)
ϕ	function of time only
τ	dimensionless time (t/τ_r)
τ_r	relaxation time (s)
η	$\tau + X$;
ζ	$\tau - X$

Dimensionless Groups

$k^{\cdot}$	reaction modulus $(k'''\tau_{mr})$
Pe_m	Peclet number (mass), ratio of superficial velocity to velocity of mass U_s/v_m
A^*	dimensionless width of the slab $a/(D\tau_r)^{1/2}$
κ	ratio of dissolution rate and velocity of mass

3.0 Introduction

The study of transient diffusion has a variety of applications such as in semiconductor device manufacturing, active solid state processes, metal treatment, coating processes, metal fusion bonding, superconducting technology, etching a substrate, optical waveguides, gas separation processes, bleaching, dyeing and chemical modification of textiles and fiber, electrolysis, simultaneous diffusion and reaction catalytic, heterogeneous and homogeneous reactions, both traditional and novel separation methods, transport in living cells, oxygen and solute transfer in tissues, dispersal of pollutants etc. The spreading of color from cupric sulfate placed at the bottom of a tall bottle filled with water is described by the process of diffusion. It is essentially a molecular phenomenon. The laws of diffusion have not been linked to what we understand about molecules and their velocities and dynamics. The phenomena of transfer of mass under the influence of a concentration driving force can be defined as diffusion. The direction of transfer is to equalize the concentration. Other driving forces that can cause diffusion, other than the concentration gradient are, the temperature difference, the concentration gradient of a second species, osmotic potential, steam,

148

centrifugation, pressure diffusion, electromotive force, surface force and surface tension gradient. The term thermophoresis refers to the processes where temperature gradient drives the diffusion. Diffusiophoresis occurs when a large drop in concentration of a second species drives the transfer. Reverse osmosis is the operation where the water is pumped across the osmotic potential through semi-permeable membrane and the desalination of sea water is effected by solvent diffusion. In sweep diffusion, steam sweeps away the solute with it. The centripetal forces in centrifugation results in different forces acting upon different masses resulting in separation. Pressure diffusion is characterized by a pressure drop in the direction of transfer. Electrolysis refers to the movement of charged particles subject to an electromotive force. The movement of species of interest on the surface of the solid is called surface diffusion. In separation by foaming the surface tension gradient is utilized.

Diffusion may be viewed as a process by which molecules intermingle as a result of their kinetic energy of random motion. Consider two containers of CO_2 gas and N_2 gas separated by a partition as shown in Figure 3.1.

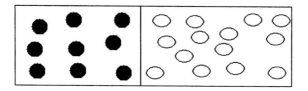

Figure 3.1 Container with a Partition Separating Nitrogen and Carbon dioxide

The molecules of both gases are in constant motion and make numerous collisions with the partition. If the partition is removed, the gases will mix due to the random velocities of their molecules. In sufficient time, a uniform mixture of CO_2 and N_2 molecules will result in the container. The speed distribution for the molecules for an ideal gas can be given by the Maxwell distribution and is;

$$P(v) = 4\pi (M/2\pi RT)^{3/2} v^2 \exp(-Mv^2/2RT) \qquad (3.1)$$

Where, Most Probable Speed, $v_p = sqrt(2RT/M)$

Mean Speed, $<v> = sqrt(8RT/M)$

Root Mean Square Speed $<v_{rms}> = sqrt(3RT/M)$ $\qquad (3.2)$

effective cross-sectional area for collision can be taken as circle of diameter 2d and the

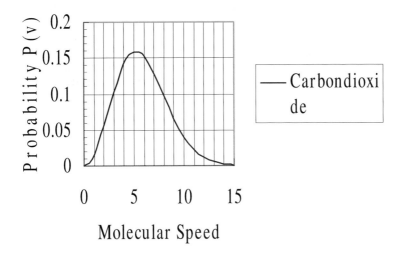

Figure 3.2 Maxwell Molecular Speed Distribution for CO$_2$

The calculation of the molecular speed depends on the molecular mass and the temperature. For mass m (amu), M (kg/mol), temperature, T (K), the three characteristic speeds can be calculated. The mean free path of the molecule can be viewed as the average distance between collisions for a gas molecule. If the molecule has a diameter d, the area itself is πd^2. itself as πd^2.

In time t, the circle would sweep out the volume ($\pi d^2 <v>t$). The number of gas molecules in the swept volume would be ($n_v \pi d^2 <v>t$). The number of collisions can thus be calculated. The mean free path of the molecule is then given by the length of the path divided by the number of collisions:

$$\lambda = <v>t/(n_v \pi d^2 <v>t) = 1/n_v \pi d^2 \qquad (3.3)$$

Refinement of the estimate may be obtained by allowing for the target molecule to have an average velocity. The resulting mean free path is;

$$\lambda = <v>t/(n_v \pi d^2 <v>t) = 1/sqrt(2)n_v \pi d^2 \qquad (3.4)$$

The number of collisions is $\sqrt{2}$ times the number with stationary targets. The number of molecules per unit volume can be determined from Avagadro's number and the ideal gas law leading to;

$$N_v = A_N P/RT$$

and $\lambda = RT/(sqrt(2) \pi d^2 A_N P)$ \qquad (3.5)

Kinetic theory can be used to predict the mass diffusivity D_{AB} for binary mixtures of nonpolar gases within about 5 %. In a similar derivation to that in section 1.1 here the definition of mass flux is revisited and the damped wave mass diffusion and relaxation equation can be shown to arise from a molecular standpoint from the accumulation of molecules near the surface. Consider a large body of gas containing two molecular species A and B with both species having the same mass, m and size and shape. Good examples of such systems are isotopes of Uranium, $U^{235}F_6$ and $U^{238}F_6$. Let the diameter of the molecular sphere be given by d. The wall collision frequency per unit area in stationary gas is given by;

$$Z = \tfrac{1}{4} \, n \, <u> \tag{3.6}$$

The molecules reaching any plane in the gas have had their last collision at a distance a from the plane,

$$a = 2/3\lambda \tag{3.7}$$

Inorder to determine the diffusivity D_{AB}, consider the motion of species A in the y direction under a concentration gradient $\partial x_A / \partial y$ when the mixture moves at a finite velocity v_y^* throughout. The temperature and total molar concentration c are assumed constant. *The molar flux, N_{Ay} of species A, across any plane of constant y is found by counting the molecules of A that cross unit area of the plane in the positive y direction and subtracting the number that cross in the negative y direction.* Thus,

$$N_{Ay} \;=\; 1/N^* \, \{ n x_A v_y^* |_y \;+\; \tfrac{1}{4}\, x_A <u>|_{y\text{-}a} \;-\; \tfrac{1}{4}\, x_A <u>|_{y+a} \} \tag{3.8}$$

Assuming that the concentration profile, $x_A(y)$ is essentially linear for a distance of several mean free paths. Thus,

$$x_A |_{y\text{-}a} \;=\; x_A|_y \;-\; 2/3\, \lambda\, x_A |_{y\text{-}a} \;=\; x_A|_y \;-\; 2/3\, \lambda\, \partial x_A / \partial y$$

$$x_A |_{y+a} \;=\; x_A|_y \;+\; 2/3\, \lambda\, \partial x_A / \partial y \tag{3.9}$$

Substituting the above two Eqs. into Eq. [3.8], and denoting $N_A + N_B = cv^*$;

$$N_{Ay} \;=\; x_A(N_{Ay} + N_{By}) - 1/3\, c <u> \lambda\, x_A |_{y\text{-}a}\, \partial x_A / \partial y \tag{3.10}$$

This equation corresponds to the y component of Fick's law with the following approximate value for D_{AB};

$$D_{AB} \;=\; 1/3 <u> \lambda \tag{3.11}$$

The Fick's law is then;

$$N_{Ay} \;=\; x_A(N_{Ay} + N_{By}) - D_{AB}\, \partial x_A / \partial y \tag{3.12}$$

In the mid 1800s Adolf Fick introduced two differential equations that quantitated the phenomenon of molecular diffusion. The Fick's first law states that the flux J of a component of concentration C across a membrane of unit area in a predefined plane is proportional to the concentration differential across that plane and is expressed as;

$$J = -D \nabla C \qquad (3.13)$$

Where the differential operator, ∇, is given by, $\nabla = i\partial/\partial x + j\partial/\partial y + k\partial/\partial z$ such that grad C is given by, $\nabla C = i\partial C/\partial x + j\partial C/\partial y + k\partial C/\partial z$. For the special case of diffusion in 1 dimension Eq. [3.13] reduces to;

$$J = -D \, \partial C/\partial x \qquad (3.14)$$

Fick's second law is derived by combining the first law with a mass balance. Consider in 1 dimension a small slice of thickness Δx at a distance x from the origin across a constant cross-sectional area of A. A mass balance in the thin slice can be written as;

$$\{J \, A|_x \, \_ \, J \, A|_{x+\Delta x} \}\Delta t \quad = \quad A\Delta x \, \Delta C \qquad (3.15)$$

$$\text{in - out} \pm \text{reaction} \quad = \quad \text{accumulation}$$

Dividing throughout Eq. [3.15] by $\Delta x \Delta t$ and taking the limits as Δx and Δt going to zero;

$$-\partial J/\partial x = \partial C/\partial t \qquad (3.16)$$

Combining Eqs.[3.16] with Eq. [3.14] the governing equation becomes;

$$D\partial^2 C/\partial x^2 = \partial C/\partial t \qquad (3.17)$$

Eq. [3.17] is the Fick's second law of diffusion. Fick was born on September 3rd 1829, the youngest of five children. He was drawn to mathematics in high school by the work of Poisson. An older brother persuaded him to go into medicine. Carl Ludwig tutored A. Fick at Marlburg. They believed that medicine should have a basis in mathematics. Fick's education continued in Berlin where he did a considerable amount of clinical work. He received his degree in 1852. His thesis, dealt with the visual errors caused by astigmatism. Ludwig brought Fick along as a prosecutor. The majority of Fick's scientific accomplishments were investigations of physiology. He did outstanding work in mechanics as applied to the functioning of muscles, in hydrodynamics and hemorheology and in the visual and thermal functioning of the human body. He drew analogies from Fourier's law of heat conduction and Ohm's law of electricity. He found second derivatives difficult to measure. He was able to establish a steady-state concentration gradient in a cylindrical cell containing crystalline Sodium Chloride at the bottom and a large volume of water on top.

In Eq. [3.11], *the molar flux, N_{Ay} of species A, across any plane of constant y is found by counting the molecules of A that cross unit area of the plane in the positive y direction and subtracting the number that cross in the negative y direction.* This may be good at steady state. During transient state processes there is the depletion or accumulation of molecules near the surface that gets neglected. Thus the damped wave mass diffusion and relaxation equation can be obtained by writing the *the molar flux, N_{Ay} of species A, across any plane of constant y is found by counting the molecules of A that cross unit area of the plane in the positive y direction and subtracting the number* *that cross in the negative y direction and by subtracting the depletion of molecules at the surface at the limit of zero slice thickness.*

$$N_{Ay} = 1/N^* \{nx_A v_y^*|_y + \tfrac{1}{4} x_A <u>|_{y-a} - \tfrac{1}{4} x_A <u>|_{y+a}$$
$$- \tau_{mr} \partial/\partial t \, (\tfrac{1}{4} x_A <u>|_y)\} \qquad (3.18)$$

Assuming that the concentration profile, $x_A(y)$ is essentially linear for a distance of several mean free paths. Thus,

$$x_A|_{y-a} = x_A|_y - 2/3 \, \lambda \, \partial x_A/\partial y \qquad (3.19)$$

$$x_A|_{y+a} = x_A|_y + 2/3 \, \lambda \, \partial x_A/\partial y \qquad (3.20)$$

Substituting Eqs.[3.19, 3.20] into Eq. [3.18], and denoting $N_A + N_B = cv^*$;

$$N_{Ay} = x_A(N_{Ay} + N_{By}) - 1/3 \, c <u> \lambda \, x_A|_{y-a} \, \partial x_A/\partial y - \tau_{mr} \partial N_{Ay}/\partial t$$

One way of accounting for the storage term is the damped wave mass diffusion and relaxation equation with the following approximate value for D_{AB};

$$D_{AB} = 1/3 <u> \lambda$$

$$N_{Ay} = x_A(N_{Ay} + N_{By}) - D_{AB} \, \partial x_A/\partial y - \tau_{mr} \partial N_{Ay}/\partial t \qquad (3.21)$$

The J flux when the convection is zero is then;

$$J = -D_{AB} \, \partial C_A/\partial y - \tau_{mr} \partial J/\partial t \qquad (3.22)$$

Where τ_{mr} is the mass relaxation time.

Worked Example 3.1

Write an expression for the J flux including the depletion of molecules at the surface. Deduce that the time rate of change of concentration at the plane y can be another way of accounting for the depletion at the surface. Contrast the resulting constitutive generalized Fick's law of diffusion to that obtained as the damped wave diffusion and relaxation equation.

$$J_{Ay} = 1/N^* \{ \tfrac{1}{4} x_A <u>|_{y-a} - \tfrac{1}{4} x_A <u>|_{y+a} - \tau_{mr} (2a\ A)\partial/\partial t \, (\tfrac{1}{4} x_A <u>|_y)\}$$

$$= -1/3 \, c <u> \lambda \partial x_A/\partial y \qquad - \tau_{mr} <u> \partial C/\partial t \qquad (3.23)$$

Assuming that the speed of the mass propagation given by $\sqrt{(D_{AB}/\tau_{mr})}$ can be taken as the average velocity of the molecule $<u>$;

$$=-D_{AB} \partial C/\partial y \quad - \quad sqrt(D\tau_{mr}) \, \partial C/\partial t \tag{3.24}$$

In one dimension when combined with the mass balance equation the governing equation will become;

$$\partial C/\partial t \; = \; D_{AB}\partial^2 C/\partial y^2 \; + \; sqrt(D\tau_{mr}) \, \partial^2 C/\partial t \, \partial y \tag{3.25}$$

In dimensionless form the governing equation would become;

$$\partial u/\partial \tau \; = \; \partial^2 u/\partial X^2 \; + \; \partial^2 u/\partial \tau \, \partial X \tag{3.26}$$

where,

$$u \; = (C - C_s)/(C_0 - C_s); \; \tau = t/\tau_{mr}; \; X = x/sqrt(D_{AB}\tau_{mr}) \tag{3.27}$$

In order to understand the ramifications of the governing equation given in Eq. [3.32] consider a finite slab at an initial concentration of C_0 at zero time. For times greater than zero the surface concentration is given a step change in concentration to C_s.

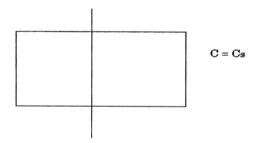

$$C = Cs$$

Figure 3.3 Finite Slab at Constant Wall Concentration

Eq. [3.26] is solved for by the method of separation of variables.

$$\text{Let } u = V(\tau)\phi(X) \tag{3.28}$$

$$V'/V \; - V' \, \phi'/V\phi \; = \; \phi''/\phi \; = -c^2$$

$$V'/V = -c^2 = \; -\phi''/(\phi' - \phi)$$

$$\text{Or } V = c_1 exp(-c^2\tau) \tag{3.29}$$

$$\phi'' \; - c^2\phi' + \; c^2\phi$$

$$\text{or } \phi = \; exp(+Xc^2/2) \, (c_2 \, exp(+Xc^2 \, sqrt(1 - 4/c^2) + c_3 exp(-Xc^2 sqrt(1 - 4/c^2)) \tag{3.30}$$

c can be chosen as less than 2 to obtain the eigen values and the Fourier series. For c less than two taking the real parts;

or ϕ = exp(+Xc^2/2)Real [(c$_2$ exp(iXc^2 sqrt(4/c^2 – 1)) + c$_3$exp(-iXc^{2}sqrt(4/c^2 – 1))

or ϕ = exp(+Xc^2/2) { c$_2$ Cos(Xc^2 sqrt(4/c^2 – 1)) + c$_3$ Cos(Xc^{2}sqrt(4/c^2 – 1))}   (3.31)

The boundary condition at the center of the slab can be seen to be obeyed. When X = X$_a$, u = 0.

$$0 = exp(X_a c^2/2) c_4 \ Cos(Xc^2 \ sqrt(4/c^2 – 1))$$

X$_a$ c^2 sqrt(4/c^2 – 1) = (2n-1)π/2.

Or c^2 = 2 - ½ sqrt(16 – (2n-1)π^{2}Dτ/a^2), for large a,   n = 1,2,....   (3.32)
Thus,

$$u = \Sigma c_n \ exp(-c^2(\tau - X/2) \ Cos(Xc^2 \ sqrt(4/c^2 – 1))$$ (3.33)

c$_n$ can be solved for from the initial condition and using the principle of orthogonality and can be shown to be 4 (-1)$^{n+1}$/ (2n-1)π.

3.1 Finite Speed Diffusion with Fast Chemical Reaction in a Semi-Infinite Catalyst

In many problems the diffusing solutes react rapidly with surrounding material. For example in the dyeing of wool, the dye can react quickly with the wool as it diffuses into the fiber. It is of interest to study the effect of such a rapid chemical reaction. The chemical reaction can radically change the process by reducing the apparent diffusion coefficient and mass relaxation time and increasing the interfacial flux of solute. Consider a semi-infinite medium at an initial concentration of zero. For times greater than zero a step change in concentration is effected at one of the surfaces. The species reacts by a first order reaction at it comes in contact with the solid medium. The mass balance equation can be written as;

$$-\partial J/\partial x - k''' \ C = \partial C/\partial t$$ (3.34)

where k''' is the first order reaction rate constant. Combining Eq. [3.34] with Eq. [3.22] the governing equation is obtained. Eq. [3.22] is differentiated wrt to x to give;

$$\partial J/\partial x = -D_{AB}\partial^2 C/\partial x^2 - \tau_{mr} \ \partial^2 J/\partial t\partial x$$

Eq. [3.34] is differentiated wrt to t to give;

$$-\partial^2 J/\partial x \partial t - k''' \, \partial C/\partial t = \partial^2 \, C/\partial t^2$$

Eliminating $\partial^2 J/\partial x \partial t$ between the two Eqs. above,

$$-k'''C - \partial C/\partial t = -D_{AB}\partial^2 C/\partial x^2 + \tau_{mr} (k''' \, \partial C/\partial t + \partial^2 \, C/\partial t^2)$$

$$D_{AB}\partial^2 C/\partial x^2 = \tau_{mr} \, \partial^2 \, C/\partial t^2 + (1 + k''' \, \tau_{mr}) \, \partial C/\partial t + k''' \, C \qquad (3.35)$$

Eq. [3.35] can be made dimensionless by the following substitutions;

$$k^* = (k''' \tau_{mr}) \, ; \quad u = C/C_s; \quad \tau = t/\tau_{mr}; \quad X = x/\text{sqrt}(D\tau_{mr}) \qquad (3.36)$$

$$\partial^2 u/\partial X^2 = \partial^2 u/\partial \tau^2 + (1 + k^*) \, \partial u/\partial \tau + k^* u \qquad (3.37)$$

The Eq. [3.37] can be solved by the method of relativistic transformation of coordinates. The damping term is first removed by a $u = w \exp(-n\tau)$ substitution.

$$\partial^2 u/\partial \tau^2 = +n^2 w \, -2n w_\tau + w_{\tau\tau}$$

$$\partial^2 w/\partial X^2 = (n^2 - n(1 + k^*))w + w_\tau (1 + k^* \, -2nw) + w_{\tau\tau} + k^* w$$

Choosing $n = (1 + k^*)/2$ Eq. [3.37] becomes;

$$\partial^2 w/\partial X^2 = -w(1 - k^*)^2/4 + w_{\tau\tau} \qquad (3.38)$$

Let the transformation be;

$$\eta = \tau^2 - X^2 \text{ for } \tau > X$$

Eq. [3.38] becomes;

$$4\tau^2\partial^2 w/\partial\eta^2 + 2\partial w/\partial\eta \quad -w(1 - k^*)^2/4 = 4X^2\partial^2 w/\partial\eta^2 - 2\partial w/\partial\eta \qquad (3.39)$$

$$\text{or } \eta^2 \, \partial^2 w/\partial\eta^2 + \eta\partial w/\partial\eta \, - w\eta(1 - k^*)^2/16 = 0 \qquad (3.40)$$

Comparing Eq. [3.40] with the generalized Bessel equation [A.30]. $a = 1$; $b = 0$; $c = 0$; $s = \frac{1}{2}$; $d = -1/16(1 - k^*)^2$. The order $p = 0$. $\text{sqrt}(d)/s = \frac{1}{2} i(1 - k^*)$ and is imaginary. Hence the solution is;

$$w = c_1 I_0 \, \frac{1}{2} |1 - k^*| \, \text{sqrt}(\tau^2 - X^2) + c_2 K_0 \, \frac{1}{2} |1 - k^*| \, \text{sqrt}(\tau^2 - X^2)$$

c_2 can be seen to be zero from the condition that at $\eta = 0$ W is finite.

$$w = \exp(-\tau(1 + k^*)/2) \, I_0 \, [\frac{1}{2} |1 - k^*| \, \text{sqrt}(\tau^2 - X^2)] \qquad (3.41)$$

From the boundary condition at $X = 0$,

$$1 = \exp(-\tau(1 + k^*)/2) \, c_1 \, I_0 \, \frac{1}{2} \, [|1 - k^* \tau|] \qquad (3.42)$$

c_1 can be eliminated between Eqs. [3.41, 3.42] to yield;

$$u = I_0 [\tfrac{1}{2} |1 - k^*| \, sqrt(\tau^2 - X^2)] / I_0 \, \tfrac{1}{2} [|1 - k^*|\tau] \qquad (3.43)$$

This is valid for $\tau > X$, $k^* \neq 1$. For $X > \tau$,

$$u = J_0 [\tfrac{1}{2} |1 - k^*| \, sqrt(X^2 - \tau^2)] / I_0 \, \tfrac{1}{2} [\, |1 - k^*|\tau] \qquad (3.44)$$

At the wave front , $\tau = X$, $u = exp(-\tau(1 + k^*)/2)) = exp(-X(1 + k^*)/2)$. The mass inertia can be calculated from the first zero of the Bessel function at 2.4048. Thus,

$$\tau_{inertia} = sqrt(X_p^2 - 23.1323/|1 - k*|^2) \qquad (3.45)$$

The concentration at an interior point in the semi-infinite medium is shown in Figure 3.4. Three regimes can be identified. During the first regime of mass inertia there is no transfer of mass upto a certain threshold time at the interior point $X_p = 10$. The second regime is given by Eq. [3.62] represented by a modified Bessel function of the first kind and zeroth order. The rise in dimensionless concentration proceeds from the dimensionless time 2.733 upto the wave front at $X_p = 10.0$. The third regime is given by Eq. [3.61] and represents the decay in time of the dimensionless concentration. It is given by the modified Bessel composite function of the first kind and zeroth order. In Figure 3.5 is shown the three regimes of the concentration when $k^* = 2.0$. It can be seen from Figure 3.5 that the mass inertia time has increased to 8.767. The rise is nearly a jump in concentration at the interior point $X_p = 10.0$. When k* = 0.25 as shown in Figure 3.6 the inertia time is 7.673. In Figure 3.6 the three regimes for the case when $k^* = 0.0$ is plotted. In Table 3.1 the mass inertia time for various values of k^* for the interior point $X_p = 10.0$ is shown. k^* needs to be sufficiently far from 1 to keep the inertia time positive.

Table 3.1 Mass Inertia Time vs k^* for Interior Point $X_p = 10.0$

S.No.	k* (k'''τ_{mr})	Mass Inertia Time (t/τ_{mr})
1.	0.01	8.741
2.	0.1	8.452
3.	0.25	7.673
4.	0.3	7.266
5.	0.4	5.979
6.	0.5	2.733
7.	1.75	7.673
8.	2.0	8.767
9.	4.0	9.871
10.	8.0	9.976
11.	25.0	9.998
12.	10.0	10.0

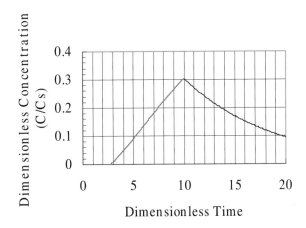

Figure 3.4   **Dimensionless Concentration at a Interior Point X_p = 10 in a Semi-infinite Medium during Simultaneous Reaction and Diffusion. $k^* = 0.5$.**

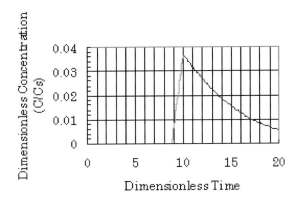

Figure 3.5   **Dimensionless Concentration at a Interior Point X_p = 10 in a Semi-infinite Medium during Simultaneous Reaction and Diffusion. $k^* = 2.0$.**

158

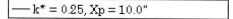

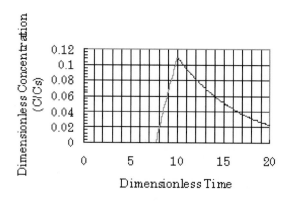

Figure 3.6 **Dimensionless Concentration at a Interior Point X_p = 10 in a Semi-infinite Medium during Simultaneous Reaction and Diffusion. k^* = 0.25**

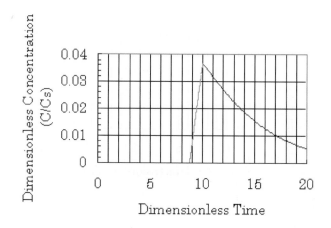

Figure 3.7 **Dimensionless Concentration at a Interior Point X_p = 10 in a Semi-infinite Medium during Simultaneous Reaction and Diffusion. k^* = 0.0**

The steady state solution for Eq. [3.37] can be written as;

$$u^{ss} = \exp(-(k^*)^{1/2}X)$$

Worked Example 3.2

The analysis discussed in the above section is valid for $k^* \neq 1$. Derive a suitable expression for $k^* = 1$. The governing equation given in Eq. [3.37] is rewritten for $k^* = 1$ as follows;

$$\partial^2 u/\partial X^2 = \partial^2 u/\partial \tau^2 + 2 \partial u/\partial \tau + u \qquad (3.46)$$

Obtaining the Laplace transforms;

$$d^2\underline{u}/dX^2 = s^2\underline{u} + 2s\,\underline{u} + \underline{u}$$

$$\underline{u} = c_1\exp(X\sqrt{s^2 + 2s + k^*}) + c_2 \exp(-X\sqrt{s^2 + 2s + k^*}) \qquad (3.47)$$

From the boundary condition at $X = \infty$, $c_1 = 0$. From the Boundary condition at $X = 0$, $u = 1$, c_2 can be seen to be $1/s$. Thus.,

$$\underline{u} = 1/s \exp(-X\sqrt{s^2 + 2s + 1}) = 1/s \exp(-X(s+1)) \qquad (3.48)$$

Inversion of the Laplace transform yields,

$$= \exp(-X)\, S_X(\tau) = 0 \quad \text{when } 0 < \tau < X$$

$$= \exp(-X)\, S_X(\tau) = \exp(-X) \text{ when, } \tau > X \qquad (3.49)$$

Worked Example 3.3

Obtain the solution to Eq. [3.37] by the method of Laplace transforms. The governing equation given in Eq. [3.37] is written as;

$$\partial^2 u/\partial X^2 = \partial^2 u/\partial \tau^2 + (1 + k^*)\, \partial u/\partial \tau + k^* u$$

Obtain the Laplace transforms of Eq. [3.37] is;

$$d^2\underline{u}/dX^2 = s^2\underline{u} + (1 + k^*)\, s\,\underline{u} + k^*\,\underline{u} \qquad (3.50)$$

$$\underline{u} = c_1 \exp(X\mathrm{sqrt}(s^2 + (1 + k^*)s + k^*)) + c_2 \exp(-X\mathrm{sqrt}(s^2 + (1 + k^*)s + k^*))$$

From the boundary condition at $X = \infty$, $c_1 = 0$. From the Boundary condition at $X = 0$, $u = 1$, c_2 can be seen to be $1/s$. Thus,

$$\underline{u} = 1/s \exp(-X\mathrm{sqrt}(s^2 + (1 + k^*)s + k^*))$$

$$\underline{u} = 1/s \exp(-X\mathrm{sqrt}((s + (1 + k^*)/2)^2 + k^* - (1 + k^*)^2/4)) \tag{3.51}$$

Define a $H = \int \underline{u}\, dX$

$$H = -1/s (s + a)^2 + k^* - a^2)^{1/2} \exp(-X((s + a)^2 + k^* - a^2)^{1/2}) \tag{3.52}$$

Where $a = (1 + k^*)/2$. Inversion of H results in;

$$H = -\int_0^\tau \exp(-ap)\, J_0 (k^* - a^2)^{1/2}(p^2 - X^2)^{1/2} dp \tag{3.53}$$

This is for $\tau > X$ and it is 0 for $0 < \tau < X$.

$$H = -\int_X^\tau \exp(-ap)\, J_0 (k^* - a^2)^{1/2}(p^2 - X^2)^{1/2} dp \tag{3.54}$$

$u = \partial H/\partial X$. Using Leibniz's rule ;

$$u = X\int_X^\tau \exp(-ap)\, [J_1 (k^* - a^2)^{1/2}(p^2 - X^2)^{1/2}]/[p^2 - X^2)^{1/2} dp \tag{3.55}$$
$$+ \exp(-aX)$$

Where $a = (1 + k^*)/2$. This is for $\tau > X$ and it is 0 for $0 < \tau < X$.

Worked Example 3.4

Consider a semi-infinite medium at a initial concentration of zero. For times greater than zero a step change in concentration is effected at one of the surfaces. The species reacts by a zeroth order reaction at it comes in contact with the solid medium. Discuss the concentration profile as a function of space and time.

The mass balance equation can be written as;

$$-\partial J/\partial x - k''' = \partial C/\partial t$$

where k''' is the zeroth order reaction rate constant. Combining Eq. [3.56] with the damped wave diffusion and relaxation equation the governing equation is obtained. The wave diffusion equation is differentiated wrt to x to give;

$$\partial J/\partial x = -D_{AB}\partial^2 C/\partial x^2 - \tau_{mr}\,\partial^2 J/\partial t\partial x$$

The mass balance equation is differentiated wrt to t to give;

$$-\partial^2 J/\partial x\partial t = \partial^2 C/\partial t^2$$

Eliminating $\partial^2 J/\partial x\partial t$ between the above two Eqs.

$$-k''' - \partial C/\partial t = -D_{AB}\partial^2 C/\partial x^2 + \tau_{mr}\,(\partial^2 C/\partial t^2)$$

$$D_{AB}\partial^2 C/\partial x^2 = \tau_{mr}\,\partial^2 C/\partial t^2 + \partial C/\partial t + k''' \tag{3.56}$$

Eq. [3.56] can be made dimensionless by the following substitutions;

$$k^* = (k'''\tau_{mr})\,;\ u = C/C_s;\ \tau = t/\tau_{mr};\ X = x/sqrt(D\tau_{mr})$$

$$\partial^2 u/\partial X^2 = \partial^2 u/\partial \tau^2 + \partial u/\partial \tau + k^* \tag{3.57}$$

Obtaining Laplace transforms of Eq. [3.57],

$$d^2 \underline{u}/dX^2 - \underline{u}(s^2 + s) - k^*/s = 0$$

$$\text{Let } \underline{v} = \underline{u}(s^2 + s) + k^*/s$$

Then , $1/(s^2 + s)\,d^2\underline{v}/dX^2 - \underline{v} = 0$

$$\text{Then, } \underline{v} = c_1 exp(+Xsqrt(s + s^2)) + c_2\, exp(-Xsqrt(s + s^2)) \tag{3.58}$$

The solution to the second order linear ODE with constant coefficients can be obtained. From the boundary condition at $X = \infty$, it can be seen that c_1 is zero. From the boundary condition at $X = 0$, $u = 1$

$$\text{or } \underline{u} = 1/s \text{ and } \underline{v} = (s+1) + k^*/s$$

So, $\underline{v} = ((s + 1) + k^*/s)\, exp(-Xsqrt(s + s^2))$

$$\underline{u} = (1/s + k^*/(s^2(s+1)))\, exp(-Xsqrt(s + s^2)) - k^*/(s^2(s+1)) \tag{3.59}$$

Inversion of Eq. [3.59] from the tables in Mickley, Sherwood and Reed [1957],

$$u = X\!\int_X^\tau exp(-p/2)\, [I_1\ 1/2(p^2 - X^2)^{1/2}]/[\,p^2 - X^2)^{1/2}dp \quad + exp(-X/2)$$

162

$$-k \int_X^\tau (\tau - p) \exp(-p) dp$$

$$+ \int_X^\tau \exp(-p/2) I_0 1/2(p^2 - X^2)^{1/2} (\tau-p)\exp(-(\tau-p)/2) [I_0 1/2 (\tau-p) + I_1 (1/2(\tau-p)dp \tag{3.60}$$

This is valid for $\tau > X$.   For $0 < \tau < X$, u = 0.

3.2 Finite Speed Diffusion with Fast Chemical Reaction in Infinite Catalyst in Cylindrical Coordinates

Consider a cylindrical catalyst medium at a initial reactant concentration of zero. The surface of the solid non-catalyst cylinder is maintained at a constant concentration of C_s for times greater than zero. The mass propagative velocity is given as the square root of the ratio of the binary diffusivity and mass relaxation time, $V_m = \text{sqrt}(D/\tau_{mr})$. The two time conditions, initial and final and the two boundary conditions are;

$$t = 0, \ r > R, \ C = 0 \tag{3.61}$$

$$t > 0, \ r = R, \ C = C_s \tag{3.62}$$

$$r = \infty, \ t > 0, \ C = 0 \tag{3.63}$$

The governing equation in concentration is obtained by eliminating the second cross-derivative of mass flux wrt to r and t between the damped wave mass diffusion and relaxation equation and the mass balance equation in cylindrical coordinates. Considering a cylindrical shell of thickness Δr,

$$\Delta t[(2\pi rL \ J_r \ - \ 2\pi(r + \Delta r)L \ J_{r + \Delta r}) \ - \ 2\pi Lr\Delta r \ k'''C] = 2\pi Lr\Delta r \ \Delta C$$

In the limit of Δr, Δt going to zero, the mass balance equation in cylindrical coordinates becomes;

$$-1/r \partial(rJ_r)/\partial r - k'''C \ = \ \partial C/\partial t \tag{3.64}$$

The non-Fourier damped wave equation is;

$$J_r \ = \ -D_{AB}\partial C/\partial r \ - \ \tau_{mr} \ \partial J_r/\partial t \tag{3.65}$$

Multiplying Eq. [3.65] by r and differentiating wrt to r and then dividing by r,

$$(1/r)\partial(rJ_r)/\partial r \ = \ (-D_{AB} /r)\partial(r\partial C/\partial r)/\partial r \ - \ (\tau_{mr}/r) \ \partial^2(rJr)/\partial t\partial r$$

Differentiating Eq. [3.64] wrt t.

$$-1/r \partial^2 (rJ_r)/\partial t \partial r - k''' \partial C/\partial t = \partial^2 C/\partial t^2$$

Substituting the above two Eqs. into Eq. [3.64] the governing equation in concentration is obtained as;

$$k''' C + \partial C/\partial t \,(\, 1 + k''' \tau_{mr}) + \tau_{mr} \, \partial^2 C/\partial t^2 = D_{AB}/r \, \partial(\, r \partial C/\partial r)/\partial r \qquad (3.66)$$

Obtaining the dimensionless variables,

$$u = (\, C/C_s); \quad \tau = (t/\tau_{mr}); \quad X = r/\sqrt{(D_{AB} \, \tau_{mr})} \quad k^* = k''' \, \tau_{mr} \quad ;$$

The governing equation in the dimensionless form can be written as;

$$k^* u + \partial u/\partial \tau \,(\, 1 + k^*) + \partial^2 u/\partial \tau^2 = \partial^2 u/\partial X^2 + 1/X \, \partial u/\partial X \qquad (3.67)$$

The damping term is removed from the governing equation. This is done realizing that the transient temperature decays with time in an exponential fashion. The other reason for this manoeuvre is to study the wave equation without the damping term. Let $u = w \exp(-n\tau)$. Choosing $n = (1 + k^*)/2$ Eq. [3.66] becomes;

$$1/X \, \partial w/\partial X + \partial^2 w/\partial X^2 = -w(1 - k^*)^2/4 + w_{\tau\tau} \qquad (3.68)$$

Eq. [3.68] can be solved by using the method of relativistic transformation of coordinates. Consider the transformation variable η for $\tau > X$ as; $\eta = \tau^2 - X^2$

$$\partial w/\partial \tau = (\partial w/\partial \eta \,)2\tau; \quad \partial^2 w/\partial \tau^2 = (\partial^2 w/\partial \eta^2 \,)4\tau^2 + 2(\partial w/\partial \eta \,)$$

In a similar fashion,

$$\partial w/\partial X = (\partial w/\partial \eta \,)2X \quad ; \quad \partial^2 w/\partial X^2 = (\partial^2 w/\partial \eta^2 \,)4X^2 + 2(\partial w/\partial \eta \,)$$

Plugging the above two Eqs. into Eq. [3.68]

$$(\partial^2 w/\partial \eta^2 \,)4(\, \tau^2 - X^2 \,) + 6(\partial w/\partial \eta \,) \quad -w(1 - k^*)^2/4 = 0 \qquad (3.70)$$

$$\eta^2 \partial^2 w/\partial \eta^2 + 3/2 \eta \partial w/\partial \eta - \eta w(1 - k^*)^2/16 = 0 \qquad (3.71)$$

Comparing Eq. [3.71] with the generalized Bessel equation as given in Eq. [2.96], the solution is;

$$a = 3/2; \; b = 0; \; c = 0; \; d = -1(1 - k^*)^2/16; \; s = ½. \text{ The order p of the}$$
solution is then $p = 2 \sqrt{(1/16)} = ½$.

$$W = c_1 I_{1/2} (\,|1 - k^*|/2 \sqrt{(\tau^2 - X^2)}/(\tau^2 - X^2)^{1/4} + c_2 I_{-1/2} |1 - k^*|/2 \sqrt{(\tau^2 - X^2)}/(\tau^2 - X^2)^{1/4} \qquad (3.72)$$

c_2 can be seen to be zero as W is finite and not infinitely large at $\eta = 0$. An approximate solution can be obtaining by eliminating c_1 between Eq. [3.62] and the above equation. It can be noted that this is a mild function of time, however. As the general solution of PDE consists of n arbitrary functions when the order of the PDE is n compared with n arbitrary constants for ODE. From the boundary condition at, $X = X_R$,

$$1 = \exp(-\tau/2) c_1 I_{1/2} (|1 - k^*|/2 \; sqrt(\tau^2 - X_R^2)/(\tau^2 - X_R^2)^{1/4} \tag{3.73}$$

$$\text{or } u = [(\tau^2 - X_R^2)^{1/4}/(\tau^2 - X^2)^{1/4}][I_{1/2} (|1 - k^*|/2 sqrt(\tau^2 - X^2) / I_{1/2} ((|1 - k^*|/2 sqrt(\tau^2 - X_R^2))] \tag{3.74}$$

In terms of elementary functions, Eq. [3.74] can be written as;

$$u = [(\tau^2 - X_R^2)^{1/2}/(\tau^2 - X^2)^{1/2}][Sinh(1/2(|1 - k^*| \; sqrt(\tau^2 - X^2) / Sinh(|1 - k^*|/2 sqrt(\tau^2 - X_R^2)] \tag{3.75}$$

In the limit of X_R going to zero, the expression becomes;

$$\text{or } u = [\tau/(\tau^2 - X^2)^{1/2}][Sinh(|1 - k^*|/2 \; sqrt(\tau^2 - X^2) / Sinh(|1 - k^*|/2\tau)] \tag{3.76}$$

This is for $\tau > X$, $k^* \neq 1$. For $X > \tau$, $k^* \neq 1$

$$u = [(\tau^2 - X_R^2)^{1/4}/(X^2 - \tau^2)^{1/4}][J_{1/2} (1/2 (|1 - k^*|sqrt(X^2 - \tau^2)/ I_{1/2} (1/2 sqrt(\tau^2 - X^2)] \tag{3.77}$$

Eq. [3.77] can be written in terms of trigonometric functions as;

$$u = [[(\tau^2 - X_R^2)^{1/4}/(X^2 - \tau^2)^{1/4}][Sin(1/2 (|1 - k^*|sqrt(X^2 - \tau^2)/ Sinh(1/2 (|1 - k^*|sqrt(\tau^2 - X^2)] \tag{3.78}$$

In the limit of X_R going to zero, the expression becomes;

$$\text{or } u = [\tau/(X^2 - \tau^2)^{1/2}][Sin((|1 - k^*|/2 sqrt(X^2 - \tau^2)/ Sinh((|1 - k^*|\tau/2)] \tag{3.79}$$

The dimensionless concentration at a point in the medium at $X_p = 7$ for example is considered and shown in Figure 3.7 for $k^* = 2.5$. Three different regimes can be seen. The first regime is that of the thermal lag and consists of no change from the initial concentration. The second regime is when the dimensionless concentration rises. The third regime is the decay with time of the concentration.

$$\text{or } \tau_{lag} = sqrt(X_p^2 - 4\pi^2/(|1 - k^*|^2) = 5.61 \text{ when } X_p = 7 \tag{3.80}$$

For times greater than the time lag and less than X_p the dimensionless concentration is given by Eq. [3.128]. For dimensionless times greater than 7, the dimensionless concentration is given by Eq. [3.125]. For distances *closer to the surface compared with* $2\pi/|1 - k^*|$ the time lag will be zero. The dimensionless concentration is shown for an exterior point $X_p = 7$ for different values of k^* in Figures 3.8 – 3.11. The mass inertia time lag for various values of k^* is shown in Table 3.2. The steady state solution for Eq. [3.67] can be written as;

$$u^{ss} = K_0(k^*)^{1/2} X/K_0(k^*)^{1/2} X_R \tag{3.81}$$

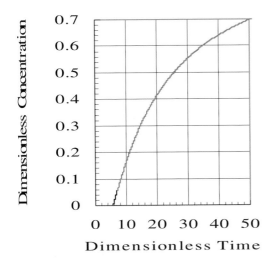

——— "X > tou; Xp = 7" ——— tou > X

Figure 3.8 Simultaneous Simple Reaction and Damped Wave Diffusion and Relaxation for $k^* = 0.05$ at a Point $X_p = 7$ in a Infinite Cylinder

Table 3.2 Mass Inertia Time vs k^* for Exterior Point $X_p = 7.0$ in a Infinite Cylinder during Simultaneous Reaction and Wave Diffusion and Relaxation

S.No.	$k^* (k'''\tau_{mr})$	Mass Inertia Time (t/τ_{mr})
1.	0.01	2.96
2.	0.05	2.3
3.	0.1	0.511
4.	2.0	3.1
5.	2.5	5.61
6.	3.5	6.533
7.	4.0	6.68
8.	5.0	6.82
9.	10.0	6.97

166

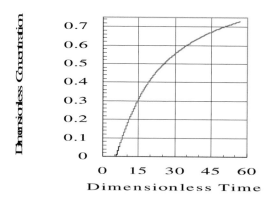

**Figure 3.9 Dimensionless Concentration Profile at a Point $X_p = 7$
in a Infinite Cylinder with Reaction and Wave Diffusion**

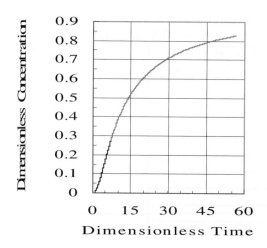

**Figure 3.10 Dimensionless Concentration Profile at a Point $X_p = 7$ in a Infinite
Cylinder during Simultaneous Simple Reaction and Damped Wave Diffusion**

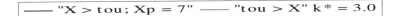

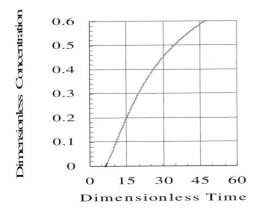

**Figure 3.11 Dimensionless Concentration Profile at a Point X_p = 7
in a Infinite Cylinder during Simultaneous Simple Reaction and
Damped Wave Diffusion and Relaxation for k^* = 3.0**

3.3 Finite Speed Diffusion with Fast Chemical Reaction in Infinite Catalyst in Spherical Coordinates

Consider a spherical catalyst medium at an initial reactant concentration of zero. The surface of the solid non-catalyst sphere is maintained at a constant concentration of C_s for times greater than zero. The mass propagative velocity is given as the square root of the ratio of the binary diffusivity and mass relaxation time, $V_m = sqrt(D/\tau_{mr})$. The two time conditions, initial and final and the two boundary conditions are;

$$t = 0, \ r > R, \ C = 0 \tag{3.82}$$

$$t > 0, \ r = R, \ C = C_s \tag{3.83}$$

$$r = \infty, \ t > 0, \ C = 0 \tag{3.84}$$

The governing equation in concentration is obtained by eliminating the second cross-derivative of mass flux wrt to r and t between the damped wave mass diffusion and relaxation equation and the mass balance equation in cylindrical coordinates. Considering a spherical shell of thickness Δr,

$$\Delta t[(\ 4\pi r^2 J_r\ -\ 4\pi(r + \Delta r)^2\ J_{r + \Delta_r}\ )\ -\ 4\pi r^2 \Delta r\ k'''C] = 4\pi r^2 \Delta r\ \Delta C \tag{3.85}$$

In the limit of Δr, Δt going to zero, the mass balance equation in spherical coordinates becomes;

$$-1/r^2 \partial(r^2 J_r)/\partial r - k'''C\ =\ \partial C/\partial t \tag{3.86}$$

The non-Fourier damped wave equation is;

$$J_r\ =\ -D_{AB}\partial C/\partial r\ -\ \tau_{mr}\ \partial J/\partial t \tag{3.87}$$

Multiplying Eq. [3.87] by r^2 and differentiating wrt to r and then dividing by r^2,

$$(1/r^2)\partial(r^2 J_r\ )/\partial r\ =\ (-D_{AB}/r^2)\partial(r^2\partial C/\partial r)/\partial r\ -\ (\tau_{mr}/r^2)\ \partial^2(r^2 Jr)/\partial t\partial r$$

Differentiating Eq. [3.8] wrt t.

$$-1/r^2 \partial^2(r^2 Jr)/\partial t\partial r - k'''\partial C/\partial t\ =\ \partial^2 C/\partial t^2$$

Substituting the above two Eqs. into Eq. [3.87] the governing equation in concentration is obtained as;

$$k'''C\ +\partial C/\partial t\ (\ 1 + k'''\tau_{mr}) + \tau_{mr}\ \partial^2 C/\partial t^2\ =\ D_{AB}/r^2\partial(\ r^2\partial C/\partial r)/\partial r \tag{3.88}$$

Obtaining the dimensionless variables,

$$u = (\ C/C_s);\ \tau = (t/\tau_{mr});\ X = r/\text{sqrt}(D_{AB}\ \tau_{mr})\ \ k^* = k'''\ \tau_{mr}\ ;$$

The governing equation in the dimensionless form can be written as;

$$k^*u + \partial u/\partial \tau\ (\ 1 + k^*)\ +\ \partial^2 u/\partial \tau^2\ =\ \partial^2 u/\partial X^2\ + 2/X\ \partial u/\partial X \tag{3.89}$$

The damping term is removed from the governing equation. This is done realizing that the transient concentration decays with time in an exponential fashion. The other reason for this manoeuvre is to study the wave equation without the damping term. Let $u = w\exp(-n\tau)$. Choosing $n = (1 + k^*)/2$ Eq. [3.89] becomes;

$$2/X\ \partial w/\partial X + \partial^2 w/\partial X^2\ =\ -w(1 - k^*)^2/4\ + w_{\tau\tau} \tag{3.91}$$

Eq. [3.91] can be solved by using the method of relativistic transformation of coordinates. Consider the transformation variable η for $\tau > X$

$$\eta\ =\ \tau^2 - X^2$$

As shown in the above sections, Eq. [3.91] becomes;

$$(\partial^2 w/\partial\eta^2)4(\tau^2 - X^2) + 8(\partial w/\partial\eta) - w(1 - k^*)^2/4 = 0 \qquad (3.92)$$

$$4\eta^2\partial^2 w/\partial\eta^2 + 8\eta\partial w/\partial\eta - \eta w(1 - k^*)^2/4 = 0$$

or

$$\eta^2\partial^2 w/\partial\eta^2 + 2\eta\partial w/\partial\eta - \eta w(1 - k^*)^2/16 = 0 \qquad (3.93)$$

Comparing Eq. [3.93] with the generalized Bessel equation as given in Eq. [2.96], the solution is;

$a = 2; b = 0; c = 0; d = -1(1 - k^*)^2/16; s = \frac{1}{2}$. The order p of the solution is then $p = 2\sqrt{1/4} = 1$

$$W = c_1 I_1 (|1 - k^*|/2 \sqrt{(\tau^2 - X^2)}/(\tau^2 - X^2)^{1/2} + c_2 K_1 |1- k^*|/2 \sqrt{(\tau^2 - X^2)}/(\tau^2 - X^2)^{1/2}$$
$$(3.94)$$

c_2 can be seen to be zero as W is finite and not infinitely large at $\eta = 0$. An approximate solution can be obtaining by eliminating c_1 between Eq. [3.83] and the above equation. It can be noted that this is a mild function of time, however. As the general solution of PDE consists of n arbitrary functions when the order of the PDE is n compared with n arbitrary constants for ODE. From the boundary condition at, $X = X_R$,

$$1 = \exp(-\tau/2) c_1 I_1(|1 - k^*|/2 \sqrt{(\tau^2 - X_R^2)}/(\tau^2 - X_R^2)^{1/2} \qquad (3.95)$$

or $u = [(\tau^2 - X_R^2)^{1/2}/(\tau^2 - X^2)^{1/2}][I_1(|1 - k^*|/2\sqrt{(\tau^2 - X^2)} / I_1((|1 - k^*|/2\sqrt{(\tau^2 - X_R^2)})]$
$$(3.96)$$

In the limit of X_R going to zero, the expression becomes;

or $u = [\tau/(\tau^2 - X^2)^{1/2}][I_1(|1 - k^*|/2\sqrt{(\tau^2 - X^2)} / I_1((|1 - k^*|\tau/2)]$
$$(3.97)$$

This is for $\tau > X$, $k^* \neq 1$. For $X > \tau$, $k^* \neq 1$

$u = [[(\tau^2 - X_R^2)^{1/2}/(X^2 - \tau^2)^{1/2}][J_1(1/2 (|1 - k^*|\sqrt{(X^2 - \tau^2)})/ I_1 (|1 - k^*|/2\sqrt{(\tau^2 - X^2)})]$
$$(3.98)$$

In the limit of X_R going to zero, the expression becomes;

or $u = [\tau/(X^2 - \tau^2)^{1/2}][J_1(1/2 (|1 - k^*|\sqrt{(X^2 - \tau^2)})/ I_1(\tau(|1 - k^*|/2)]$
$$(3.99)$$

Seventeen terms were taken in the series expansion of the modified Bessel composite function of the first kind and first order and the Bessel composite function of the first kind and first order respectively and the results plotted in Figures 3.12-3.15 for a given $Xp = 9$ and different values of k^* using a Miocrosoft Excel spreadsheet on a Pentium IV desktop microcomputer. Three regimes can be identified. The first regime is that of the thermal lag and consists of no change from the initial concentration. The second regime is when

$$\tau_{lag}^2 = X^2 - (7.6634)^{\wedge}2$$

or $\tau_{lag} = \text{sqrt}(X_p^2 - 7.6634^2/(|1 - k^*|^2)) = 4.72$, when $X_p = 9$, $k^* = 0.0$.

The first zero of $J_1(x)$ occurs at $x = 3.8317$. The 7.6634 is twice the first root of the Bessel function of the first order and first kind. For times greater than the time lag and less than X_p the dimensionless concentration is given by Eq. [3.160]. For dimensionless times greater than X_p, the dimensionless concentration is given by Eq. [3.158]. For distances closer to the surface compared with $7.6634\text{sqrt}(\alpha\tau_r)$ the thermal lag time will be zero. The ballistic term manifests as a thermal lag at a given point in the medium. The mass inertia time for various values of k^* is given in Table 3.3.

Table 3.3 Mass Inertia Time vs k^* for Exterior Point
$X_p = 9.0$ in a Infinite Sphere during Simultaneous
Reaction and Wave Diffusion and Relaxation

S.No.	k^* ($k'''\tau_{mr}$)	Mass Inertia Time (t/τ_{mr})
1.	0.01	4.59
2.	0.05	3.99
3.	0.1	2.91
4.	2.0	4.72
5.	2.5	7.41
6.	3.5	8.46
7.	5.0	8.79
8.	10.0	8.96

The steady state solution for the concentration for Eq. [3.89] can be seen to be;

$$u^{ss} = (X_R/X)^{1/2} I_{-1/2} [(k^*)^{1/2} X] / I_{-1/2} [(k^*)^{1/2} X_R] \qquad (3.100)$$

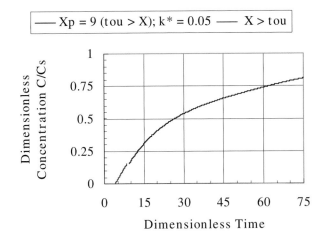

Figure 3.12   Dimensionless Concentration Profile at a Point X_p = 9 in a Infinite Sphere during Simultaneous Simple Reaction and Damped Wave Diffusion and Relaxation for k^{*} = 0.05

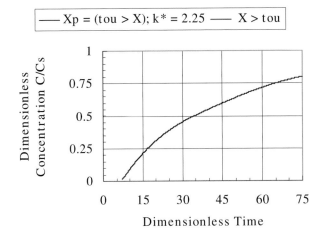

Figure 3.13   Dimensionless Concentration Profile at a Point X_p = 9 in a Infinite Sphere during Simultaneous Simple Reaction and Damped Wave Diffusion and Relaxation for k^{*} = 2.25

172

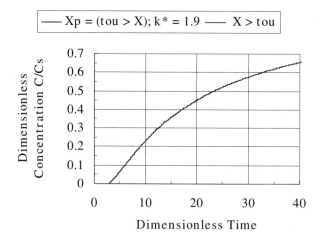

Figure 3.14 **Dimensionless Concentration Profile at a Point X_p = 9 in a Infinite Sphere during Simultaneous Simple Reaction and Damped Wave Diffusion and Relaxation for k^* = 1.9**

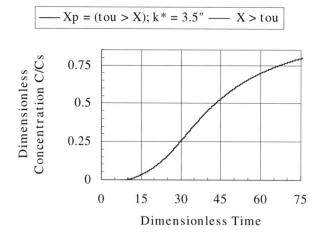

Figure 3.15 **Dimensionless Concentration Profile at a Point X_p = 9 in a Infinite Sphere during Simultaneous Simple Reaction and Damped Wave Diffusion and Relaxation for k^* = 3.5**

3.4 Finite Slab at Constant Wall Concentration

Consider a finite slab of width 2a with an initial concentration at C_0. The sides of the slab are maintained at constant concentration C_s. The governing equation can be obtained by eliminating J_x between Eqs. (1) and the equation from mass balance, $-\partial J/\partial x - k'''C = \partial C/\partial t$, where k''', is the simple first order reaction rate constant. This is achieved by differentiating Eq.(1) with respect to x and the balance equation with respect to t and eliminating the second cross derivative of J with respect to x and time. Thus,

$$\tau_r \, \partial^2 C/\partial t^2 + (1 + k^*)\partial C/\partial t + k'''C = D \, \partial^2 C/\partial x^2 \qquad (3.101)$$

$$\text{Let} \quad \tau = t/\tau_r; X = x/\sqrt{D\tau_r}; k^* - (k'''\tau_r)$$

The governing equation in the dimensionless form is then;

$$k^* C + \partial C/\partial \tau \, (1 + k^*) + \partial^2 C/\partial \tau^2 = \partial^2 C/\partial X^2 \qquad (3.102)$$

The solution can be assumed to consist of a steady state part and a transient part, i.e., $C = C^t + C^{ss}$. The steady state part and boundary conditions can be selected in such as fashion that the transient portion becomes homogeneous.

$$\partial^2 C^{ss}/\partial X^2 = k^* C^{ss} \qquad (3.103)$$

The Boundary Conditions are;

$$X = 0, \ \partial C^{ss}/\partial X = 0 \qquad (3.104)$$

$$X = \pm a/\sqrt{D\tau_r}, \ C^{ss} = C_s \qquad (3.105)$$

The solution for Eq. [3.103] can be written in terms of the hyperbolic sine and cosine functions as;

$$C^{ss} = A_1 \text{Sinh} \, ((k^*)^{1/2}X) + A_2 \text{Cosh} \, ((k^*)^{1/2}X) \qquad (3.106)$$

Applying the boundary condition given in Eq. [3.104] it can be seen that $A_1 = 0$. Solving for A_2 from Eq. [3.105].

$$C^{ss}/C_s = \text{Cosh} \, ((k^*)^{1/2}X) / \text{Cosh} \, ((k^*)^{1/2}X_a) \qquad (3.107)$$

The equation and time and space conditions for the transient portion of the solution can be written as;

$$\partial^2 C^t/\partial \tau^2 + (1 + k^*)\partial C^t/\partial \tau + k'''C^t = D \, \partial^2 C^t/\partial X^2 \qquad (3.108)$$

$$\text{Initial Condition}: \tau = 0, \quad C^t = C_0 \qquad (3.109)$$

$$\text{Final Condition:} \ \tau = \infty, \quad C^t = 0 \qquad (3.110)$$

The Boundary Conditions are now homogeneous after the expression of the result as a sum of steady state and transient parts and are;

$$\partial C^t / \partial X = 0, X = 0 \tag{3.111}$$

$$X = X_a, C^t = 0 \tag{3.112}$$

The solution is obtained by the method of separation of variables. Initially the damping term is eliminated using a substitution such as $C^t = W \exp(-n\tau)$. Eq. [3.108] then becomes at $n = \frac{1}{2}(1 + k^*)$;

$$W_{xx} = -(|1 - k^*|)^2 W/4 + W_{\tau\tau} \tag{3.113}$$

Eq. [3.113] can be solved by the method of separation of variables.

$$\text{Let } W = g(\tau) \phi(X)$$

Eq. [3.113] becomes,

$$g(\tau) \phi^{\cdot}(X) = -(|1 - k^*|)^2 g(\tau) \phi(X)/4 + g^{\cdot}(\tau) \phi(X) \tag{3.114}$$

$$\text{Or } \phi^{\cdot}(X)/\phi(X) = -(|1 - k^*|)^2/4 + g^{\cdot}(\tau)/g(\tau) = -\lambda_n^2 \tag{3.115}$$

The space domain solution is;

$$\phi(X) = c_1 Sin(\lambda_n X) + c_2 Cos(\lambda_n X) \tag{3.116}$$

From the boundary conditions,

At $X = 0$, $\partial \phi / \partial X = 0$, it can be seen that, $c_1 = 0$ (3.117)

$$\phi(X) = c_1 Cos(\lambda_n X) \tag{3.118}$$

From the boundary Condition given by Eq. [3.105],

$$0 = c_1 Cos(\lambda_n X_a) \tag{3.119}$$

$$(n + \frac{1}{2})\pi = \lambda_n X_a \tag{3.120}$$

$$\lambda_n = (n + 1/2)\pi \, sqrt(D\tau_r)/a, \ n = 0,1,2,3... \tag{3.121}$$

The time domain solution would be,

$$g = c_3 \exp(sqrt((|1 - k^*|)^2/4 - \lambda_n^2) \, \tau) + c_4 \exp(-sqrt((|1 - k^*|)^2/4 - \lambda_n^2)\tau) \tag{3.122}$$

From the FINAL condition given by Eq.[3.110], not only the transient concentration has to decay out to 0 but also the wave concentration. As, $W = C^t \exp(\tau/2)$, at time infinity the transient concentration is zero and as any number multiplied by zero is zero even if it is infinity, $W = 0$ at the final condition. Applying this condition in the solution Eq.[3.122] it can be seen that $c_3 = 0$. Thus,

$$C^t = \sum_0^\infty C_0 \, 2(-1)^n /(n+1/2)\pi \, \exp(-\tau/2) \, \exp(-\mathrm{sqrt}((\,|1-k^*|\,)^2/4 - \lambda_n^2)\tau)\mathrm{Cos}(\lambda_n X) \tag{3.123}$$

λ_n is described by Eq. [3.121]. C_n can be derived using the orthogonality property and can be shown to be $C_0 \, 2(-1)^n /(n+1/2)\pi$. It can be seen that the model solutions given by Eq. [3.123] is bifurcated. i.e., the characteristics of the function change considerably when a parameter such as the width of the slab is varied. Here a decaying exponential becomes exponentially damped cosinous. This is referred to as subcritical damped oscillatory behaviour. For a $< \pi \, \mathrm{sqrt}(D \, \tau_r)/(|\, 1 - k^*|\,)$ even for n =1, all the terms in the infinite series will pulsate. This when the argument within the square root sign in the exponentiated time domain expression becomes negative and the result becomes imaginary. Using Demovrie's theorem and taking real part for small width of the slab,

$$C^t = \sum_0^\infty c_n \exp(-\tau/2) \, \mathrm{Cos}(\mathrm{sqrt}((\lambda_n^2 - |1-k^*|\,)^2/4)\tau)\mathrm{Cos}(\lambda_n X) \tag{3.124}$$

3.5 Finite Cylinder at Constant Wall Concentration

Consider a finite cylinder of radius R with an initial concentration of C_0. The sides of the cylinder are maintained at a finite concentration of C_s. The governing equation from mass balance which is, $(-1/r\partial(rJ)/\partial r - k'''C = \partial C/\partial t\,)$ is achieved by differentiating Eq.(1) with respect to r and the balance equation with respect to t and eliminating the second cross derivative of J with respect to r and time. Then the equation is non-dimensionalized to give;

$$\partial^2 C/\partial\tau^2 + (1+k^*)\partial C/\partial\tau + k^*C = (\partial^2 C/\partial X^2 + 1/X \, \partial C/\partial X\,) \tag{3.125}$$

$$\text{where } \tau = t/\tau_r \quad ; \quad X = r/\mathrm{sqrt}(D\,\tau_r); \quad k^* = k''' \, \tau_r \tag{3.126}$$

The initial condition

$$\tau = 0, \; Vr, \; C = C_0 \tag{3.127}$$

$$\tau > 0, \; X = 0, \; \partial C/\partial X = 0 \tag{3.128}$$

$$\tau > 0, \; X = R/\mathrm{sqrt}(D/\tau_r), \; C = C_s \tag{3.129}$$

The solution can be expressed as a sum of a steady state part and a transient part.

$$\text{i.e., } C = C^{ss} + C^t. \tag{3.130}$$

The steady state part is given by;

$$k* \, C^{ss} \quad = (\partial^2 \, C^{ss} / \partial X^2 + 1/X \; \partial \, C^{ss} / \partial X) \tag{3.131}$$

This can be recognized as a modified Bessel differential equation of the zeroth order;

$$C^{ss} / C_s = \; I_0 \, (k*^{1/2} \, X) \, / \; I_0 \, (k*^{1/2} \, X_s) \tag{3.132}$$

The transient portion of the concentration is governed by;

$$\partial^2 \, C^t / \partial \tau^2 + (1 + k*) \partial \, C^t / \partial \tau + k* \, C^t \quad = (\partial^2 \, C^t / \partial X^2 + 1/X \; \partial \, C^t / \partial X) \tag{3.133}$$

The solution is obtained by the method of separation of variables. Initially the damping term is removed using a $C^t = W \exp(-n\tau)$ substitution. Eq. (3.196) becomes at $n = (1 + k*)/2$

$$W_{xx} + W_X / X = W \, (- (1 - k*)^2/4 \,) + \; W_{\tau\tau} \tag{3.134}$$

The solution to Eq. [3.197] is obtained by the method of separation of variables

$$\text{Let } W = g(\tau) \, \phi \, (X) \tag{3.135}$$

$$\text{Then, } (\phi" + \phi'/X \,)/\phi \; = \; g"/g - |1 - k*|^2/4 = -\lambda_n^{\,2} \tag{3.136}$$

$$\text{Or } X^2 \, \phi" + X \, \phi' + X^2 \lambda_n^{\,2} \, \phi = 0 \tag{3.137}$$

This can be recognized as the Bessel equation:

$$V = c_1 \, J_0 \, (\lambda_n X) \tag{3.138}$$

c_2 can be seen to be zero as ϕ is finite at $X = 0$ from boundary condition . From the boundary condition a the surface with the non-homogeneity attributed to the steady state portion of the solution;

$$C^t = \; 0 = c_1 \, J_0 \, (\lambda_n \, R/\text{sqrt}(D \, \tau_r \,)) \tag{3.139}$$

$$\text{or } \lambda_n = \; \text{sqrt}(D \, \tau_r \,)/R(2.4048 + (n-1)\pi \,), \text{ where } n = 1,2.3... \tag{3.140}$$

$$\text{Now, } \phi"/\phi \; = \; |1 - k*|^2/4 \; - \lambda_n^{\,2} \tag{3.141}$$

$$\text{Or } \phi = A \exp(-\text{sqrt}((1 - k*)^2/4 \; - \lambda_n^{\,2} \,)\tau + B \exp(+\text{sqrt}((1 - k*)^2/4 \; - \lambda_n^{\,2} \,)\tau \tag{3.142}$$

B can be seen to be zero as the system decays to steady state . $W = C^t \exp(1+k*)\tau/2$. At infinite time the value of the transient concentration is zero and the product of 0 and infinity, is zero and hence W is zero and so B has to be set to 0.

$$\text{For } |1 - k*|/2 \quad < \lambda_n \tag{3.143}$$

$$\text{Or } R < \; \text{sqrt}(D \, \tau_r \,)(4.8096)/|1 - k*| \tag{3.144}$$

When Eq. [3.144] is satisfied the solution in time domain becomes periodic. Thus the general solution is bifurcated depending on the value of R with respect to the binary diffusivity and relaxation time.

$$C^t = \sum_1^\infty A_n \exp(-1/2 - \mathrm{sqrt}((1 - k^*)^2/4 \; -\lambda_n^2)\tau \, J_0(\lambda_n X) \qquad (3.145)$$

Where A_n can be obtained from the principle of orthogonality and the initial condition;

$$A_n = C_0 \int_0^{XR} J_0(\lambda_n X)dX \; / \int_0^{XR} J_0^2(\lambda_n X) \, dX \qquad (3.146)$$

The nature of Eq. [3.145] is bifurcated. For values of $R < \mathrm{sqrt}(D \, \tau_r)(4.8096)/|1 - k^*|$

$$C^t = C_0 \sum_1^\infty A_n \exp(-1/2\tau) \, \mathrm{Cos}\tau(\mathrm{sqrt}(\lambda_n^2 - (1 - k^*|^2/4) \, J_0 A_n X) \qquad (3.147)$$

where A_n, is given by Eq. [3.146] and λ_n given by Eq. [3.140]

3.6 Finite Sphere with Constant Wall Concentration

Consider a finite sphere of radius R with an initial concentration of C_0. The sides of the sphere are maintained at a finite concentration of C_s. The governing mass balance equation from mass balance and the constitutive relation for wave diffusion and relaxation which can be written after non-dimensionalization as;

$$X^2\partial^2 C /\partial X^2 + 2X\partial C /\partial X \; - \; k^* C = \partial^2 C/\partial\tau^2 + (1 + k^*)\partial C/\partial\tau \qquad (3.148)$$

$$\text{Where, } \tau = t/\tau_r ; X = r/\mathrm{sqrt}(D \tau_r); k^* = (k'''\tau_r) \qquad (3.149)$$

The four conditions in space and time are;

$$\tau = 0, \; Vr, \; C = C_0 \qquad (3.150)$$

$$\tau > 0, \; X = 0, \; \partial C/\partial X = 0 \qquad (3.151)$$

$$\tau > 0, \; X = R/\mathrm{sqrt}(D\tau_r) \quad C = C_s \qquad (3.152)$$

$$\tau = \infty, \; C = C_s \qquad (3.153)$$

The solution to the concentration can be expressed as a sum of a transient portion and a steady state portion in order to ensure that the transient boundary values are homogeneous.

$$\text{Let, } C = C^t + C^{ss} \qquad (3.154)$$

The steady state part and boundary conditions can be selected in such a fashion that the transient portion becomes homogeneous. The steady state portion of the concentration and the boundary Conditions are;

$$X^2 \partial^2 C^{ss}/\partial X^2 + 2X \partial C^{ss}/\partial X - k^* C^{ss} = 0 \qquad (3.155)$$

$$X = 0, \quad \partial C^{ss}/\partial X = 0 \qquad (3.156)$$

$$X = R/\text{sqrt}(D\tau_r), \quad C^{ss} = C_s \qquad (3.157)$$

Eq. [3.155] can be recognized as a Bessel differential equation. The solution is;

$$C^{ss} = c_1 I_{1/2}(Xk^{*1/2})/X^{1/2} + c_2 I_{-1/2}(Xk^{*1/2})/X^{1/2} \qquad (3.158)$$

Since the concentration is finite at $X = 0$, c_2 is set to zero. Solving for c_1 from the boundary condition given the solution at steady state is;

$$C^{ss}/C_s = X_R^{1/2} I_{1/2}(Xk^{*1/2})/ I_{1/2}(X_R k^{*1/2})X^{1/2} \qquad (3.159)$$

The solution of the transient part is the rest of the problem:

$$(1 + k^*) \partial C^t/\partial \tau + \partial^2 C^t/\partial \tau^2 + k^* C^t = \partial^2 C^t/\partial X^2 + 2/X \partial C^t/\partial X \qquad (3.160)$$

The four conditions become;

$$\text{Initial Condition}: C^t = C_0 \qquad (3.161)$$

$$\text{Final Condition}: C^t = 0 \qquad (3.162)$$

$$X = 0, \quad C^t = \text{finite} \qquad (3.163)$$

$$X = X_R, \quad C^t = 0 \qquad (3.164)$$

The solution is obtained by the method of separation of variables. Initially the damping term is removed using a $C^t = W \exp(-n\tau)$ substitution. Eq. [3.160] becomes at $n = (1 + k^*)/2$

$$\partial^2 W/\partial X^2 + 2/X \partial W/\partial X = - W (1 - k^*)^2/4 + \partial^2 W/\partial \tau^2 \qquad (3.165)$$

The solution to Eq. [3.164] can be obtained by the method of separation of variables.

Let $W = g(\tau) \phi(X)$

$$\text{Then, } (\phi'' + 2\phi'/X)/\phi = g''/g - |1 - k^*|^2/4 = -\lambda_n^2 \qquad (3.166)$$

$$\text{Or } X^2 \phi'' + 2X \phi' + X^2 \lambda_n^2 \phi = 0 \qquad (3.167)$$

This can be recognized as the Bessel equation of the 1/2th order:

$$V = c_1 X^{-1/2} J_{1/2} (\lambda_n X) \qquad (3.168)$$

c_2 can be seen to be zero as ϕ is finite at $X = 0$ from boundary condition given in Eq. [3.163].

$$0 = c_1 J_{1/2} (\lambda_n R/sqrt(D \tau_r))/X^{1/2} \qquad (3.169)$$

$$\text{or } \lambda_n = n\pi \, sqrt(D\tau_r)/R \quad \text{where } n = 0, 1, 2,3... \qquad (3.170)$$

$$\text{Now, } \phi''/\phi = |1 - k^*|^2/4 - \lambda_n^2 \qquad (3.171)$$

$$\text{or} \quad \phi = A \exp (-sqrt((1 - k^*)^2/4 - \lambda_n^2)\tau + B \exp (sqrt((1 - k^*)^2/4 - \lambda_n^2)\tau \qquad (3.172)$$

B can be seen to be zero as the system decays to steady state from the FINAL condition given in Eq. [3.162]. $W = u\exp(1+k^*)\tau/2$. At infinite time the value of u is zero and the product of 0 and infinity, is zero and hence W is zero and so B has to be set to 0.

$$C^t = \Sigma_0^\infty A_n \exp (-(1+ k^*)/2 - sqrt((|1 - k^* |)^2/4 - \lambda_n^2)\tau J_{1/2} (\lambda_n X)/X^{1/2} \qquad (3.173)$$

Where A_n can be obtained from the principle of orthogonality at the initial condition.

$$C_0 = \Sigma_0^\infty A_n J_{1/2} (\lambda_n X)/sqrt(X) \qquad (3.174)$$

Multiplying both sides of the equation by $J_{1/2} (\lambda_m X)$ and integrating between the limits of 0 and $R/ sqrt(D\tau_r)$ and using the principle of orthogonality;

$$A_n = (C_0)\int_0^{XR} J_{1/2} (\lambda_n X)/sqrt(X)dX / \int_0^{XR} J_{1/2}^2 (\lambda_n X)/sqrt(X)dX \qquad (3.175)$$

For terms in the infinite series where the $(\lambda_n > |1 - k^*|/2)$ the exponentiated time domain terms become negative within the square root and using the De Movrie's theorem the real part gives the $Cos ((\lambda_n^2 - (|1 - k^*|)^2/4)^{1/2}\tau)$.

3.7 Critical Radii of Nuclear Fuel Rod Neither Less than the Cycling Limit Nor Greater than the Shape Limit

In autocatalytic reactions such as during nuclear fission and metabolic pathway the neutrons can be studied by a first order reaction. The mass balance in a long cylindrical rod with first order auto catalytic reaction can be written as;

$$1/r \, \partial (r J_r)/ \partial r + \partial C/\partial t + k^- C = 0 \qquad (3.176)$$

In this study the concentration profile during simultaneous diffusion and auto catalytic reaction is explored by the method of separation of variables using the modified Fick's law accounting for the finite speed of mass propagation. Consider a long cylindrical rod at zero initial concentration of autocatalytic reactant A. The surface of the rod is maintained at a constant concentration C_s for times greater than zero. The mass propagative velocity is given as the square root of the ratio of binary diffusivity and relaxation time.

$$V_m = sqrt(D/\tau_r) \tag{3.177}$$

The initial condition

$$t = 0, \ Vr, \ C = 0 \tag{3.178}$$

$$t > 0, \ r = 0, \ \partial C/\partial r = 0 \tag{3.179}$$

$$t > 0, \ x = R, \ C = C_s \tag{3.180}$$

The governing equation can be obtained by eliminating J_r between Eq. [3.176] and the damped wave diffusion and relaxation equation. This is achieved by eliminating the second cross derivative of J with respect to r and time. Thus,

$$\tau_r \, \partial^2 C/\partial t^2 + \partial C/\partial t \ (1 - k''\tau_r) = D \, \partial^2 C/\partial r^2 + D/r \, \partial C/\partial r + k'' C \tag{3.181}$$

where k'' is the first order autocatalytic reaction rate with the units of sec^{-1}. Obtaining the dimensionless variables ;

$$u = (C - C_s)/(C_s); \ \tau = t/\tau_r; \quad X = r/sqrt(D\tau_r); \ k^* = k''\tau_r$$

The governing equation in the dimensionless form is then;

$$\partial u/\partial \tau \ (1 - k^*) + \partial^2 u/\partial \tau^2 = \partial^2 u/\partial X^2 + 1/X \, \partial u/\partial X + k^* u \tag{3.182}$$

The solution is obtained by the method of separation of variables. First the damping term is removed by the substitution, $u = exp(-n\tau) \, W(X, \tau)$, Eq. [3.182] becomes,

$$exp(-n\tau) \, W (-n(1 - k^*) + n^2 - k^*) + exp(-n\tau) \, \partial W/\partial \tau ((1 - k^*) - 2n)$$
$$+ exp(-n\tau) \, \partial^2 W/\partial \tau^2 = exp(-n\tau)(\partial^2 W/\partial X^2 + 1/X \, \partial W/\partial X \tag{3.183}$$

By choosing $n = (1 - k^*)/2$, the damping component of the equation is removed and Eq. (10) becomes;

$$-W/4(1 + k^*)^2/4 + \partial^2 W/\partial \tau^2 = \partial^2 W/\partial X^2 + 1/X \, \partial W/\partial X \tag{3.184}$$

The method of separation of variables can be used to obtain the solution of Eq. [3.184].

$$\text{Let } W = V(\tau)\, \phi\,(X) \tag{3.185}$$

Substituting Eq. [3.185] into Eq. [3.184], and separating the variables that are a function of X only and τ only;

$$\phi'' + \phi'/X + \lambda^2\, \phi = 0 \tag{3.186}$$

$$V''/V = (1 + k^*)^2/4 - \lambda^2 \tag{3.187}$$

The solution for Eq. [3.184] is the Bessel function of zeroth order and first kind;

$$\phi = c_1\, J_0\,(\lambda\, X) + c_2\, Y_0\,(\lambda\, X) \tag{3.188}$$

It can be seen that $c_2 = 0$ as the concentration is finite at $X = 0$. Now from the BC at the surface,

$$\phi = c_1\, J_0\,(\lambda\, R/\sqrt{D\tau_r}) = 0$$

$$\text{or } \lambda_n = \sqrt{D\tau_r}\,/R\,(2.4048 + (n-1)\pi),\ \text{for } n = 1, 2, 3, .. \tag{3.189}$$

The solution for Eq. [3.187] is the sum of two exponentials. The term containing the positive exponential power exponent will drop out as with increasing time the system may be assumed to reach steady state. At steady state or infinite time $W = u\exp(\tau/2)$, becomes zero multiplied with infinity which is zero. Thus,

$$V = c_4\, \exp(-\tau\, \sqrt{((1 + k^*)^2/4 - \lambda_n^2)}) \tag{3.190}$$

$$\text{Or, } u = \sum_1^\infty c_n\, J_0\,(\lambda_n\, X)\, \exp(-\tau\,(1 + k^*)/2 - \sqrt{((1 + k^*)^2/4 - \lambda_n^2)}) \tag{3.191}$$

The c_n can be solved for from the initial condition by using the principle of orthogonality for Bessel functions. At time is zero the LHS And RHS are multiplied by $J_0\,(\lambda_m\, X)$. Integration between the limits of 0 and R is performed. When n is not m the integral is zero from the principle of orthogonality. Thus when $n = m$,

$$c_n = -\int_0^R J_0\,(\lambda_n\, X) \;/\; \int_0^R J_0^2\,(\lambda_n\, X) \tag{3.192}$$

It can be noted from Eq. [3.191] that when

$$(1 + k^*)^2/4 < \lambda_n^2 \tag{3.193}$$

the solution will be periodic with respect to time domain. This can be obtained by using De Movries theorem and obtaining the real part to $\exp(-i\tau\, \sqrt{(\lambda_n^2 - (1 + k^*)^2/4)})$. Also it can be shown that for terms in the infinite series after n or for

$$n > 1 + R\,(1 + k^*)/\pi\, \sqrt{D\tau_r} - 2.4048/\pi \tag{3.194}$$

the contribution to the solution will be periodic. Thus a bifurcated solution is obtained. Also from Eq. [3.191] it can be seen that all terms in the infinite series will be periodic. i.e, even for n =1 when,

$$R < 4.8096 \sqrt{D\tau_r} / (1 + k^*) \tag{3.195}$$

$$u = \sum_1^\infty c_n J_0(\lambda_n X) \exp(-\tau(1 + k^*)/2)\cos(\tau\sqrt{\lambda_n^2 - (1 + k^*)^2/4}) \tag{3.196}$$

The steady state solution of Eq. [3.182] can be obtained as follows after re-defining $u^* = C/C_s$.

$$\partial^2 u^*/\partial X^2 + 1/X \ \partial u^*/\partial X + k^* u^* = 0 \tag{3.197}$$

$$X^2 \partial^2 u^* /\partial X^2 + X \ \partial u^*/\partial X + X^2 k^* u^* = 0 \tag{3.198}$$

Eq. [3.198] can be recognized as the Bessel equation. The solution,

$$u^* = c_1 J_0(X\sqrt{k^*}) + c_2 Y_0(X\sqrt{k^*}) \tag{3.199}$$

It can be seen that $c_2 = 0$ as the concentration is finite at $X = 0$. The boundary condition for surface concentration is used to obtain c_1. Thus

$$c_1 = 1/ J_0(R\sqrt{k^*}/D\tau_r) \tag{3.200}$$

Thus,

$$u^* = J_0(X\sqrt{k^*}) /J_0(R\sqrt{k^*}/D\tau_r) \tag{3.201}$$

The surface to volume ratio needs to be maintained high. It can be seen that there exist a critical value of R above which the rate at which the neutrons are produced in the nuclear reaction is larger than the rate at which it is removed by diffusion. This will lead to a runaway condition in autocatalytic reactions. At the critical value,

$$(2R \ h \ \pi) \ \partial u/ \partial X \ C_s \ D/\sqrt{D\tau_r} = (\pi R^2 h) (k''' C) \tag{3.202}$$

$$\text{or} \quad R_{crit} = 4 \sqrt{D/k'''} \ J_1(R \sqrt{k'''/D}) /J_0(R \sqrt{k'''/D}) \tag{3.203}$$

Considering the average reaction rate instead,

$$R_{crit} = 2\sqrt{D/k'''} \ J_1(R \sqrt{k'''/D}) \tag{3.204}$$

Thus the transient concentration profile in a long cylindrical rod is obtained for simultaneous diffusion and autocatalytic reaction using the modified Fick's diffusion law. A lower limit on the radius is obtained to avoid cycling of concentration in the time domain during transience and an upper limit on the radius is obtained to avoid runaway condition from geometric shape effect. A lower limit on the radius of the long cylindrical rod subject to simultaneous autocatalytic reaction and diffusion exist and found to be $4.8096 \sqrt{D\tau}/(1 + k^{\cdot} \tau_r)$. A upper limit on the radius of the long cylindrical rod subject to simultaneous autocatalytic reaction and diffusion exist and found to be $2\sqrt{D/k'''} \ J_1$ (R

$\sqrt{(k'''/D)}$. The exact solution for transient concentration profile for simultaneous autocatalytic reaction and finite speed diffusion is derived by the method of separation of variables. It is a bifurcated solution. For certain values of λ_n the time portion of the solution is cosinous and damped and for others it is an infinite series of Bessel function of the first kind and zeroth order and decaying exponential in time. The upper limit on the radius was derived from the steady state solution by using the average concentration in the rod as well as the maximum reaction rate at the center of the rod. The surface to volume ratio or shape factor is an important consideration in the design of nuclear fuel rod.

3.8 Zone of Zero Transfer in a Ziegler Natta Polymerization Catalyst

In the gas solid fluidized bed Ziegler Natta coordination polymerization process to manufacture polypropylene often times small particle size catalyst is used. With reports of finite speed heat conduction and relaxation being significant by analogy there is increasing concern that the transient concentration can be affected by the finite speed diffusion and relaxation phenomena. This is very little attention paid to this in the literature. Fick's law of mass diffusion is valid only for low rate steady state transfer process. Physically, there exists a time where the linear mass diffusion relationship is not valid. This is because the Fick's law implicitly assumes an infinite speed of propagation of mass in any media. In reality, mass travels at a large but finite velocity. In this section an attempt is made to obtain an exact solution for the finite slab subject problem subject to an exponential decay of surface concentration in time at either of its boundaries using the method of transformation to the wave coordinates after removal of the damping term. Exponential decay in time at the boundary can be found for simple first order reaction at the surface of the solid. Consider a finite slab with an initial concentration at C_0. The initial condition is given as follows

$$t = 0, \ \forall x, \ C = C_0 \qquad (3.205)$$

Boundary Conditions in Space

$$t > 0, \ x = \pm a, \ C = C_0 + (C_s - C_0) \exp(-t/2\tau_r) \qquad (3.206)$$

$$t > 0, \ x = 0, \ \partial C/\partial x = 0 \qquad (3.207)$$

The fourth and final condition in time,

$$t = \infty, \ \forall x, \ C = C_0 \qquad (3.208)$$

A surface reaction is imposed on either ends of a finite slab for times greater than zero. The exponential decay of concentration serves as a time varying boundary condition. At infinite time the concentration is unchanged from the initial concentration. The governing equation can be obtained by a 1 dimensional mass balance [in – out + reaction = accumulation]. This is achieved by eliminating J_x between the damped wave diffusion and relaxation equation and the equation from mass balance [$-\partial J/\partial x = \partial C/\partial t$]. This is achieved by differentiating the damped wave equation with respect to x and the mass balance equation with respect to t and eliminating the second cross derivative of J with respect to r and time. Thus,

$$\tau_r \, \partial^2 C/\partial t^2 + \partial C/\partial t = D \, \partial^2 C/\partial x^2 \qquad (3.209)$$

Obtaining the dimensionless variables ;

$$u = (C - C_0)/(C_s - C_0); \ \tau = t/\tau_r \ ; \ X = x/sqrt(D\tau_r) \qquad (3.210)$$

The governing equation in the dimensionless form is then;

$$\partial u/\partial \tau + \partial^2 u/\partial \tau^2 = \partial^2 u/\partial X^2 \qquad (3.211)$$

Initially the damping term is eliminated using a $u = W \exp(-n\tau)$ substitution. Eq. [3.211] then becomes, At $n = \frac{1}{2}$;

$$W_{xx} = -W/4 + W_{\tau\tau} \qquad (3.213)$$

Eq. [3.213] can be solved by the following transformations to the wave coordinates or canonical form. Let the transformation be as follows;

$$\eta = \tau + X \qquad (3.214)$$

$$\xi = \tau - X \qquad (3.215)$$

Then equation [3.213] becomes;

$$\partial^2 W/\partial \tau^2 = \partial^2 W/\partial \eta^2 + \partial^2 W/\partial \xi^2 + 2\partial^2 W/\partial \xi \partial \eta \qquad (3.216)$$

$$\partial^2 W/\partial X^2 = \partial^2 W/\partial \eta^2 + \partial^2 W/\partial \xi^2 - 2\partial^2 W/\partial \xi \partial \eta \qquad (3.217)$$

$$4 \, \partial^2 W/\partial \xi \partial \eta = W/4 \qquad (3.218)$$

At this stage Laplace transforms is obtained in the ξ domain;

$$d\underline{W}/d\eta = \underline{W}/16/s \qquad \text{or the solution is}$$

$$\underline{W} = C' \exp \eta/16s \qquad (3.219)$$

At, $\xi = 0$, $\tau = X$. This is at the wave front. The time taken at a point x in the medium to see the disturbance at the boundary traveling with a speed of $sqrt[D/\tau_r]$. Only for times greater than this the problem exists. For times equal to and less than this time, the initial concentration remain and thus $W = u = 0$. Using the boundary condition given by Eq. [3.206] $W = \exp[-\tau/2]\exp[\tau/2] = 1$ at $\eta = a/\sqrt{D\tau_r} + \tau$.

$$1/s = C' \exp((a^* + \tau)/16s) \qquad (3.220)$$

where $a^* = a/\sqrt{D\tau_r}$

$$\text{or } C' = 1/s \exp(-(a^* + \tau)/16s) \qquad (3.221)$$

$$\text{Thus, } \underline{W} = 1/s \exp(\eta - \tau - a^*)/16s = 1/s \exp(X - a^*)/16s \qquad (3.222)$$

The inversion of Eq. [3.222] is within the tables available [Mickley, Sherwood and Reed, 1957]. So,

$$u = \exp(-\tau/2) J_0 (\tfrac{1}{2} \text{sqrt}((a^* - X)(\tau - X))) \qquad (3.223)$$

Eq. [3.223] can be rewritten for $X > \tau$ as;

$$u = \exp[-\tau/2] I_0 [\tfrac{1}{2} \text{sqrt}[[a^* - X][X - \tau]]] \qquad (3.224)$$

For $\tau > X$, $\qquad u = \exp(-\tau/2) J_0 (\tfrac{1}{2} \text{sqrt}((a^* - X)(\tau - X))) \qquad (3.225)$

It can be seen that at infinite time the expression becomes zero as specified. The expression is valid only for times greater than the penetration time $t > x/\text{sqrt}[D/\tau_r]$. For $X = a^*$, the imposed boundary condition $u = \exp[-\tau/2]$. By symmetry the solution from $-a \leq x \leq 0$ can be written for the concentration. So at the wave front $u = \exp[-\tau/2]$. The Bessel composite function of the first kind and zeroth order has many zeros and can be seen to be damped oscillatory. In order to amplify the wavy nature of the solution the exponential decay boundary condition was considered in this study.The plane of zero transfer can be identified and is when,

$$-(a^* + \tau)X + X^2 + a^*\tau = 23.132 \qquad (3.226)$$

$$\text{Thus } X = ((\tau + a^*) \pm \text{sqrt}((a^* + \tau)^2 - 4(a^*\tau - 23.132)))/2$$

There are two roots to the quadratic equation. The root occurs before the wavefront. This is an interesting result. This is similar to the Krogh solution of the anorexic region in tissue at steady state with Michael Menton kinetics and cylindrical geometry with two layers. It can be seen that for sufficient a^* there will be z zone near the center of the slab where there will be zero transfer. Not only is the boundary condition met there is a zone where the species will not be present.

The solution to the hyperbolic diffusion and relaxation equation in a finite slab with surface reaction at either of the boundaries was obtained by double transformation, of the governing equation, first into wave coordinates and then into Laplace domain . The solution is a product of decaying exponential in time and Bessel composite function of the first kind and zeroth order. At certain values of the composite function the value of the dimensionless concentraion becomes zero. This is identified as the plane of null transfer. This was found to occur before the wave front. Upon solving for the first zero of the Bessel composite function the spatio-temporal conditions when the zero condition happens is given by equation [3.226]. This is a quadratic equation with two roots. Both

186

roots seems significant and laden with meaning. Thus the Cattaneo and Vernotte equation can be used as a add-on to the Fick's law to describe shorter time scale events especially in a more physically realistic manner. A zone was identified in the center of the slab for sufficient width of the slab where the spcies will not migrate at all. Inorder to avoid this the slab has to be made small enough. However too small will result in subcritical damped oscillations as shown earlier [Sharma, 2003]. Therefore an optimal size of the catalyst need be identified for efficient operation.

The dimensionless concentration for a typical value of $a^* = 15$ and $\tau = 5$, as a function of the dimenionless distance in the half-width from 0 to a^* is shown in Figure 3.16. It can be seen that the zone of zero transfer exists between -3.06 to + 3.06 in dimensionless distance. In addition a maxima can be found in the expression for the concentration. This means that a backflux of mass is indicated. When exponential decay is faster compared with the diffusion and relaxation times, the surface concentration can become lower than the concentration in the interior of the slab close to the surface. This can cause a flow of mass in the direction toward the surface of the slab.

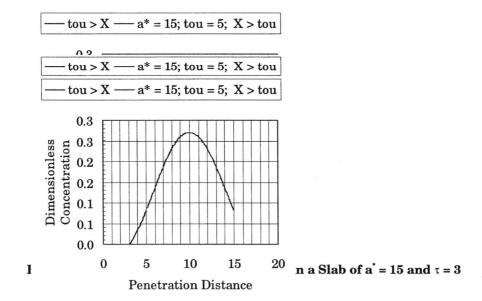

I   n a Slab of $a^* = 15$ and $\tau = 3$

Worked Example 3.5

In the analysis of the above section consider a first order reaction in the bulk of the catalyst and obtain the spatiotemporal patterns of the concentration in the slab.

The species reacts by a first order reaction at it comes in contact with the solid medium in addition to the surface reaction. The mass balance equation can be written as;

$$-\partial J/\partial x - k''' C = \partial C/\partial t \qquad (3.227)$$

where k''' is the first order reaction rate constant. Combining Eq. [3.227] with the damped wave diffusion and relaxation equation the governing equation is obtained. The damped wave equation is differentiated wrt to x to give;

$$\partial J/\partial x = -D_{AB}\partial^2 C/\partial x^2 - \tau_{mr} \partial^2 J/\partial t\partial x \qquad (3.228)$$

Eq. [3.227] is differentiated wrt to t to give;

$$-\partial^2 J/\partial x\partial t - k''' \partial C/\partial t = \partial^2 C/\partial t^2 \qquad (3.229)$$

Eliminating $\partial^2 J/\partial x\partial t$ between Eqs.[3.228, 3.229] ;

$$-k'''C - \partial C/\partial t = -D_{AB}\partial^2 C/\partial x^2 + \tau_{mr} (k''' \partial C/\partial t + \partial^2 C/\partial t^2) \qquad (3.230)$$

$$D_{AB}\partial^2 C/\partial x^2 = \tau_{mr} \partial^2 C/\partial t^2 + (1 + k''' \tau_{mr}) \partial C/\partial t + k''' C \qquad (3.231)$$

Eq. [3.231] can be made dimensionless by the following substitutions;

$$k^* = (k'''\tau_{mr}) ; \quad u = (C - C_0)/(C_s - C_0); \quad \tau = t/\tau_{mr}; \quad X = x/sqrt(D\tau_{mr}) \qquad (3.232)$$

$$\partial^2 u/\partial X^2 = \partial^2 u/\partial \tau^2 + (1 + k^*) \partial u/\partial \tau + k^* u \qquad (3.233)$$

The damping term is first removed by a $u = w \exp(-n\tau)$ substitution. Choosing n $= (1 + k^*)/2$ Eq. [3.233] becomes;

$$\partial^2 w/\partial X^2 = -w(1 - k^*)^2/4 + w_{\tau\tau} \qquad (3.234)$$

The initial condition is given as follows

$$t = 0, Vx, C = C_0 \qquad (3.235)$$

Boundary Conditions in Space

$$t > 0, \quad x = \pm a, \quad C = C_0 + (C_s - C_0) \exp(-(1 + k^* t/2\tau_r) \qquad (3.236)$$

$$t > 0, \quad x = 0, \quad \partial C/\partial x = 0 \qquad (3.237)$$

The fourth and final condition in time,

$$t = \infty, Vx, C = C_0 \qquad (3.238)$$

Let the transformation be as follows;

$$\eta = \tau + X \qquad (3.239)$$

$$\xi = \tau - X \qquad (3.240)$$

Then Eq. [3.234] becomes;

$$\partial^2 W/\partial \tau^2 = \partial^2 W/\partial \eta^2 + \partial^2 W/\partial \xi^2 + 2\partial^2 W/\partial \xi\partial\eta \qquad (3.241)$$

$$\partial^2 W/\partial X^2 \quad = \partial^2 W/\partial\eta^2 + \quad \partial^2 W/\partial\xi^2 \quad - 2\partial^2 W/\partial\xi\partial\eta \tag{3.242}$$

$$4\,\partial^2 W/\partial\xi\partial\eta \quad = W/4|1-k^*|^2 \tag{3.243}$$

At this stage Laplace transforms is obtained in the ξ domain;

$$d\underline{W}/d\eta \;=\underline{W}/16/|1-k^*|^2 s \quad \text{or the solution is}$$

$$\underline{W} \;=\; C' \exp\eta/\,[|1-k^*|^2 16s] \tag{3.244}$$

At, $\xi = 0$, $\tau = X$. This is at the wave front. The time taken at a point x in the medium to see the disturbance at the boundary traveling with a speed of sqrt[D/ τ_r]. Only for times greater than this the problem exists. For times equal to and less than this time, the initial concentration remain and thus $W = u = 0$. Using the boundary condition given by equation $W = \exp[-(1 + k^*)\tau/2]\exp[(1 + k^*)\tau/2] = 1$ at $\eta = a/\sqrt{D\tau_r} + \tau$.

$$1/s \;=\; C' \exp(\,(a^* + \tau)/\,(|1-k^*|^2 16s) \tag{3.245}$$

where $a^* = a/\sqrt{D\tau_r}$

$$\text{or } C' \;=\; 1/s\,\exp(-((a^* + \tau)/(|1-k^*|^2 16s) \tag{3.246}$$

Thus, $\underline{W} = 1/s\,\exp[\eta\text{- }\tau\text{ - }a^*\,]/16s = 1/s\,\exp[X\text{- }a^*\,]/\,[|1-k^*|^2 16s] \tag{3.247}$

The inversion of equation [3.247] is within the tables available [Mickley, Sherwood and Reed, 1957]. So,

$$u \;=\; \exp(-\tau/2)\,J_0\,(1/(2\,|1-k^*|)\mathrm{sqrt}((a^* \text{ - } X)(\tau \text{ - } X)) \tag{3.248}$$

Eq. [3.348] can be rewritten for $X > \tau$ as;

$$u \;=\; \exp(\tau/2)\,I_0\,(1/(2|1-k^*|)\,\mathrm{sqrt}((a^*\text{- }X)(X\text{ - }\tau)) \tag{3.249}$$

For $\tau > X$,

$$u \;=\; \exp(-\tau/2)\,J_0\,(1/(2|1-k^*|]\mathrm{sqrt}((a^*\text{- }X)(\tau\text{ - }X)) \tag{3.250}$$

The first zero of the Bessel function of the zeroth order occurs at 2.4048. When the argument in the composite Bessel function given in Eq. [3.250] reaches 2.4048 interesting things begin to happen. Thus when,

$$23.132\,|1-k^*|^2 \;=\; X^2 - X(a^* + \tau) + a^*\tau \tag{3.251}$$

$$\text{or } X^2 - X(a^* + \tau) + a^*\tau\text{ - }23.132\,|1-k^*|^2 \tag{3.252}$$

The roots of the quadratic equation given in Eq. [3.252] are;

$$[(a^* + \tau) \pm \text{sqrt} ((a^* + \tau)^2 - 4 a^* \tau + 92.528 |1 - k^*|^2]/2 \qquad (3.253)$$

Worked Example 3.6

The surface of a spherical drop of radius R, dissolves at a constant rate at D^S for times greater than zero. Obtain the spatio-temporal concentration in the surrounding medium due to migration of species by damped wave diffusion and relaxation.

The initial condition in the surrounding medium is;

$$t = 0, \qquad C = C_0 \qquad (3.254)$$

The final condition in the surrounding medium is;

$$t = \infty, \ C = C_s \qquad (3.255)$$

For times greater than zero,

$$r = \infty, \ C = C_0 \qquad (3.256)$$

From the moving boundary,

$$r = R - D^s t, \ C = C_s \qquad (3.257)$$

The governing equation in concentration is obtained by eliminating the second cross-derivative of mass flux wrt to r and t between the damped wave mass diffusion and relaxation equation and the mass balance equation in spherical coordinates. Considering a spherical shell of thickness Δr,

$$\Delta t[(4\pi r^2 J_r - 4\pi (r + \Delta r)^2 J_{r+\Delta r})] = 4\pi r^2 \Delta r \ \Delta C \qquad (3.258)$$

In the limit of Δr, Δt going to zero, the mass balance equation in spherical coordinates becomes;

$$-1/r^2 \partial(r^2 J_r)/\partial r = \partial C/\partial t \qquad (3.259)$$

The damped wave mass diffusion and relaxation equation written in the r direction only is;

$$J_r = -D_{AB}\partial C/\partial r - \tau_{mr} \partial J_r/\partial t \qquad (3.260)$$

Multiplying Eq. [3.260] by r^2 and differentiating wrt to r and then dividing by r^2,

$$(1/r^2)\partial(r^2 J_r)/\partial r = (-D_{AB}/r^2)\partial(r^2\partial C/\partial r)/\partial r - (\tau_{mr}/r^2) \partial^2(r^2 Jr)/\partial t\partial r$$

Differentiating Eq. [3.259] wrt t.

$$-1/r^2 \partial^2(r^2 Jr)/\partial t \partial r \ = \ \partial^2 C/\partial t^2$$

Substituting the above two Eqs. into Eq. [3.259] the governing equation in concentration is obtained as;

$$\partial C/\partial t + \tau_{mr} \ \partial^2 C/\partial t^2 \ = \ D_{AB}/r^2 \partial(\ r^2 \partial C/\partial r)/\partial r \qquad (3.261)$$

Obtaining the dimensionless variables,

$$u = (\ C - C_0)/C_s - C_0); \quad \tau = (t/\tau_{mr}); \quad X = r/sqrt(D_{AB}\ \tau_{mr}); \quad k^* = k''' \ \tau_{mr} \qquad (3.262)$$

The governing equation in the dimensionless form can be written as;

$$\partial u/\partial \tau \ + \ \partial^2 u/\partial \tau^2 \ = \ \partial^2 u/\partial X^2 \ + 2/X \ \partial u/\partial X \qquad (3.263)$$

The damping term is removed from the governing equation. This is done realizing that the transient concentration decays with time in an exponential fashion. The other reason for this manoeuvre is to study the wave equation without the damping term. Let $u = w \exp(-n\tau)$. Choosing $n = 1/2$ Eq. [3.263] becomes;

$$2/X \ \partial w/\partial X + \partial^2 w/\partial X^2 \ = \ -w/4 \ + w_{\tau\tau} \qquad (3.265)$$

Eq. [3.265] can be solved by using the method of relativistic transformation of coordinates. Consider the transformation variable for $\tau > X$ η as;

$$\eta \ = \ \tau^2 - X^2 \qquad (3.266)$$

As shown in the above sections, Eq. [3.265] becomes;

$$(\partial^2 w/\partial \eta^2)4(\ \tau^2 - X^2) \ + 8(\partial w/\partial \eta\ ) \ -w/4 = 0 \qquad (3.267)$$

$$4\eta^2 \partial^2 w/\partial \eta^2 + 8\eta \partial w/\partial \eta \ - \eta w/4 \ = 0 \qquad (3.268)$$

or

$$\eta^2 \partial^2 w/\partial \eta^2 + 2\eta \partial w/\partial \eta \ - \eta w/16 \ = 0 \qquad (3.269)$$

Comparing Eq. [3.269] with the generalized Bessel equation as given in Eq. [2.96], the solution is;

$$a = 2; b = 0; \ c = 0; \ d = -1/16; \ s = \tfrac{1}{2}$$

The order p of the solution is then $p = 2 \ sqrt(1/4) = 1$

Or

$$W \ = \ c_1 I_1(\ |1/2 \ sqrt(\tau^2 - X^2)/(\tau^2 - X^2)^{1/2} \ + \ c_2 K_1 \ 1/2 \ sqrt(\tau^2 - X^2)/(\tau^2 - X^2)^{1/2}$$

$$u \ = c_1 \exp(-\tau/2)I_1(\ |1/2 \ sqrt(\tau^2 - X^2)/(\tau^2 - X^2)^{1/2} \qquad (3.270)$$

c_2 can be seen to be zero as W is finite and not infinitely large at $\eta = 0$. An approximate solution can be obtaining by eliminating c_1 between the above equation and the equation from the boundary condition.

$$At\ X = X_R - tD^s/sqrt(D\tau_r) = X_R - \kappa\tau \tag{3.271}$$

$$u = 1\ or\ W = exp(\tau/2) \tag{3.272}$$

where $\kappa = D^s/sqrt(D/\tau_r) = $ Dissolution Rate/velocity of mass. Thus,

$$1 = exp(-\tau/2)c_1\ I_{1/}(|1/2\ sqrt(\tau^2 - (X_R^2 + \kappa^2\tau^2 - 2\kappa\tau X_R)/(\tau^2 - (X_R^2 + \kappa^2\tau^2 - 2\kappa\tau X_R))^{1/2} \tag{3.273}$$

Dividing Eq. [3.270] by Eq.[3.273]

$$u = [(\tau^2 - (X_R^2 + \kappa^2\tau^2 - 2\kappa\tau X_R))^{1/2} / (\tau^2 - X^2)^{1/2}][I_1(1/2\ sqrt(\tau^2 - X^2)/ I_{1/}(|1/2\ sqrt(\tau^2 - (X_R^2 + \kappa^2\tau^2 - 2\kappa\tau X_R)] \tag{3.274}$$

It can be noted that this is a mild function of time, however. As the general solution of PDE consists of n arbitrary functions when the order of the PDE is n compared with n arbitrary constants for ODE. In the limit of X_R going to zero, the expression becomes for $\tau > X$, $\kappa \neq 1$;

$$or\ u = [\tau(1 - \kappa^2)^{1/2}/(\tau^2 - X^2)^{1/2}][I_1(1/2sqrt(\tau^2 - X^2) / I_1[\tau(1 - \kappa^2)^{1/2}/2)] \tag{3.275}$$

For $X > \tau$, $\kappa \neq 1$

$$u = [\tau^2 - (X_R^2 + \kappa^2\tau^2 - 2\kappa\tau X_R))^{1/2} / (X^2 - \tau^2)^{1/2}][J_1(1/2\ sqrt(X^2 - \tau^2)/$$
$$I_{1/}(|1/2\ sqrt(\tau^2 - (X_R^2 + \kappa^2\tau^2 - 2\kappa\tau X_R)] \tag{3.276}$$
In the limit of X_R going to zero, the expression becomes;

$$or\ u = [\tau(1 - \kappa^2)^{1/2}/(X^2 - \tau^2)^{1/2}][J_1(1/2\ sqrt(X^2 - \tau^2)/ I_1(\tau((1 - \kappa)^{1/2})/2)] \tag{3.277}$$

Worked Example 3.7

Consider a finite slab of width 2a at an initial concentration of C_0 subject to a step change in concentration at either of its boundaries to C_s for times greater than zero. Using the method of Laplace transforms discuss the spatiotemporal concentration in the slab. Compare this solution that obtained for damped wave heat conduction and relaxation in Chapter 2. What is the surface flux ? What happens to the poor convergence at short times of the Fourier series solution as shown in Chapter 1 using the damped wave mass diffusion and relaxation equation ?

The governing equation for the 1 dimensional transient mass diffusion and relaxation without any chemical reaction can be written as shown in the above sections as;

$$\partial u/\partial \tau \ + \ \partial^2 u/\partial \tau^2 \ = \ \partial^2 u/\partial X^2 \qquad (3.278)$$

where, $\quad u = (C - C_0)/(C_s - C_0); \quad \tau = (t/\tau_{mr}); \quad X = r/sqrt(D_{AB} \ \tau_{mr})$ $\qquad (3.279)$

The two conditions in time and space domains respectively are as follows:

$$X = \pm a/sqrt(D_{AB} \ \tau_{mr}) \ = \pm X_a, \ u \ = 1$$

$$X = 0, \quad \partial u/\partial X = 0 \qquad (3.280)$$

$$\tau = 0, \quad u = 0 \qquad (3.281)$$

$$\tau = \infty, \quad u = 1 \qquad (3.282)$$

Obtaining the Laplace transform of Eq. [3.278]

$$\underline{u} (s + s^2) \ = \ d^2\underline{u}/dX^2 \qquad (3.283)$$

Solving the second order ODE with constant coefficients,

$$\underline{u} \ = A \exp(X(s + s^2)^{1/2}) \ + \ B \exp(-X(s + s^2)^{1/2})$$

From the boundary condition given by Eq. [3.280],

$$0 \ = (s + s^2)^{1/2} (A \exp(X(s + s^2)^{1/2}) \ - \ B \exp(X(s + s^2)^{1/2})) \qquad (3.284)$$

So A = B.

$$\underline{u} \ = A (\exp(X(s + s^2)^{1/2}) \ + \ \exp(-X(s + s^2)^{1/2}) \qquad (3.285)$$

From the Boundary Condition given in Eq. [3.359]

$$1/s \ = \ B \ CoshX_a(s + s^2)^{1/2}) \quad \text{or } B = 1/[s \ CoshX_a(s + s^2)^{1/2})] \qquad (3.386)$$

Thus, $\qquad \underline{u} \ = \ (\exp(X(s + s^2)^{1/2}) \ + \ \exp(-X(s + s^2)^{1/2})/[s \ CoshX_a(s + s^2)^{1/2})]$
$$\qquad (3.287)$$

The dimensionless heat flux in Laplace domain is given by;

$$\underline{q}^* \ = \ -1/(s+1)d\underline{u}/dX \qquad (3.288)$$

Plugging Eq. [3.287] in Eq. [3.288].

$$\underline{q}^* \ = \ -[(\exp(X(s + s^2)^{1/2}) \ - \ \exp(-X(s + s^2)^{1/2} \ 1/[s(s+1)]^{1/2}]/[Cosh(X_a(s + s^2)^{1/2})]$$
$$\qquad (3.289)$$

For small finite slabs the hyperbolic cosine function can be approximated by Taylor series expansion and with a truncation error of $O(X_a^5)$.

$$\text{Cosh}\,(X_a(s+s^2)^{1/2}) \;=\; 1 \;+\; X_a^3\,(s+s^2)^{3/2}/3! \tag{3.290}$$

The inversion of Eq. [3.289] after approximating the $\exp+X(s+s^2)$ as 1 and looking up the tables for the inversion of $\exp(-X(s+s^2)^{1/2}/[s(s+1)]^{1/2}$ then;

$$q^* \;=\; \exp(-\tau/2)\,[I_0(1/2\,(\tau^2-X^2)^{1/2}) + I_0(\tau/2)\,] \tag{3.291}$$

Eq. [3.291] is well bounded. This is a more realistic solution compared with the singularity of infinite flux seen in the surface flux of the Fourier series solution as shown in Chapter 1. The added terms that were truncated will give smaller contributions to the solution.

Worked Example 3.8

A container containing perfume was opened in the shopping mall in King of Prussia, at Pennsylvannia in USA. Write the governing equation for transient damped wave mass diffusion and relaxation in three dimensions using spherical coordinates. Consider an infinite sphere. Discuss the spatiotemporal concentration in 3 dimensions. What approximations are needed ?

The governing equation for the concentration when the mass balance equation and the constitutive damped wave diffusion and relaxation equation are combined and written after modification of the equation given in Cussler [1997] in three dimensions is as follows;

$$\tau_r\,\partial^2 C/\partial t^2 \;+\; \partial C/\partial t \;=\; 2D/r\partial C/\partial r \;+\; D\partial^2 C/\partial r^2 \;\; D\text{Cot}\theta/r^2\partial^2 C/\partial\theta^2 \;+\; D/r^2\text{Sin}^2\theta\,\partial^2 C/\partial\phi^2 \tag{3.292}$$

Let $u = C/C_s$; $X = r/\text{sqrt}(D\tau_r)$; $\tau = t/\tau_r$;

Then, $\quad \partial^2 u/\partial\tau^2 + \partial u/\partial\tau = 2/X\partial u/\partial X + \partial^2 u/\partial X^2 + 1/X^2\partial^2 u/\partial\theta^2 + 1/X^2\text{Sin}^2\theta\,\partial^2 u/\partial\phi^2 + \text{Cot}\theta/X^2\partial u/\partial\theta \tag{3.293}$

After removing the damping term by a $u = W\exp(-\tau/2)$ substitution,

$$\partial^2 W/\partial\tau^2 \;-\; W/4 \;=\; 2/X\partial W/\partial X \;+\; \partial^2 W/\partial X^2 \;+\; 1/X^2\partial^2 W/\partial\theta^2 \;+\; 1/X^2\text{Sin}^2\theta\,\partial^2 W/\partial\phi^2 + \text{Cot}\theta/X^2\partial W/\partial\theta \tag{3.294}$$

194

The "creeping transfer" assumption can be used and Eq. [3.294] can be rewritten as;

$$\partial^2 W/\partial \tau^2 - W/4 = 2/X \partial W/\partial X + \partial^2 W/\partial X^2 + 1/X^2 \partial^2 W/\partial \theta^2 + 1/X^2 Sin^2\theta \; \partial^2 W/\partial \phi^2 \quad (3.295)$$

$$Let \; \xi = \theta X, \; Then \; 1/X^2 \partial^2 W/\partial \theta^2 = \partial^2 W/\partial \xi^2 \quad (3.296)$$

$$\psi = \phi X Sin\theta, \; Then, \; 1/X^2 Sin^2\theta \; \partial^2 W/\partial \phi^2 = \partial^2 W/\partial \psi^2 \quad (3.297)$$

Eq. [3.296] then becomes;

$$\partial^2 W/\partial \tau^2 - W/4 = 2/X \partial W/\partial X + \partial^2 W/\partial X^2 + \partial^2 W/\partial \xi^2 + \partial^2 W/\partial \psi^2 \quad (3.298)$$

Consider the transformation, $\quad \eta = \tau^2 - X^2 - \xi^2 - \psi^2$

$$\partial W/\partial \tau = 2\tau \partial W/\partial \eta$$

$$\partial^2 W/\partial \tau^2 = 4\tau^2 \; \partial^2 W/\partial \eta^2 + 2\partial W/\partial \eta$$

$$\partial^2 W/\partial X^2 = 4X^2 \; \partial^2 W/\partial \eta^2 - 2\partial W/\partial \eta$$

$$2/X \partial W/\partial X = -4\partial W/\partial \eta$$

$$\partial^2 W/\partial \xi^2 = 4\xi^2 \; \partial^2 W/\partial \eta^2 + 2\partial W/\partial \eta$$

$$\partial^2 W/\partial \psi^2 = 4\psi^2 \qquad \partial^2 W/\partial \eta^2 - 2\partial W/\partial \eta$$

Eq. [3.298] becomes,

$$(\partial^2 w/\partial \eta^2)4(\tau^2 - X^2 - \xi^2 - \psi^2) + 12(\partial w/\partial \eta) - w/4 = 0 \quad (3.299)$$

$$or \; \eta^2(\partial^2 w/\partial \eta^2)) + 3\eta(\partial w/\partial \eta) - w\eta/16 = 0 \quad (3.300)$$

Comparing Eq. [3.300] with the generalized Bessel equation as given in Eq. [2.96], the solution is;
$$a = 3; b = 0; \; c = 0; \; d = -1/16; \; s = \tfrac{1}{2}. \text{ The order p of the solution is}$$
then $p = 2 \; sqrt(1) = 2$

$$W = c_1 I_2(1/2 \; sqrt(\tau^2 - X^2 - \xi^2 - \psi^2)/(\tau^2 - X^2 - \xi^2 - \psi^2) +$$
$$c_2 K_2 \; sqrt(\tau^2 - X^2 - \xi^2 - \psi^2)/(\tau^2 - X^2 - \xi^2 - \psi^2) \quad (3.301)$$

c_2 can be seen to be zero as W is finite and not infinitely large at $\eta = 0$. An approximate solution can be obtaining by eliminating c_1 between the above equation and the equation from the boundary condition.

$$1 = exp(-\tau/2) \; c_1 I_2(\tau/2)/(\tau^2) \quad (3.302)$$

Thus, $\quad u = [(\tau^2)/(\tau^2 - X^2 - \xi^2 - \psi^2)]\, I_2(1/2\ \text{sqrt}(\tau^2 - X^2 - \xi^2 - \psi^2)/\, I_2(\tau/2) \qquad (3.303)$

This is for $\tau > \text{sqrt}(X^2 - \xi^2 - \psi^2)$

For $X^2 + \xi^2 + \psi^2 > \tau^2$, in a similar fashion,

$\quad u = [(\tau^2)/(X^2 + \xi^2 + \psi^2 - \tau^2]\, J_2(1/2\ \text{sqrt}(X^2 + \xi^2 + \psi^2 - \tau^2)/\, I_2(\tau/2) \qquad (3.304)$

On examining Eq. [3.304] it can be seen that the Bessel function of the second order and first kind will go to zero at some value of η. The first root of the Bessel function occurs when

$$\tfrac{1}{2}(X^2 + \xi^2 + \psi^2 - \tau^2)^{1/2} = 5.1356 \qquad (3.305)$$

$$\text{Or} \quad X^2 + \theta^2 X^2 + \phi^2 Sin^2\theta X^2 - \tau^2 = 105.498 \qquad (3.306)$$

When an exterior point in the infinite sphere is considered a lag time can be calculated prior to which there is no mass transfer to that point. After the lag time there exists two regimes. One is described by Eq. [3.304] and the third regime is described by Eq. [3.302]. Thus,

$$\tau_{lag} = \text{sqrt}(\,X_p^2\,(1 + \theta_p^2 + \phi_p^2 Sin^2\theta p - 105.498) \qquad (3.307)$$

Worked Example 3.9

For the problem given in Worked Example 3.8 derive the solution relaxing the creeping transfer assumption. Consider an infinite sphere. Discuss the spatiotemporal concentration in 3 dimensions. What approximations are needed?

The governing equation in spherical coordinates in three dimensions after non-dimensionalizing and removal of the damping term is found to be;

$$\partial^2 W/\partial\tau^2 - W/4 = 2/X\partial W/\partial X + \partial^2 W/\partial X^2 + 1/X^2\partial^2 W/\partial\theta^2 + 1/X^2 Sin^2\theta\ \partial^2 W/\partial\phi^2 + Cot\theta/X^2\partial W/\partial\theta \qquad (3.308)$$

$$\text{Let } \xi = \theta X, \quad \text{Then } 1/X^2\partial^2 W/\partial\theta^2 = \partial^2 W/\partial\xi^2 \qquad (3.309)$$

$$\psi = \phi X Sin\theta, \text{ Then, } 1/X^2 Sin^2\theta\ \partial^2 W/\partial\phi^2 = \partial^2 W/\partial\psi^2 \qquad (3.310)$$

$Cot\theta$ for small θ will approximate to a $1/Sin\theta$ which can be written as $1/\theta$
Eq. [3.308] then becomes;

$$\partial^2 W/\partial\tau^2 - W/4 = 2/X\partial W/\partial X + \partial^2 W/\partial X^2 + \partial^2 W/\partial\xi^2 + \partial^2 W/\partial\psi^2 + 1/\xi\partial W/\partial\xi \quad (3.311)$$

Consider the transformation, $\quad \eta = \tau^2 - X^2 - \xi^2 - \psi^2 \quad\quad (3.312)$

$$\partial W/\partial\tau = 2\tau\partial W/\partial\eta$$

$$\partial^2 W/\partial\tau^2 = 4\tau^2 \partial^2 W/\partial\eta^2 + 2\partial W/\partial\eta$$

$$\partial^2 W/\partial X^2 = 4X^2 \partial^2 W/\partial\eta^2 - 2\partial W/\partial\eta$$

$$2/X\partial W/\partial X = -4\partial W/\partial\eta$$

$$\partial^2 W/\partial\xi^2 = 4\xi^2 \partial^2 W/\partial\eta^2 + 2\partial W/\partial\eta$$

$$\partial^2 W/\partial\psi^2 = 4\psi^2 \quad\quad \partial^2 W/\partial\eta^2 - 2\partial W/\partial\eta$$

Eq. [3.311] becomes,

$$(\partial^2 w/\partial\eta^2)4(\tau^2 - X^2 - \xi^2 - \psi^2) + 14(\partial w/\partial\eta) \quad\quad -w/4 = 0 \quad\quad (3.313)$$

or $\eta^2(\partial^2 w/\partial\eta^2)) + 7/2\eta(\partial w/\partial\eta) - w\eta/16 = 0 \quad\quad (3.314)$

Comparing Eq. [3.314] with the generalized Bessel equation as given in Eq. [2.96], the solution is;

$$a = 7/2; \, b = 0; \, c = 0; \, d = -1/16; \, s = \tfrac{1}{2}$$

The order p of the solution is then $p = 5/2$

Or $\quad W = c_1 I_{5/2}(1/2 \text{ sqrt}(\tau^2 - X^2 - \xi^2 - \psi^2)/(\tau^2 - X^2 - \xi^2 - \psi^2)^{5/4}$
$\quad\quad\quad + c_2 I_{-5/2} 1/2\text{sqrt}(\tau^2 - X^2 - \xi^2 - \psi^2)/(\tau^2 - X^2 - \xi^2 - \psi^2)^{5/4} \quad\quad (3.315)$

c_2 can be seen to be zero as W is finite and not infinitely large at $\eta = 0$. An approximate solution can be obtaining by eliminating c_1 between the above equation and the equation from the boundary condition at the point source at the center of the sphere, i.e., at the perfume container.

$$1 = \exp(-\tau/2) c_1 I_{5/2}(\tau/2)/(\tau^{5/2}) \quad\quad (3.316)$$

Thus, $u = [(\tau^{5/2})/ (\tau^2 - X^2 - \xi^2 - \psi^2)^{5/4}] I_{5/2}(1/2 \text{ sqrt}(\tau^2 - X^2 - \xi^2 - \psi^2)/ I_{5/2}(\tau/2) \quad\quad (3.317)$

This is for $\tau > \text{sqrt}(X^2 - \xi^2 - \psi^2)$

For $X^2 + \xi^2 + \psi^2 > \tau^2$, in a similar fashion,

$$u = [(\tau^{5/2})/ (X^2 + \xi^2 + \psi^2 - \tau^2)^{5/4}] J_{5/2}(1/2 \text{ sqrt}(X^2 + \xi^2 + \psi^2 - \tau^2)/ I_{5/2}(\tau/2) \quad\quad (3.318)$$

Worked Example 3.10

The rate of permeation of a particular medicine in the human body is of interest in drug delivery systems. The dissolution of the drug is often times governed by mass diffusion and relaxation. The time scales involved with these systems are short that the ballistic term may become significant. The objective of drug delivery system design is to increase the amount of drug dissolved. When a pill is taken, after a said period of time the dissolution of pill will reach a steady state. The time required to reach this steady supply of drug is of interest. It is assumed that the dissolution of this pill is controlled by diffusion into the stagnant contents of the human anatomy. The dissolution is diffusion controlled and the surroundings are stagnant.

A mass balance on the spherical shell around the pill can be written and when combined with the damped wave diffusion and relaxation equation can be written as;

$$\tau_r \partial^2 C/\partial t^2 + \partial C/\partial t = D/r^2 \, \partial/\partial r(r^2 \, \partial C/\partial r) \qquad (3.319)$$

$$\text{Let } u = (C - C_0)/(C_{sat} - C_0); \; ; \; \tau = t/\tau_r; X = r/sqrt(D\tau_r) \qquad (3.320)$$

Eq. [3.319] becomes;

$$\partial^2 u/\partial \tau^2 + \partial u/\partial \tau = 1/X^2 \, \partial/\partial X(X^2 \, \partial u/\partial X) \qquad (3.321)$$

The time and space conditions can be written as;

$$\tau = 0, \; u = 0 \qquad (3.322)$$

$$\tau = \infty, \; u = 1 \qquad (3.323)$$

$$\tau > 0, \; X = X_{R0}, \; u = 1 \qquad (3.324)$$

$$X = \infty, \; u = 0 \qquad (3.325)$$

Consider the substitution, $V = u/X$. Eq. [3.321] becomes,

$$\partial^2 V/\partial \tau^2 + \partial V/\partial \tau = 2V/X^2 + 4/X\partial V/\partial X + \partial^2 V/\partial X^2 \qquad (3.327)$$

The damping term can be removed by a $u = wexp(-n\tau)$ substitution. As shown in the preceding sections for $n = \frac{1}{2}$, Eq. [3.327] becomes.

$$\partial^2 W/\partial \tau^2 - W/4 = 2W/X^2 + 4/X\partial W/\partial X + \partial^2 W/\partial X^2 \qquad (3.328)$$

Let $\eta = \tau^2 - X^2$

The term $2W/X^2$ can be neglected for large X. W is small for large r as $u = W\exp(-\tau/2)/r$. As shown in the above section Eq. [3.328] for large X can be written as;

Now, $4/X\partial W/\partial X \quad = -8\partial W/\partial\eta$

$$4\eta\partial^2 W/\partial\eta^2 + 12\partial W/\partial\eta \qquad - W/4 \; = 0 \tag{3.329}$$

$$\text{Or } \eta^2\partial^2 W/\partial\eta^2 + 3\eta\partial W/\partial\eta \qquad - \eta W/16 \; = 0 \tag{3.330}$$

Comparing Eq. [3.330] with the generalized Bessel equation as given in Eq. [2.96], the solution is;
$$a = 3; b = 0; \; c = 0; \; d = -1/16; \; s = \tfrac{1}{2}$$

The order p of the solution is then $p = 2 \, \text{sqrt}(1) = 2$

$$\text{Or} \qquad W \; = \; c_1 I_2(1/2 \, \text{sqrt}(\tau^2 - X^2)/(\tau^2 - X^2) + c_2 K_2 \, 1/2\text{sqrt}(\tau^2 - X^2)/(\tau^2 - X^2) \tag{3.331}$$

c_2 can be seen to be zero as W is finite and not infinitely large at $\eta = 0$.

$$V \; = \exp(-\tau/2) \, c_1 I_2(1/2 \, \text{sqrt}(\tau^2 - X^2)/(\tau^2 - X^2) \tag{3.332}$$

An approximate solution can be obtaining by eliminating c_1 between the above equation and the equation from the boundary condition.

$$1/X_{R0} \; = \; \exp(-\tau/2) \, c_1 I_2(1/2\text{sqrt}(\tau^2 - X_{R0}^{\;2})/(\tau^2 - X_{R0}^{\;2}) \tag{3.333}$$

Thus for $\tau > X$

$$V \; = \; (1/X_{R0}) \, [(\tau^2 - X_{R0}^{\;2})/ (\tau^2 - X^2)] \, [I_2(1/2 \, \text{sqrt}(\tau^2 - X^2)/ I_2 1/2(\text{sqrt}(\tau^2 - X_{R0}^{\;2}) \tag{3.334}$$

For $X > \tau$,

$$u \; = \; (X/X_{R0}) \, [(\tau^2 - X_{R0}^{\;2})/ (X^2 - \tau^2)] \, J_2(1/2 \, \text{sqrt}(X^2 - \tau^2) / I_2 1/2\text{sqrt}(\tau^2 - X_{R0}^{\;2}) \tag{3.335}$$

On examining Eq. [3.335] it can be seen that the Bessel function of the second order and first kind will go to zero at some value of η. The first root of the Bessel function occurs when
$$\tfrac{1}{2}(X^2 - \tau^2)^{1/2} = 5.1356 \tag{3.336}$$

$$\text{Or } X^2 - \tau^2 = 105.498 \tag{3.337}$$

When an exterior point in the infinite sphere is considered a lag time can be calculated prior to which there is no mass transfer to that point. After the lag time there exists two regimes. One is described by Eq. [3.335] and the third regime is described by Eq. [3.334]. Thus,

$$\tau_{lag} = sqrt(X_p^2 - 105.498)$$

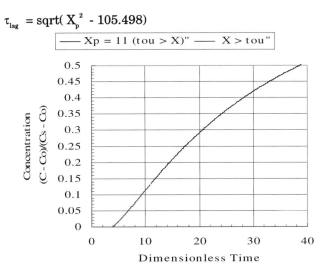

Figure 3.17 Three Regimes of Dimensionless Concentration at a Exterior Point from a Pill

Worked Example 3.11

The EPA mandate for California Phase II regulations on automobile engines from 2006 and beyond require that the sulfur content of gasoline to be less than 40 ppm. The organic sulfides content in the gasoline have to be reduced to a few parts per billion. The existing technology for catalytic converter and tailpipe emissions has to be refined further. A process development team at a major corporation making catalysts has designed a bed of Palladium to adsorb the organic sulfides present in the gasoline. Derive the concentration profile at steady state.

Applications such as the chemisorption of organic sulfides onto an adsorbent bed of Palladium to effect desulfurization of automobile fuel gasoline involve simultaneous molecular diffusion and convection and simple reaction. The thiopenes, sulfides and disulfides present in gasoline prior to operation in an internal combustion engine can be treated in a chemisorption bed of Palladium or a mixture of Nickel and Palladium. Ascertaining the time scale of scrutiny the wave diffusion and relaxation may become important. Chemisorption can be accounted for by a first order reaction rate. Usually the reaction results in a net removal and a net minus sign are used in the mass balance equation to denote the reaction rate. When they are is an autocatalytic effect in the adsorbent a net generation is denoted by a plus sign.

Let $u = y/y^*$ where y^* is the equilibrium adsorbent at the entrance of the adsorbent. The governing equation at steady state for autocatalytic chemisorption can be written as;

$$d^2u/dX^2 + Pe_m \, du/dX + k^* u = 0 \qquad (3.339)$$
where,
$$X = x/(D\tau_r)^{1/2}; \quad Pe_m = v/v_m = v/(D/\tau_r)^{1/2}; \quad k^* = (k'''\tau_r) \qquad (3.340)$$

The boundary conditions for an infinite adsorbent bed can be written as:

$$X = 0, \ u = 1 \qquad (3.341)$$

$$X = \infty, \ u = 0 \qquad (3.342)$$

Pe_m is the Peclet number (mass) and gives the ratio of the convective flow velocity to the speed of mass diffusion and relaxation. k^* is the dimensionless reaction rate constant which is a product of the first order autocatalytic reaction rate and the mass relaxation time. X is the dimensionless distance. The solution to the second order ODE with constant coefficients in Eq. [3.339] after setting one of the constants of integration to zero from the boundary condition at infinite adsorbent depth, can be written as;

$$u = c_1 \exp(-Pe_m X/2) \exp[-X/2 \, sqrt(Pe_m^2 - 4k^*)] \qquad (3.343)$$

c_1 can be seen to be 1 from the boundary condition given by Eq. [3.341], when, $Pe_m < 2k^*$, the concentration can be seen to cosinous.

$$u = \exp(-PeX/2)\cos[X/2\sqrt{(4k^* - Pe^2)}] \qquad (3.344)$$

A critical length beyond which the concentration will be zero can be calculated. This is when,

$$X_{crit} = \pi/\sqrt{(4k^* - Pe^2)} \qquad (3.345)$$

If the chemisorption rate is a net removal of the species then the governing equation can be written as,

$$d^2u/dX^2 + Pe\, du/dX - k^* u = 0 \qquad (3.346)$$

The solution for the concentration profile will then be given by;

$$u = \exp(-PeX/2)\exp[-X/2\sqrt{(Pe^2 + 4k^*)}] \qquad (3.347)$$

Worked Example 3.12

In the dissolving pill problem consider all three dimensions of the spherical coordinates. Use the $V = u/r$ substitution if necessary. Discuss the spatiotemporal concentration in the infinite sphere.

The governing equation for the concentration when the mass balance equation and the constitutive damped wave diffusion and relaxation equation are combined and written after modification of the equation given in Cussler [1997] in three dimensions is as follows;

$$\text{Let } u = (C - C_0)/(C_s - C_0); \quad X = r/\sqrt{(D\tau_r)}; \quad \tau = t/\tau_r; \qquad (3.348)$$

Then the governing equation in three dimensions in spherical coordinates can be written as;

$$\partial^2u/\partial\tau^2 + \partial u/\partial\tau = 2/X\partial u/\partial X + \partial^2u/\partial X^2 + 1/X^2\partial^2u/\partial\theta^2 + 1/X^2\sin^2\theta\, \partial^2u/\partial\phi^2 + \text{Cot}\theta/X^2\partial u/\partial\theta \qquad (3.349)$$

Consider the substitution, $V = u/X$. Eq. [3.448] becomes,

$$\partial^2V/\partial\tau^2 + \partial V/\partial\tau = 2V/X^2 + 4/X\partial V/\partial X + \partial^2V/\partial X^2 + 1/X^2\partial^2V/\partial\theta^2 + 1/X^2\sin^2\theta\, \partial^2V/\partial\phi^2 + \text{Cot}\theta/X^2\partial V/\partial\theta \qquad (3.351)$$

$$\partial^2V/\partial\tau^2 + \partial V/\partial\tau = 2V/X^2 + 4/X\partial V/\partial X + 1/X\partial^2V/\partial X^2 + 1/X^2\partial^2V/\partial\theta^2 + 1/X^2\sin^2\theta\, \partial^2V/\partial\phi^2 + \text{Cot}\theta/X^2\, \partial V/\partial\theta \qquad (3.352)$$

The damping term can be removed by a $V = w\exp(-n\tau)$ substitution. As shown in the preceding sections for $n = \frac{1}{2}$, Eq. [3.352] becomes.

$$\partial^2 W/\partial \tau^2 - W/4 \quad = \quad 2W/X^2 + 4/X\partial W/\partial X + \partial^2 W/\partial X^2 + 1/X^2 \partial^2 W/\partial \theta^2 + 1/X^2 Sin^2\theta$$
$$\partial^2 W/\partial \phi^2 \quad + Cot\theta/X^2 \partial W/\partial \theta \tag{3.352}$$

For small θ,

$$\partial^2 W/\partial \tau^2 - W/4 = 2W/X^2 + 4/X\partial W/\partial X + \partial^2 W/\partial X^2 + 1/X^2 \partial^2 W/\partial \theta^2 + 1/X^2 Sin^2\theta \, \partial^2 W/\partial \phi^2$$
$$+ 1/\theta X^2 \, \partial W/\partial \theta \tag{3.353}$$

$$\text{Let } \xi = \theta X, \text{ Then } 1/X^2 \partial^2 W/\partial \theta^2 = \partial^2 W/\partial \xi^2 \tag{3.354}$$

$$\psi = \phi X Sin\theta, \text{ Then, } 1/X^2 Sin^2\theta \, \partial^2 W/\partial \phi^2 = \partial^2 W/\partial \psi^2 \tag{3.355}$$

Eq. [3.353] then becomes for large X,

$$\partial^2 W/\partial \tau^2 - W/4 = 4/X\partial W/\partial X + \partial^2 W/\partial X^2 + \partial^2 W/\partial \xi^2 + \partial^2 W/\partial \psi^2 + 1/\xi \partial W/\partial \xi \tag{3.356}$$

Consider the transformation, $\eta = \tau^2 - X^2 - \xi^2 - \psi^2$

As shown in the analysis in Worked Example 3.10 the derivatives in Eq. [3.356] in 4 variables become converted into 1 variable,

$$(\partial^2 w/\partial \eta^2)4(\tau^2 - X^2 - \xi^2 - \psi^2) + 18(\partial w/\partial \eta) \quad -w/4 = 0 \tag{3.358}$$

$$\text{or } \eta^2(\partial^2 w/\partial \eta^2)) + 9/2\eta(\partial w/\partial \eta) - w\eta/16 = 0 \tag{3.359}$$

Comparing Eq. [3.359] with the generalized Bessel equation as given in Eq. [2.96], the solution is;
$$a = 9/2; \; b = 0; \; c = 0; \; d = -1/16; \; s = \tfrac{1}{2}$$

The order p of the solution is then $p = 7/2$

$$W = c_1 I_{7/2}(1/2 \, sqrt(\tau^2 - X^2 - \xi^2 - \psi^2)/(\tau^2 - X^2 - \xi^2 - \psi^2) + c_2 I_{-7/2} \, sqrt(\tau^2 - X^2 - \xi^2 - \psi^2)/(\tau^2 - X^2 - \xi^2 - \psi^2) \tag{3.360}$$

$$\text{Or} \quad u = X exp(-\tau/2) \, c_1 I_{7/2}(1/2 \, sqrt(\tau^2 - X^2 - \xi^2 - \psi^2)/(\tau^2 - X^2 - \xi^2 - \psi^2) \tag{3.361}$$

c_2 can be seen to be zero as W is finite and not infinitely large at $\eta = 0$. An approximate solution can be obtaining by eliminating c_1 between the above equation and the equation from the boundary condition. The equation from the boundary condition can be written as;

$$1 = X_{R0} \, exp(-\tau/2) \, c_1 I_{7/2}(1/2 \, sqrt(\tau^2 - X_{R0}^2)/(\tau^2 - X_{R0}^2) \tag{3.362}$$

Dividing Eq. [3.361] by Eq. [3.360],

$$u = (X/X_{R0}) \, [(\tau^2 - X_{R0}^2)/(\tau^2 - X^2 - \xi^2 - \psi^2)] I_{7/2}(1/2 \, sqrt(\tau^2 - X^2 - \xi^2 - \psi^2)/I_{7/2}(1/2 sqrt(\tau^2 - X_{R0}^2) \tag{3.363}$$

For small X,

$$u = (X/X_{R0}) [(\tau^2 - X_{R0}{}^2)/(X^2 + \xi^2 + \psi^2 - \tau^2] J_{7/2}(1/2 \text{ sqrt}(X^2 + \xi^2 + \psi^2 - \tau^2)/I_{7/2}(1/2\text{sqrt}(\tau^2 -X_{R0}{}^2) \quad (3.364)$$

In the creeping mass transfer limit Eq. [3.356] can be approximated as;

$$\partial^2 W/\partial\tau^2 - W/4 = 4/X\partial W/\partial X + \partial^2 W/\partial X^2 + \partial^2 W/\partial\xi^2 + \partial^2 W/\partial\psi^2 \quad (3.365)$$

After the transformation the PDE with 4 variables is converted to a Bessel equation in 1 variable:

$$(\partial^2 w/\partial\eta^2)4(\tau^2 - X^2 - \xi^2 - \psi^2) + 16(\partial w/\partial\eta) \qquad -w/4 = 0 \quad (3.366)$$

$$\text{or } (\partial^2 w/\partial\eta^2)\eta^2 + 4\eta(\partial w/\partial\eta) -\eta w/16 = 0 \quad (3.367)$$

The order of the Bessel solution for Eq. [3.367] can be calculated by comparing Eq. [3.367] with the generalized Bessel equation given in Eq. [2.96] and the solution is; a = 4; b = 0; c = 0; d = -1/16; s = ½ . The order p of the solution is then p = 3

$$W = c_1 I_3(1/2 \text{ sqrt}(\tau^2 - X^2 - \xi^2 - \psi^2)/(\tau^2 - X^2 - \xi^2 - \psi^2) + c_2 K_3 \text{ sqrt}(\tau^2 - X^2 - \xi^2 - \psi^2)/(\tau^2 - X^2 - \xi^2 - \psi^2) \quad (3.368)$$

$$\text{Or} \quad u = X\exp(-\tau/2) c_1 I_3(1/2 \text{ sqrt}(\tau^2 - X^2 - \xi^2 - \psi^2)/(\tau^2 - X^2 - \xi^2 - \psi^2) \quad (3.369)$$

c_2 can be seen to be zero as W is finite and not infinitely large at $\eta = 0$. An approximate solution can be obtaining by eliminating c_1 between the above equation and the equation from the boundary condition. The equation from the boundary condition can be written as;

$$1 = X_{R0} \exp(-\tau/2) c_1 I_3(1/2 \text{ sqrt}(\tau^2 - X_{R0}{}^2)/ (\tau^2 - X_{R0}{}^2) \quad (3.370)$$

Dividing Eq. [3.369] by Eq. [3.370],

$$u = (X/X_{R0}) [(\tau^2 - X_{R0}{}^2)/(\tau^2 - X^2 - \xi^2 - \psi^2)] I_3(1/2 \text{ sqrt}(\tau^2 - X^2 - \xi^2 - \psi^2)/I_3(1/2\text{sqrt}(\tau^2 - X_{R0}{}^2) \quad (3.371)$$

For small X,

$$u = (X/X_{R0}) [(\tau^2 - X_{R0}{}^2)/(X^2 + \xi^2 + \psi^2 - \tau^2] J_3(1/2 \text{ sqrt}(X^2 + \xi^2 + \psi^2 - \tau^2)/I_3(1/2\text{sqrt}(\tau^2 - X_{R0}{}^2) \quad (3.372)$$

In the limit of zero radius of the dissolving pill,

$$u = (X/X_{R0}) [\tau^2/(\tau^2 - X^2 - \xi^2 - \psi^2)] I_3(1/2 \text{ sqrt}(\tau^2 - X^2 - \xi^2 - \psi^2)/I_3(\tau/2) \quad (3.373)$$

For small X,

$$u = (X/X_{R0})\ [(\tau^2)/(\ X^2 + \xi^2 + \psi^2 - \tau^2]\ J_3(1/2\ sqrt(X^2 + \xi^2 + \psi^2 - \tau^2)/I_3(\tau/2) \qquad (3.374)$$

The solution is in terms of a Bessel composite function of the third order and first kind for small X and a modified Bessel composite function of the third order and first kind for times greater than X. The first root of the Bessel function of the third order was calculated by using 17 terms of the series expansion of the Bessel function in a Pentium IV microprocessor using a Microsoft Spreadsheet upto 4 decimal places. The root was found to be 6.3802.

$$\tfrac{1}{2}(X^2 + \xi^2 + \psi^2 - \tau^2\ )^{1/2} = 6.3802 \qquad (3.375)$$

$$\text{Or}\quad X^2 + \xi^2 + \psi^2 - \tau^2 = 162.828 \qquad (3.376)$$

When an exterior point in the infinite sphere is considered a lag time can be calculated prior to which there is no mass transfer to that point. After the lag time there exists two regimes. One is described by Eq. [3.374] and the third regime is described by Eq. [3.373]. Thus,

$$\tau_{lag} = sqrt(\ X_p^{\ 2} + \xi_p^{\ 2} + \psi_p^{\ 2} - 162.828) \qquad (3.377)$$

3.9 Diffusion, Relaxation and Convection

Convection occurs in parallel to diffusion and relaxation in problems such as fast evaporation. Considering the time scales of scrutiny, the damped wave diffusion and relaxation may be of increased relevance. Consider the fast evaporation by diffusion and convection. The mass balance equation in 1 dimension can be written as;

$$-\partial J_z/\partial z - V_z \partial C/\partial z = \partial C/\partial t \qquad (3.378)$$

The velocity V_z is taken in the direction of diffusion. The damped wave diffusion and relaxation equation can be written as;

$$J_z = -D\partial C/\partial z - \tau_{mr}\partial J_z/\partial t \qquad (3.379)$$

Inorder to eliminate the J between Eqs. [3.378, 3.379], Eq. [3.378] is differentiated wrt t to give;

$$-\partial^2 J_z/\partial z\partial t - V_z\partial^2 C/\partial z\partial t = \partial^2 C/\partial t^2 \qquad (3.380)$$

Differentiating Eq. [3.380] wrt z,

$$\partial J_z/\partial z = -D\partial^2 C/\partial z^2 - \tau_{mr}\partial^2 J_z/\partial t\partial z \qquad (3.381)$$

Eliminating the second cross derivative of the flux J wrt time and position,

$$\tau_{mr} V_z \partial^2 C/\partial z \partial t + \tau_{mr} \partial^2 C/\partial t^2 + V_z \partial C/\partial z + \partial C/\partial t = D\, \partial^2 C/\partial z^2 \qquad (3.382)$$

Obtaining the dimensionless form by the following substitutions;

$$X= z/sqrt(D\tau_r); \quad \tau = t/\tau_r; \quad u = (C- C_s)/(C_0 - C_s); \quad k^* = k'''\tau_{mr}; \quad Pe_m = V_z/sqrt(D/\tau_{mr}) \qquad (3.383)$$

Eq. [3.382] becomes;

$$Pe_m \partial^2 u/\partial X \partial \tau + \partial^2 u/\partial \tau^2 + Pe_m \partial u/\partial X + \partial u/\partial \tau = \partial^2 u/\partial X^2 \qquad (3.384)$$

Obtaining the Laplace transform of the Eq. [3.384],

$$d^2 \underline{u}/dX^2 - (s + 1)Pe_m d\, \underline{u}/dX - \underline{u}\,(s^2 + s) = 0 \qquad (3.385)$$

The solution to the second order ODE in Laplace domain can be written as;

$$\text{``}b^2 - 4ac\text{''} = (s+1)^2 + 4s(s+1) = (s+1)(5s + 1) \qquad (3.386)$$

$$= 5\,[(\,s + 3/5)^2\, -4/25) \qquad (3.387)$$

$$\underline{u} = A\, exp(X(s+1)Pe_m/2)exp(+Xsqrt[(s+1)(5s+1)] + \\ Bexp(X(s+1)Pe_m/2)exp(-Xsqrt[(s+1)(5s+1)] \qquad (3.388)$$

At $X = \infty$, $u = 0$ and hence $A = 0$. From the boundary condition at $X = 0$, $\underline{u} = 1/s$. Thus,

$$\underline{u} = 1/sexp(X(s+1)Pe_m/2)exp(-X5^{1/2}\, sqrt[(s+3/5)^2 - 4/25] \qquad (3.389)$$

Multiplying both the numerator and denominator of Eq. [3.389] by sqrt((s + 3/5)2 – 4/25) and expressing the transformed expression as a product of h(s)g(s)M(s)

$$h(\tau) = L^{-1} h(s) = L^{-1}\, (1/\, sqrt[(s+3/5)^2 - 4/25]exp(-X5^{1/2}\, sqrt[(s+3/5)^2 - 4/25]$$

$$= exp(-3\tau/5)/\, I_0\, 2/5sqrt(\tau^2 - 5X^2) \qquad (3.390)$$

$$L^{-1} g(s) = L^{-1}(+1/5\, sqrt((s + 3/5)^2 - 4/25)) = 1/\,5exp(-3\tau/5)I_0\,(2\tau/5) \qquad (3.391)$$

$$([(s+3/5)^2 - 4/25]\ \text{can be approximated as} = 1/5$$

$$L^{-1} M(s) = L^{-1}\, 1/s\, exp(X(s+1) = exp(X)S_X(\tau) = exp(X), \quad \tau > 0 \qquad (3.392)$$

Thus using the convolution theorem;

$$\tau \qquad q$$

206

$$u = \int_0 (\int_0 \exp(X) \exp(-3p5) I_0 (2/5) dp) \quad \exp(-3(\tau-q)/5)/ I_0 2/5 \mathrm{sqrt}((\tau - q)^2 - 5X^2) dq$$
(3.393)

The evaporation rate at the boundary can be given by the flux at $X = 0$. Thus,

$$J^* = (5s + 1)^{1/2}/(s+1)^{1/2}/s + 1/s$$
(3.394)

$$= [1 + 4s/(s+1)]^{1/2}/s + 1/s$$
(3.395)

Expanding by a binomial infinite series;

$$= 2/s + \tfrac{1}{2} 4/(s+1) - 1/4 \, 16s/(s+1)^2 + \ldots$$
(3.396)

Inversion of each term in the infinite series,

$$= 2 + 2 \exp(-\tau) - 4 \exp(-\tau)(1 - \tau) + \ldots$$
(3.397)

Worked Example 3.13

The sequential separation of whey proteins is often effected by radial chromatography [Mozaffar, Salah, Saxena and Miranda, 2000]. In 1994 the USA exports of dairy products amounted to $717 million. A part of the success is due to the improved cattle breeding and milk processing capabilities. Cheese is made from the milk of various mammals. The basic components of milk contain fat, protein, lactose, ash, enzymes and vitamins. Two major categories of proteins are cassein and whey proteins. Cassein is a colloid and whey is soluble. In the cheese industry, two types of precipitation techniques are most commonly used to separate the total milk proteins separated into caseins and whey proteins, i.e, rennet precipitation and acid precipitation. In rennet precipitation, rennin is added to warm milk at 30-35 ^{0}C, precipitating the caseins and leaving the whey proteins in solution. Whey produced by this method is referred to as "sweet whey." Acid precipitation is carried out at the isoelectric point of milk (i.e., 4.7) through the use of acid to precipitate out casein and leave the whey in solution. Whey produced by this method is referred to as "acid whey." The choice of the method used to make the curd depends on the desired cheese product. Although the whey is discarded during the cheesemaking process, it has significant value as a source of nutrition. Unless whey is used to produce other products (including feed), it is a liability for cheese factories. The high biological oxygen demand (BOD) of whey presents disposal problems. Thus, large quantities of whey are used in candies and special cheese products. In addition, large quantities of dry and concentrated forms of whey are also used in dry and concentrated forms that are mixed with food and feeds. Whey may also be condensed and spray dried by the roller process for other uses. Various whey drinks have been developed, including wines, carbonated beverages, buttermilk substitutes, whey-fruit flavored drinks, and whey-tomato drinks. Whey is also used to produce whey butter, soups, protein hydrolysates, cheeses, processed cheese foods and spreads, bakery products, infant foods, geriatric foods, hydrolyzed lactose syrup, pills, riboflavin concentrates, alcohol (e.g., butyl alcohol), methane, acetone, spirit vinegar, food acidulant, resins, coatings, tanning, acrylic plastics, and biomass. In their undenatured, soluble form, whey proteins are also useful as binders in extruded vegetable or animal protein foods. Whey contains various proteins (e.g., .beta.-lactoglobulin and .alpha.-lactalbumin), lactose, soluble minerals, water-soluble vitamins, and enzymes.

The cheese industry produces large amounts of whey, much of which is used to make whey protein concentrate Commercial-scale fractionation of different whey proteins has been hampered by the lack of an economical fractionation technology. It would be desirable, therefore, to provide a method for the continuous and sequential separation of various proteins from whey in a simple one or two step separation process. None of the existing methods provide a product which can be easily, economically, and efficiently utilized as a supplement for infant formulas, fat substitutes or other commercially important products. A need remains in the industry for economical methods to process whey products, including processes and methods that are designed so as to permit reutilization of buffers and constituents utilized during the purification of whey compounds. In addition, methods are needed to increase the efficiency of whey processing procedures, including methods designed to reduce the time necessary to obtain a final product. Sepragen Corp. San Jose, CA, USA, provides a process for the sequential separation of at least five different proteins from whey and incorporating these separated whey proteins into pharmaceutical and food formulations. The process of the invention is directed to the continuous, sequential separation of whey proteins by chromatography, comprising adsorbing the proteins in liquid whey on a suitable separation medium packed in a chromatographic column and sequentially eluting immunoglobulin (e.g, IgG), .beta.-lactoglobulin (.beta.-Lg), .alpha.-lactalbumin (.alpha.-La), bovine serum albumin (BSA), and lactoferrin (L-Fe) fractions with buffers at suitable pH and ionic strength. Even though both axial and *radial flow chromatography* may be utilized, a horizontal flow column is particularly suitable. The liquid whey is selected from the group consisting of pasteurized sweet whey, pasteurized acid whey, non-pasteurized acid whey, and whey protein concentrate. Discuss the spatio-temporal concentration in the radial flow chromatographic column.

The convection effects due to radial flow also have to be taken into account compared with the analysis done in earlier section. The mass balance on a thin cylindrical slice including the radial convective effects can be written as;

$$-1/r\partial(rJ_r)/\partial r - V_r\partial(rC)/\partial r - k'''C = \partial C/\partial t \qquad (3.398)$$

where k''' is the first order rate constant for irreversible adsorption. V_r is the radial velocity. The damped wave diffusion and relaxation equation can be written is;

$$J_r = -D_{AB}\partial C/\partial r - \tau_{mr}\,\partial J_r/\partial t \qquad (3.399)$$

In order to eliminate J_r between Eqs. [3.398, 3.399], Eq. [3.398] is differentiated wrt to t and Eq. [3.399] wrt r after multiplying by r

$$1/r\partial^2(rJ_r)/\partial r\partial t - V_r/r\partial^2(rC)/\partial r\partial t - k'''\partial C/\partial t = \partial^2 C/\partial t^2 \tag{3.400}$$

The damped wave diffusion and relaxation equation can be written is;

$$1/r\partial(rJ_r)/\partial r = -D_{AB}/r\,\partial/\partial r(r\,\partial C/\partial r) - \tau_{mr}/r\partial^2(rJr)/\partial t\partial r \tag{3.401}$$

Eliminating $1/r\partial^2(rJ_r)/\partial r\partial t$, between Eqs.[3.400, 3.401],

$$D_{AB}/r\,\partial/\partial r(r\,\partial C/\partial r) - V_r/r\partial(rC)/\partial r - k'''C = (1 + k'''\tau_{mr})\partial C/\partial t + \tau_{mr}(V_r/r\partial^2(rC)/\partial r\partial t + \partial^2 C/\partial t^2) \tag{3.402}$$

Obtaining the dimensionless form by the following substitutions;

$$X = r/\sqrt{(D\tau_r)}; \quad \tau = t/\tau_r; \quad u = (C/C_s); \quad k^* = k'''\tau_{mr}; \quad Pe_m = V/\sqrt{(D/\tau_{mr})} \tag{3.403}$$

$$1/X\,\partial/\partial X(X\,\partial u/\partial X) - Pe_m/X\partial(Xu)/\partial X - k^*u = (1 + k^*)\partial u/\partial \tau + Pe_m/X\partial^2(uX)/\partial X\partial \tau + \partial^2 u/\partial \tau^2 \tag{3.404}$$

The damping term can be removed using a $u = W\exp(-n\tau)$ substitution;

$$\partial u/\partial \tau = -nW\exp(-n\tau) + \exp(-n\tau)\partial W/\partial \tau$$

$$\partial^2 u/\partial \tau\partial X = -n\partial W/\partial X\exp(-n\tau) + \exp(-n\tau)\partial^2 W/\partial \tau\partial X$$

$$\partial(uX)/\partial \tau = -n(WX)\exp(-n\tau) + \exp(-n\tau)X\partial W/\partial \tau + \exp(-n\tau)W$$

$$\partial^2(uX)/\partial X\partial \tau = -n(W)\exp(-n\tau) - nX\partial W/\partial X\exp(-n\tau) + \exp(-n\tau)\partial W/\partial \tau + \exp(-n\tau)X\partial^2 W/\partial X\partial \tau + \exp(-n\tau)\partial W/\partial X$$

$$\partial^2 u/\partial \tau^2 = +n^2 W\exp(-n\tau) + -2n\exp(-n\tau)\partial W/\partial \tau + \exp(-n\tau)\partial^2 W/\partial \tau^2$$

Substituting the above, Eq. [3.404] becomes;

$$[1/X + nPe_m - Pe_m/X]\,\partial W/\partial X + \partial^2 W/\partial X^2 - Pe_m/X\partial(XW)/\partial X$$
$$= (1 + k^* - 2n + Pe_m/X)\partial W/\partial \tau + Pe_m/X\,X\partial^2 W/\partial X\partial \tau +$$
$$W(-n(1 + k^*) + n^2 + k^* + nPe_m/X) + \partial^2 W/\partial \tau^2 \tag{3.405}$$

For small Pe_m/X Eq. [3.405] can be written as;

$$[1/X + nPe_m]\,\partial W/\partial X + \partial^2 W/\partial X^2 - Pe_m/X\partial(XW)/\partial X$$
$$= (1 + k^* - 2n)\partial W/\partial \tau + Pe_m/X\,X\partial^2 W/\partial X\partial \tau + W(-n(1 + k^*) + n^2 + k^*) + \partial^2 W/\partial \tau^2 \tag{3.406}$$

By choosing $n = (1 + k^*)/2$, Eq. [3.514] becomes;

$$\partial^2 W/\partial X^2 + 1/X\partial W/\partial X + (1 + k^*)/2Pe_m\,\partial W/\partial X - Pe_m\partial W/\partial X - Pe_m W/X$$

$$= Pe_m/X \partial^2 W/\partial\tau\partial X \quad -W/4 |1 - k^*|^2 \quad + \partial^2 W/\partial\tau^2 \tag{3.407}$$

Eq. [3.407] can be solved by the method of separation of variables;

Let $W = \phi(X)V(\tau)$,

$$\phi''/\phi \quad + \phi'/\phi \ (1/X) \ + (1 + k^*)Pe_m/2(\phi'/\phi) \ - Pe_m\phi'/\phi \ - Pe_m/X \quad = Pe_m /XV'\phi'/V\phi \ - 1/4|1 - k^*|^2$$
$$+ \quad V''/V \tag{3.408}$$

$$\phi''/\phi \quad + \phi'/\phi \ [(1/X)\{1 \ - Pe_m - \ Pe_mV'/V\} + \ Pe_m(k^* + 1)/2\}] - Pe_m/X$$
$$= V''/V \ = \ 1/4|1 - k^*|^2 - \lambda_n^2 \tag{3.409}$$

The time domain solution can seen to be;

$$V = c_1\exp(+\tau\mathrm{sqrt}(1/4|1 - k^*|^2 - \lambda_n^2) + c_2\exp(-\tau\mathrm{sqrt}(1/4|1 - k^*|^2 - \lambda_n^2)) \tag{3.410}$$

At infinite time the concentration cannot be infinity and hence c_1 has to be set to zero.

$$V = c_2\exp(-\tau\mathrm{sqrt}(1/4|1 - k^*|^2 - \lambda_n^2)) \tag{3.411}$$

Then,
$$V'/V = -\mathrm{sqrt}(1/4(|1 - k^*|^2 - \lambda_n^2)) \tag{3.412}$$

The space domain solution can be obtained

$$X^2\phi'' + \phi' \ [X[1 - \ Pe_m + Pe_m \ \mathrm{sqrt}1/4(|1 - k^*|^2 - \lambda_n^2] + XPe_m/2 \ (1 +k^*)] \ - \phi \ [X^2(1/4|1 - k^*|^2 - \lambda_n^2)$$
$$- Pe_mX] \ = 0 \tag{3.413}$$

Comparing Eq. [3.413] with the generalized Bessel equation given in Eq. [2.96],

$$a = 1 - Pe_m + Pe_m \ \mathrm{sqrt}1/4(|1 - k^*|^2 - \lambda_n^2]$$

$$r = 1; \quad b = +Pe_m/4 \ (1 + k^*)$$

$$1 - a - r = Pe_m - Pe_m \ \mathrm{sqrt}1/4(|1 - k^*|^2 - \lambda_n^2] - 1 = 0$$

$$s = 1/2 ; \ d = Pe_m ; \ c = 0$$

Thus λ_n can be chosen from the relation given in Eq. [3.413]

$$1/4(|1 - k^*|^2 - \lambda_n^2 = (Pe_m - 1)^2 /Pe_m^2 \tag{3.414}$$

$$[1 + 1/(Pe_m(k^*-1)/2 - 1) \ ]^2 = 1/4(|1 - k^*|^2 - \lambda_n^2)) \tag{3.415}$$

$$\text{or } \lambda_n = \mathrm{sqrt}[\tfrac{1}{4}|1 - k^*|^2 - (Pe_m -1)^2/Pe_m^2] \tag{3.416}$$

$$a = 1 - Pe_m + Pe_m (Pe_m -1)/Pe_m = 0 \tag{3.417}$$

The order of the Bessel equation is thus; $p = 2/2 = 1$. The solution is given by;

$$\phi = \exp(-X(1 + k^*)Pe_m/4)[\ c_1\ J_1\ (2X|Pe_m|^{1/2} + c_2\ Y_1(2X|Pe_m|^{1/2}]$$ (3.418)

The value of b differs between the two coefficients of the Bessel differential equation. The solution is approximately obtained. c_2 can be seen to be zero from the condition that W is finite at $X = 0$. Thus,

$$\phi = \exp(-X(1 + k^*)Pe_m/4)[\ c_1\ J_1\ (2X|Pe_m|^{1/2}$$ (3.419)

$$u = c\ \exp(-(1 + k^*)\tau/2)\ \exp(-\tau sqrt(1/4|1 - k^*|^2 - \lambda_n^2))\ \exp(-X(1 + k^*)Pe_m/4)[\ c_1\ J_1 (2X|Pe_m|^{1/2}$$ (3.420)

There exists a non-homogeneity at the boundary condition at $X = 0$. This can be solved by the method of superposition and by expressing the solution as a sum of a transient part and a steady state part. At steady state, Eq. [3.488] becomes;

$$1/X\,\partial/\partial X(X\,\partial u/\partial X) - Pe_m/X\partial(Xu)/\partial X - k^* u = 0$$ (3.421)

$$\partial^2 u/X^2 + 1/\ X\partial u/\partial X - Pe_m\ u/X - Pe_m\partial u/\partial X - k^* u = 0$$ (3.422)

For small Pe_m/X
$$X^2\ \partial^2 u/X^2 + X(\ 1 - XPe_m)\partial u/\partial X - u[X^2\ k^* + X\ Pe_m] = 0$$ (3.423)

Comparing Eq. [3.539] with the generalized Bessel equation given in Eq. [2.96] Approximately,
$$a = 1;\ \ r = 1;\ \ 2b = -Pe_m$$

$$c = 0;\ \ s = 1;\ \ 1 - a - r = -1;\ \ -b(1-a-r) = Pe_m/2\ ;\ \ d = -k^{*1/2}$$

The order of the Bessel solution can be calculated as $p = 0$. Thus approximately,

$$u = \exp(-Pe_m X/2)\ [c_1\ I_0(Xk^{*1/2}) + c_2\ K_0(Xk^{*1/2})$$ (3.424)

The value of b differs in the two coefficients of the Bessel differential equation. As a first pass, analysis the variation is neglected. From the condition that the concentration is finite at zero X c_2 can be seen to be zero. For $X = 0$,

$$1 = c_1$$

Thus at steady state,
$$u = \exp(-Pe_m X/2)\ I_0(Xk^{*1/2})$$ (3.425)

Now the transient portion can have the boundary condition at $X = 0$, $u = 0$. Eq. [3.425] can be seen to obey the boundary condition at $X = 0$ for the transient portion of the solution. As the $J_1(x)$ goes to zero at $x = 3.8317$, beyond a certain distance the concentration will be zero. This critical distance is when,

$$R > 1.916 \text{sqrt}(D\tau/Pe_m) \qquad (3.426)$$

The design of the radial flow chromatography column greater than the R calculated by Eq. [3.426] will not result in any further separation. This is the critical length of diffusion.

Worked Example 3.14

The diffusion coefficient is sometimes measured using transient concentration profiles in an infinite couple. Two bars are different concentrations are joined together for times greater than zero. The concentration is raised to facilitate diffusion and at the end of the experiment the bars are quenched and then the compositions are measured as a function of position. This analysis is made by grinding off small amounts of bar and determining the composition by a series of wet chemical tests. Recent changes to the method include the use of electron microprobe. Derive the concentration profiles in both the bars.

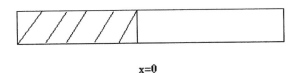

x=0

Figure 3.18 **Infinite Couple to Measure Diffusion Coefficients. Two Bars of Different Concentrations are Joined Together at Zero Time**

Assume that the two bars are made of the same material. The initial concentration in slab 1 may be taken as C_{1s} and the initial concentration in slab two may be taken as C_{2s}.

For timed greater than zero the two bars are joined together. The interface concentration at steady state will be at the average of the two concentrations and be given by;

$$t = \infty, \ C_i = (C_{1s} + C_{2s})/2 \tag{3.427}$$

Studies have shown that the interface concentration will reach equilibrium by an exponential process. Thus for the first bar,

$$C_i = A + B \exp(-t/\tau_r)) \tag{3.428}$$

$$\text{For } t = 0, \ C_{1s} = A + B \tag{3.429}$$

$$\text{For } t = \infty, \ A = (C_{1s} + C_{2s})/2 \tag{3.430}$$

$$\text{So, } B = C_{1s} - (C_{1s} + C_{2s})/2 \tag{3.431}$$

Thus,

$$C_i = (C_{1s} + C_{2s})/2 + [C_{1s} - (C_{1s} + C_{2s})/2]\exp(-t/2\tau_r) \tag{3.432}$$

For the second bar,

$$C_i = (C_{1s} + C_{2s})/2 + [C_{2s} - (C_{1s} + C_{2s})/2]\exp(-t/2\tau_r) \tag{3.433}$$

The governing equation in 1 dimension including the effects of damped wave diffusion and relaxation can be written as;

$$\tau_r \, \partial^2 C/\partial t^2 + \partial C/\partial t = D \, \partial C/\partial x^2 \tag{3.434}$$

Define the dimensionless variables in the first bar as;

$$u = (C - (C_{1s} + C_{2s})/2)/(C_{1s} . (C_{1s} + C_{2s})/2); \ \tau = t/\tau_r; \ X = x/\text{sqrt}(D\tau_r) \tag{3.436}$$

The governing equation becomes;

$$\partial^2 u/\partial \tau^2 + \partial u/\partial \tau = \partial^2 u/\partial X^2 \tag{3.437}$$

The time and space conditions become;

$$\tau = 0, \ u = 1 \tag{3.438}$$

$$\tau = \infty, \ u = 0 \tag{3.439}$$

$$X = 0, \ u = \exp(-\tau/2) \tag{3.440}$$

Let $V = 1 - u$. Then,

$$\partial^2 V/\partial \tau^2 + \partial V/\partial \tau = \partial^2 V/\partial X^2 \tag{3.442}$$

The time and space conditions become;

$$\tau = 0, \quad V = 0 \tag{3.443}$$

$$\tau = \infty, \quad V = 1 \tag{3.444}$$

$$X = 0, \quad V = 1 - \exp(-\tau/2) \tag{3.445}$$

The damping term can be removed from Eq. [3.442] by a $V = W\exp(-n\tau)$ substitution. For $n = \frac{1}{2}$, Eq. [3.442] becomes,

$$\partial^2 W/\partial \tau^2 - W/4 \quad = \partial^2 W/\partial X^2 \tag{3.446}$$

Let the transformation be as follows;

$$\eta \quad = \tau + X \tag{3.447}$$

$$\xi \quad = \tau - X \tag{3.448}$$

Then equation [3.446] becomes;

$$\partial^2 W/\partial \tau^2 \quad = \partial^2 W/\partial \eta^2 + \quad \partial^2 W/\partial \xi^2 \quad + 2\partial^2 W/\partial \xi \partial \eta \tag{3.449}$$

$$\partial^2 W/\partial X^2 \quad = \partial^2 W/\partial \eta^2 + \quad \partial^2 W/\partial \xi^2 \quad - 2\partial^2 W/\partial \xi \partial \eta \tag{3.450}$$

$$4 \, \partial^2 W/\partial \xi \partial \eta \quad = W/4 \tag{3.451}$$

At this stage Laplace transforms is obtained in the ξ domain;

$$d\underline{W}/d\eta \quad = \underline{W}/16/s \quad \text{or the solution is}$$

$$\underline{W} = C' \exp \eta/16s \tag{3.452}$$

At, $\xi = 0$, $\tau = X$. This is at the wave front. There exists an inertia time taken at a point x in the medium to see the disturbance at the boundary traveling with a speed of $\mathrm{sqrt}[D/\tau_r]$. Only for times greater than this the problem exists. For times equal to and less than this time, the initial concentration remain and thus $W = V = 0$. Using the boundary condition, $W = \exp[-\tau/2]\exp[\tau/2] = (\exp(\tau/2) - 1)$ at $\eta = a/\sqrt{D\tau_r} + \tau$.

$$1/(s - 1/2) - 1/s \quad = C' \exp((a^* + \tau)/16s) \tag{3.453}$$

where $a^* = a/\sqrt{D\tau_r}$

$$\text{or } C' = (1/(s-1/2) - 1/s) \exp[-[[a^* + \tau]/16s] \tag{3.454}$$

Thus, $\underline{W} = 1/s \exp[\eta - \tau - a^*]/16s = (1/(s-1/2) - 1/s)\exp[X - a^*]/16s \tag{3.455}$

The inversion of equation [3.576] is within the tables available [Mickley, Sherwood and Reed, 1957]. So,

$$u = \int_0^\tau \exp(\tau-p)/2) \exp[-p/2] J_0 [\frac{1}{2} \mathrm{sqrt}[[a^* - X][p - X]]dp \quad + \quad \exp[-\tau/2] J_0 [\frac{1}{2} \mathrm{sqrt}[[a^* - X][\tau - X]] \tag{3.456}$$

The first root of the Bessel function of the zeroth order and first kind occurs at 2.4048. Thus after a critical length the concentration in the bar will be unchanged from the initial condition. This needs to be factored into the analysis. Such a critical length can be calculated as shown in the previous analysis on zone of zero transfer in a catalyst pellet in the above sub-section.

Summary

The study of diffusion has a variety of applications. Diffusion can be viewed as a intermingling of molecules due to random motion. The Maxwell speed distribution of the molecules, root mean square averaged velocity of the molecule for a monatomic gas is discussed. Kinetic theory is utilized to obtain the mean free path of the molecule. The depiction of Fick's first law in terms of kinetic theory is revisited. It is noted that the accumulation of molecules at the incremental volume is often neglected. The linear concentration gradient is needed for the expression of mass flux to reduce to the expression of Fick's first law. The accumulation term inclusion can result in the expression for mass flux can reduce to the damped wave diffusion and relaxation equation. An alternate expression for Fick's law is obtained as a worked example to account for the depletion of molecules at the surface. The finite speed diffusion and relaxation with the velocity of mass equal to $\sqrt{D/\tau_r}$ is used to represent the spatio-temporal concentration in a semi-infinite catalyst during simultaneous diffusion and fast reaction. The governing equation is a hyperbolic PDE. The damping term is removed by a $u = W\exp(-n\tau)$ substitution. The resulting PDE in two variables is reduced to a Bessel differential equation in 1 variable using a transformation such as $\eta = \tau^2 - X^2$.

The spatio-temporal concentration is given approximately in three regimes. The first regime is one of zero transfer and also refers to the mass inertia. Thus $\tau_{inertia} = \sqrt{X_p^2 - 23.1323/|1 - k^*|^2}$ for an interior point X_p in the medium and k^* is the dimensionless first order rate constant. The second regime is given by a ratio of the Bessel composite function of the first kind and zeroth order and a modified Bessel function in time of the first kind and zeroth order. The third regime is given by a ratio of the modified Bessel composite function of the first kind and zeroth order and the modified Bessel function of the first kind and zeroth order. The mass inertia time is tabulated as a function of the dimensionless first order rate constant. The three regimes are shown in Figures 3.4 – 3.7 for various values of k^*. A separate expression for the transient concentration profile is derived for $k^* = 1$. The governing equation is solved by an aliter, by the method of Laplace transforms. The solution is obtained using the Leibniz's rule for differentiating an integral. The analysis is repeated for a zeroth order chemical reaction using the method of Laplace transforms. The problem of fast reaction and simultaneous diffusion and relaxation is studied for an infinite cylindrical catalyst in cylindrical coordinated. The radial effects are accounted for and the solution is sought by the method of relativistic transformation of coordinates. Three regimes are identified in the spatio-temporal concentration. For a exterior point in the infinite cylindrical catalyst, the mass inertia time was shown to be $\tau_{lag} = $ sqrt($X_p^2 - 4\pi^2/(|1 - k^*|^2)$). The spatiotemporal concentration in the second regime is given as a ratio of the Bessel composite function of the half order and first kind and the modified Bessel function of time of the half order and first kind. The third regime is given by a ration of the modified Bessel function of the half order and first kind. . The mass inertia time is tabulated as a function of the dimensionless first order rate constant. The three regimes are shown in Figures 3.8 – 3.11 for various values of k^*. The problem of fast reaction and simultaneous diffusion and relaxation is studied for an infinite spherical catalyst in spherical coordinates. The radial effects are accounted for and the solution is sought by the method of relativistic transformation of coordinates. Three regimes are identified in the spatio-temporal concentration. For an exterior point in the infinite cylindrical catalyst, the mass inertia time was shown to be $\tau_{lag} = $ sqrt($X_p^2 - 7.6634^2/(|1 - k^*|^2)$). The spatiotemporal concentration in the second regime is given as a ratio of the Bessel composite function of the first order and first kind and the modified Bessel function of time of the first order and first kind. The third regime is given by a ration of the modified Bessel function of the first order and first kind. The mass inertia time is tabulated as a function of the dimensionless first order rate constant. The three regimes are shown in Figures 3.12 – 3.15 for various values of k^*.

The transient concentration profile in a finite slab, cylinder and sphere using the damped wave diffusion and relaxation equation are derived using the method of separation of variables. An infinite series solution is provided. The FINAL condition in time domain is used to keep the solution within the bounds of the second law of thermodynamics. The boundary condition is used is the constant wall concentration. The non-homogeneity in the boundary condition was removed by a superposition of steady state and transient state solutions.

During autocatalytic reactions such as the simultaneous diffusion, relaxation and reaction of neutrons during fission in nuclear power plant two limits of importance were derived. One is the lower cycling limit the size below which subcritical damped oscillations in concentration can be found. An upper shape limit is obtained, above which the rate of generation of neutrons is much greater than the rate of removal by mass transfer leading to the runaway reaction.

A zone of zero transfer is identified in a Zeigler Natta catalyst. The governing equation was converted using transformation of wave coordinates to the canonical form. This form was transformed into the Laplace domain and upon substituting for the boundary conditions was inverted to give a Bessel function of the first kind and zeroth order. The first zero of the Bessel function at 2.4048 marked the line of zero transfer of mass. This problem was repeated for a simultaneous reaction of the first order and diffusion and relaxation.

The spatio-temporal concentration in the exterior of a dissolving drop is obtained by the method of relativistic transformation of coordinates. The method of Laplace transforms was used to obtain the transient concentration profile obtained in a finite catalyst slab. The 4 spherical coordinates were considered in a problem of opening the perfume in a shopping mall. Three regimes for the solution were obtained. The first inertial regime of zero transfer. The second regime represented by a critical Bessel composite function of the second order and first kind and a third regime where the critical part is a modified Bessel composite function of the second order and first kind. The creeping transfer assumption normally used for a sphere is used and the analysis repeated. The Palladium adsorber to remove the organic sulfides in automobile gasoline was analyzed at steady state. The dissolving pill problem was discussed in spherical coordinates with a u/r = V substitution. The solution to this problem in 4 spherical coordinates with the substitution resulted in a 3 regime solution. The critical portion of the second regime contained a Bessel composite function of the third order and first kind. The first root of the third order Bessel function was solved for using a Microsoft spreadsheet on Pentium IV microcomputer and the mass inertia time calculated. The use of the creeping transfer assumption is evaluated. The order of the Bessel function result obtained was 7/2.

The convection effects along with wave diffusion and relaxation was discussed. The method of Laplace transforms was used to analyze the 1 dimensional wave diffusion and relaxation along with convective effects. A problem in radial flow chromatography was studied to separate whey proteins. The governing equation in cylindrical coordinated in 1 dimension was reduced to a Bessel differential equation. The results for steady state and transient state are provided. The critical term in the solution is the Bessel function of the first order and first kind. In the cylindrical slice considered prior to obtaining the mass balance equation the variation of cross-sectional area in the convection term was also considered and not neglected. The measurement of diffusion coefficient using infinite couple was discussed.

Exercises

1. *Interstitial Mechanism of Movement of Atoms*

The atomic motions in crystal lattice can be used to make rough estimates of diffusion coefficients. The theory of Face Centered Cubic, FCC, is used to count the atom movement by an interstitial mechanism. The diffusion coefficients obtained using this theory is given by Franklin [1975] and Stark [1976].

$$D = a_0^2 N \omega$$

a_0 is the spacing between atoms, N is the fraction of sites vacant in the crystal and ω is the jump frequency, the number of jumps per unit time from one position to the next. Values of interatomic spacing are obtained from crystallographic data and the fraction N can be estimated from the Gibbs free energy of mixing. Consider in one dimension the net diffusion flux of atoms from z to z + Δz.

A(Net Flux J_1)   = 4Nω A(number of atoms/area)$|_z$

                   - 4Nω A(number of atoms/area)$|_{z+\Delta z}$

                   - A Δz (rate of accumulation of atoms/volume)

Dividing throughout by A and obtaining the limit as Δz goes to zero. The factor of 4 reflects the fact that the FCC structure has 4 sites into which jumps can occur. The number of atoms per unit area can be related to the concentration,

$$C_1|_z = C_1|_{z+\Delta z} - a_0/2 \; \partial C_1/\partial z$$

Assuming a linear concentration gradient.

$$J_1 = -4N\omega a_0^2/4 \; \partial C_1/\partial z \quad - 4N\omega \; a_0^3/8 \; \partial C/\partial t$$

$$= -D\partial C_1/\partial z - D \, a_0/2 \; \partial C/\partial t$$

Obtain the governing equation in 1 dimension.

2. *Zeroth Order Reaction*

Obtain the spatio-temporal concentration in a semi-infinite medium in a) 1 dimension and b) three dimensions in cartesian coordinates during simultaneous fast reaction and diffusion and relaxation. Assume a zeroth order reaction and use the method of relativistic transformation of coordinates.

3. *Method of Laplace Transforms*

Obtain the spatio-temporal concentration in a infinite cylindrical catalyst during simultaneous fast reaction and wave diffusion and relaxation. Use the method of Laplace transforms. How does your result compare with the result obtained by method of relativistic transformation of coordinates.

4. *Constant Wall Flux during Simultaneous Reaction and Diffusion and Relaxation*

Consider a finite slab at constant wall flux. Consider simultaneous fast reaction and wave diffusion and relaxation and obtain the transient concentration profile by method of separation of variables.

5. *Spatio-Temporal Concentration Profile in Cone*

Consider a cone with the apex at a concentration C_a. Obtain the spatio-temporal concentration profile in the cone. Use the method of relativistic transformation of coordinates.

6. *Pulse Condition in Infinite Medium*

A pulse of concentration is injected at the center of an infinite medium. Consider the damped wave diffusion and relaxation away from a sharp pulse of solute. Express the pulse as a dirac delta function of $\delta(z)$. Compare the results with the Gaussian profile obtained for the same problem using the Fick parabolic diffusion equation.

7. *Parabolic and Hyperbolic*

Derive the transient concentration profile for a semi-infinite medium in 1 dimension in cartesian coordinates using the damped wave diffusion and relaxation using the Fick parabolic diffusion equation. Express the results in terms of % change.

8. *Chemical Treatment of Wood*

In chemical treatment of wood, the diffusion of a solute into the cylinder is of interest. The cylinder initially contains no solute. At time zero, the cylinder is suddenly immersed in a well-stirred solution that is of such enormous volume that its solute concentration is constant. Obtain the spatio-temporal concentration in the cylinder using a) Fick parabolic diffusion equation. b) damped wave diffusion and relaxation equation.

9. *Wave Diffusion and Relaxation and Convection in a Moving Film*

Consider a problem where the diffusion direction and the direction of convection flow are in perpendicular directions. Consider the diffusion across a thin, moving liquid film. The concentrations on both sides of this film are fixed by electrochemical reactions, but the film itself is moving steadily. Assume that the liquid solution is dilute. The liquid is the only resistance to mass transfer. This implies that the electrode reactions are fast. Mass transport is by diffusion in the z direction and by convection in the x direction. Convection can be neglected in the z direction. Obtain the transient concentration profile using the damped wave diffusion and relaxation equation.

10. *Wave Diffusion in Moving Film*

A thin liquid film flows slowly and without ripples down a flat surface. One side of the film wets the surface and the other side is in contact with a gas which is sparingly soluble in the liquid. How much gas dissolves in the liquid.

$$-\partial J/\partial z = V_x \partial C/\partial z + \partial C/\partial t$$

Using the Fick's second law for the diffusion flux the governing equation for the concentration of the solute gas in the liquid is given by;

$$D_{AB} \partial^2 C/\partial z^2 = V_x \partial C/\partial z + \partial C/\partial t$$

Using the damped wave diffusion and relaxation equation the governing equation for the solute gas in the liquid is given by.

$$D\partial^2 C/\partial z^2 \quad -V_x\partial C/\partial z \ = \ + \ \tau_r \, (V_x\partial^2 C/\partial z\partial t \ + \partial^2 C/\partial t^2) + \partial C/\partial t$$

Obtain the spatio-temporal concentration profile using a) the governing equation derived from Fick's second law and b) the governing equation derived from damped wave diffusion and relaxation equation.

11. *Evaporation*

Consider the fast unsteady evaporation by damped wave diffusion and relaxation and convection. Consider a very long capillary filled with solvent gas. As the liquid evaporated the liquid gas interface recedes. Obtain the transient concentration profile of the vapor in the gas phase.

12. *Desalination*

In some parts of the world, whole cities are going to have to be abandoned because of prolonged drought. Inexhaustible supply of water is the sea. Many desalination plants operate on the concept of Reverse Osmosis. Other methods include, electro dialysis, ion exchange, evaporation, flash distillation, adsorption, pervaporation and extraction. Water obtained has to possess the minerals that are needed for human consumption. Potable water can be obtained by reverse osmosis in mobile units. The semi-permeable membrane used is polyamide. It needs to be (1) Chlorine and bifouling resistant. The currently used elements have a spiral construct. They are reported to have several deficiencies. These are poor chemical stability to oxidants, chlorine and possess high fouling rates. New polyamide membrane is made from all trans stereoisomer of cyclopentanetetracaraboxylic acid chloride by interfacial reaction of m-phenylenediamine, trimeosyl acid chloride. A copolymer/blend design may be used to prepare a thin film membrane. It has to possess good balance of properties. Porosity of the membrane is a salient consideration. The control of hydrophilicity by tweaking the molecular structure is possible. The novel membrane can have a composite construction. Two films of complex polymer are formed. A salt passage can then be defined. Water may flow to a central tube through mesh of elements and ooze out as pure water. The flow is maintained in the state of turbulence. Brine is recovered as by product.

In a typical plant, 99.3 % of dissolved solids are rejected. Usually sea water comprises of 3.6% NaCl and is reduced by desalination to 400- 500 ppm.

Fouling of membrane may be caused by mineral disposition, organic growth and formation of concentration polarization layer. Sodium hexametaphosphate is used as inhibitor. Low flux rate, backwashing can improve the performance. In this process, the efficiency is high, fouling is reduced, and no dosing of mineral minerals is necessary. Turbulence is induced by a pump with an inlet pressure of 50 – 65 bars and a pressure drop of 1-2 bars. The turbulence keeps the concentration polarization layer thickness low. The concentration polarization layer thickness can be described using the product of mass flux at the interface of the membrane. Discuss the spatio-temporal concentration profile.

13. *Coextrusion*

In the manufacture of the casings of the solid rocket motor SRM, the material requirements are bifunctional. They have to have high hoop strength on one side and high ablation resistance on the other. In order to prepare such materials the technology of coextrusion is utilized. In a twin-screw extruder both the materials are co-extruded together. During the residence time of the polymers in the extruder, the interdiffusion of either material in the other occurs. Calculate the interlayer thickness as a function of the extruder residence time, diffusivities of the two materials, mass relaxation times of the two materials. The damped wave diffusion and relaxation equation in cylindrical coordinates can be used to obtain the governing equation and the method of relativistic transformation of coordinates as the solution approach.

14. *Transformation $\eta = (\tau^2 - X^2)^{1/2}$*

Consider an infinite catalyst in cylindrical coordinates as discussed in Section 3.2. Use the transformation, $\eta = (\tau^2 - X^2)^{1/2}$ after removing the damping term in the governing equation in cylindrical coordinates. In 1 dimension. Compare the Bessel differential equation and solution obtained with the ones obtained in the section 3.2.

4.0 Transient Momentum Transfer and Relaxation

Nomenclature

A	area of the plate (m^2)
A_{tube}	area of the tube (m^2)
C_p	heat capacity (J/kg/K)
d	diameter (m)
D	diameter of the tube (m)
f	friction factor ($F_k = A K_e f$)
F_k	drag force (N)
g	acceleration due to gravity (m/s^2)
H	hematocrit, the red blood cell volume fraction
$I_0(x)$	modified Bessel function of the zeroth order and first kind
$I_1(x)$	modified Bessel function of the first order and first kind
$I_2(x)$	modified Bessel function of the second order and first kind
$J_0(x)$	Bessel function of the zeroth order and first kind
$J_1(x)$	Bessel function of the first order and first kind
$J_2(x)$	Bessel function of the second order and first kind
$J_3(x)$	Bessel function of the third order and first kind
$K_0(x)$	modified Bessel function of the zeroth order and second kind
$K_1(x)$	modified Bessel function of the first order and second kind
$K_2(x)$	modified Bessel function of the second order and second kind
$K_3(x)$	modified Bessel function of the third order and second kind
K_e	characteristic energy per unit volume (J/m^3)
k	thermal conductivity (W/m/K)
P	pressure (N/m^2)
U_t	terminal settling velocity of sphere (m/s)
u	dimensionless velocity (v_z/v_{mom})
v_x	x component velocity (m/s)
v_y	y component velocity (m/s)
v_z	z component velocity (m/s)
V	velocity of fluid (m/s)
V_{mom}	velocity of momentum sqrt(γ/τ_r)
W	wave velocity
$Y_0(x)$	Bessel function of the zeroth order and second kind
$Y_1(x)$	Bessel function of the first order and second kind
$Y_2(x)$	Bessel function of the second order and second kind
$Y_3(x)$	Bessel function of the third order and second kind
Z	dimensionless distance (z/sqrt($\gamma\tau_r$)

Greek

β	Substitution constant
μ	viscosity of fluid (kg/(m s))
ρ	density of fluid (kg/m^3)
κ	permeability of porous medium (m^2)
γ	kinematic viscosity (m^2/s)
ψ	stream function
τ_{mom}	relaxation time of momentum
τ_{xx}	viscous normal stress on the x face
τ_{yx}	viscous tangential stress on the y face
τ_{yy}	viscous normal stress on the y face
τ_{zy}	viscous tangential stress on the z face
τ_{zz}	viscous normal stress on the z face
τ_{zx}	viscous tangential stress on the x face

Subscripts

cs	control surface
cv	control volume

Dimensionless Groups

Acc	Accumulation number ($g\tau_{mom}/V$)
C_D	Drag coefficient $4gD(\rho_s - \rho)/(3\rho v_\infty^2)$
F	Froude number ($v/(gl)^{1/2}$)
F*	dimensionless force, ($F/AV\rho\, v_{mom}$)
Ma	Mach number (V/V_{sound})
Mom	Momentum Number ($g\tau_{mom}/V_{mom}$)
Osc	Oscillation number ($g\tau_{mom}^2/L$)
P*	Dimensionless pressure ($p/\rho v_R v_{mom}$)
Pe$_{mom}$	Peclect (momentum) Pe$_{mom}$ $= v_{rb}/v_{mom}$
Pb	Permeability number, Pb $= (\gamma\tau_{mom}/\kappa)$,
Pr	Prandtl number ($C_p\mu/k$)
Re	Reynolds number ($\rho Vd/\mu$)
τ_{zx}^*	Dimensionless shear stress ($\tau_{zx}/$ sqrt($\rho V^2\, \mu/\tau_{mom}$))
θ	Substitution variable $\theta = Z + (1 - \beta)^{1/2}\tau$
ψ	Transformation variable, $\psi = \theta/$sqrt($4\beta\tau$)
η	Tansformation variable, $\eta = \tau^2 - Z^2$

4.1 34 Different Flow Types

A fluid is a gas or liquid that flows when subjected to sufficient shear stress. Shear force is the tangential component of force field and divided by the area normal to the force is the average shear stress over the area. Shear stress at a point is the limiting value of shear force to area as the area is reduced to a point. 34 different flow types can be identified.

Osborn Reynolds (1883)   presented his experimental investigation of the circumstances which determine whether the motion of water shall be direct or sinuous and of the laws of resistance in parallel channels to the Royal society 122 years ago. To this day the dimensionless group ($\rho V d/\mu$) named after him called the Reynolds number is used extensively. It gives the ratio of the inertia forces and viscous forces and is used to delineate *laminar flow* from *turbulent flow*.

A glass tube was mounted horizontally with one end in a tank and a valve on the opposite end. A smooth bell mouth entrance was attached to the upstream end with a dye jet so arranged that a fine stream of dye could be injected at any point in front of the bellmouth. Reynolds took the average velocity V as characteristic velocity and the diameter of tube as characteristic length. For small flows the dye stream moved as a straight line through the tube indicating that the flow was laminar. As the flow rate was increased, the Reynolds' number increased since d, ρ, μ were held constant and V was directly proportional to the rate of flow. With increasing discharge a condition was reached at which the dye stream wavered and then suddenly broke up and was diffused throughout the tube. The nature of flow had changed to turbulent flow with its violent interchange of momentum that had completely disrupted the orderly movement of laminar flow. By careful manipulation of variables, Reynolds was able to obtain a value of Re = 12,000 before turbulence sets in. Later investigators obtained the value of 40,000 using the original equipment used by Reynolds. They let the water stand in the tank for several days before the experiments and by taking precautions to avoid vibration of the water or equipment. These numbers are referred to the upper critical Reynolds number. Starting with turbulent flow in glass tube, Reynolds found that it was always laminar when the velocity is reduced to enable Re < 2000. This is the lower critical Reynolds number. With the usual piping installation the flow will change from laminar to turbulent in the range of Reynolds numbers from 2000- 4000. Reynolds number may be interpreted as the ratio of the bulk transfer of momentum to the momentum by shear stress.

Heleshaw [1898] refers to two-dimensional laminar flow between closely spaced plates. Laminar flow is defined as flow in which the fluid moves in layers, or laminas, one layer gliding smoothly over an adjacent layer with only a molecular interchange of momentum. Turbulent flow, however, has very erratic motion of fluid particles with a vibrant transverse interchange of momentum. Reynolds number calculations have been popular with many a successful practioner and withstood the test of time for over 12 decades.

In 1904, Prandtl [1952] presented the concept of the boundary layer. It provides the important link between ideal fluid flow and real fluid flow. For fluids with small viscosity, the effect of internal friction in a fluid is appreciable only in a narrow region surrounding the fluid boundaries. From this premise, the flow outside the narrow region near the solid boundaries may be considered as ideal flow or potential flow. Relations within the boundary layer region can be computed from the general equation for viscous fluid. The momentum equation permits the developing of approximate equation for boundary layer growth and drag. When motion is started in a fluid having small viscosity the flow is irrotational initially. The fluid at the boundaries has zero velocity relative to the boundaries. As a result, there is a steep velocity gradient from the boundary into the flow. The velocity gradient in a real fluid sets up near the boundary shear forces that reduce the flow relative to the boundary. That fluid layer which has had its velocity affected by the boundary shear is called the *boundary layer*.

The velocity in the boundary layer approaches the velocity in the main flow asymptotically. The boundary layer is very thin at the upstream end of a streamlined body at rest in an otherwise uniform flow. As this layer moves along the body, the continual action of shear stress tends to slow down additional fluid particles, causing the thickness of the boundary layer to increase with distance from the upstream point. The fluid in the layer is also subjected to a pressure gradient determined from the potential flow, that increased the momentum of the layer if the pressure decreases downstream and decreases its momentum if the pressure increases downstream (adverse pressure gradient). The flow outside the boundary layer may also bring momentum into this layer. For smooth upstream boundaries, the boundary layer starts out as a laminar boundary layer in which the fluid particles move in smooth layers. As the *laminar boundary layer* increases in thickness, it becomes unstable and finally transforms into a turbulent region in which the fluid particles move in zig zag paths although their velocity has been reduced by the action of viscosity at the boundary. Where the boundary layer has become turbulent, there is still a very thin layer next to the boundary that has laminar motion. It is called *laminar sublayer*.

A*diabatic flow* is that flow during which no heat is transferred to or from the fluid. *Isentropic flow* is reversible, adiabatic and frictionless in nature. *Steady flow* is said to occur when conditions such as velocity, temperature at a certain point is invariant in time. When the conditions of flow do change with time the flow is said to be unsteady or *transient*. When all the points in the flow field have the same velocity then the flow is said to be in *plug or uniform flow*. *Vortex flow* or rotational flow is said to occur when fluid particles exhibit rotation about any axis. When the fluid within the region has no rotation the flow is described as *irrotational flow*. *One dimensional flow* neglects variations or changes in velocity, pressure, temperature, concentration etc, transverse to the main flow direction. When there is no change in flow normal to the planes of flow along identical path the flow is described as *two dimensional flow*. *Three dimensional flow* is the generalized description of flow and described by the u, v, w components of the velocity vector as a function of space coordinates, x, y, z and t.

A *streamline* is the imaginary continuous line drawn through the fluid so that it has the direction of the velocity vector at every point. A stream tube or stream filament is a tube of small or large cross section and of any convenient cross sectional shape that is entirely bounded by streamlines. A stream tube can be visualized as an imaginary pipe in the mass of flowing fluid through the walls of which no net flow is occurring. A path line is the path followed by a material element of fluid. When flow is steady stream line and pathline coincide. In transient flow, pathline generally do not coincide with the steam line.

A dye or smoke is frequently injected into a fluid in order to trace its subsequent motion. The resulting dye or smoke trials are called streak lines. For steady fluids, streamlines, pathlines and streaklines are coincidental. In two dimensional flow, flow stream lines are contours of the stream function. Streamlines in two dimensional flow can be obtained by injecting fire bright particles such as aluminum dust into the fluid, brilliantly lighting one plane and taking a photograph of the streaks made in a short time interval. Tracing on the picture continuous lines that have the direction of the streaks at every point portrays the streamlines for either steady or unsteady flow. Flow patterns may be detected using laser interferometers and Wollaston prism. The tracer particles are illuminated by creating laser sheet and photographs reveal the streamlines, when a sphere settles in a fluid, for example.

Incompressible flow, is said to occur when during study the density is not changed. *Compressible flow* is when the density changes during flow is more than 5%. Equation of state in addition to the equation of continuity, equation of mass, equation of momentum and equation of energy need be considered. The Mach number is obtained by taking the ratio of the velocity of fluid to the velocity of sound. When Ma < 1 the flow is said to be *subsonic*, and for Ma > 1 the flow is said to be *supersonic*. When Ma = 1 the flow is said to be *sonic or critical*. *Isothermal compressible flow* is often encountered in long transport lines where there is sufficient heat transfer to maintain constant temperature. *Annular flow* is found to happen in a cylindrical annulus. Flow can be classified as *rapid* or *tranquil*. When flow occurs at low velocities so that a small disturbance can travel upstream conditions, it is said to be in tranquil flow conditions. Upstream are affected by downstream conditions and the flow is controlled by the downstream conditions. The delineating dimensionless group is the Froude number, F $(v/(gl)^{1/2})$ for the tranquil flow F < 1. When flow occurs at such high velocities that a small disturbance such as an elementary wave is swept downstream, the flow is described as shooting or rapid (F > 1). Small changes in downstream condition do not effect any change in upstream condition. When flow is such that the velocity is just equal to the velocity of an elementary wave the flow is said to be critical (F = 1). *Subcritical* refers to tranquil flow at velocities less than critical and *supercritical* corresponds to rapid flows when velocities are greater than the critical point. Time dependent flow is a function of history of fluid.

Knudsen flow is said to occur when the mean free path of the molecule was greater than the width of the channel and the process is described by the pressure and temperature of the system. *Ballistic or relaxational flow* is said to occur when the accumulation of momentum is higher than an exponential rise and when the width of the channel is small the velocity of the fluid exhibits *subcritical damped oscillations*. Oscillations exist in *pulsatile flow* for example in the inhale and exhale of Oxygen and Carbon Dioxide. *Radial flow or squeeze flow* is said to happen when the r component of the velocity becomes a salient consideration.

The *Rayleigh Bernard* instabilities arise due to natural convection and the *Marangoni flow* is said to happen on account of thermocapillary stress. When chemical reactions take place during flow the condition is described as *reacting flow*. *Capillary flow* can be said to occur of blood in arteries and veins. Sub-atmospheric pressure condition leads to *vacuum flow*. *Tangential flow* emanates from moving circular objects. *Slip flow* is the transition between molecular and viscous flow. The slip boundary condition permits flow at the wall of the container. *Two phase flow* refers to the flow of more than one fluid such as gas-solid or liquid-gas etc.

4.2 Damped Wave Momentum Transfer and Relaxation

Newton's law of viscosity relates the shear stress to the shear rate with the constant of proportionality, the viscosity of the fluid, which offers resistance to flow.

$$\tau_{yx} = -\mu \partial v / \partial y \qquad (4.1)$$

where τ_{yx} is the shear stress, $\partial v / \partial y$ is the velocity gradient, and μ is the absolute viscosity. The negative sign in Eq. [4.1] is written to normalize the momentum transfer direction. Consider a plate at $y = l$, pulled at a constant velocity, V atop a stationary liquid. The layer of the liquid adjacent to the plate is also subjected to motion. The layer adjoint the bottom surface is stationary. The velocity gradient can be calculated as V/l. This multiplied by the absolute viscosity gives the shear stress in magnitude. If the force acting on the plate is F and the area of the plate, A then

$$\tau_{yx} = F/A \quad -\mu V/l = -\gamma (V \rho / l) \qquad (4.2)$$

The RHS of Eq. [4.2] represents the rate of momentum transfer. γ is the kinematic viscosity with units of m^2/s. The direction of momentum transfer in the downward direction, from atop the liquid towards the origin. Hence in order to render $F/A = \tau$ positive, the negative sign is added to Eq. [4.1]. Consequently, τ may be viewed as the momentum flux in the y direction. Besides the momentum flux and velocity gradient have to have opposite signs to stay within the bounds of the second law of thermodynamics. Momentum transfers from the high velocity region to a low velocity region by molecular transfer and the other direction is not allowed. The shear stress expression when combined with the equation of momentum results in a PDE that can be solved and the solution expressed as an infinite Fourier series. The singularity in Fourier series representation can be addressed by the use of the damped wave momentum and relaxation equation.

$$\tau_{yx} = -\mu \partial v / \partial y - \tau_r \partial \tau_{yx} / \partial t \qquad (4.3)$$

The damped wave momentum transfer and relaxation equation can arise from the accumulation term in the theory of kinetic theory of gases and derivation of physical properties of monatomic gases from molecular properties. From a molecular view, the viscosity can be derived and the momentum transport mechanism can be illustrated [Bird, Stewart and Lightfoot, 1960]. This derivation is revisited here. Consider molecules to be rigid, non-attracting spheres of mass m and diameter, d. The gas is assumed to be at rest and the molecular motion is considered. The following results of kinetic theory for a rigid sphere dilute gas in which temperature, pressure and velocity gradients are small are used;

Mean Molecular Speed $\quad <u> \; = \;$ sqrt($8\,\kappa T/\pi m$) $\hspace{3cm}$ (4.4)

Wall Collision Frequency $\quad Z \quad = \; \frac{1}{4}\,n'<u>$ $\hspace{3cm}$ (4.5)
per unit area

Mean free path $\hspace{2cm} \lambda \quad = \;$ 1/sqrt(2)/($\pi d^2 n'$) $\hspace{2cm}$ (4.6)

The molecules reaching any plane in the gas have on an average had their last collision at a distance a from the plane, where

$$a \quad = \; 2/3\,\lambda \hspace{3cm} (4.7)$$

In order to determine the viscosity of a dilute monatomic gas consider the gas when it flows parallel to the x axis with a velocity gradient $\partial v_x/\partial\partial z$. Assuming the relations for the mean free path of the molecule, wall collision frequency, distance to collision, mean velocity of molecule are good during the non-equilibrium conditions, the flux of momentum in the x direction across any plane z is found by summing the x momenta of the molecules that cross in the positive y direction and subtracting the x momenta of those that cross in the opposite direction. Thus,

$$\tau_{zx}$$
$$= \; Z\,mv_x\big|_{z-a} \; - \; Z\,mv_x\big|_{z+a} \hspace{3cm} (4.8)$$

It may be assumed that the velocity profile is essentially linear for a distance of several mean free paths. Molecules have velocity representative of their last collision. Accordingly.

$$v_x\big|_{z-a} \hspace{1cm} = \hspace{1cm} v_x\big|_z \; - \; 2/3\,\lambda\,\partial v_x/\partial z \hspace{2cm} (4.9)$$

$$v_x\big|_{z+a} \hspace{1cm} = \hspace{1cm} v_x\big|_z \; + \; 2/3\,\lambda\,\partial v_x/\partial z \hspace{2cm} (4.10)$$

Substituting Eqs. [4.9, 4.10] into Eq. [4.8],

$$\tau_{zx} \hspace{2cm} = \hspace{1.5cm} -1/3n\;m\;<u>\;\lambda\;dv_x/dz \hspace{2cm} (4.11)$$

Eq. [4.11] corresponds to the Newton's Law of viscosity with the viscosity given by;
$$\mu \hspace{2cm} = \hspace{1cm} 1/3\,\rho\,<u>\,\lambda \hspace{3cm} (4.12)$$

Apparently Maxwell obtained the above relation in 1860. It can be seen that prior to writing Eq. [4.8] the *accumulation of momentum* was neglected. This may be a good assumption at steady state but not at short time transient events. Thus considering a time increment t* the momentum is the momentum of molecules in minus momentum of molecules out minus the accumulation of momentum in the incremental volume under consideration, near the surface. The accumulation of momentum may be written in terms of,

$$\tau_{zx} = \; Z\,mv_x\big|_{z-a} \; - \; Z\,mv_x\big|_{z+a}$$
$$- \; t^*\partial/\partial t\{\,Z\,mv_x\big|_{z-a} \; - \; Z\,mv_x\big|_{z+a}\,\} \hspace{2cm} (4.13)$$

where t* is some characteristic time constant. To simplify matters Eq. [4.13] is used in Eq. [4.8] to give,

$$\tau_{zx} = -\mu dv_x/dz - t^* \partial\tau_{zx}/\partial t \tag{4.14}$$

t^* can be taken as the relaxation time of momentum, τ_{mom}. The equation of conservation of mass for any fluid can be derived as follows;

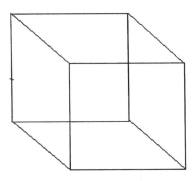

Figure 4.1 Region of Volume $\Delta x \, \Delta y \, \Delta z$ Fixed in Space through which Fluid is Moving

Consider a stationary volume element $\Delta x \, \Delta y \, \Delta z$ through which the fluid is flowing.

(rate of mass in)   -   (rate of mass out) $\pm$ (reaction rates)
                $=$ (rate of mass accumulation)   (4.15)

For the case when there is no chemical reaction,

$$\begin{aligned}
&\Delta x\Delta z \,(\rho\, v_y|_y - \rho\, v_y|_{y+\Delta y}) + \\
&\Delta x\Delta y \,(\rho\, v_z|_z - \rho\, v_z|_{z+\Delta z}) + \\
&\Delta y\Delta z \,(\rho\, v_x|_x - \rho\, v_x|_{x+\Delta x}) \qquad = \qquad \Delta x \, \Delta y \, \Delta z \, \partial\rho/\partial t
\end{aligned} \tag{4.16}$$

Dividing throughout the Eq. [4.16] by $\Delta x \, \Delta y \, \Delta z$ and taking the limits as the increments in the three directions, Δx, Δy, Δz go to zero.

$$\partial\rho/\partial t = -(\,\partial(\rho v_x)/\partial x + \partial(\rho v_y)/\partial y + \partial(\rho v_z)/\partial z) \tag{4.17}$$

Eq. [4.17] can be written in terms of the substantial derivative;

$$D\rho/Dt = -\rho(\partial v_x/\partial x + \partial v_y/\partial y + \partial v_z/\partial z) \tag{4.18}$$

230

Where the total derivative is given by;

$$D\rho/Dt = \partial\rho/\partial t + v_x \partial\rho/\partial x + v_y \partial\rho\partial y + v_z \partial\rho/\partial z) \tag{4.19}$$

At steady state for a fluid at constant density,

$$0 = -\rho(\partial v_x/\partial x + \partial v_y/\partial y + \partial v_z/\partial z) \tag{4.20}$$

Eq. [4.18] is the differential form of the equation of continuity. An integral form of the equation of continuity can be written as;

$$\partial/\partial t \int_{cv} \rho \, dv \quad + \int_{cs} \rho \, V \, dA = 0 \tag{4.21}$$

where, cv refers to the control volume and cs to the control surface. A control volume refers to a region in space and is useful in the analysis situations where flow occurs into and out of the space. The boundary of a control volume is its control surface. The size and shape of the control are entirely arbitrary. They can be made to coincide with solid boundaries in parts. The control volume is also referred to as an open system.

The Equation of Motion can be derived using a momentum balance on the control volume shown in Figure 4.1.

(rate of momentum in) - (rate of momentum out) + (sum of forces acting on system)
$$= \quad \text{(rate of momentum accumulation)} \tag{4.22}$$

Consider the x component of momentum into and out of volume element shown in Figure 4.1. Momentum flows into and out of the volume element by two mechanisms;

1) by convection or by bulk fluid flow
2) by molecular transfer (velocity gradients)

$\Delta y \Delta z \, (\tau_{xx}|_x - \tau_{xx}|_{x+\Delta x})$
$+ \ \Delta x \Delta y \, (\tau_{zx}|_z - \tau_{zx}|_{z+\Delta z})$
$+ \ \Delta x \Delta z \, (\tau_{yx}|_y - \tau_{yx}|_{y+\Delta y})$ = net transfer of x component momentum by molecular transfer

τ_{xx} is the normal stress on the x face and τ_{yx} the tangential stress on the y face from viscous forces. By convection,

$\Delta y \Delta z \, (\rho \, v_x^2|_x - \rho \, v_x^2|_{x+\Delta x})$
$+ \ \Delta x \Delta z \, (\rho \, v_y v_x|_y - \rho \, v_y v_x|_{y+\Delta y})$
$+ \ \Delta x \Delta y \, (\tau(\rho \, v_z v_x|_z - \rho \, v_z v_x|_{z+\Delta z})$ = net transfer of x component momentum by convection

$$\tag{4.23}$$

The sum of the external forces arise from that of hydrostatic pressure and gravity. The resultant force in x direction is,

$$\Delta y \Delta z \, (p|_x - p|_{x+\Delta x}) + \rho \, g_x \, \Delta x \Delta y \Delta z \tag{4.24}$$

$$(\Delta x \Delta y \Delta z)\partial(\rho v)/\partial t = \text{rate of accumulation of momentum} \quad (4.25)$$

Substituting Eqs.[4.23 – 4.25] into Eq. [4.22] and dividing throughout $\Delta x \Delta y \Delta z$ and obtaining the limits as Δx, Δy, Δz going to zero the x component of equation of motion of the fluid can be obtained.

$$\partial(\rho v_x)/\partial t = -[\partial (\rho v_x^2)/\partial x + \partial(\rho v_x v_y)/\partial y + \partial(\rho v_x v_z)/\partial z]$$
$$-[\partial\tau_{xx}/\partial x + \partial\tau_{yx}/\partial y + \partial\tau_{zx}/\partial z]$$
$$-\partial p/\partial x + \rho g_x \quad (4.26)$$

The equation of momentum in the x can be written in terms of the substantial derivative as;

$$\rho DV_x/Dt = -\nabla p - (\partial\tau_{xx}/\partial x + \partial\tau_{yx}/\partial y + \partial\tau_{zx}/\partial z) + \rho g_x \quad (4.27)$$

where ∇ is the vector differential operator. Adding the x component, y component and z components of momenta and use made of the substantial derivate the equation of motion including all three components, can be written as;

$$\rho DV/Dt = -\nabla p + \mu \nabla^2 V + \rho g \quad (4.28)$$

Eq.[4.27] is the Navier – Stokes equation. The finite speed of momentum transfer can be taken into account and the Navier [1821] - Stokes [1845] expression modified or extended to include the damped wave transport effects. As discussed in Chapter 1 the ballistic transport or phonon or at a molecular level another factor needs to be included in addition to the ones considered in writing the Fourier [1822] , Fick [1855] and Newton's [1623-1727] laws.

Consider the damped wave momentum and relaxation equation that contribute to the x component of the momentum. The three shear stress components are differentiated with respect to x, y, and z respectively to give;

$$\partial\tau_{xx}/\partial x = -\mu\partial^2 v_x/\partial x^2 - \tau_{mom}\partial^2\tau_{xx}/\partial t\partial x \quad (4.29)$$

$$\partial\tau_{yx}/\partial z = -\mu\partial^2 v_x/\partial y^2 - \tau_{mom}\partial^2\tau_{yx}/\partial t\partial y \quad (4.30)$$

$$\partial\tau_{zx}/\partial z = -\mu\partial^2 v_x/\partial z^2 - \tau_{mom}\partial^2\tau_{zx}/\partial t\partial z \quad (4.31)$$

Differentiating Eq. [4.27] wrt to time,

$$\rho D (\partial V_x/\partial t) /Dt = -\nabla(\partial p/\partial t) - (\partial^2\tau_{xx}/\partial x\partial t + \partial^2\tau_{yx}/\partial y\partial t + \partial^2\tau_{zx}/\partial z\partial t) \quad (4.32)$$

Adding Eq.[4.29-4.31] and substituting for the sum of the three cross-derivatives in shear stress wrt to time and x, y, and z,

$$\rho D (\partial V_x/\partial t) /Dt = -\nabla(\partial p/\partial t) + 1/\tau_{mom}(\mu \nabla^2 V + \partial\tau_{xx}/\partial x + \partial\tau_{yx}/\partial y + \partial\tau_{zx}/\partial z) \quad (4.33)$$

232

Eliminating the three stress derivatives that contribute to the x component of the momentum between Eq. [4.33] and Eq. [4.27],

$$\tau_{mom}\rho D\, \partial V_x/\partial t\, /Dt\, +\, \rho DV_x/Dt\quad +\tau_{mom}\nabla(\partial p/\partial t)\, =\quad \mu\, \nabla^2 V\, -\nabla p\, +\, \rho g_x \quad (4.34)$$

Adding the three components in the x, y and z direction the Navier-Stokes equation of motion is extended to include the fluid inertial and transient effects using the damped wave momentum transport and relaxation equation to give;

$$\tau_{mom}\rho D\, \partial V_x/\partial t\, /Dt\, +\, \rho DV/Dt\quad +\tau_{mom}\nabla(\partial p/\partial t)\, =\quad \mu\, \nabla^2 V\, -\nabla p\, +\, \rho g \quad (4.35)$$

For constant density system and with negligible viscosity, Eq. [4.35] becomes;

$$\tau_{mom}\rho D\, \partial V_x/\partial t\, /Dt\quad +\, \rho DV/Dt\quad +\tau_{mom}\nabla(\partial p/\partial t)\, =\quad -\nabla p\, +\, \rho g \quad (4.36)$$

Eq. [4.36] gives the extended Euler equation that now includes the damped wave transport and relaxation effects. Euler equation [1755] neglects the viscous effects. Each term in Eq. [4.35] have different physical meaning. The mass times acceleration on the fluid is equal to the sum of the pressure forces, gravity forces and viscous forces in the Navier-Stokes equation. To this are added two more terms. One is due to the damped wave transport and relaxation and provides an "acceleration force" that becomes important when the accumulation of momentum term becomes higher than an exponential rise in time. The other term is due to the variation of pressure with time. An additional "force" other than the pressure force in Navier-Stokes due to the dynamics of pressure is also included in Eq. [4.36]. These two terms are on account of the relaxation phenomena in the fluid. This in later analysis provides well bounded results and removes the anomalies due to an infinite speed of momentum propagation.

Worked Example 4.1

Consider a pipe filled with water upto a level H at initial time that begins to empty out using an orifice. The pressure and velocity distribution as a function of z and t in the long tube can be derived using the extended Euler equation in 1 dimension. Thus Eq. [4.36] written in the z direction in Figure 4.2 is;

$$\tau_{mom}\rho[\partial^2 v_z/\partial t^2\, +\, v_z\partial^2 v_z/\partial z\partial t\, +\, (\partial v_z/\partial t)(\partial v_z/\partial z)]\, +\, \rho\partial v_z/\partial t\, +\, v_z\partial v_z/\partial z\quad +\tau_{mom}\nabla(\partial p/\partial t)$$
$$=\quad -\partial p/\partial z\, -\, \rho g \quad (4.37)$$

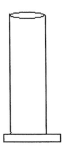

Figure 4.2 Emptying of a Filled Pipe

$$p = (H - z)\, \rho g_z + p_{atm} \tag{4.38}$$

$$-\partial p/\partial z = -\rho g \tag{4.39}$$

$$-\partial^2 p/\partial t \partial z = \rho g \partial v_z/\partial z = 0 \tag{4.40}$$

It can be assumed that during the drainage the velocity of liquid in the tube the velocity of the liquid, v_z is independent of z and is only a function of time. This is true for wide reservoirs where the height in the contianer does not change appreciably. Thus at constant density,

$$\tau_{mom}\, \partial^2 v_z/\partial t^2 + \partial v_z/\partial t + 2g = 0 \tag{4.41}$$

Let $\tau = t/\tau_{mom}$

$$\partial^2 v_z/\partial \tau^2 + \partial v_z/\partial \tau + 2g\tau_{mom} = 0$$

Let $u = v_z/g\tau_{mom}$

$$\partial^2 u/\partial \tau^2 + \partial u/\partial \tau + 2 = 0 \tag{4.43}$$

The solution to the second order ODE with constant coefficients in Eq. [4.43] is given by;

Let $V = \partial u/\partial \tau + 2$ \hfill (4.44)

$$\partial V/\partial \tau = -V; \quad -\ln V = \tau + c \tag{4.45}$$

$$V = c'\exp(-\tau) \tag{4.46}$$

$$u = c''\exp(-\tau_{efflux})\,((1 - \exp(\tau_{efflux} - \tau)) + 2(\tau_{efflux} - \tau) \tag{4.47}$$

$$\int u\, d\tau = c''\exp(-\tau_{efflux})\,((\tau + \exp(\tau_{efflux} - \tau)) + 2\tau_{efflux}\tau - \tau^2 + c' \tag{4.48}$$

$$\int u \, d\tau \ = \ 1/g\tau_{mom}{}^2 z \ + c$$

Thus,

$$z \, /g\tau_{mom}{}^2 \ = (c" \exp(-\tau_{efflux}) \, ((\ \tau + \exp(\tau_{efflux} - \tau) \,) \ + 2\tau_{efflx} \tau \ - \ \tau^2 \,) \ + c"" \quad (4.49)$$

At $\tau = 0$, $z = H$ $\qquad\qquad\qquad\qquad\qquad\qquad\qquad\qquad\qquad\qquad$ (4.50)

$\tau \ = \tau_{efflux}, \ \ z = 0$

$$c_. + c" = H/ \, g\tau_{mom}{}^2 \qquad\qquad\qquad\qquad\qquad\qquad\qquad (4.51)$$

$$c" \ = \ H*/ \, (\ 1 + (\tau_{efflux} \ + \tau_{efflux}{}^2 \,)(\ \exp(-\tau_{efflux})) \qquad\qquad (4.52)$$

Thus,

$$u \ = [H/ \, g\tau_{mom}{}^2] \ [(1 - \exp(\tau_{efflux} - \tau) \,) \ + 2(\tau_{efflux} - \tau)]/[(\ \exp(+\tau_{efflux)} + (\tau_{efflux} \ + \tau_{efflux}{}^2 \,)] \quad (4.53)$$

The above expression is derived based upon a lumped azimuthal velocity. It can be a good estimate of the space averaged velocity. The expression is independent of the height of liquid in the pipe. The efflux time can be calculated using a pseudo-steady state assumption.

A pseudo-steady state is assumed to have been reached. At steady state the Toricelli's theorem relates the efflux velocity to the height of liquid in the tube. Thus,

$$v_z \ = \text{sqrt}(2gz) \qquad\qquad\qquad\qquad\qquad\qquad (4.54)$$

This can be arrived at by using the Bernoulli equation which is the Euler equation at steady state integrated in 1 dimension [Sharma, 2003]. The time to efflux is then obtained by performing a mass balance on the contents of the pipe at constant density;

$$\text{(rate in)} \quad \text{- (rate out)} \ = \text{(rate of accumulation)} \qquad\qquad (4.55)$$

$$\text{-sqrt}(2gz) \, A_{orifice} \qquad = \ A_{tube} \, \partial z/\partial t \qquad\qquad\qquad (4.56)$$

Integrating between the limits of z ranging from H to 0 and the time from 0 to t_{efflux}.

$$t_{efflux} \ = \ 2\text{sqrt}(H/2g)(A_{tube}/A_{orifice}) \qquad\qquad\qquad (4.57)$$

$$\tau_{eflux} \ = 2\text{sqrt}(H/2g\tau_{mom}{}^2)(A_{tube}/A_{orifice}) \qquad\qquad\qquad (4.58)$$

Eq. [4.61] can be used in Eq. [4.56] as an approximation from pseudo-steady state assumption.

Worked Example 4.2

In the above worked example 4.1, consider multiple ports along the periphery of the tall tube. As the orifices are in the periphery of the tube the x component velocity may also exist. Consider the x component velocity in addition to the z component velocity in the analysis. Discuss the z component velocity. From the equation of continuity,

$$\partial v_x/\partial x + \partial v_z/\partial z = 0 \tag{4.59}$$

The x component velocity can be assumed to vary in a linear fashion. At the periphery the efflux velocity, v_{efflux}, may be assumed to be constant, and independent of z. Consider long slender tubes where the height of the fluid is much greater than the location of the multiple ports. At $x = 0$, which is at the center of the tube the x component velocity is zero. Thus,

$$v_x = c_1 x + c_2 \tag{4.60}$$

As at $x = 0$, v_x is zero, c_2 need be set to zero. At $x = D/2$, where D is the diameter of the tube, the velocity of the fluid in the x component is equal to the efflux velocity which is assumed to be independent of the height of the port,

$$v_{efflux} = c_1 D/2 \text{ or } c_1 = 2v_{efflux}/D \tag{4.61}$$

$$v_x/v_{efflux} = 2x/D \tag{4.62}$$

$$\partial v_x/\partial x = -\partial v_z/\partial z = 2v_{efflux}/D \tag{4.63}$$

Thus, $v_z = 2v_{efflux}z/D + c'$ $\tag{4.64}$

At the bottom of the tube at $z = 0$, the z component velocity is zero as all of the efflux velocity can be assumed to be in the x component. So $c' = 0$. Thus,

$$\partial v_z/\partial z = 2v_{efflux}/D \tag{4.65}$$

Writing the equation of motion in the z component at constant density,

$$\tau_{mom} [\partial^2 v_z/\partial t^2 + (\partial v_z/\partial t)(2v_{efflux}/D)] + \partial v_z/\partial t + v_z 2v_{efflux}/D \quad = \quad -2g \tag{4.66}$$

In order to convert the governing equation in v_z into dimensionless form, define the following variables.

$$\text{Let } \tau = t/\tau_{mom}; \quad u = v_z/v_{efflux}; \quad D^* = D/(v_{efflux}\,\tau_{mom}) \tag{4.67}$$

$$\partial^2 u/\partial \tau^2 + \partial u/\partial \tau (2/D^* + 1) + 2u/D^* + 2\,(Acc) = 0 \tag{4.68}$$

Eq. [4.72] is a second order ODE with constant coefficients. The homogeneous portion of the solution can be obtained first and then the particular integral for a constant is another constant can be added to the solution. In order to solve the homogeneous portion of the solution,

$$r^2 + r(2/D^* + 1) + 2/D^* = 0 \qquad (4.69)$$

The two roots are -1, $-2/D^*$

$$u = c_1 \exp(-\tau) + c_2 \exp(-2\tau/D^*) + c_3 \qquad (4.70)$$

c_3 can be seen to be zero as at infinite time u has to go to zero.

$$\int u d\tau = -c_1 \exp(-\tau) - D^*/2\, c_2 \exp(-2\tau/D^*) + c_4 = z/\tau_{mom} v_{efflux} \qquad (4.71)$$

when $\tau = 0$, $z = H$ $\qquad (4.72)$

$$-c_1 - D^*/2\, c_2 + c_4 = H \qquad (4.73)$$

$$0 = c_1 \exp(-\tau_{efflux}) - D^*/2\, c_2 \exp(-2\tau_{efflux}/D^*) + c_4 \qquad (4.74)$$

$$0 = c_1 \exp(-\tau_{efflux}) + c_2 \exp(-2\tau_{efflux}/D^*) \qquad (4.75)$$

Thus $c_4 = 0$,

$$c_2 = H/[\exp(-2\tau_{efflux}/D^* + \tau_{efflux}) - D^*/2] \qquad (4.76)$$

$$c_1 = -HD^*/2\ (\exp(-2\tau_{efflux}/D^* + \tau_{efflux}) - D^*/2] - H \qquad (4.77)$$

4.3 Flow Near a Horizontal Wall Suddenly Set in Motion

Consider a fluid with constant density, ρ and constant viscosity, μ, atop a horizontal plate. The fluid medium is assumed to be in continuum. In most cases, at the macroscopic scale, molecular structure of the fluid is not taken into account. Mass is concentrated in the nuclei of atoms and is far from uniformly distributed over the volume occupied by the liquid. Non-uniform distribution can be seen in other variable such as composition and velocity when viewed on a microscopic scale. The continuum supposition is that the behavior of fluids is the same as if they were perfectly continuous in structure and physical quantities such as mass and momentum associated with matter contained within a given volume will be regarded as being spread uniformly over that volume, instead of as is strict reality, being concentrated n a small fraction of it. This is the continuum hypothesis. Atop the horizontal plate is a semi-infinite medium of fluid. The fluid is stationary at time zero. For times greater than zero, the plate is set at a constant velocity, V. The velocity in the z direction as a function of space and time is of interest. An error function result when the Newton's law of viscosity is used for the fluid. The spatio-temporal velocity of the fluid is obtained from the damped wave momentum transfer and relaxation equation. From the equation of motion, neglecting convection effects,

$$\tau_{mom} \partial^2 v_x / \partial t^2 + \partial v_x / \partial t \quad = \quad \gamma \partial^2 v_x / \partial z^2 \; - 2g \tag{4.78}$$

Let $u = v_x / V$; $\tau = t / \tau_{mom}$; $Z = z / sqrt(\gamma \; \tau_{mom})$ (4.79)

The one time and two space conditions are

$\tau = 0, \; u = 0$ (4.80)

$Z = 0, \; u = 1$ (4.81)

$Z = \infty, \; u = 0$ (4.82)

$$\partial^2 u / \partial \tau^2 + \partial u / \partial \tau \quad = \quad \partial^2 u / \partial Z^2 \; - 2g\tau_{mom}/V \tag{4.83}$$

$[g\tau_{mom}/V]$ is a dimensionless number which represents the ratio of gravity forces to the ballistic "force" that corrects for the accumulation of momentum and can be called accumulation number.

$$\text{Thus,} \quad Acc = \quad [g\tau_{mom}/V] \tag{4.84}$$

Thus the governing equation is;

$$\partial^2 u / \partial \tau^2 + \partial u / \partial \tau \quad = \quad \partial^2 u / \partial Z^2 \; - 2Acc \tag{4.85}$$

The solution to the homogenous portion of the equation is sought by the method of transformation of coordinates as shown in earlier sections.

The damping term can be removed using a $u = Wexp(-n\tau)$ substitution. As shown in the previous chapters at $n = \frac{1}{2}$ the governing equation becomes;

$$\partial^2 W / \partial \tau^2 \; - W/4 \quad = \quad \partial^2 W / \partial Z^2 \tag{4.86}$$

Let $\eta = \tau^2 - X^2$ (4.87)

Then Eq. [4.87] becomes,

$$4\eta \; \partial^2 W / \partial \eta^2 + 4 \; \partial W / \partial \eta \; - W/4 \quad = 0 \tag{4.88}$$

Comparing the above equation to the generalized Bessel differential equation,

$$W = c_1 I_0 \tfrac{1}{2} sqrt(\tau^2 - Z^2) \quad + c_2 K_0 \tfrac{1}{2} sqrt(\tau^2 - Z^2), \quad \text{for } \tau > Z \tag{4.89}$$

The term containing $K_0 \tfrac{1}{2} sqrt(\tau^2 - X^2)$ can be dropped to zero as at the wave front the velocity is finite. At $\tau = X$,

$$u = exp(-Z/2) = exp(-\tau/2), W = 1 \tag{4.90}$$

At $Z = 0$, $\quad 1 = \quad c_1 I_0 (\tau/2)$ (4.91)

Eliminating c_1 between Eq. [4.91] and Eq. [4.89] for $\tau > X$,

$$u = I_0 \tfrac{1}{2} \operatorname{sqrt}(\tau^2 - Z^2)/(I_0(\tau/2)) \qquad (4.92)$$

For $X > \tau$, $\qquad\qquad u = J_0 \tfrac{1}{2} \operatorname{sqrt}(Z^2 - \tau^2)/(I_0(\tau/2))$

The shear stress can be given as shown in the heat transfer and mass transfer sections,

$$\tau_{zx} / \operatorname{sqrt}(\rho V^2 \, \mu/\tau_{mom}) = \exp(-\tau/2) \, I_0 1/2(\operatorname{sqrt}(\tau^2 - Z^2)) \qquad (4.93)$$

For $Z > \tau$, $\qquad \tau_{zx} / \operatorname{sqrt}(\rho V^2 \, \mu/\tau_{mom}) = \exp(-\tau/2) \, J_0 1/2(\operatorname{sqrt}(Z^2 - \tau^2)) \qquad (4.94)$

The shear force can be seen to be;

$$F/[\, AV\rho \, v_{mom}] = \exp(-\tau/2) \, I_0 1/2(\operatorname{sqrt}(\tau^2 - Z^2)) \qquad (4.95)$$

The dimensionless force exerted by the plate on the fluid can be seen to be;

$$F^* = F/AV\rho \, v_{mom} \qquad (4.96)$$

The solution exhibits some space time symmetry with respect to the negative values as well as with each other. It can be seen that for a plate at some point in the interior of the semi-infinite medium the shear force exerted by the fluid on the plate is bifurcated. In fact it has three different regimes. The first is the thermal inertia regime. In this regime there is no action of the fluid on the plate. In the second regime the shear stress is given by Eq. [4.94] which is a product of the decaying exponential and a Bessel composite function of the first kind and zeroth order. The third regime after the wave front, is represented by Eq. [4.93] and is a product of a modified composite Bessel function of the first and zeroth order. The momentum inertia can be calculated as;

$$\tau_{inertia} = \operatorname{sqrt}(Z_p^2 - 23.1323) \qquad (4.97)$$

In the first regime, the shear force may be negative should the Bessel function's negative sign have meaning. This could be the first few ripples that the plate see from the disturbance from the surface. The shear force can be in the opposite direction and eventually after the thermal time lag has elapsed the force is in the right direction. The shear stress undergoes a maximum. The second regime is a steep rise. The first regime is an inertial time of upto 3.597 in dimensionless quantities. The third regime is a tailed fall. The curvature changes from convex to concave. There is an inflection point in the third regime. There is a skew to the right and the kurtososis may be compared to the Maxwell distribution.

On examining Eq.[4.92] it can be seen that when $Z > \tau$, the expression for the dimensionless velocity becomes a Bessel composite function. This is because when $Z > \tau$, the argument in the modified Bessel composite function within the square root sign becomes negative. The square root of -1 is i. Further,

$$J_0(x) = I_0(ix) \tag{4.98}$$

Hence, Eq. [4.92] becomes for $Z > \tau$,

$$u = J_0 \tfrac{1}{2} \text{sqrt}(Z^2 - \tau^2)/(I_0(\tau/2)) \tag{4.99}$$

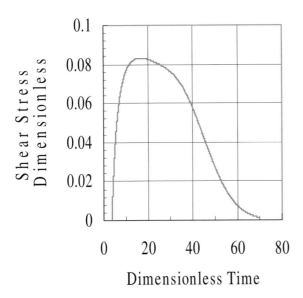

Figure 4.3   Three Regimes of Dimensionless Shear Stress in the Interior
Of a Semi-infinite Fluid Suddenly Pulled by a Plate at a
Constant Velocity from the Bottom

It is generally realized in analysis such as in boundary layer theory , that after a finite region from the moving plate the fluid will be at the initial state or will have zero velocity. The first zero of the Bessel function occurs at 2.4048. Beyond that the velocity predicted will be negative. Granted that the denominator in Eq. [4.103] will dampen the oscillations, why would the velocity of fluid be negative after a said distance from the moving flat plate at the boundary for a given time instant under consideration. Since it is damped oscillatory, the effect of the surface disturbance for distances further than sqrt(23.13 + τ^2) acts differently on account of the ballistic transport. It can be taken as zero from an analogy from heat wave conduction and relaxation or mass wave diffusion and relaxation. If it is taken as zero then, the boundary layer thickness for a given instant in time greater than zero is given by,

$$\delta(\tau) = sqrt(23.13 + \tau^2) \qquad (4.100)$$

Beyond this distance, the fluid velocity can be taken to be zero from the analogy from heat or mass diffusion and relaxation.

The model prediction of Eq. [4.99] gives negative values for velocity beyond the boundary layer thickness. Velocity is a vector. In the momentum balance equation from which the solution is derived the velocity is preserved through the analysis. Hence, a negative velocity, could mean that the velocity of the fluid is in the opposite direction compared with the velocity of the flat plate. Upto the first root of the Bessel function the second regime for the dimensionless velocity profile hold good. For a given instant of time, for values of Z smaller than the instant of dimensionless time, the third regime or the modified Bessel composite function solution is applicable. The negative values for the velocity can be due to the ballistic transport mechanism. The disturbance swims back from the region beyond the boundary layer. This is sort of the ripple effect and backflow phenomena. This needs to be borne out by experiment. It can be seen in graphical form as follows. For large values of the argument, the Bessel function can be approximated with a cosinuous function as follows;

$$u = sqrt(4/(\pi(Z^2 - \tau^2)^{1/2}))Cos[\tfrac{1}{2}(Z^2 - \tau^2)^{1/2} - \pi/4] /(I_0(\tau/2) \qquad (4.101)$$

In Figures [4.4, 4.5] are plotted the dimensionless velocity for a given instant in time ($\tau = 5$) as a function of dimensionless distance. In Figure 4.4, it can be seen that close to the flat plate the dimensionless velocity obeys Eq. [4.98] and is valid for dimensionless distances less than the instant in time under consideration. This is given by the modified composite Bessel spatio-temporal function of the first kind and zeroth order divided by the modified Bessel function in time, of the first kind and first order. For dimensionless distances greater than the time instant under consideration the Bessel composite spatio-temporal function of the first kind and zeroth order divided by the modified Bessel function in time of the first kind and zeroth order gives the dimensionless velocity profile. Beyond the first zero, of the Bessel function the solution predicts damped oscillations for the dimensionless velocity. Upto the first zero of the Bessel function the velocity of the fluid is positive. In this case, this value can be calculated to be sqrt(22.21 + 16) = 6.18. Beyond 6.18, the velocity is in the negative direction. This is the subcritical damped oscillatory regime. This needs to be verified by experiment. The ballistic transport mechanism gives credence to some wave motion for certain conditions. This was seen as model predictions in the heat and mass transfer sections as well. However, for the fluid problem it manifests as a vector and a minus sign indicates a reversal of flow in the direction opposite to the direction of movement of the flat plate.

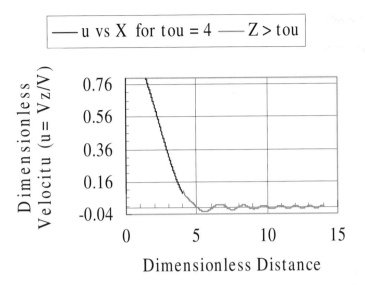

Figure 4.4 Dimensionless Velocity of Fluid in a Semi-infinite
Fluid from a Moving Flat Plate

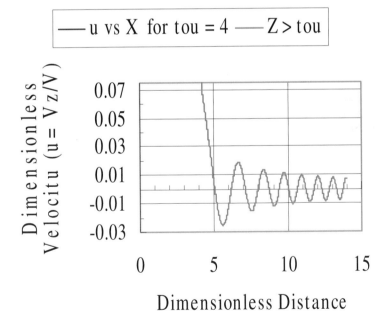

Figure 4.5 Damped Oscillatory Behavior of Dimensionless Velocity
Far from the Flat Plate in Semi-infinite Medium

Worked Example 4.3

The homogeneous portion of Eq. [4.91] is hyperbolic in nature as explained in Chapter 1.0. There are a number of solutions available in the literature for parabolic equation. Search for a substitution that can transform the hyperbolic PDE into a parabolic PDE in a form for which the solutions are available.

The governing equation for transient velocity profile can be given by;

$$\partial^2 u/\partial \tau^2 + \partial u/\partial \tau \quad = \quad \partial^2 u/\partial Z^2 \qquad (4.102)$$

The parabolic PDE that this can become after a suitable transformation is;

$$\partial u/\partial \tau \quad = \quad \beta \partial^2 u/\partial \theta^2 \qquad (4.103)$$

Let θ be a general substitution,

$$\theta = g(Z) + V(\tau) \qquad (4.104)$$

Then,

$$\partial u/\partial \tau \quad = (\partial u/\partial \theta)V'(\tau) \qquad (4.105)$$

$$\partial^2 u/\partial \tau^2 \quad = (\partial^2 u/\partial \theta^2)V'^2 + (\partial u/\partial \theta)V'' \qquad (4.106)$$

$$\partial^2 u/\partial Z^2 \quad = (\partial^2 u/\partial \theta^2)g'^2 + (\partial u/\partial \theta)g'' \qquad (4.107)$$

Substituting Eqs. [4.105-4.107] in Eq. [4.103],

$$\partial u/\partial \tau \quad = \quad \partial^2 u/\partial Z^2 - \partial^2 u/\partial \tau^2 \qquad (4.108)$$

$$= (\partial^2 u/\partial \theta^2)(g'^2 - V'^2) + (\partial u/\partial \theta)(g'' - V'') \qquad (4.109)$$

In order for Eq. [4.109] to take on the form of Eq. [4.103],

$$g'' - V'' = 0 \qquad (4.110)$$

$$g'^2 - V'^2 = \beta \qquad (4.111)$$

Thus $g' = \mathrm{sqrt}(1 + V'^2) = c$ (only then two different functions can be equal)

$$g = cZ \qquad (4.112)$$

$$g'' = 0 \text{ and } V'' = 0, \qquad (4.113)$$

$$V' = d; V = d\tau + e \qquad (4.114)$$

$$c^2 - d^2 = \beta \qquad (4.115)$$

$$d = \mathrm{sqrt}(c^2 - \beta) \qquad (4.116)$$

Hence, $g'^2 - V^2 = \beta$ (4.117)

$$\partial u/\partial \tau = \beta \partial^2 u/\partial \theta^2$$ (4.118)

The substitution, θ that made the transformation possible was,

$$\theta = c(Z + sqrt(1 - \beta/c^2)\tau + e$$ (4.119)

Assuming $c = 1$ and $e = 0$,

$$\theta = Z + \tau(1 - \beta)^{1/2}$$ (4.120)

Eq. [4.118] is a PDE in two variables. This can be converted into an ODE in one variable by the Boltzmann transformation;

Let $\psi = \theta/(4\beta\tau)^{1/2}$ (4.121)

$$\beta \partial^2 u/\partial \theta^2 = \partial^2 u/\partial \psi^2 /4\tau$$ (4.122)

$$\partial u/\partial \tau = -\psi/(2\tau)(\partial u/\partial \psi)$$ (4.123)

Substituting Eqs. [4.121, 4.123] into Eq. [4.118],

$$\partial^2 u/\partial \psi^2 = -2\psi(\partial u/\partial \psi)$$ (4.124)

Let $\partial u/\partial \psi = p$,

$$-\partial p/p = 2\psi \partial \psi$$

Integrating both sides,

$$\ln(p) = -\psi^2 + c'$$ (4.125)

$$u = \int exp(-\psi^2) \, d\psi + c''$$

$$erf(\psi) = 2/sqrt(\pi) \int exp(-\psi^2) \, d\psi$$ (4.126)

Thus $u = c_1 erf(\psi) + c_2$ (4.127)

From the boundary condition at $\psi = \infty$, where $erf(\infty) = 1$, $u = 0$.

$$c_1 = -c_2$$ (4.128)

From the boundary condition at the surface,

$$u = 1, \quad when \ Z = 0$$

$$\theta = \tau(1 - \beta)^{1/2} \ ; \ \psi = \tau^{1/2}(1 - \beta)^{1/2} /2\beta^{1/2}$$ (4.129)

$$1 = c_2 (1 - \text{erf}(\tau^{1/2}(1 - \beta)^{1/2} /2\beta^{1/2})) \tag{4.130}$$

$$u = c_2 (1 - \text{erf}(\psi)) \tag{4.131}$$

Eliminating c_2 between the two Eqs. [4.130, 4.131],

$$u = [1 - \text{erf}(\psi)]/[1 - \text{erf}(\tau^{1/2}(1 - \beta)^{1/2} /2\beta^{1/2})] \tag{4.132}$$

c_1, c_2 were solved from the space and time conditions. β can be chosen as follows;

$$1 = (1 - \beta)^{1/2} /2\beta^{1/2} \tag{4.133}$$

$$4\beta + \beta - 1 = 0 \tag{4.134}$$

$$\beta = 1/5 \tag{4.135}$$

Then, $$u = [1 - \text{erf}((5)^{1/2}Z/(4\tau)^{1/2} + \tau^{1/2})]/[1 - \text{erf}(\tau^{1/2})] \tag{4.136}$$

The parabolic PDE can be described by two space and one time condition. For a PDE of order n, n functions need to be solved for as against n constants for an ODE of order n. Although the hyperbolic PDE needs two space and two time conditions for complete description, it was converted to a parabolic PDE.

Worked Example 4.4

In the worked example 4.1 in the study of emptying a pipe filled with liquid consider a tube filled with porous packing. The Darcy's law can be used to relate the pressure gradient with the flow velocity. Consider the resulting governing equation and solve for the vertical component of the velocity of the fluid,

The equation of motion considering the ballistic transport effects for the vertical component of the velocity of the fluid can be written as;

$$\tau_{mom}\rho[\partial^2 v_z/\partial t^2 + v_z \partial^2 v_z/\partial z \partial t + (\partial v_z/\partial t)(\partial v_z/\partial z)] + \rho \partial v_z/\partial t + v_z \partial v_z/\partial z + \tau_{mom}\nabla(\partial p/\partial t)$$
$$= -\partial p/\partial z - \rho g \tag{4.137}$$

From Darcy's law,

$$v_z = - (\kappa/\mu) (\partial p/\partial z - \rho g) \tag{4.138}$$

$$-\partial^2 p/\partial t \partial z = (\mu/\kappa) \partial v_z/\partial t \tag{4.139}$$

where κ is the permeability of the porous medium. Eq. [4.137] becomes;

$$\tau_{mom}\rho[\partial^2 v_z/\partial t^2 + v_z \partial^2 v_z/\partial z \partial t + (\partial v_z/\partial t)(\partial v_z/\partial z)]$$

$$+ [\rho - \mu/\kappa\tau_{mom}]\partial v_z/\partial t + v_z\partial v_z/\partial z - \mu/\kappa v_z = 0 \qquad (4.140)$$

It can be assumed that during the drainage the velocity of liquid in the tube the velocity of the liquid, v_z is independent of z and is only a function of time. Or this can be arrived at by writing the equation of continuity. This is true for wide reservoirs where the height in the contianer does not change appreciably. Thus at constant density the equation of momentum becomes;

$$\tau_{mom} \partial^2 v_z/\partial t^2 + [1 - \gamma/\kappa\tau_{mom}] \partial v_z/\partial t - \gamma/\kappa v_z = 0 \qquad (4.141)$$

Let $\tau = t/\tau_{mom}$

$$\partial^2 v_z/\partial\tau^2 + [1 - (\gamma\tau_{mom} /\kappa)]\partial v_z/\partial\tau + - (\gamma\tau_{mom} /\kappa) v_z = 0 \qquad (4.142)$$

where,

$$Pb = (\gamma\tau_{mom} /\kappa)$$

Pb is a sort of a permeability number that gives the ratio of the kinematic viscosity times the relaxation time of momentum divided by the Darcy permeability of the medium. It may represent the ratio of viscous forces and ballistic transport "forces" to the permeability forces.

Let $u = v_z/g\tau_{mom}$

$$\partial^2 u/\partial\tau^2 + (1 - Pb)\partial u/\partial\tau - Pb\, u = 0 \qquad (4.143)$$

The solution to Eq. [4.143] which is a second order ODE with constant coefficients can be written as;

$$u = \exp(-\tau/2(1- Pb) [c_1 \exp(-\tau) + c_2 \exp(Pb\tau) \qquad (4.144)$$

From the constraint at infinite time, i.e., $u = 0$, c_2 can be seen to be zero.

The initial condition can be written assuming a pseudo-steady state and using Toricelli's theorem;

$$\mathrm{sqrt}(2gH) = c_1 \qquad (4.145)$$

Thus,

$$u = (2gH)^{1/2} \exp(-\tau(3/2- Pb/2) \qquad (4.146)$$

Worked Example 4.5

In the worked example 4.1 in the study of emptying a pipe filled with liquid consider a tube filled with porous packing and the apparatus taken in a space shuttle and into the galaxy. The Darcy's law can be used to relate the pressure gradient with the flow velocity as follows. Consider the resulting governing equation and solve for the vertical component of the velocity of the fluid,

The equation of motion considering the ballistic transport effects for the vertical component of the velocity of the fluid can be written as;

$$\tau_{mom}\rho[\partial^2 v_z/\partial t^2 + v_z \partial^2 v_z/\partial z \partial t + (\partial v_z/\partial t)(\partial v_z/\partial z)] + \rho \partial v_z/\partial t + v_z \partial v_z/\partial z + \tau_{mom}\nabla(\partial p/\partial t)$$
$$= -\partial p/\partial z - \rho g \qquad (4.147)$$

From Darcy's law in a new gravitational field in the space shuttle in the galaxy;

$$v_z = +(\kappa/\mu)(\partial p/\partial z + \rho g) \qquad (4.148)$$

where κ is the permeability of the porous medium. Eq. [4.151] becomes after neglecting the pressure changes with time;

$$\tau_{mom}\rho[\partial^2 v_z/\partial t^2 + v_z \partial^2 v_z/\partial z \partial t + (\partial v_z/\partial t)(\partial v_z/\partial z)]$$
$$+ \rho \partial v_z/\partial t + v_z \partial v_z/\partial z + \mu/\kappa v_z = 0 \qquad (4.149)$$

It can be assumed that during the drainage the velocity of liquid in the tube the velocity of the liquid, v_z is independent of z and is only a function of time. Or this can be arrived at by writing the equation of continuity. This is true for wide reservoirs where the height in the contianer does not change appreciably. Thus at constant density the equation of momentum becomes;

$$\tau_{mom}\, \partial^2 v_z/\partial t^2 + \partial v_z/\partial t + \gamma/\kappa v_z = 0 \qquad (4.150)$$

Let $\tau = t/\tau_{mom}$

$$\partial^2 v_z/\partial \tau^2 + \partial v_z/\partial \tau + (\gamma \tau_{mom}/\kappa)\, v_z = 0 \qquad (4.151)$$

where

$$Pb = (\gamma \tau_{mom}/\kappa)$$

Pb is a sort of a permeability number that gives the ratio of the kinematic viscosity times the relaxation time of momentum divided by the Darcy permeability of the medium. It may represent the ratio of viscous forces and ballistic transport "forces" to the permeability forces. Let $u = v_z/g\tau_{mom}$

$$\partial^2 u/\partial \tau^2 + \partial u/\partial \tau + Pb\, u = 0 \qquad (4.152)$$

Eq. [4.157] is a second order ODE with constant coefficients and is homogeneous. The two roots of the auxiliary equation,

$$-1/2 \pm 1/2 sqrt(1 - 4Pb^2) \qquad (4.153)$$

$$u = \exp(-\tau/2)\, [\, c_1 \exp(-\tau/2\, sqrt(1 - 4Pb^2)) + c_2 \exp(\tau/2\, sqrt(1 - 4Pb^2)) \qquad (4.154)$$

From the constraint at infinite time, i.e, $u\exp(\tau/2) = 0$, c_2 can be seen to be zero. The initial condition can be written assuming a pseudo-steady state and using Toricelli's theorem; $sqrt(2gH) = c_1$. Thus,

$$u = (2gH)^{1/2} \exp(-\tau/2)\ [\ c_1 \exp(-\tau/2\ \text{sqrt}(1 - 4Pb^2)) \quad\quad (4.155)$$

It can be seen that for small values of the permeability number,
i.e, when, Pb > ½,

$$u = (2gH)^{1/2} \exp(-\tau/2\ )\text{Cos}(\tau/2\ \text{sqrt}(4Pb^2 - 1)) \quad\quad (4.156)$$

    The positive gradient of pressure dependence of velocity of flow through a porous medium can happen in packings that have the channel size change on account of pressure. With increased pressure, when the channel size decreases with increased pressure, the flow velocity through the porous medium will decrease with increased pressure. The dimensionless velocity as a function of dimensionless time is shown in Figure 4.6, and can be seen to be subcritical damped oscillatory. After a said time the velocity changes in direction on account of the added consideration of the ballistic transport that takes into consideration of the accumulation of momentum in the momentum flux expression.

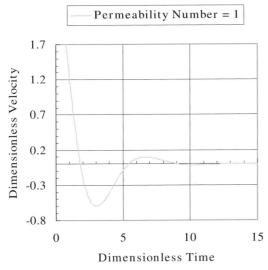

Figure 4.6 Subcritical Damped Oscillations with Positive Permeability

4.4   Shear Flow Between Two Plates Moving in Opposite Direction at Constant Velocity with Separation Distance 2a

    Consider two flat plates pulled in opposite directions at a constant velocity, V. Let the separation distance between the plates be 2a . The initial velocity of the fluid is zero.

248

Defining the axes in a fashion that the plate velocity is in the ± x direction and the shear stress acts in and imparts the momentum transfer in the z direction. The governing equation for the velocity for the fluid at constant density and viscosity neglecting pressure and gravity effects in one dimension including the ballistic transport term for correcting for the accumulation of momentum can be written as;

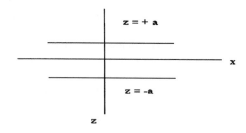

Figure 4.7 Fluid Between Two Flat Plates Moving with Constant Velocity in Opposite Directions

$$\tau_r \, \partial^2 v_x/\partial t^2 \; + \; \partial v_x/\partial t \;\; = \;\; \gamma \, \partial^2 v_x/\partial z^2 \tag{4.157}$$

$$\text{Let} \;\; \tau \; = \; t/\tau_r; \, Z \; = \; z/\mathrm{sqrt}(\gamma\tau_r); \, u \; = \; (v_x - V)/V \tag{4.158}$$

The governing equation in the dimensionless form is then;

$$\partial u/\partial\tau \;\; + \;\; \partial^2 u/\partial\tau^2 \;\; = \;\; \partial^2 u/\partial Z^2 \tag{4.159}$$

The solution can be assumed to consist of a steady state part and a transient part, i.e., $u \; = \; u^t \; + \; u^{ss}$. The steady state part and boundary conditions can be selected in such as fashion that the transient portion becomes homogeneous.

$$\partial^2 u^{ss}/\partial Z^2 \;\; = \;\; 0 \tag{4.160}$$

The Boundary Conditions are;

$$Z = 0, \; u \; = -1 \;\; \text{from symmetry} \tag{4.161}$$

$$Z \; = \; a/\mathrm{sqrt}(\gamma\tau_r), \;\; u = 0 \tag{4.162}$$

$$Z \; = -a/\mathrm{sqrt}(\gamma_r), \; u = -2 \tag{4.163}$$

Solving for Eq. [4.160],
$$u^{ss} = c_1 Z + c_2 \qquad (4.164)$$

From the boundary condition given in Eq. [4.162] the c_2 can be seen to be -1. From the boundary condition given in Eq. [4.163], c_1 can be seen to be $1/Z_a$. Eq. [4.160] is obeyed by Eq. [4.165]. Thus,
$$u^{ss} = Z/Z_a - 1 \qquad (4.165)$$

The equation and time and space conditions for the transient portion of the solution can be written as;

$$\partial^2 u^t/\partial \tau^2 + \partial u^t/\partial \tau = \partial^2 u /\partial Z^2 \qquad (4.166)$$

Initial Condition : $\tau = 0$, $u^t = -1$ $\qquad (4.167)$

Final Condition: $\tau = \infty$, $u^t = 0$ $\qquad (4.168)$

The Boundary Conditions are now homogeneous after the expression of the result as a sum of steady state and transient parts and are;

$$u^t = 0, Z = 0 \qquad (4.169)$$

$$Z = \pm Z_a , u^t = 0 \qquad (4.170)$$

The solution is obtained by the method of separation of variables. Initially the damping term is eliminated using a substitution such as $u^t = W \exp(-n\tau)$. Eq. (4.166) then becomes at $n = \frac{1}{2}$

$$W_{xx} = - W /4 + W_{\tau\tau} \qquad (4.171)$$

Eq. (4.171) can be solved by the method of separation of variables.

$$\text{Let } W = g(\tau) \phi (Z) \qquad (4.172)$$
Eq. (4.171) becomes,

$$g(\tau) \phi^{''} (Z) = - g(\tau) \phi(Z)/4 + g^{''}(\tau) \phi(Z) \qquad (4.173)$$

$$\text{Or } \phi^{''} (Z)/ \phi(Z) = -1/4 + g^{''}(\tau)/ g(\tau) = -\lambda_n^2 \qquad (4.174)$$

The space domain solution is;
$$\phi(Z) = c_1 Sin(\lambda_n Z) + c_2 Cos(\lambda_n Z) \qquad (4.175)$$
From the boundary conditions,
$$X = 0, \quad u = 0, \text{ it can be seen that }, c_2 = 0 \qquad (4.176)$$

$$\phi(Z) = c_1 Sin(\lambda_n Z) \qquad (4.177)$$

From the boundary condition given by Eq. (4.170),

$$0 = c_1 Sin(\lambda_n Z_a) \tag{4.178}$$

$$n\pi = \lambda_n Z_a \tag{4.179}$$

$$\lambda_n = n\pi sqrt(\gamma\tau_r)/a \ , \tag{4.180}$$

$$a = sqrt(\gamma\tau_r)n\pi/\lambda_n \tag{4.181}$$

Since a is a non-zero quantity n can take on the values, 1, 2, 3........
The time domain solution would be,

$$g = c_3 exp(sqrt(1/4 - \lambda_n^2) \tau)+ c_4 exp(-sqrt(1/4 - \lambda_n^2)\tau) \tag{4.182}$$

From the FINAL condition given by Eq.(4.168), not only the transient velocity has to decay out to 0 but also the wave velocity. As, $W = u^t exp(\tau/2)$, at time infinity the transient velocity is zero and as any number multiplied by zero is zero even if it is infinity, $W = 0$ at the final condition. Applying this condition in the solution Eq.(4.182) it can be seen that $c_3 = 0$.   Thus,

$$u^t = \sum_1^\infty -2(1 - (-1)^n) /n\pi \, exp(-\tau/2) \, exp(-sqrt(1/4 - \lambda_n^2)\tau)Sin(\lambda_n Z) \tag{4.183}$$

λ_n is described by Eq. [4.180]. C_n can be derived using the orthogonality property and can be shown to be $-2(1 - (-1)^n)/n\pi$. It can be seen that the model solutions given by Eq. (4.183) is bifurcated. i.e., the characteristics of the function change considerably when a parameter such as the separation distance of the plates are varied. Here a decaying exponential becomes exponentially damped cosinous. This is referred to as subcritical damped oscillatory behaviour.

For $a < 2\pi sqrt(\gamma\tau_r)$, all the terms in the infinite series will pulsate. This when the argument within the square root sign in the exponentiated time domain expression becomes negative and the result becomes imaginary. Using Demovrie's theorem and taking real part for small width of the slab,

$$u^t = \sum_1^\infty c_n exp(-\tau/2) \, Cos(sqrt((\lambda_n^2 -1/4)\tau)Sin(\lambda_n Z) \tag{4.184}$$

At $Z = Z_a/2$, the dimensionless velocity,

$$= \sum_1^\infty c_n exp(-\tau/2) \, Cos(sqrt((\lambda_n^2 -1/4)\tau) \tag{4.185}$$

This is shown in Figures 4.7 and 4.8. The maximum velocity can be seen to be subcritical damped oscillatory. The oscillations are overdamped by the decaying exponential in time term.

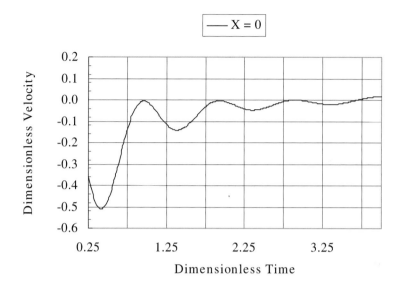

**Figure 4.8 Subcritical Damped Oscillations in a Fluid
Between Two Moving Plates in Opposite Direction
$Z = a/2(\gamma\tau_r)$**

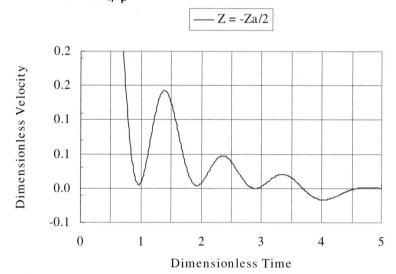

**Figure 4.9 Subcritical Damped Oscillations in a Fluid
Between Two Moving Plates in Opposite Direction
$Z = -a/2\sqrt{(\gamma\tau_{mom})}$**

252

Worked Example 4.6

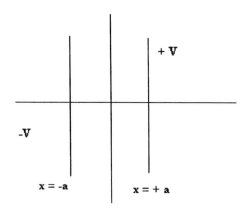

Figure 4.10 Fluid in-between Two Vertical Plates Moving in Opposite Directions with Constant Velocity

Consider the additional forces of pressure and gravity and modify the governing equation for the fluid between two moving plates in the opposite direction. Assume a permeability law between pressure gradient and flow velocity where the as the velocity increases the pressure gradient is lower. This can be seen during elutriation in gas-solid flow and pneumatic conveying under certain conditions. Discuss the dimensionless velocity variation spatio-temporally. Assume that the two plates are moved in opposite directions along the z axis this time. Thus the shear stress or momentum transfer is in the x direction, the horizontal axis. Let the separation between the two vertical plates be given by 2a.

The governing equation including the pressure gradient and gravity forces, and after writing the pressure gradient in terms of the velocity of flow and neglecting the changes in pressure gradient with time can be written as;

$$\partial^2 v_z/\partial \tau^2 + \partial v_z/\partial \tau + (\gamma \tau_{mom}/\kappa)\, v_z = \gamma\, \partial^2 v_z/\partial z^2 \qquad (4.186)$$

$$\text{Let } \tau = t/\tau_r;\, Z = z/\text{sqrt}(\gamma \tau_r);\, u = (v_z - V)/V \qquad (4.187)$$
$$Pb = (\gamma \tau_{mom}/\kappa)$$

The governing equation in the dimensionless form is then;

$$\partial u/\partial \tau + \partial^2 u/\partial \tau^2 + Pb\, u + Pb = \partial^2 u/\partial Z^2 \qquad (4.188)$$

The solution can be assumed to consist of a steady state part and a transient part, i.e., $u = u^t + u^{ss}$. The steady state part and boundary conditions can be selected in such as fashion that the transient portion becomes homogeneous.

$$\partial^2 u^{ss}/\partial Z^2 + Pb + Pb\, u^{ss} = 0 \tag{4.189}$$

The Boundary Conditions are;

$$Z = 0, \ u = -1 \ \text{ from symmetry} \tag{4.190}$$

$$Z = R/\text{sqrt}(\gamma\tau_r), \ u^{ss} = 0 \tag{4.191}$$

$$Z = -R/\text{sqrt}(\gamma\tau_r), \ u^{ss} = -2 \tag{4.192}$$

Solving for Eq. [4.189],

$$u^{ss} = c_1 \text{Sin}\,(Pb^{1/2}Z) + c_2 \text{Cos}(Pb^{1/2}Z) + c_3 \tag{4.193}$$

From the boundary condition given in Eq. [4.190]

$$c_2 + c_3 = -1 \tag{4.194}$$

$$0 = c_1 \text{Sin}(Pb\, Z_R) + c_2 (\text{Cos}(Pb^{1/2}Z_R) - 1) + 1 \tag{4.195}$$

$$-2 = -c_1 \text{Sin}(Pb\, Z_R) + c_2 (\text{Cos}(Pb^{1/2}Z_R) - 1) + 1 \tag{4.196}$$

$$\text{or } c_1 = -c_2(\text{Cot}(Pb^{1/2}Z_R) - 1/\text{Sin}(Pb^{1/2}Z_R) + 1/\text{Sin}(Pb^{1/2}Z_R) \tag{4.197}$$

$$c_2 = -2/(\text{Sin}(Pb^{1/2}Z_R) - 1/\text{Sin}^2(Pb^{1/2}Z_R)) \tag{4.198}$$

The boundary condition given in Eq. [4.190] assumes that the viscous effects predominate over the gravitational and pressure effects. The equation and time and space conditions for the transient portion of the solution can be written as;

$$\partial^2 u^t/\partial\tau^2 + \partial u^t/\partial\tau + Pb\, u^t = \partial^2 u^t/\partial Z^2 \tag{4.199}$$

$$\text{Initial Condition}: \ \tau = 0, \quad u^t = -1 \tag{4.200}$$

$$\text{Final Condition}: \ \tau = \infty, \ u^t = 0 \tag{4.201}$$

The Boundary Conditions are now homogeneous after the expression of the result as a sum of steady state and transient parts and are;

$$= 0 \qquad\qquad u^t = 0, Z \tag{4.202}$$

$$Z = \pm Z_a, \ u^t = 0 \tag{4.203}$$

The solution is obtained by the method of separation of variables. Initially the damping term is eliminated using a substitution such as $u^t = W \exp(-n\tau)$. Eq. (4.210) then becomes at $n = \frac{1}{2}$

$$W_{zz} = (Pb - \tfrac{1}{4})W/4 \; + \; W_{\tau\tau} \tag{4.204}$$

For large permeability numbers, Eq. [4.204] can be transformed into a Bessel equation by the following substitution;

For $\tau > Z$, and $Pb > \frac{1}{4}$, $\quad \eta = \tau^2 - Z^2$ (4.205)

As shown in earlier sections, Eq. [4.204], is transformed into;

$$4\eta W_{\eta\eta} + 4 W_\eta + (Pb - \tfrac{1}{4})W = 0 \tag{4.206}$$

The solution is a Bessel composite function and can be written by;

$d = 1/4(Pb - \frac{1}{4})$

$$W = c \, J_0 \, \text{sqrt}(\, (Pb - \tfrac{1}{4})(\, \tau^2 - Z^2) \tag{4.207}$$

$$u = c \exp(-\tau/2) \, J_0 \, \text{sqrt}(\, (Pb - \tfrac{1}{4})(\tau^2 - Z^2) \,) \tag{4.208}$$

From the boundary condition,

$$-1 = c \exp(-\tau/2) \, J_0 \, (\tau\text{sqrt}(Pb - \tfrac{1}{4})) \tag{4.209}$$

Eliminating c between Eqs. [4.221, 4.220],

$$u = -J_0 \, \text{sqrt}(\, (Pb - \tfrac{1}{4})(\tau^2 - Z^2) \,) / J_0(\tau\text{sqrt}(Pb - \tfrac{1}{4})) \tag{4.210}$$

It can be seen that Eq. [4.210] describes the velocity profile in-between vertical plates considering the viscous, gravitational, permeability effects. The spatio-temporal velocity is given by a Bessel composite function of the first kind and zeroth order. This is expected to be valid for permeability numbers greater than $\frac{1}{4}$. The expression exhibits space symmetry and is subcritical damped oscillatory. The Bessel function can be approximated as ;

$$u = -(\tau/(\tau^2 - Z^2)^{1/4})\text{Cos}(\text{sqrt}(\, (Pb - \tfrac{1}{4})(\tau^2 - Z^2) \, - \pi/4)/\text{Cos}(\tau\text{sqrt}(Pb - \tfrac{1}{4} \, - \pi/4))$$
$$\tag{4.211}$$

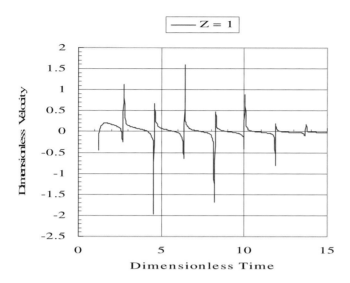

Figure 4.11 Transient Velocity of Viscous Fluid Flow in Porous Medium With a Positive Permeability Coefficient Between Vertical Plates

4.5 Transient Laminar Flow in a Circular Conduit

Consider the laminar flow in a circular pipe of narrow dimensions in transience. The damped wave momentum transfer and relaxation equation is written as;

$$\tau_{xy} + \tau_{mom}\partial\tau/\partial t = -\mu\,\partial v/\partial r \tag{4.212}$$

The governing equation for the axial velocity as a function of the radial direction can be written after combining the modified Newton's law of viscosity including the relaxation term with the momentum balance equation to yield:

$$\partial v/\partial t + \tau_{mom}\,\partial^2 v/\partial t^2 = \Delta P/L\rho + v/r\,\partial/\partial r(r\,\partial v/\partial r) \tag{4.213}$$

Defining the dimensionless variables;

$$u = v/v_{max};\ X = r/sqrt(v\tau_{mom});\ \tau = t/\tau_{mom};\ P^* = \Delta P\tau_{mom}/L\rho v_{max} \tag{4.214}$$

The dimensionless governing equation then becomes;

$$\partial u/\partial \tau + \partial^2 u/\partial \tau^2 \quad = \partial^2 u/\partial X^2 + 1/X \, \partial u/\partial X + P^* \tag{4.215}$$

The solution is assumed to be a sum of two parts, i.e., the steady state and transient part; Let $u = u^{ss} + u^t$. Then the governing equation becomes;

$$d^2 u^{ss}/dX^2 + 1/X \, du^{ss}/dX + P^* = 0 \tag{4.216}$$

$$\partial u^t/\partial \tau + \partial^2 u^t/\partial \tau^2 \quad = \partial^2 u^t/\partial X^2 + 1/X \, \partial u^t/\partial X \tag{4.217}$$

Integrating the equation for the steady state component of the velocity wrt X;

$$X \, \partial u^{ss}/\partial X = -P^* X^2/2 + C' \tag{4.218}$$

At $X = 0$, the gradient of the velocity is zero as a condition of extremama can be expected from symmetry considerations. So it can be seen that $C' = 0$. Integrating the resulting equation again wrt X,

$$u^{ss} = -P^* X^2/4 + D' \tag{4.219}$$

From the boundary condition at $X = X_R$

$$D' = P^* X_R^2/4 \quad \text{and} \tag{4.220}$$

$$u^{ss} = P^* X_R^2/4(\nu \tau_{mom}) \, (1 - r^2/R^2) \tag{4.221}$$

The above relation is the Poiseuille distribution. The above relation is the Hagen [1839] -Poisuelle [1841] flow in laminar pipes at steady state where P^* is given by Eq. [4.226]. The rest of the problem is obtaining the transient part. First the damping term is removed by a substitution, $u^t = W \exp(-n\tau)$. At $n = \frac{1}{2}$ the governing equation reduces to ;

$$-W/4 + \partial^2 W/\partial \tau^2 = \partial^2 W/\partial X^2 + 1/X \, \partial W/\partial X \tag{4.223}$$

The method of separation of variables can be used to obtain the exact solution to the above equation. The boundary and time conditions for the transient portion of the velocity are then;

$$\tau = 0, \quad u^t = 1 \tag{4.224}$$

$$\tau = \text{infinity}, u^t = 0 \tag{4.225}$$

$$\tau > 0, \quad X = X_R, \quad u^t = 0 \tag{4.226}$$

$$X = 0, \text{ symmetry considerations} \tag{4.227}$$

Let $W = V(\tau)\phi(X)$ \hfill (4.228)

The wave equation becomes;

$$V''/V - 1/4 = (\phi'' + \phi'/X)/\phi = -\lambda_n^2 \tag{4.229}$$

Thus,

$$X^2\phi'' + X\phi' + X^2\lambda_n^2\phi = 0 \tag{4.230}$$

This can be recognized as a Bessel equation of the first order [Mickley, Sherwood, Reed, 1955] and the solution can be written as;

$$\phi = c_1 J_0(\lambda_n X) \tag{4.231}$$

c_2 can be seen to be zero as ϕ is finite at $X = 0$. From the boundary condition at $X = R/\text{sqrt}(\nu\tau_{mom})$;

$$c_1 J_0(\lambda_n X_R) = 0 \tag{4.232}$$

$$\lambda_n = \text{sqrt}(\nu\tau_{mom})/R \, (2.4048 + (n-1)\pi) \tag{4.233}$$
$$n = 1, 2, 3....$$

Now the time domain part of the wave is obtained by solving the second order ODE.

$$V''/V - 1/4 = -\lambda_n^2 \qquad (4.234)$$

Thus,

$$uexp(\tau/2) = W = (c_3 \exp(+\tau sqrt(1/4 - \lambda_n^2)) + c_34exp(-\tau sqrt(1/4 - \lambda_n^2))) \phi \qquad (4.235)$$

At infinite times the RHS becomes infinitely large. The LHS is zero multiplied by infinity and is zero. At steady state the velocity is bounded and hence the constant c_3 is set to zero. It can be seem that for small channel dimensions,

i.e., when $R < 4.8096 \ sqrt(\nu\tau_{mom})$, the solution is periodic wrt to time. The general solution for such cases can be written as;

$$u = \sum_1^\infty C_n \exp(-\tau/2) \ Cos \ (\tau \ sqrt(\lambda_n^2 - 1/4)) \ J_0(\lambda_n X) \qquad (4.236)$$

The initial condition can be taken to be maximum velocity in plug flow. So the initial superficial velocity essentially at plug flow becomes a periodic profile as in Hagen-Poiseulle flow when channels are formed. The transient portion is governed for small channels by the generalized Netwon's law of viscosity. So the initial condition is;

$$1 = \sum_1^\infty C_n J_0(\lambda_n X) \qquad (4.237)$$

The constant can be solved for by the orthogonality property.

$$C_n = -\int_0^{XR} J_0(\lambda_m X)dX / \int_0^{XR} J_0^2(\lambda_m X)dX \qquad (4.238)$$

The maximum transient velocity is given by that at the center of the circular tube;

$$u_{max} = \sum_1^\infty C_n \exp(-\tau/2) \ Cos \ (\tau \ sqrt(\lambda_n^2 - 1/4)) \qquad (4.239)$$

Worked Example 4.7

Consider the oscillations in a U tube manometer. Use the additional ballistic term and discuss the velocity and the height in the manometer.

Eq. [4.36] written in the z direction in the U tube manometer and integrated with respect to z between the two points in the manometer 1 and 2 on either sides, is as follows;

$$\tau_{mom} L\, \partial^2 v_z/\partial t^2 \ + L \partial v_z/\partial t \ = \ -2gz \qquad (4.240)$$

$\partial v_z/\partial t$ is independent of z, L, where L is the length of the column,

$$P_1 = P_2 \qquad (4.241)$$

$$V_{z1} = -V_{z2} \ \text{(from continuity)} \qquad (4.242)$$

$$V_{z1}^{\ 2} = V_{z2}^{\ 2} \qquad (4.243)$$

Writing v_z as dz/dt , Eq. [4.240] can be written as;

$$\tau_{mom}\, \partial^3 z/\partial t^3 \ + \partial^2 z/\partial t^2 + \ 2gz/L \ = 0 \qquad (4.244)$$

Let the Oscillation number, Osc $= (g\tau_{mom}^{\ 2}/L)$; $\tau = t/\tau_{mom}$; $Z = z/L$ (4.245)
Eq. [4.256] becomes,

$$\partial^3 Z/\partial \tau^3 \ + \partial^2 Z/\partial \tau^2 + \ (2Osc)\, Z \ = 0 \qquad (4.246)$$

The third order ODE with constant coefficients is homogeneous and can be solved for as follows.

$$r^3 \ + r^2 \ + 2Osc \ = 0 \qquad (4.247)$$

Eq. [4.247] can be compared with the general form of the cubic equation,

$$r^3 \ + a_2\, r^2 + \ a_1 r \ + a_0 \ = 0 \qquad (4.248)$$

Let e and f be defined as;

$$e = 1/3\, a_1 - 1/9 a_2^{\ 2} \ = \ -1/9 \qquad (4.249)$$

$$f = 1/6(\, a_1 a_2 - 3a_0) - 1/27 a_2^{\ 3} \ = -(Osc + 1/27) \qquad (4.250)$$

where $a_0 = 2Osc$; $a_1 = 0$; $a_2 = 1$ (4.251)

Consider $e^3 + f^2 > 0$ (4.252)

This is when,

$$Osc > -2/27 \qquad (4.253)$$

Oscillation number is the ratio of the gravitational force divided by the relaxational frequency normalized by the length of the column in the U tube. The oscillation number will always be positive and hence the Eqs. [4.267, 4.268] will be valid for real systems. In such cases, the cubic equation solution results in 1 real root and two imaginary roots. Let,

$$s_1 = (f + (e^3 + f^2)^{1/2})^{1/3} = (Osc[(1 + 2/27Osc)^{1/2} - 1] - 1/27)^{1/3} \qquad (4.254)$$

$$s_2 = (f - (e^3 + f^2)^{1/2})^{1/3} = -(Osc[(1 + 2/27Osc))^{1/2} - 1] + 1/27)^{1/3} \qquad (4.255)$$

The roots of the cubic are then,

$$r_1 = (s_1 + s_2) - a_2/3 \qquad (4.256)$$

$$r_2 = 1/2(s_1 + s_2) - a_2/3 + isqrt(3)/2 (s_1 - s_2) \qquad (4.257)$$

$$r_3 = 1/2(s_1 + s_2) - a_2/3 + isqrt(3)/2 (s_1 - s_2) \qquad (4.258)$$

Thus for finite Oscillation number the displacement will pulsate.

$$Z = c_1 exp(r_1 t) + c_2 exp(r_2 t) + c_3 exp(r_3 t) \qquad (4.259)$$

The imaginary roots can be seen to predict the oscillations that are subcritical damped oscillatory. Using the De Movrie's theorem and obtaining the real parts the term that contributes the subcritical damped oscillations can be written as;

$$Z = C'exp(-\tau (s_1 + s_2)/2 - 1/3)Cos(\tau ((s_1 + s_2)/2 - 1/3)) \qquad (4.260)$$

C' can solved for using the initial condition for velocity. The other conditions are the velocity at zero time and infinite time to solve for the other two integration constants.

4.6 Flow of Blood in Capillaries

Blood is a dispersed fluid system and consists of cells and and plasma. Major proteins found in the blood are albumin, globulin, and fibrinogen. Albumin has a major role in regulating the pH and the colloid osmotic pressure. Alpha and beta globulins are involved in solute transport and the gamma globulins are the antibodies that fight infection and form the basis of the humoral component of the immune system.

Fibrinogen has a major role in blood clotting. Serum is the fluid remaining afer the blood is allowed to clot. The three main cells present in the blood are the RBCs, red blood cells or erythrocytes and occupies 95% of the cellular component of the blood. Their have a major role in the transport of oxygen by haemoglobin contained within the RBCs. The density of RBC is higher than that of plasma. The red blood cell volume fraction is called the hematocrit and typically varies between 40 – 50%. The true hematocrit (H) is about 96% of the measured hematocrit (Hct). RBCs can form stacked coin like structures called rouleaux and rouleaux can clump to form the aggregates. They break up in conditions of high shear or increased volumetric flow rate.

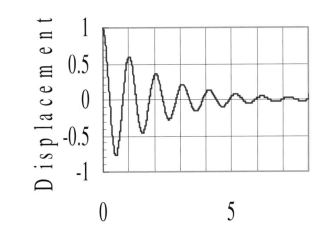

Figure 4.12 Subcritical Damped Oscillations in a U Tube Manometer Including the Ballistic Transport Term

About 5% of the blood consists of the platelets. They are responsible for the blood coagulation and haemostasis. The leukocytes, or WBCs, the white blood cells form the basis of the cellular component of the immune system. The effect of blood platelets and WBCs on the flow characterises of the blood, on account of their low volume fraction can be expected to be low. Rheology is the study of the flow behaviour of fluids. The shear stress and shear rate relationship of blood can be written in terms of the damped wave momentum transfer and relaxation equation. For non-Newtonian fluid the linear term of the velocity gradient becomes non-linear. Here an apparent viscosity is considered in-place of the Newtonian viscosity

The Fahraeus- Lindquist effect (Gaehtgens, 1980) and the Fahraeus effect (Barbee and Cokelet, 1971a, 1971b) were two anomalous effects that were detected during tube flow of blood at high shear rates, greater than 100 sec^{-1}. Blood flows into smaller and smaller vessels from the artery to the vein. If the hematocrit of the blood in the artery is given by H_F, the hematocrit of the blood in the vessel is of interest H_T, and the discharge hematocrit, H_D. In tubes with diameters from about 15 to 500 μm it is found that the tube hematocrit, H_T, is actually less than that of the discharge hematocrit, ($H_D = H_F$, at steady state). This is called the Fahreus effect. For tubes less than 15 microns in size the ration of H_T to H_D increases with the increase in capillary diameter. One attributable reason for this is plasma skimming. When the tube size is smaller than 15 microns, the small vessel draws blood from a much larger vessel and it becomes physically very difficult for the RBC to even enter the vessel. The orientation of the branch as well as that of the RBC will affect the tube hematocrit as well as the resulting discharge hematocrit. The Fahraues effect was explained to using in vivo and in vitro experiments. The cells are unevenly distributed across the tube cross-section. The RBCs accumulate in the axis of the wall and a thin cell free layer called the plasma layer forms near the wall. The thickness of the plasma layer depends on the tube thickness and the hematocrit and is of the order of a few microns.

The marginal zone theory proposed by Haynes (1960) can explain the Fahraeus Lindquist effect. Tube flow of blood at high shear rates (> 100/sec) exhibits the dependence of diameter on the viscosity effect is the Fahreus –Lindquist effect. When the diameter of the tube decreases below 500 microns upto 4 – 6 microns, the viscosity of the blood also decreases. The marginal zone theory may be used to characterize the effect from 4-6 microns to 500 microns in tune diameter. An expression is obtained for the apparent viscosity in terms of the plasma layer thickness, tube diameter and the hematocrit.

The bold flow within a tube or vessel is divided into two regions; a central core that contains the cells with a viscosity, μ_c, and the cell free marginal or plasma layer that consists only of plasma with a thickness of δ, and a viscosity equal to that of the plasma denoted by μ_p. In each region the flow is considered to be Newtonian and at steady state. For the core region, the governing equation may be written neglecting the ballistic term as;

$$\tau_{rz} = (\Delta P)r/2L = -\mu_c \partial v_z^c / \partial r \qquad (4.261)$$

The boundary conditions can be written as;

$$r = R - \delta, \quad \tau_{rz} \text{ (core)} = \tau_{rz} \text{ (plasma)} \qquad (4.262)$$

$$r = 0, \quad \partial v_z^c / \partial r = 0 \tag{4.263}$$

The first boundary conditions stems from the continuity of the transfer of momentum across the interface between the core and plasma layer and the second boundary condition derives from the fact that the axial velocity would be a maximum at the center of the tube from symmetry arguments. In a similar vein, for the plasma layer the governing equation and boundary conditions can be written as;

$$\tau_{rz} = (\Delta P)r/2L = -\mu_p \partial v_z^p / \partial r \tag{4.264}$$

$$r = R, \quad v_z^p = 0 \tag{4.265}$$

$$r = R - \delta, \quad v_z^p = v_z^c \tag{4.266}$$

The boundary condition given Eq. [4.265] comes from the zero velocity condition at the wall and the boundary condition given in Eq. [4.266] is that the velocity need be the same at the interface of the two phases. Eq. [4.261] and [4.264] may be integrated and the core and plasma flow rates given by the following;

$$Q_p = \pi(\Delta P)/8\mu_p L \left(R^2 - (R - \delta)^2 \right)^2 \tag{4.267}$$

$$Q_c = \pi(\Delta P) R^2/8\mu_p L \left((R - \delta)^2 - (1 - /\sigma^2)(R - \delta)^4 / R^2 - \mu_p/\mu_c (R - \delta)^4 / 2R^2 \right) \tag{4.268}$$

The total discharge rate of the blood would equal the sum of the flow rates in the core and plasma regions and is given by;

$$Q = \pi(\Delta P) R^4/8\mu_p L \left(1 - (1 - \delta/R)^4 (1 - \mu_p/\mu_c) \right) \tag{4.269}$$

Eq. (4.269) can be used to fit the apparent viscosity data and obtain values of the plasma layer thickness and the core hematocrit as a function of the tube diameter. A relation between the core hematocrit, H_c and the feed hematocrit, H_F and the thickness of the plasma layer is needed. An equation is needed to describe the dependence of the blood viscosity on the hematocrit since the value of the H_c will be larger than H_F because of the axial accumulation on the RBCs. This relative increase in the core hematocrit will make the equation in the core have a higher viscosity than the blood in the feed. The following equation by Charm and Kurland (1974) may be used to express the dependence of the viscosity of blood at high shear rates on the hematocrit and temperature;

$$\mu = \mu_p (1/(1 - \alpha H)) \tag{4.270}$$

$$\text{or } \alpha H = 1 - \mu_p/\mu \tag{4.271}$$

$$\text{where, } \alpha = 0.070 \exp(2.49 H + 1107/T \exp(-1.69H)) \tag{4.272}$$

where the temperature is in Kelvin. Theses equations are valid to a hematocrit of 0.6 with a stated accuracy of 10 %. If a subscript occurs on the viscosity then the corresponding values of H and α in the above equations will carry the same subscript. If $\sigma = 1 - \delta/R$ then the solution for the plasma layer thickness are implicit and require solution of two equations and two unknowns:

$$\mu_{app}/\mu_F = (1 - \alpha_F H_F)/(1 - \sigma^4 \alpha_C H_C) \tag{4.273}$$

$$H_C/H_F = 1 + (1 - \sigma^2)^2/\sigma^2 (2(1 - \sigma^2) + \sigma^2 \mu_p/\mu_c) \tag{4.274}$$

An explicit expression for the plasma layer thickness is desirable and a method is developed as follows..

Equation [4.272] is examined using a spreadsheet. It can be observed that αH is linear with H at a given temperature for the range of hematocrit for which the Charm and Kurland expression is valid (Figure 4.13). Thus for the core and tube ;

$$\alpha_c H_c = m H_c + C \tag{4.275}$$

$$\alpha_T H_T = m H_T + C \tag{4.276}$$

The slope and intercept can be obtained by a least squares regression between the αH and H as given by Eq. [4.284]. From a balance of cells in the two phases in the tube it can be seen that;

$$H_T = \sigma^2 H_c \tag{4.277}$$

Dividing Eqs. [4.275] by [4.276],

$$\alpha_c H_c / \alpha_T H_T = m H_c + C / m H_T + C \tag{4.278}$$

From Eq. [4.278] it can be seen that;

$$\alpha_T H_T = 1 - \mu_p/\mu_{app} \tag{4.279}$$

With minor rearrangement it can be seen that;

$$\alpha_c H_c \sigma^4 = 1 - \mu_p/\mu_{app} \tag{4.282}$$

or $\alpha_c H_c \sigma^4 = \alpha_T H_T = \alpha_T \sigma^2 H_c$

$$\sigma = \sqrt{\alpha_T/\alpha_c} \tag{4.281}$$

Plugging Eq. [4.273] and Eq. [4.269] into Eq. [4.270];

$$\sigma^4 = (m H_T + C)/(m H_T/\sigma^2 + C) \tag{4.282}$$

with, $\sigma^2 = p$

the quadratic can be solved;

$$C\, p^2 \,+\, (m\, H_r - 1)p \,+ C = 0 \qquad (4.283)$$

Thus an explicit expression for the plasma layer thickness in terms of the tube hematocrit is developed. The tube hematocrit can be read from the linear regression developed between the αH and H at a given temperature once the apparent viscosity of the tube is known.

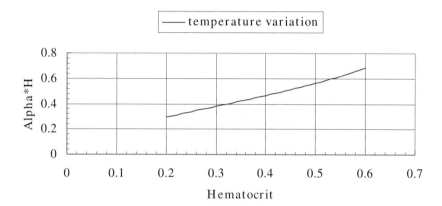

Figure 4.13 Temperature Variation Parameter of the Hematocrit During Blood Flow in Capillaries

4.7. Tangential Flow Induced by a Rotating Cylinder

The velocity distributions and pressure distributions during the tangential laminar flow of an in incompressible fluid induced by a sudden rotating cylinder at constant velocity is examined. The rotation of the cylinder of radius R, is at a tangential velocity of V_θ. The radial and azimuthal velocity components are zero. From Table C.2, Eqs. [C.8-C.10] from the equation of motion can be written as;

$$\rho\tau_{mom} [\partial^2 v_\theta/\partial t^2 + \quad + v_\theta/r \, \partial^2 v_\theta/\partial t\partial\theta \,] + \quad \rho\partial v_\theta/\partial t \quad = + \ \mu\partial/\partial r(1/r \,\partial/\partial r(r v_\theta)) \qquad (3.284)$$

θ component

r component

$$\rho\tau_{mom} [- 2/rv_\theta \, \partial v_\theta/\partial t] + + \tau_{mom} \partial^2 p/\partial t \, \partial r + - \rho v_\theta^2/r = -\partial p/\partial r \qquad (4.285)$$

z component

$$0 = -\partial p/\partial z + + \rho g_z \qquad (4.286)$$

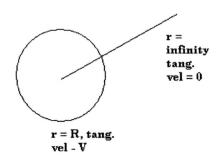

**r =
infinity
tang.
vel = 0**

**r = R, tang.
vel - V**

Figure 4.14 Tangential Flow Past a Cylinder

It can be seen that the pressure gradient in the z direction does not change with time and hence $\tau_{mom}\partial^2 p/\partial t \, \partial z$ is zero. Once the v_θ is solved for from Eq. [4.284] the radial pressure distribution can then be calculated using Eq. [4.285]. The tangential velocity is assumed to be a function of only time and the radial space coordinate. Eq. [4.284] is made dimensionless by the following substitutions;

$$u = v_\theta/V_\theta; \quad X = r/\sqrt{\gamma\tau_{mom}}; \quad \tau = t/\tau_{mom};$$

$$\partial^2 u/\partial \tau^2 + \partial u/\partial \tau = \partial/\partial X(1/X \, \partial/\partial X(uX)) \qquad (4.287)$$

Let uX = V

$$\partial^2 V/\partial \tau^2 + \partial V/\partial \tau = -1/X \, \partial V/\partial X + \partial^2 V/\partial X^2 \qquad (4.288)$$

Eq. [4.288] can be solved by a similar method shown in section 2.2 for the temperature distribution in a infinite medium heated by a hot cylindrical surface.

The damping term is removed from the governing equation. This is done realizing that the transient velocity decays with time in an exponential fashion. The other reason for this manoeuvre is to study the wave equation without the damping term. Let $V = w\exp(-n\tau)$. By choosing $n = \frac{1}{2}$, the damping component of the equation is removed; Thus for $n = \frac{1}{2}$

$$-w/4 + \partial^2 w/\partial \tau^2 = \partial^2 w/\partial X^2 - 1/X \, \partial w/\partial X \quad (4.289)$$

Eq. [4.297] can be solved by using the method of relativistic transformation of coordinates. Consider the transformation variable η for $\tau > X$

$$\eta = \tau^2 - X^2 \quad (4.290)$$

$$\partial w/\partial \tau = (\partial w/\partial \eta)2\tau \quad (4.291)$$

$$\partial^2 w/\partial \tau^2 = (\partial^2 w/\partial \eta^2)4\tau^2 + 2(\partial w/\partial \eta) \quad (4.292)$$

In a similar fashion,

$$\partial w/\partial X = (\partial w/\partial \eta)2X \quad (4.293)$$

$$\partial^2 w/\partial X^2 = (\partial^2 w/\partial \eta^2)4X^2 + 2(\partial w/\partial \eta) \quad (4.294)$$

Plugging Eq. [4.291, 4.294] into Eq. [4.289]

$$(\partial^2 w/\partial \eta^2)4(\tau^2 - X^2) + 2(\partial w/\partial \eta) -w/4 = 0 \quad (4.295)$$

$$4\eta^2 \partial^2 w/\partial \eta^2 + 2\eta \partial w/\partial \eta - \eta w /4 = 0 \quad (4..296)$$

or

$$\eta^2 \partial^2 w/\partial \eta^2 + 1/2\eta \partial w/\partial \eta - \eta w/16 = 0 \quad (4.297)$$

Comparing Eq. [4.296] with the generalized Bessel equation as given in Eq. [A.30], the solution is; $a = 1/2$; $b = 0$; $c = 0$; $d = -1/16$; $s = \frac{1}{2}$. The order p of the solution is then $p = 2 \, \text{sqrt}(1/16) = \frac{1}{2}$.

$$W = c_1 I_{1/2} (1/2 \, \text{sqrt}(\tau^2 - X^2)(\tau^2 - X^2)^{1/4} + c_2 I_{-1/2} (1/2 \, \text{sqrt}(\tau^2 - X^2)(\tau^2 - X^2)^{1/4} \quad (4.298)$$

c_2 can be seen to be zero as W is finite and not infinitely large at $\eta = 0$. An approximate solution can be obtaining by eliminating c_1 between the equation derived from the boundary condition at $X = X_R$ and the above equation. It can be noted that this is a mild function of time, however. As the general solution of PDE consists of n arbitrary functions when the order of the PDE is n compared with n arbitrary constants for ODE. From the boundary condition at, $X = X_R$,

$$X = \exp(-\tau/2) c_1 I_{1/2} (1/2 \, \text{sqrt}(\tau^2 - X_R^2)(\tau^2 - X_R^2)^{1/4} \quad (4.299)$$

or $u = [(\tau^2 - X^2)^{1/4}/ [(\tau^2 - X_R^2)^{1/4}][I_{1/2} (1/2 \, \text{sqrt}(\tau^2 - X^2) / I_{1/2} (1/2 \, \text{sqrt}(\tau^2 - X_R^2)]$

$$\quad (4.300)$$

In terms of elementary functions, Eq. [4.300] can be written as;

or $u = [(\tau^2 - X^2)^{1/4}/ [(\tau^2 - X_R^2)^{1/4}][\text{Sinh}(1/2 \, \text{sqrt}(\tau^2 - X^2) / \text{Sinh}(1/2 \, \text{sqrt}(\tau^2 - X_R^2)]$

$$\quad (4.301)$$

In the limit of X_R going to zero, the expression becomes for $\tau > X$;

or $u = [(\tau^2 - X^2)^{1/2}/\tau^{1/2}][\operatorname{Sinh}(1/2 \operatorname{sqrt}(\tau^2 - X^2)) / \operatorname{Sinh}(\tau/2)]$ (4.302)

For $X > \tau$,

$u = [(X^2 - \tau^2)^{1/4} /[(\tau^2 - X_R^2)^{1/4}][J_{1/2} (1/2 \operatorname{sqrt}(X^2 - \tau^2))/ I_{1/2} (1/2 \operatorname{sqrt}(\tau^2 - X^2))]$ (4.303)

Eq. [4.303] can be written in terms of trigonometric functions as;

$u = [(X^2 - \tau^2)^{1/4} /[(\tau^2 - X_R^2)^{1/4}] [\operatorname{Sin}(1/2 \operatorname{sqrt}(X^2 - \tau^2))/ \operatorname{Sinh}(1/2 \operatorname{sqrt}(\tau^2 - X^2))]$ (4.304)

Three different regimes can be seen. The first regime is that of the thermal lag and consists of no change from the initial velocity. The second regime is when

$\tau_{lag}^2 = X^2 - 4\pi^2$

or $\tau_{lag} = \operatorname{sqrt}(X_p^2 - 4\pi^2) = 3.09$ when $X_p = 7$. (4.305)

For times greater than the time lag and less than X_p the dimensionless velocity is given by Eq. [4.300]. For dimensionless times greater than 7, for example, the dimensionless velocity is given by Eq. [4.303]. For distances *closer to the surface compared with 2π the time lag will be zero*. The radial pressure distribution can be estimated from the following equation;

$\rho [- 2/rv_\theta \, \partial v_\theta/\partial\tau] + \partial^2 p/\partial\tau\partial r + - \rho v_\theta^2/r = -\partial p/\partial r$ (4.306)

It can be seen that for distances greater than a certain distance for an instant of time, the pressure will be pulsational.

Worked Example 4.8

In writing Eq. [4.290], the assumption that is valid at large distances was made. Obtain the governing equation from the tangential shear term and obtain the solution for the above problem by relaxing the assumption at large distances.

θ component

$$\rho\tau_{mom} \, [\partial^2 v_\theta/\partial t^2 + \quad + \; v_\theta/r \; \partial^2 v_\theta/\partial t \partial\theta \;] + \; \rho\partial v_\theta/\partial t \; = -1/r^2 \, (\partial/\partial r(r^2\tau_{,\theta}) \tag{4.307}$$

The tangential shear stress in terms of the tangential velocity can be written as;

$$\tau_{,\theta} = -\mu \, (r \; \partial/\partial r \; (v_\theta/r)) \tag{4.308}$$

The RHS of Eq. [4.317] will then become;

$$\mu/r^2(\partial/\partial r(r^3 \; \partial/\partial r \; (v_\theta/r)) \,) \tag{4.309}$$

Let $V = u/X$ where u, X and τ are defined using the dimensionless variables in Eq. [4.307]

$$\partial^2 V/\partial\tau^2 + \partial V/\partial\tau \; = \; 1/X^3 \; \partial/\partial X \, (X^3 \; \partial V/\partial X) = 3/X\partial V/\partial X + \partial^2 V/\partial X^2 \tag{4.310}$$

The damping term is removed from the governing equation. This is done realizing that the transient velocity decays with time in an exponential fashion. The other reason for this manoeuvre is to study the wave equation without the damping term. Let $V = w\exp(-n\tau)$. By choosing $n = \frac{1}{2}$,   the damping component of the equation is removed;
Thus for $n = \frac{1}{2}$

$$-w/4 \; + \; \partial^2 \, w/\partial\tau^2 \; = \; \partial^2 \, w/\partial X^2 + 3/X \; \partial w/\partial X \tag{4.311}$$

Eq. [4.311] can be solved by using the method of relativistic transformation of coordinates. Consider the transformation variable η for $\tau > X$ as;

$$\eta \; = \; \tau^2 - X^2 \tag{4.312}$$

$$\partial w/\partial\tau \; = \; (\partial \, w/\partial\eta \,)2\tau \tag{4.313}$$

$$\partial^2 \, w/\partial\tau^2 \; = \; (\partial^2 \, w/\partial\eta^2 \,)4\tau^2 \; + \; 2(\partial \, w/\partial\eta \,) \tag{4.314}$$

In a similar fashion,

$$\partial \, w/\partial X = \; (\partial \, w/\partial\eta \,)2X \tag{4.315}$$

$$\partial^2 \, w/\partial X^2 \; = \; (\partial^2 \, w/\partial\eta^2 \,)4X^2 \; + \; 2(\partial \, w/\partial\eta \,) \tag{4.316}$$

Plugging Eq. [4.316, 4.35] into Eq. [4.311]

$$(\partial^2 \, w/\partial\eta^2 \,)4(\, \tau^2 - X^2 \,) \; + \; 10(\partial \, w/\partial\eta \,) \; -w/4 = 0 \tag{4.317}$$

$$4\eta^2\partial^2 \, w/\partial\eta^2 \; + 10\eta\partial \, w/\partial\eta \; - \; \eta w \; /4 \; = 0 \tag{4..318}$$

or

$$\eta^2\partial^2 \, w/\partial\eta^2 \; + 5/2\eta\partial \, w/\partial\eta \; - \; \eta w/16 \; = 0 \tag{4.319}$$

270

Comparing Eq. [4.319] with the generalized Bessel equation as given in Eq. [A.30], the solution is;

$a = 5/2$; $b = 0$; $c = 0$; $d = -1/16$; $s = ½$. The order p of the solution is then $p = 3/2 \sqrt{(1/16)} = ½$.

Or $\quad W = c_1 I_{3/2} (1/2 \, \text{sqrt}(\tau^2 - X^2)/(\tau^2 - X^2)^{3/4} + c_2 I_{-3/2} (1/2 \, \text{sqrt}(\tau^2 - X^2)/(\tau^2 - X^2)^{3/4}$

c_2 can be seen to be zero as W is finite and not infinitely large at $\eta = 0$. An approximate solution can be obtaining by eliminating c_1 between Eq. [4.331] and the above equation. It can be noted that this is a mild function of time, however. As the general solution of PDE consists of n arbitrary functions when the order of the PDE is n compared with n arbitrary constants for ODE. From the boundary condition at, $X = X_R$,

$$1/X = \exp(-\tau/2) \, c_1 I_{1/2} (1/2 \, \text{sqrt}(\tau^2 - X_R^2)/(\tau^2 - X_R^2)^{3/4} \qquad (4.320)$$

or $\text{u} = [(\tau^2 - X_R^2)^{3/4}/(\tau^2 - X^2)^{3/4}][I_{3/2} (1/2 \, \text{sqrt}(\tau^2 - X^2) / I_{3/2} (1/2 \, \text{sqrt}(\tau^2 - X_R^2)] \qquad (4.321)$

This is for $\tau > X$
For $X > \tau$,

$\text{u} = [(\tau^2 - X_R^2)^{3/4}/(X^2 - \tau^2)^{3/4}][J_{3/2} (1/2 \, \text{sqrt}(X^2 - \tau^2))/ I_{3/2} (1/2 \, \text{sqrt}(\tau^2 - X^2)] \qquad (4.322)$

Worked Example 4.9

In writing Eq. [4.290], allowance was made for the angular velocity. Using a shell balance obtain the governing equation from the tangential shear term and obtain the solution for the above problem by using a similar expression for the shear force compared with the cartesian coordinates.

Consider a thin cylindrical shell from a distance r from the center of the axis. The shear force that pulls the liquid along with it is a product of the tangential shear stress, $\tau_{r\theta}$ multiplied the area of the shell through which the force is transmitted, i.e., $2\pi rL$. Thus the accumulation of momentum is given by the rate of change of shear stress, ignoring the gravity and pressure forces in the tangential velocity component;

$$-1/r\partial(r\tau_{r\theta})/\partial r = \rho \, \partial v_\theta/\partial t \qquad (4.323)$$

The tangential shear stress in terms of the tangential velocity and accounting for the damped wave ballistic term can be written as;

$$\tau_{r\theta} = -\mu (r \, \partial/\partial r (v_\theta/r)) - \tau_{mom} \, \partial\tau_{r\theta}/\partial t \qquad (4.324)$$

Inorder to eliminate the tangential shear stress and obtain a governing equation in terms of the tangential velocity Eq. [4.324] is multiplied by r and differentiated wrt r

$$1/r\partial(r\tau_{r\theta})/\partial r = -\mu/r\ \partial/\partial r(r^2\ \partial/\partial r\ (v_\theta/r))\qquad -\ \tau_{mom}/r\partial^2(r\tau_{r\theta})/\partial t\partial r \qquad (4.325)$$

Differentiating Eq. [4.334] wrt t

$$-1/r\partial^2(r\tau_{r\theta})/\partial r\partial t\ = \rho\ \partial^2 v_\theta/\partial t^2 \qquad (4.326)$$

$$\mu/r\ (\partial/\partial r(r^2\ \partial/\partial r\ (v_\theta/r))\ ) \qquad = \rho\ \tau_{mom}\partial^2 v_\theta/\partial t^2\ +\ \rho\partial v_\theta/\partial t \qquad (4.327)$$

Let $u = v_\theta/V_0$; $X = r/\sqrt{(\gamma\tau_{mom})}$; $\tau = t/\tau_{mom}$; $\qquad (4.328)$

$$1/X\ (\partial/\partial X(X^2\ \partial/\partial X\ (u/X))\ ) \qquad = \partial^2 u/\partial\tau^2\ +\ \rho\partial u/\partial\tau \qquad (4.329)$$

Let $V = u/X$

$$\partial^2 V/\partial\tau^2\ +\quad \partial V/\partial\tau\quad = 1/X^2\ \partial/\partial X\ (X^2\ \partial V/\partial X) = 2/X\partial V/\partial X\quad +\ \partial^2 V/\partial X^2 \quad (4.330)$$

The damping term is removed from the governing equation. This is done realizing that the transient velocity decays with time in an exponential fashion. The other reason for this manoeuvre is to study the wave equation without the damping term. Let $V = w\exp(-n\tau)$. By choosing $n = \frac{1}{2}$,   the damping component of the equation is removed; Thus for $n = \frac{1}{2}$

$$-w/4\ +\ \partial^2\ w/\partial\tau^2\ =\ \partial^2\ w/\partial X^2 + 2/X\ \partial w/\partial X \qquad (4.331)$$

Eq. [4.331] can be solved by using the method of relativistic transformation of coordinates.   Consider the transformation variable η as;

$$\eta\ =\ \tau^2 - X^2 \qquad (4.332)$$

Eq. [4.331] becomes

$$(\partial^2\ w/\partial\eta^2\ )4(\ \tau^2 - X^2\ )\ +\ 10(\partial\ w/\partial\eta\ )\ -w/4 = 0 \qquad (4.333)$$

$$4\eta^2\partial^2\ w/\partial\eta^2\ +\ 8\eta\partial\ w/\partial\eta\ -\ \eta w\ /4\ = 0 \qquad (4..334)$$

or $\eta^2\partial^2\ w/\partial\eta^2\ +2\eta\partial\ w/\partial\eta\ -\ \eta w/16 = 0 \qquad (4.335)$

Comparing Eq. [4.347] with the generalized Bessel equation as given in Eq. [A.30], the solution is; $a = 2; b = 0;\ c = 0;\ d = -1/16;\ s = \frac{1}{2}$. The order p of the solution is then p $= 1\ \mathrm{sqrt}(1/16) = \frac{1}{2}$

Or $W\ =\ c_1 I_1\ (1/2\ \mathrm{sqrt}(\tau^2 - X^2)/(\tau^2 - X^2)^{1/2}\ +\ c_2 I_{-1}(1/2\ \mathrm{sqrt}(\tau^2 - X^2)/(\tau^2 - X^2)^{1/2}$

$$(4.336)$$

c_2 can be seen to be zero as W is finite and not infinitely large at $\eta = 0$. An approximate solution can be obtaining by eliminating c_1 between the equation from the boundary condition at the surface and the above equation. It can be noted that this is a mild function of time, however. As the general solution of PDE consists of n arbitrary functions when the order of the PDE is n compared with n arbitrary constants for ODE. From the boundary condition at, $X = X_R$,

$$1/X = \exp(-\tau/2)\, c_1 I_1(1/2\, \text{sqrt}(\tau^2 - X_R^2)/(\tau^2 - X_R^2)^{1/2} \tag{4.337}$$

$$\text{or } u = [(\tau^2 - X_R^2)^{1/2}/(\tau^2 - X^2)^{1/2}][\, I_1\,(1/2\, \text{sqrt}(\tau^2 - X^2)\, /\, I_1\,(1/2\, \text{sqrt}(\tau^2 - X_R^2)] \tag{4.338}$$

This is for $\tau > X$. For $X > \tau$,

$$u = [[(\tau^2 - X_R^2)^{1/2}/(X^2 - \tau^2)^{1/2}][\, J_1\,(1/2\, \text{sqrt}(X^2 - \tau^2\,)/\, I_1\,(1/2\, \text{sqrt}(\tau^2 - X^2)] \tag{4.339}$$

4.8.   Transient Flow Past a Sphere

Consider a solid sphere settling in an infinite fluid at terminal settling velocity. Of interest is the transient velocity in the tangential direction of the sphere as a function of r.   Defining the stream function as follows and neglecting the ϕ and r component velocities;

$$v_\theta = 1/r\text{Sin}\theta\, \partial\psi/\partial r \tag{4.340}$$

The θ component of the velocity considering only its dependence in r direction becomes;

$$\tau_{mom}[\partial^3\psi/\partial t^2\partial r\,] + \quad \partial^2\psi/\partial t\partial r + = \quad \gamma[1/r\partial/\partial r(r^2\, \partial/\partial r\, 1/r\, \partial\psi\, /\partial r\,)\,] \tag{4.341}$$

Let $\tau = t/\tau_{mom}$; $X = r/\text{sqrt}(\gamma\tau_{mom})$; $\tag{4.342}$

Then, $v_\theta = 1/[X\text{Sin}\theta\, \gamma\tau_{mom}]\, \partial\psi/\partial X$ $\tag{4.343}$

$$\partial^3\psi/\partial\tau^2\partial X \quad + \quad \partial^2\psi/\partial\tau\partial X \quad = \quad 1/X\partial/\partial X(X^2 \ \partial/\partial X 1/X\partial\psi/\partial X)$$

$$-\partial^2\psi/\partial X^2 + X \ \partial^2\psi/\partial X^2 + \partial\psi/\partial X$$

$$=\partial^3\psi/\partial X^3 \qquad (4.344)$$

Integrating wrt X,

$$\partial^2\psi/\partial\tau^2 \quad + \quad \partial\psi/\partial\tau \quad = \quad \partial^2\psi/\partial X^2 \qquad (4.345)$$

Let $\psi = \chi \exp(-\tau/2)$. This will remove the damping term in the governing equation to give;

$$\partial^2\chi/\partial\tau^2 \ - \ \chi/4 \quad = \quad \partial^2\chi/\partial X^2 \qquad (4.346)$$

Using the substitution as shown before;

$$\eta = \tau^2 - X^2 \qquad (4.347)$$

for, $\tau > X$

The governing equation without the damping term becomes;

$$4\eta \ \partial^2\chi/\partial\eta^2 \ + \ 4 \ \partial\chi/\partial\eta \ - \ \chi/4 = 0 \qquad (4.348)$$

$$\text{or } \eta^2\partial^2\chi/\partial\eta^2 + \eta\partial\chi/\partial\eta \ - \ \eta\chi/16 = 0 \qquad (4.349)$$

Comparing Eq. [4.349] with the generalized Bessel equation the solution can be written as;

$$\chi = c_1 \exp(-\tau/2) \ I_0(1/2\text{sqrt}(\tau^2 - X^2))$$

$$v_\theta = 1/[\gamma\tau_{mom} \text{Sin}\theta] \ c_1 \exp(-\tau/2) \ I_1(1/2\text{sqrt}(\tau^2 - X^2))/(\tau^2 - X^2)^{1/2} \qquad (4.350)$$

From the boundary condition at r = R,

$$v_\theta = U_t = 1/[\gamma\tau_{mom} \text{Sin}\theta] \ c_1 \exp(-\tau/2) \ I_1(1/2\text{sqrt}(\tau^2 - X_R^2))/(\tau^2 - X_R^2)^{1/2} \qquad (4.351)$$

For $X > \tau$,

$$v_\theta/U_t = [(\tau^2 - X_R^2)^{1/2}/(X^2 - \tau^2)^{1/2}] \ J_1(1/2\text{sqrt}(X^2 - \tau^2)/I_1(1/2\text{sqrt}(\tau^2 - X_R^2) \qquad (4.352)$$

Eliminating c_1 between the two Eqs. [4.351, 4.350], for $\tau > X$,

$$v_\theta/U_t = [(\tau^2 - X_R^2)^{1/2}/(\tau^2 - X^2)^{1/2}] I_1(1/2\text{sqrt}(\tau^2 - X^2))/I_1(1/2\text{sqrt}(\tau^2 - X_R^2)$$

Worked Example: 4.10

Consider the radial flow between two concentric spheres of an incompressible, isothermal liquid. The transient velocity distribution is examined using the damped wave momentum transfer and relaxation equation. Let the radii of the two spheres be R and mR respectively. The governing equation for the radial component of the velocity can be written from the Eq. [C.11] in Appendix C as;

$$\tau_{mom} [1/r^2 \, \partial^2(r^2v_r)/\partial t^2 + v_r/r^2 \, \partial^2(r^2v_r)/\partial t \partial r] + [(\tau_{mom}/r^2 \, \partial(r^2v_r)/\partial t + v_r)(1/r^2 \, \partial(r^2v_r)/\partial r) + \tau_{mom}\partial^2 p/\partial t \, \partial r + \rho/r^2 \, \partial(r^2 v_r)/\partial t = -1/\rho \partial p/\partial r + \gamma[1/r^2 \, \partial^2(r^2v_r)/\partial r^2] \quad (4.353)$$

Inorder to obtain the dimensionless form of the governing equation the substitution given in Eq. [4.355] is used and after neglecting the nonlinear term using the creeping flow assumption;

$$1/X^2 \, \partial^2(X^2u)/\partial\tau^2 + [1/X^2 \, \partial(X^2u)/\partial\tau + u](1/X^2 \, \partial(X^2u)/\partial X) + \partial^2 P^*/\partial\tau \, \partial X + 1/X^2 \, \partial(X^2 u)/\partial\tau$$
$$= -\partial P^*/\partial X + [1/X^2 \, \partial^2(X^2u)/\partial X^2] \quad (4.354)$$

where, $P^* = p/\rho v_R v_{mom}$ $u = (v_r - v_R)/v_R$ $\quad\quad\quad\quad\quad$ (4.355)

The space boundary conditions can be written as; $r = R$, $v_r = v_R$ $\quad\quad$ (4.356)

From the equation of continuity for a constant density system,

$$1/r^2 \, \partial(\rho r^2 v_r)/\partial r = 0 \quad\quad\quad\quad\quad\quad\quad\quad (4.357)$$

$$r^2 v_r = c_1 = R^2 v_R \quad\quad\quad\quad\quad\quad\quad (4.358)$$

The velocity at mR is then,

$$v_{mR} = v_R/m^2 \quad\quad\quad\quad\quad\quad\quad\quad\quad (4.359)$$

Thus, $u^{ss} = (R/r)^2 - 1$ $\quad\quad\quad\quad\quad\quad\quad\quad$ (4.360)

$$= X_R^2/X^2 - 1 \quad\quad\quad\quad\quad\quad\quad (4.361)$$

The time conditions are;

$$\tau = 0, \, u = -1 \quad\quad\quad\quad\quad\quad (4.362)$$

Let the velocity consist of a steady state and a transient component,

$$u = u^{ss} + u^{t} \tag{4.363}$$

$$(2X_R^2/X^3 + 2/X) + 2/X^2 = -\partial P^*/\partial X \tag{4.364}$$

$$P^* = -X_R^2/X^2 + 2\ln X - 2/X + c_3 \tag{4.365}$$

$$X = 0, P^* \text{ is finite and therefore } c_3 = 0 \tag{4.366}$$

$$u^{ss} = (X_R^2/X^2) \tag{4.367}$$

$$X^2 u^{ss} = X_R^2 \tag{4.368}$$

Thus $P^* = c_2$

The transient dimensionless velocity can be written as;

$$1/X^2 \, \partial^2(X^2u)/\partial\tau^2 + [1/X^2 \, \partial(X^2u)/\partial\tau + u)(1/X^2 \, \partial(X^2u)/\partial X)$$
$$+ 1/X^2 \, \partial(X^2 u)/\partial\tau = + [1/X^2 \, \partial^2(X^2u)/\partial X^2] \tag{4.369}$$

Let $V = X^2u$ (4.370)

$$\partial^2V/\partial\tau^2 + X^2 (\partial V/\partial\tau + V)\partial V/\partial X) + \partial V/\partial\tau = \partial^2V/\partial X^2 \tag{4.371}$$

The method of separation of variables can be used to solve the above equation;

Let $V = g(X) \, \theta(\tau)$ (4.372)

$$\theta'' g + X^2 [\theta' g + g\theta] g'\theta + g\theta' = g''\theta \tag{4.373}$$

Dividing by $g\theta$ throughout,

$$\theta''/\theta + \theta'/\theta \quad = g''/g - X^2[\theta' + \theta] g' = -\lambda_n^2 \tag{4.374}$$

$[\theta' + \theta]$ can be set to zero to obtain a separation of the time and space variables. From this constraint,

$$\theta' + \theta \quad = c \tag{4.375}$$

Then, $\quad \theta''/\theta \quad = 1 - \lambda_n^2 \tag{4.376}$

Eq. [4.376] can be used to obtain the θ. The θ obtained from this constraint may not meet the $\theta' + \theta = 0$ requirement. Hence the solution is an approximation.

$$g''/g \quad = -\lambda_n^2 \tag{4.377}$$

Thus, $\quad \theta = c_1 \exp(\tau \mathrm{sqrt}(1 - \lambda_n^2)) + c_2 \exp(-\tau \mathrm{sqrt}(1 - \lambda_n^2)) \tag{4.378}$

At steady state the velocity profile is given by Eq. [4.376]. Hence, c_1 can be taken as zero as θ is not infinite at infinite time.

$$\theta = c_2 \exp(-\tau \mathrm{sqrt}(1 - \lambda_n^2)) \tag{4.379}$$

Inorder to solve for the constants c_3 and c_4 redefine X as follows;

$$Y = X - X_R \tag{4.380}$$

Eq. [4.377] then becomes, $\quad g''(Y)/g \quad = -\lambda_n^2 \tag{4.381}$

$$g = c_3 \mathrm{Sin}(\lambda_n Y) + c_4 \mathrm{Cos}(\lambda_n Y) \tag{4.382}$$

At $Y = 0$, $u^t = g\theta = 0$, $\tag{4.383}$

Hence, $c_4 = 0 \tag{4.384}$

$$g = c_3 \operatorname{Sin} (\lambda_n Y) \tag{4.385}$$

$$\text{At } Y = Y_{mR}, \ u^t = u - u^{ss} = 1/m^2 - 1 - 1/m^2 + 1 = 0 \tag{4.386}$$

$$\text{Hence, } \lambda_n = n\pi/Y_{mR}, \quad \text{for, } n = 1,2,3.... \tag{4.387}$$

$$= (\gamma\tau_{mom})^{1/2} n\pi/mR \tag{4.388}$$

Thus,

$$V = \sum_{n=1}^{\infty} c_n \exp(-\tau\operatorname{sqrt}(1 - \lambda_n^2))\operatorname{Sin} (\lambda_n (X - X_R) \tag{4.389}$$

C_n can be solved using the orthogonal property and the initial condition and shown to be; $2(1 - (-1)^n)/n\pi$. It can be seen that when,

$$(\gamma\tau_{mom})^{1/2} \pi/mR > 1 \tag{4.390}$$

$$\text{or } mR < \pi(\gamma\tau_{mom})^{1/2} \pi \tag{4.391}$$

the solution is given after using the De Movrie's theorem and obtaining the real parts,

$$X^2 u = V = \sum_{n=1}^{\infty} c_n \operatorname{Cos}(\tau\operatorname{sqrt}(1 - \lambda_n^2))\operatorname{Sin} (\lambda_n (X - X_R)) \tag{4.392}$$

The maximum velocity is obtained when the $\lambda_n X = \pi/2$. The velocity is sustained periodic. When the velocity changes in sign the radial velocity becomes inward in direction. This occurs when the distance between the spheres are small. The energy for the oscillations are provided by the kinetic energy from the inflow of the fluid from the surface at R. The pressure drop at steady state can be calculated from Eq. [4.369].

Worked Example 4.11

Consider a room cooled by a ceiling fan. Obtain the tangential and azimuthal velocity of the air blown from the fan at transient state. Idealize the room into a cylinder.

Let the axes be drawn with the origin at the fan and the + z axis pointing downward. From the equation of motion considering only the tangential velocity as a function of r and z as given in Eq. [C.9] in Appendix C;

θ Component

$$\rho\tau_{mom}\ [\partial^2 v_\theta/\partial t^2 + \quad v_\theta/r\ \partial^2 v_\theta/\partial t\partial\theta\ ] + \rho[\ (\tau_{mom}\ \partial v_z/\partial t + v_z)(\partial v_\theta/\partial z)]$$
$$+ \tau_{mom}\partial^2 p/\partial t\partial\theta + \rho\partial v_\theta/\partial t$$
$$= + \mu[\partial/\partial r(1/r\ \partial/\partial r(rv_\theta) + \quad + (\partial^2 v_\theta/\partial z^2)] \tag{4.393}$$

Obtaining the dimensionless form of the governing equation by using the following substitutions;

$$\text{Let } \tau = t/\tau_{mom};\ X = r/\text{sqrt}(\gamma\tau_{mom});\ u = v_\theta/V_\theta;\qquad ; Z = z/\ \text{sqrt}(\gamma\tau_{mom}) \tag{4.394}$$

$$\partial^2 u/\partial\tau^2 + \quad uV_\theta/X\ \partial^2 u/\partial\tau\partial\theta \quad + \partial u/\partial\tau$$
$$= [\partial/\partial X(1/X\ \partial/\partial X(Xu) + (\partial^2 u/\partial Z^2)] \tag{4.395}$$

$$\text{Let } Xu = V \tag{4.396}$$

$$\partial^2 V/\partial\tau^2 + \quad uV_\theta/X\ \partial^2 V/\partial\tau\partial\theta \quad + \partial V/\partial\tau$$
$$= X[\partial/\partial X(1/X\ \partial V/\partial X) + 1/X(\partial^2 V/\partial Z^2)] \tag{4.397}$$

$$= -1/X\ \partial V/\partial X + (\partial^2 V/\partial X^2) + \quad (\partial^2 V/\partial Z^2)] \tag{4.398}$$

Neglecting the bulk flow effects,

$$\partial^2 V/\partial\tau^2 + \quad \partial V/\partial\tau \quad = -1/X\ \partial V/\partial X + (\partial^2 V/\partial X^2) + \quad (\partial^2 V/\partial Z^2)] \tag{4.399}$$

The damping term can be removed by a u = Wexp(-τ/2). Eq. [4.412] becomes ;

$$\partial^2 W/\partial\tau^2 - W/4 \quad = -1/X\ \partial W/\partial X + (\partial^2 W/\partial X^2) + \quad (\partial^2 W/\partial Z^2)] \tag{4.400}$$

Consider the substitution,

$$\eta = \tau^2 - X^2 - Z^2 \tag{4.401}$$

Eq. [4.400] then becomes;

$$4\eta\ \partial^2 W/\partial\eta^2 + 4\partial W/\partial\eta - W/4 = 0 \tag{4.402}$$

Multiplying throughout by η, and dividing by 4,

$$\eta^2\, \partial^2 W/\partial\eta^2 + \eta\, \partial W/\partial\eta - \eta W/16 = 0 \qquad (4.403)$$

Comparing Eq. [4.403] with the generalized Bessel equation the solution can be written as; $a = 1$; $b = 0$; $d = -1/16$; $s = \frac{1}{2}$; $c = 0$. Then, $p = 0$

$$W = c_1\, I_0\,(1/2\ \mathrm{sqrt}(\tau^2 - X^2 - Z^2)) + c_2\, K_0(1/2\ \mathrm{sqrt}(\tau^2 - X^2 - Z^2)) \qquad (4.404)$$
When $\eta = 0$, W is finite and hence $c_2 = 0$.

$$Xu = c_1\, \exp(-\tau/2)\, I_0\,(1/2\ \mathrm{sqrt}(\tau^2 - X^2 - Z^2)) \qquad (4.405)$$

From the boundary condition at $X, Z = 0$

$$X\,(1) = c_1\, \exp(-\tau/2)\, I_0\,(\tau/2) \qquad (4.406)$$

c_1 can be eliminated by dividing Eq. [4.405] by Eq. [4.406] to give,

$$u = I_0\,(1/2\ \mathrm{sqrt}(\tau^2 - X^2 - Z^2))\, /I_0(\tau/2) \qquad (4.407)$$

For, $X^2 + Z^2 > \tau^2$
$$u = J_0\,(1/2\ \mathrm{sqrt}(X^2 + Z^2 - \tau^2))\, /I_0(\tau/2) \qquad (4.408)$$

The steady state solution can be obtained as;

$$0 = -1/X\, \partial V/\partial X + (\partial^2 V/\partial X^2) + (\partial^2 V/\partial Z^2) \qquad (4.409)$$

The equation can be solved by separation of variables;

Let $V = g(X)\, \theta\,(Z)$ $\qquad\qquad (4.410)$

$$0 = -1/X\, g'\, \theta + g''\theta + g\theta'' \qquad (4.411)$$

Dividing throughout by $g\theta$,

$$-1/X \, g'/g + g''/g = \theta''/\theta = +\lambda_n^2 \tag{4.412}$$

$$\theta = c_1 \exp(\lambda_n Z) + c_2 \exp(-\lambda_n Z) \tag{4.413}$$

Far from the fan, the velocity can be expected to be zero, hence c_1 can be set to zero.

$$X^2 g'' - Xg' - X\lambda_n^2 g = 0 \tag{4.414}$$

Comparing Eq. [4.414] to the generalized Bessel equation [A.30];
$b = 0$; $c = 0$; $d = -\lambda_n^2$; $a = -1$; $s = \frac{1}{2}$. $p = 2$

$$g = c_1 X I_2 (X^{1/2}/2) + c_2 X K_2(X^{1/2}/2) \tag{4.415}$$

c_2 can be set to zero from the condition that the velocity is finite at $X = 0$.

Thus, $\quad Xu = \exp(-\lambda_n Z) \, c_1 X I_2 (X^{1/2}/2) + c_2 X K_2(X^{1/2}/2) \tag{4.416}$

So u $\quad = \exp(-\lambda_n Z) \, c_1 I_2 (X^{1/2}/2) + c_2 K_2(X^{1/2}/2) \tag{4.417}$

c_1 can be set to zero from the condition that the velocity is zero at $X = \infty$.

So u $\quad = \exp(-\lambda_n Z) \, c_2 K_2 (X^{1/2}/2) \tag{4.418}$

From the boundary condition at the tip of the fan blade,

$X = X_R$, $Z = 0$, $u = 1 \tag{4.419}$

$$1 = c_1' K_2 (X_R^{1/2}/2) \tag{4.420}$$

Eliminating c_1' between the two Eqs. [4.418, 4.420]

$$u = \exp(-\lambda_n Z) K_2 (X^{1/2}/2)/ K_2 (X_R^{1/2}/2)] \tag{4.421}$$

This expression is valid only for $\quad X > X_R$. The solution is a modified Bessel function of the second order and second kind.

Worked Example 4.12

Consider the outward radial squeeze flow between two parallel circular disks. A potential application is in a lubricant system consisting of two circular disks between which a lubricant flows radially. The flow takes place because of a pressure drop Δp between the inner and outer radii r_1 and r_2. Perform the analysis including the transient effects.

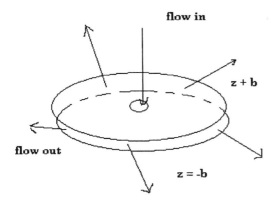

Figure 4.15 Radial Outflow between two Circular Parallel Disks

The equation of motion for v_r from Appendix C as given in Eq. [C.8] can be written as;

$$\tau_{mom} [1/r\partial^2(rv_r)/\partial t^2 + (v_r/r)\partial^2(rv_r)/\partial t\partial r] + [(\tau_{mom}/r\partial(rv_r)/\partial t + v_r)(1/r\partial(rv_r)/\partial r)$$
$$+ (\tau_{mom} \partial v_\theta/\partial t + v_\theta)(1/r^2\partial(rv_r)/\partial\theta) + (\tau_{mom} \partial v_z/\partial t + v_z)(\partial v_r/\partial z)]$$
$$+ \tau_{mom}\partial^2 p/\partial t \partial r + 1/r\partial(rv_r)/\partial t$$
$$= -\partial p/\partial r + \mu[\partial/\partial r(1/r \partial/\partial r(rv_r) + (\partial^2 v_r/\partial z^2)] \tag{4.422}$$

Considering only the radial component of velocity from the equation of continuity as given in Eq. [C.3]

$$1/r \ \partial(\rho r v_r)/\partial r = 0 \tag{4.423}$$

Obtaining the dimensionless form of the Eq. [4.422] using the following substitutions;

dimensionless velocity,	$u = v/v_{ref}$	(4.424)
dimensionless time,	$\tau = t/\tau_{mom}$	(4.425)
dimensionless distance,	$X = z/sqrt(\gamma\tau_{mom})$	(4.426)
dimensionless radius	$Y = r/sqrt(\gamma\tau_{mom})$	(4.427)
dimensionless pressure,	$P^* = p/\rho v_{rb}v_{mom}$	(4.428)
Peclet (momentum)	$Pe_{mom} = v_{rb}/v_{mom}$	(4.429)

$$1/Y\partial^2(Yu)/\partial\tau^2 + \partial^2 P^*/\partial\tau\partial Y + 1/Y\partial(Yu)/\partial\tau + uPe_{mom}(1/Y\partial(Yu)/\partial Y) = -\partial P^*/\partial Y$$
$$+ \partial^2 u/\partial X^2$$

(4.430)

Let $Yu = V$ (4.431)

Then Eq. [4.430] becomes,

$$\partial^2 V/\partial\tau^2 + + \partial V/\partial\tau - \partial^2 V/\partial X^2 = -Y\partial P^*/\partial Y - Y\partial^2 P^*/\partial\tau\partial Y$$

(4.432)

Eq. [4.432] is obeyed when the RHS and LHS go to zero or constant.

The solution of $\partial^2 V/\partial\tau^2 + \partial V/\partial\tau - \partial^2 V/\partial X^2 = 0$ is sought as follows,

Let $V = V^{ss} + V^t$ (4.433)

The solution is assumed to consist of a steady state and transient part. At steady state,

$$\partial P^*/\partial Y = 1/Y(\partial^2 V^{ss}/\partial X^2)$$

(4.434)

Integrating with respect to Y

$$\Delta P^*/\ln(Y_2/Y_1) = \partial^2 V^{ss}/\partial X^2$$

(4.435)

or $V^{ss} = \Delta P^*/\ln(Y_2/Y_1) X^2/2 + c_1 X + c_2$ (4.436)

At $X = 0$, $\partial u/\partial X = 0$ (4.437)

So c_1 can be set to zero.

Solving for c_2 from the boundary condition at $X = \pm X_b$, $u = 0$ (4.438)

$$V^{ss} = (\Delta P^* (X^2 - X_b^2)/2\ln(Y_2/Y_1)$$ (4.439)

The transient part of the solution can be obtained by solving the following equation;

$$\partial^2 V/\partial\tau^2 + + \partial V/\partial\tau = \partial^2 V/\partial X^2$$ (4.440)

The damping term can be removed by a $V = W\exp(-\tau/2)$. Eq. [4.471] becomes as shown in the above sections;

$$\partial^2 W/\partial\tau^2 - W/4 \quad = Y\partial^2 W/\partial X^2$$ (4.441)

The solution can be obtained by the method of separation of variables;

Let $W = f(\tau)g(X)$ (4.442)

Eq. [4.441] then becomes,

$$f''/f - \tfrac{1}{4} = g''/g = -\lambda_n^2$$ (4.443)

$$g = c_1 Sin(\lambda_n X) + c_2 Cos(\lambda_n X)$$ (4.444)

From the boundary condition at $X = 0$, i.e, $g' = 0$, it can be seen that c_1 can be set to zero.

$$g = c_2 Cos(\lambda_n X)$$ (4.445)

From the boundary condition at $X = \pm X_b$, $g = 0$,

$$\lambda_n X_b = (2n-1)\pi/2, \quad n = 1,2,3.....$$ (4.446)

$$f = c_3 \exp{+\tau}sqrt(1/4 - \lambda_n^2) + c_4 \exp{-\tau}sqrt(1/4 - \lambda_n^2)$$ (4.447)

At infinite time, $u = 0$ and $V = uX = 0$. W is $V\exp(\tau/2)$ and is zero at infinite time. Hence, c_3 need to be set to zero. Thus,

$$uY = V = \sum_1^\infty c_n \, \text{Cos}((2n-1)\pi z/2b)\exp\text{-}\tau\text{sqrt}(1/4 - \lambda_n^2) \qquad (4.448)$$

c_n can be solved from the initial condition which is when the fluid is at the reference velocity. Cn can be shown to be $4(-1)^{n+1}/(2n-1)\pi$.

For large values of λ_n, the characteristic nature of the solution changes from decaying exponential to subcritical damped oscillatory. Even for $n = 1$, when,

$$b < \pi(\gamma\tau_{mom})^{1/2} \qquad (4.449)$$

Worked Example 4.13

In worked example 4.3 the hyperbolic PDE was converted to a parabolic PDE by a general substitution to obtain the dimensionless velocity profile in a semi-infinite body of fluid, set in motion by a horizontal plate moving at constant velocity. Use the Boltzmann transformation and convert the hyperbolic PDE into a PDE with two variables, one in time and the other in the similarity variable. Show that the separation of variables technique can be used to separate the two functions as long as the Z in non zero in the similarity variable.

The governing equation can be written as;

$$\partial^2 u/\partial\tau^2 + \partial u/\partial\tau = \partial^2 u/\partial Z^2 \qquad (4.450)$$

The Boltzmann transformation can be written as;

$$\chi = Z/2 \, \tau^{1/2} \qquad (4.451)$$

Then Eq. [4.450] can be rewritten as;

$$\partial^2 u/\partial\tau^2 = \partial^2 u/\partial Z^2 \quad - \partial u/\partial\tau \qquad (4.452)$$

The RHS of Eq. [4.450] alone in subject to the transformation given by Eq. [4.451] as is done in several parabolic PDE problems. The RHS then becomes;

$$\tfrac{1}{4}\tau(\ \partial^2 u/\partial\chi^2 + 2\chi \, \partial u/\partial\chi\ ) = \partial^2 u/\partial\tau^2 \qquad (4.453)$$

Eq. [4.453] can be solved by the method of separation of variables;

Let $u = g(\chi) V(\tau)$ (4.454)

For non-zero Z, the variables χ and τ can be treated as separate variables and separated out as follows;

$$(g'' + 2\chi g')/g = 4\tau V''/V = -c$$ (4.455)

where c is an arbitrary constant;

$$\tau^2 d^2V/d\tau^2 + c/4\tau V = 0$$ (4.456)

Comparing Eq. [4.455] with the generalized Bessel equation as shown in Eq. [A.30],

$$
\begin{aligned}
a &= 0 \\
b &= 0 \\
c &= 0 \\
d &= c/4 \\
s &= \tfrac{1}{2} \\
p &= 2. \ \tfrac{1}{2} = 1
\end{aligned}
$$
(4.457)

The solution to Eq. [4.456] can be written as;

$$V = c' \tau^{1/2} J_1 (c/2 \ \tau^{1/2}) + c'' \ \tau^{1/2} Y_1 (c/2 \ \tau^{1/2})$$ (4.458)

At time zero, V is finite and hence $c'' = 0$.

$$g'' + 2\chi g' + cg = 0$$ (4.459)

Eq. [4.459] can be solved by the power series method a variation of the method of Frobenius.

Let $g = \sum_0^\infty c_n \chi^n$ (4.460)

$g' = \sum_1^\infty nc_n \chi^{n-1}$ (4.461)

$g'' = = \sum_2^\infty n(n-1) c_n \chi^{n-2}$ (4.462)

$$\sum_2^\infty n(n+1) c_n \chi^{n-2} + 2\sum_1^\infty nc_n \chi^n + c\sum_0^\infty c_n \chi^n = 0$$ (4.463)

Equating the coefficients of like powers,

$cC_0 + 2.3 C_2$ (χ^0) (4.464)

$C_2 = -cC_0/6$

$$cC_1 + 2C_1 + 12\, C_3 \quad = 0 \qquad\qquad (\chi^1) \qquad\qquad (4.465)$$

$$C_3 \quad = \quad -C_1(c + 2)/12 \qquad\qquad (4.466)$$

$$cC_3\ \text{-}6C_3\ + 1.2.C_1$$

$$C_2(c + 4) + \ 4.3\, C_4\ = 0 \qquad\qquad (\chi^2) \qquad\qquad (4.467)$$

$$C_4\ = -C_2(c + 4)/4.3 \qquad = +c(c + 4)C_0/24 \qquad\qquad (4.468)$$

Thus,

$$u\ = c'\, \tau^{1/2}\, J_1\, (c/2\, \tau^{1/2})\ \ (C_0\ + C_1\chi\ \text{-}\ cC_0/2\, \chi^2\ +\ C_1(c + 2)/12\, \chi^3\ + c(c+ 4)C_0/24\ \text{-..})$$
$$(4.469)$$

The constants C_0 and C_1 can be solved from the boundary conditions at $Z = 0$, and the shear stress at $Z = 0$.

Worked Example 4.14

Consider the oscillations of a fluid in a tube of radius R about a mean position where an oscillating pressure gradient is imposed on the system using the momentum transfer and relaxation equation. The pressure gradient imposed is periodic wrt time with frequency ω.

$$-\partial p/\partial z\ =\ a_0\ \text{Real Part of } \exp(i\omega t) \qquad\qquad (4.470)$$

The equation of motion for the vertical component of the velocity in the pipe from the Appendix is given from Eq. [C. 10]:

$$\rho\tau_{mom}\ [\partial^2 v_z/\partial t^2\ +\ + v_z\partial^2 v_z/\partial t\partial z\ ] \qquad\qquad (4.471)$$
$$+\ (\tau_{mom}\ \partial v_z/\partial t\ + v_z)(\ \partial v_z/\partial z)]$$
$$+\ \tau_{mom}\partial^2 p/\partial t\ \partial z\ + \rho\partial v_z/\partial t$$
$$=\ -\partial p/\partial z\ +\ \mu[\partial/\partial r(r\partial v_z/\partial r)\ +(\partial^2 v_z/\partial z^2)]\ \ + \rho g_z$$

From the equation of continuity, neglecting the radial and angular component of velocities and considering a incompressible fluid,

$$\partial(\rho v_z)/\partial z\ = 0 \qquad\qquad (4.472)$$

and $v_z\ =\ \phi(r)$

Hence the equation of motion can be written after non-dimensionalizing and neglecting the gravity effects and keeping the tube horizontal as;

$$\tau = t/\tau_{mom} \; ; \; P^* = p/\rho v_{mom}v_{ref} \; ; \; u = -(v_z - v_{zmax})/v_{ref}; \; X = r/\text{sqrt}(\gamma\tau_{mom}) \qquad (4.473)$$

$$\partial^2 u/\partial\tau^2 + \; \partial^2 P^*/\partial X\partial\tau \; + \partial u/\partial\tau \; = \; -\partial P^*/\partial Z \; + \; \partial/\partial X(X\partial u/\partial X) \qquad (4.474)$$

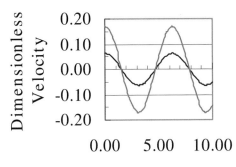

$$—— X = 0.1 —— X = 0.2$$

Dimensionless Velocity

0.20
0.10
0.00
-0.10
-0.20

0.00   5.00   10.00

Dimensionless Time

Figure 4.16   Dimensionless Velocity Response to a Oscillating Pressure Gradient Imposed on Tube Flow by Momentum Transfer and Relaxation Equation

The non-homogeneity in the boundary condition can be removed by supposing that the solution can consist of two parts, one transient and the other steady state part.

Let $u = \phi(X) \exp(i\omega^*\tau) + u^{ss}$ \hfill (4.475)

$$\Delta P^*/L^* = f(X) \exp(i\omega^*\tau)$$
Where $\omega^* = \omega\tau_{mom}$ \hfill (4.476)

The steady state part of the solution can be written as;

$$\partial P^{*}/\partial Z \quad = \quad \partial/\partial X(X\partial u/\partial X) \tag{4.477}$$

Integrating both sides wrt to Z

$$\Delta P^{*}/L \quad = \quad \partial/\partial X(X\partial u/\partial X) \tag{4.478}$$

Integrating wrt X,

$$C' + X \Delta P^{*}/L \quad = \quad X\partial u/\partial X \tag{4.479}$$

$$C' = 0 \text{ as at } X = 0, \partial u/\partial X \quad = 0 \tag{4.480}$$

$$C'' + X\Delta P^{*}/L \quad = u^{ss} = \quad v_{zmax}/v_{ref} \tag{4.481}$$

$$\text{At } X = X_{R,} \tag{4.482}$$

$$C'' = v_{zmax}/v_{ref} - X_{R}\Delta P^{*}/L \tag{4.483}$$

$$u^{ss} = v_{zmax}/v_{ref} - (X_{R} - X) f(X) Cos(\omega^{*}\tau) \tag{4.484}$$

The transient part of the velocity profile can be solved for as follows by assuming that the velocity also has a periodic component with the same frequency, ω;

$$-\omega^{*2} \phi + \phi i\omega^{*} + i\omega^{*} f = -f' + \phi' i\omega^{*} - X\phi'' \omega^{*2} \tag{4.485}$$

Assuming f and ϕ are the same;

$$X^{2} \phi'' + X/\omega^{*2} \phi' + X(i/\omega^{*} - 1)\phi = 0 \tag{4.486}$$

Comparing Eq. [4.519] with the generalized Bessel equation given by Eq. [A.30];

$a = 1/\omega^{*2}$
$b = 0$

$c = 0$

$s = \frac{1}{2}$ $\qquad\qquad$ (4.487)

$d = i/\omega^* - 1$

$p = (\omega^{*2} - 1)/\omega^{*2}$

For high dimensionless frequency the order of the Bessel solution can be taken as 1. The solution can be presented as;

$$\phi = c'X^{1/2} \ I_1 (2 (i/\omega^* - 1)^{1/2}X^{1/2} \ + c'' X^{-1} I_{-1} (2 (i/\omega^* - 1)^{1/2}X^{1/2} \qquad (4.488)$$

c'' can be set to be zero because ϕ is finite at $X = 0$.

$$\phi = c'X^{1/2} \ J_1 (2 (1/i\omega^* + 1)^{1/2}X^{1/2}) \qquad (4.489)$$

c' can be solved from the boundary condition of zero velocity at $r = R$ or $X = X_R$

$$-v_{zmax}/v_{rel} = (c'X_R^{1/2} J_1 (2 (1/i\omega^* + 1)^{1/2} X_R^{1/2}) \qquad (4.490)$$

$$c' = -v_{zmax}/v_{rel}X_R^{1/2} J_1 (2 (1/i\omega^* + 1)^{1/2} X_R^{1/2}))$$

$$(v_{zmax} - v_z)/v_{zmax} = (X/X_R)^{1/2} J_1 (2 (1/i\omega^* + 1)^{1/2}X^{1/2})Cos(\omega^*\tau) \qquad (4.491)$$

At high frequency,

$$(v_{zmax} - v_z)/(v_{zmax}) = (X/X_R)^{1/2} J_1(2X^{1/2})/J_1(2X_R^{1/2})Cos(\omega^*\tau) \qquad (4.492)$$

4.9 Friction Factors

Many engineering flow problems fall under the categories of flow in circular pipes and flow past spherical objects. The peace pipeline project discussed is to construct a pipeline to meter out oil from the gulf region such as from Iran to N.Delhi, India through a host of countries, such as Iran, Turkmenistan, Afghanisthan, Pakisthan and India.   Later in the section   the optimal number of pipes needed is derived. Examples of flow in pipes involve piping oil in pipes, flow of water in channels, extrusion of polymer throug die, flow of fluid through a filter, pulsatile flow from the lungs to the nostril,   blood flow through the capillaries, and flow during reverse osmosis in desalination. Examples of flow around submerged objects are the flow of air around the airplane wing, motion of fluid around particles, fluid flow in fluidized bed combustors, reactors, and heat exchangers circulating fluidized beds.   In such problems the relationship between the pressure drop and the volume rate of flow. In flow past submerged objects the drag force is important. Sometimes experimental data is utilized to obtain correlations for the drag coefficients for the appropriate geometry and flow situation. In this section, the notion of friction factor is introduced and charts can be found in handbooks of chemical engineering, bioengineering, mechanical engineering, civil engineering. Often times friction factors are defined for steady state scenarios. In this section after introducing the concept of friction factor, during the transient flow due to momentum transfer and relaxation the friction factor and drag forces are discussed.

A force is exerted by the flowing fluid on the solid surface it is in contact with.   This has been calculated at steady state, incompressible flow under various conditions.   This force consists of two parts;   i) that would act even when fluid is stationary and 2) additional force associated with the kinetic behavior of the fluid.

$$F_k = A K_e f \tag{4.493}$$

A is the characteristic area, K a characteristic energy per unit volume and a dimensionless quantity, f, called the friction factor. For circular tubes, of radius R and length L, f is defined as;

$$F_k = (2\pi RL)(1/2\rho <v>^2) f \tag{4.494}$$

For a fully developed pipe flow, a force balance on the fluid between 0 and L in the direction of flow yields;

$$F_k = \pi R^2 (\Delta p + \rho g (h_0 - h_L)) = \pi R^2 (\Delta P) \tag{4.495}$$

Comparing Eqs. [4.494,4.495],

$$f = \frac{1}{4} D/L \ (\Delta P/(1/2\rho <v>^2)) \tag{4.496}$$

This is usually referred to as the fanning friction factor. For flows around submerged objects, the characteristic area A is usually taken to be the area obtained by the projecting the solid on to a plane perpendicular to the velocity of approach of the fluid. Then, k is taken to be $1/2\rho \ v_\infty^2$, where v_∞, is the approach velocity of the fluid at large distance from the object. Thus for flow past a sphere at steady state,

$$F_k = (\pi R^2)(1/2\rho \ v_\infty^2) \ f \tag{4.497}$$

The resultant force of gravity and buoyancy driving the motion of the sphere is given by;

$$F_s = \pi D^3/6 \ (\rho_s - \rho)/\rho \tag{4.498}$$

Comparing Eqs. [4.597, 4.598] for the net force to be zero as it is at the terminal settling velocity of the fluid;

$$f = 4/3 \ gD/v_\infty^2 \ (\rho_s - \rho)/\rho \tag{4.499}$$

The friction factor Eq. [4.529] is referred to as drag coefficient, and represented by C_D. For a long smooth horizontal pipe of length L, at steady state for fluid with constant ρ and μ. The force exerted by the fluid on the inner pipe wall for either laminar or turbulent flow is;

$$F_k = \int_0^L \int_0^{2\pi} (-\mu \ \partial v_z/\partial r)|_R \ Rd\theta \ dz \tag{4.500}$$

Comparing Eqs. [4.499, 4.500]

$$f = \frac{\int_0^L \int_0^{2\pi} (-\mu \ \partial v_z/\partial r)|_R \ Rd\theta \ dz}{(\pi RL) \ \rho <v^2>} \tag{4.501}$$

$$f = \phi \ (Re, L/D) \tag{4.502}$$

Thus the friction factor is a function of the Reynolds number and the L/D ratio.

For laminar, steady, incompressible flow in circular pipe the Hagen – Poiseulle flow distribution is given by;

$$v^{ss} = \Delta P/4\mu LR^2(1 - r^2/R^2) \qquad (4.503)$$

Substituting Eq. [4.533] in Eq. [4.531] f can be calculated as;

$$f = 16/Re \qquad (4.504)$$

where $Re = \rho<v_z>2R/\mu$

Eq. [4.504] has been found to be valid for Reynolds number less than 2100. For turbulent flow in smooth circular tube, the friction factor is given by the Blausius' formula [1938];

$$f = 0.0791/Re^{1/4} \qquad (4.505)$$

The Blausis formula is valid for Reynolds number less than 10^5. The friction factor for during flow between two plates moving in opposite directions when governed by the momentum transfer and relaxation equation is given by;

For $a < 2\pi sqrt(\gamma\tau_r)$,

$$u^t = \sum_1^\infty c_n exp(-\tau/2) Cos(sqrt((\lambda_n^2 -1/4)\tau)Sin(\lambda_n Z) \qquad (4.506)$$

$$\partial u^t/\partial Z\big|_b = \sum_1^\infty \lambda_n (-1)^n c_n exp(-\tau/2) Cos(sqrt((\lambda_n^2 -1/4)\tau) \qquad (4.507)$$

The definition used for the friction factor at steady state is retained and,

$$f = -v_{max}/sqrt(\gamma\tau_{mom}) \mu\partial u/\partial Z/(\rho/2 <v>^2) \qquad (4.508)$$

$$<u^t> = \sum_1^\infty -(1 - (-1)^n)^2 exp(-\tau/2) Cos(sqrt((\lambda_n^2 -1/4)\tau)/n^2\pi^2 \qquad (4.509)$$

Combining Eqs. [4.537 – 4.539]

$$f = -v_{max}/sqrt(\gamma\tau_{mom}) \mu\partial u/\partial Z/(\rho/2 <v>^2) \qquad (4.510)$$

Defining, $Re = (\rho <v>b/\mu)$

$$f = (2/Re) \frac{\sum_1^\infty -2 (1 - (-1)^n)(-1)^n exp(-\tau/2) Cos(sqrt((\lambda_n^2 -1/4)\tau)}{\sum_1^\infty -(1 - (-1)^n)^2 exp(-\tau/2) Cos(sqrt((\lambda_n^2 -1/4)\tau)/n^2\pi^2}$$

$$(4.511)$$

For example taking the first few terms in the infinite series in Eq. [4.512]

$$f = (2/Re) \frac{[Cos(sqrt((\lambda_1^2 - 1/4)\tau) + Cos(sqrt((\lambda_3^2 - 1/4)\tau) ...]}{[\ 1/\pi^2\ Cos(sqrt((\lambda_1^2 - 1/4)\tau) + 1/9\pi^2\ Cos(sqrt((\lambda_3^2 - 1/4)\tau) + ..]}$$

(4.513)

For the flow around the sphere, the friction factor can be shown to be;

$$f = 24/Re$$ (4.514)

This can be derived from the Stokes law. This has been found valid for Re < 0.1 in the creeping flow regime.

Summary

34 different flow types were identified and compared with each other. These are, laminar, turbulent, laminar boundary layer, laminar sublayer, adiabatic, isentropic, steady, transient, pug/uniform, irrotational, vortex, dimensional, potential, incompressible, compressible, subsonic, supersonic, transonic, supercritical, subcritical, critical, Knudsen, ballistic/relaxational, radial, Marangoni, annular, rapid, capillary, reacting, vacuum, tangential, slip, pulsatile, two phase flows.

The damped wave momentum transfer and relaxation equation was introduced. The equation was examined from a molecular standpoint and the accumulation term in the definition of momentum flux was identified as room for the introduction of the dynamic component of the shear stress. The velocity of momentum transfer is given by $v_{mom} = (\gamma/\tau_{mom})^{1/2}$. The relaxation time of momentum and the dynamic component of the shear stress were introduced. The differential form of equation of continuity was derived using a mass balance on a cubic control volume in cartesian coordinates. The substantial or total derivative was defined. An integral equation of continuity was derived. The equation for the x component of momentum was derived using the shell balance and equating the sum of all the external forces acting on the control volume of the fluid element plus the momentum in and momentum out to the accumulation of momentum in the control volume.

The Navier-Stokes equation was derived by extending the derivation to all the three cartesian components of the momentum of the fluid. The Navier-Stokes equation was modified using the finite speed momentum transfer and relaxation equation for the fluid. Two additional terms were introduced; one to denote the rate of acceleration of the fluid and the other the ballistic contribution of the pressure forces in transience. When the apple fell from the tree, Newton deduced his second law that related the external forces to the mass times acceleration of the object. As the apple falls from a initial zero velocity it accelerates due to a resultant force of gravity, buoyancy and drag until it reaches the terminal settling velocity where the sum of the forces are zero. The acceleration is the ratio of the sum of the resultant forces to the mass of the apple changes with the motion of the apple and the rate of acceleration exists and seldom accounted for. By analogy, during the transience of the fluid motion the rate of acceleration of the fluid can be accounted for by using the finite speed momentum transfer and relaxation equation. The extended Euler equation was written including the finite speed momentum transfer and relaxation effects.

The extended Euler equation was used to predict the pressure and velocity distributions in the fluid when a filled tube is emptied. Another worked example with multiple ports was considered and two components of the velocity were considered. The flow near a horizontal plate suddenly pulled at a constant velocity was derived using the transient damped wave momentum transfer and relaxation equation. An accumulation number is defined. The momentum inertia was calculated. The solution for the velocity was obtained by the method of relativistic transformation of coordinates. The dimensionless shear stress was obtained by the method of Laplace transforms. The solution is a modified composite Bessel function of the first kind and zeroth order and decaying exponential in time for large times and a composite Bessel function of the first kind and zeroth order and decaying exponential in time. The time taken for a point in the semi-infinite medium to see the momentum disturbance at the surface was shown to be $\tau_{inertia} = sqrt(Z_p^2 - 23.1323)$. The boundary layer thickness for a given instant in time greater than zero is given by, $\delta(\tau) = sqrt(23.13 + \tau^2)$. Beyond this distance, the fluid velocity can be taken to be zero from the analogy. The ripple effect and the backflow were calculated.

A general substitution, $\theta = Z + \tau(1 - \beta)^{1/2}$ was shown to transform the hyperbolic PDE into a parabolic PDE. The resulting PDE can be solved using the Botlzmann transformation. The emptying pipe problem was solved with a porous packing and use made of the Darcy's law. A positive permeability dependence with pressure as in the case of elutriating particles was shown to lead to subcritical damped oscillations in the velocity of the fluid. A permeability dimensionless number is defined. The shear flow between two plates moving in opposite directions at constant velocity was solved for using the damped wave momentum transfer and relaxation equation. For a < $2\pi sqrt(\gamma\tau_r)$, the velocity of the fluid is found to be subcritical damped oscillatory. In Figure 4.11 is shown the oscillations of the fluid in shear flow between two vertical plates moving in opposite directions at constant velocity. The positive permeability coefficient in the pressure drop and velocity relationship is assumed.

The transient laminar flow in a circular pipe was derived by the method of separation of variables. When R < $4.8096 \, sqrt(v\tau_{mom})$, subcritical damped oscillations in the fluid velocity were predicted. The oscillations in a manometer were derived using the ballistic accumulation term and by solving a third order equation. For a finite oscillation number the displacement was shown to pulsate.

The flow rate and pressure drop relationships for blood flow in capillaries were discussed using the marginal zone theory. The plasma and cell component of the fluid have separate relationships.

The tangential flow induced by a rotating cylinder was derived using the damped wave momentum transfer and relaxation equation. Transient flow past a sphere was revisited with the hyperbolic PDE. The radial flow between two concentric spheres was derived. The air cooled by a ceiling fan was modeled using a two –component governing equation. The radial outflow from two circular disks was studied. The oscillations in a tube upon imposing a oscillating pressure gradient were studied.

Friction factors for transient problems were introduced.

Exercises

1.0   *Momentum Transfer to Two Fluids from Flat Plate*
A flat plate is moved suddenly in-between two fluids of kinematic viscosity, γ_1, and γ_2 and relaxation times τ_{mom1} and τ_{mom2} respectively. The fluid at the two surfaces that bound the fluids are stationary. Compute force exerted on the plate and obtain the distance of the plate from the wall for balance.

Show that the governing equations for the two fluid may be written as

$$\partial^2 u / \partial \tau^2 + \partial u / \partial \tau \quad = \quad \partial^2 u / \partial X^2$$

$$\kappa \partial^2 u / \partial \tau^2 + \partial u / \partial \tau \quad = \quad \beta \partial^2 u / \partial X^2$$

where $u = v_x/V$; $\tau = t/\tau_{mom}$; $X = z/(\gamma \tau_{mom1})$; $\beta = \gamma_2/\gamma_1$; $\kappa = \tau_{mom2}/\tau_{mom1}$

The non-homogeneity in the boundary condition can be removed By superposing the steady state and transient solutions:

Let the distance of the plate from the top surface be b and from the bottom surface be a.

Let $u = u^{ss} + u^t$

Show that at steady state,

$$u^{ss} = 1 - X/X_b \quad \text{(fluid on top of the plate)}$$

$$u^{ss} = X/X_a + 1 \text{ (fluid below the plate)}$$

After removing the damping terms show that the governing equations for the transient component of the fluid atop the plate and for the fluid below the plate can be written after a $u = w\exp(-\tau/2)$ substitution;

$$\partial^2 w/\partial\tau^2 - w/4 = \partial^2 w/\partial X^2$$

$u = w\exp(-\tau/2\kappa)$ substitution the governing equation for the fluid below

$$\partial^2 w/\partial\tau^2 - w/4\kappa^2 = (\beta/\kappa)\partial^2 w/\partial X^2$$

The space and time conditions are;

$$X = 0, u^t = 0$$

$X = X_a$, $u^t = 0$; $X = X_b$, $u^t = 0$ $\tau = 0$, $u = 0$. Use the method of separation of variables to solve for the transient component of the dimensionless velocity. Show that $u^t = \sum_1^\infty c_n \exp(-\tau/2)\exp(-\tau\sqrt{1/4 - \lambda_n^2}))\text{Sin}(\lambda_n X)$. where $\lambda_n = n\pi/X_b$, $n = 1,2,3....$ and where $c_n = 2(1 - (-1)^n)/n\pi$

Further for small distance between the plate and the bounded surface,
Show that $u^t = \sum_1^\infty c_n \exp(-\tau/2)\text{Cos}(\tau\sqrt{\lambda_n^2 - 1/4})\text{Sin}(\lambda_n X)$
In a similar fashion for the fluid below the plate,

Show that $u^t = \sum_1^\infty c_n \exp(-\tau/2)\exp(-\tau\sqrt{1/4\kappa^2 - \zeta_n^2}))\text{Sin}((\kappa/\beta)^{1/2}\zeta_n X)$

where $\zeta_n = n\pi/X_a$, $n = 1,2,3...$

Where $c_n = 2(1 - (-1)^n)/n\pi$
Further for small distance between the plate and the bounded surface,
Show that $u^t = \sum_1^\infty c_n \exp(-\tau/2)\text{Cos}(\tau\sqrt{\zeta_n^2 - 1/4\kappa^2})\text{Sin}((\kappa/\beta)^{1/2}\lambda_n X)$

Derive the force on the plate during transient flow and equate the contributions of the fluid on top of the plate and from the bottom of the plate. Make use of the integrating factor if necessary. Is the force oscillatory for small separation distances between the plate and the bounded surface.

2. *Microlayer Composites*

Microlayer composites were coextruded upto 3713 alternating layers. The Interdiffusion of two miscible layers of polycarbonate and copolyester was Studied at temperatures from 200 – 230 ^{0} C at the polymer laboratory at Case Western Reserve university. Extend the analysis in problem 1 to n layers.

3. *Layer Rearrangement*

From the analysis in Problem 1 can the layer re-arrangement be predicted ? What secondary flows can be predicted ?

4. *Toricelli's Theorem for n Fluids*

Consider n layers of n different fluids on top of each other in a vertical container. Extent Toricelli's theorem to obtain the efflux velocity of the fluid

from the bottom of the container. Derive the azimuthal velocity as a function of space and time.

5. *Layered Flow in a Slit*

In section 4.6 flow of blood in a circular pipe for the plasma layer and core Layer was studied and the discharge rate as a function of the intrinsic viscosity of the l=plasma and core layers and the thickness of the plasma layer radius of the capillary was derived. Derive the discharge rates as a function of pressure drop and other parameters of flow for the slit flow limit.

plasma layer

core layer

Slit Flow Limit of the Plasma and Core Layers in the Blood Flow in Capillary

Let the width of the flat plates be 2W and the area of the cross-section be A, the Thickness of the core layer be $2\delta_c$. The boundary conditions are;

$$x = 0, \quad \partial v_x^c/\partial z = 0$$
$$x = W, \quad v_x^P = 0$$

$$x = \delta_c, \quad \tau_{zx}^c = \tau_{zx}^c \; ; \; v_x^P = v_x^c$$

Show that at steady state;

$$\partial p/\partial x \;\; = \;\; \mu_c(\partial^2 v_x^c/\partial z^2) \;\; = c_1 = \; -(p_0 - p_L)/L$$

$$v_x^c = (\Delta p/2\mu_c L)(W^2 - z^2) - \; \Delta p \delta_c^2/2L(1/\mu_p - 1/\mu_c)$$

$$v_x^P = (\Delta p/2\mu_p L)(W^2 - z^2)$$

Show that the average flow rates in the core and plasma layers can be calculated as;

$$\langle v_x^c \rangle \; = 2/\delta_c^2 \!\! \int_0^{\delta_c} \!\! z v_x^c \, dz = \;\; (\Delta p/\mu_c L)(W^2/2 - \delta_c^2/4) - \; \Delta p \delta_c^2/L(1/\mu_p - 1/\mu_c)$$

$$Q^c = \; (\Delta p W \delta_c^3/\mu_c L)\,(2(W/\delta_c)^2 - 1) - \; 4\Delta p W(\delta_c^3)/L(1/\mu_p - 1/\mu_c)$$

$$Q^P = \; 4W(\Delta p/\mu_p L)(W^2 - \delta_c^2)^2/(W + \delta_c)$$

6. *Transient Layered Flow*

For the geometry and space conditions as shown in Problem 5 derive the transient pressure and velocity distributions as a function of z and time. What is the critical thickness prior to the onset of subcritical oscillations in the flow rate.

7. *Momentum Transfer to Two Fluids from a Vertical Plate*

Repeat the analysis in problem 1 for vertical plate. B) Study the response to a oscillating velocity introduced by the vertical plate.

8. *Coaxial Flow*

Consider the coaxial flow between two cylinders. The free stream velocity approaching the coaxial cylinders is constant at V. Develop the transient velocity profile in the annulus using the damped wave momentum transfer and relaxation equation. What is the average velocity ? Where is the location of maximum velocity ? Obtain the pressure drop vs discharge rate relationships at steady state and transient state. Use the method of separation of variables and let the inner and outer radii be R And κR respectively and develop the conditions where subcritical oscillations in the velocity can be observed. What is the force exerted by the fluid on the surface. Defining the friction factor, f, as;

$$F_k = f (1/2\rho <v>^2) A_k$$

Obtain the friction factor at a) steady state and b) transient state for large Pipes and c) transient state for small pipes. A_k may be taken as the wetted surface area which is $2\pi LR (1 + \kappa)$.

9.0 *Bulk Flow Effect*

Consider a 1 dimensional flow due to a constant pressure drop along with bulk flow. Show that the governing equation can be written after neglecting the viscous effects as;

$$\tau_{mom} (\partial^2 v_x/\partial t^2 + v_x \partial^2 v_x/\partial t \partial x) + (\tau_{mom} \partial v_x/\partial t + v_x)(\partial v_x/\partial x) + \partial v_x/\partial t$$
$$= \Delta p/\rho L$$

At steady state,

$$v_x^2/2 = c + x(\Delta p/L)$$

Let $v_x = v_x^s + v_x^t$

$\tau = t/\tau_{mom}; \quad u = v_x/V_{ref}; \quad X = x/V\tau_{mom}: \quad P^* = \Delta p/(\rho V_{ref} L/\tau_{mom})$

Show that the transient portion of the solution will obey;

$$(\partial^2 u/\partial \tau^2 + u\partial^2 u/\partial \tau \partial X) + (\partial u/\partial \tau + u)(\partial u/\partial X) + \partial u/\partial \tau = 0$$

Let $u = V(\tau)g(X)$

$$V"g + Vg\, V'g' + g'V(V'g + Vg) + V'g = 0$$

$$V"/V + V'/V /(2V' + V) = -g' = c^2$$

$$g(x) = -c^2 x + d$$

$$V" + V'(1 - 2c^2 V) - V^2 c^2 = 0$$

Seek a solution for V by the method of Frobenius.

10. *Friction Factor in Radial Flow in Concentric Spheres*

In worked example 4.10, the radial flow between two concentric spheres of an incompressible, isothermal liquid was derived. The transient velocity distribution is examined using the damped wave momentum transfer and relaxation equation. Let the radii of the two spheres be R and mR respectively. The governing equation for the radial component of the velocity can be written from the Eq. [C.11] in Appendix C. The velocity was assumed to consist of a steady state and transient part. From the steady state part of the solution to the velocity profile obtain the friction factor as a function of Reynolds number for laminar flow. The transient velocity profile was derived by the method of separation of variables. Obtain the friction factor for transient flow for large spheres and small spheres.

11. *Conical Thrust Bearing*

A conical thrust bearing idealized as cone of vertex angle 2θ maximum cone radius R rests and revolves over a uniform fluid layer of thickness, δ, at a constant angular velocity ω. Derive the transient and steady state velocity and obtain expressions for torque required and the rates of heat dissipation in the bearing at steady state and transient state using the damped wave momentum transfer and relaxation equation.

12. *Falling Ball Viscometer*

In the falling ball viscometer, the shear rate is given by the terminal settling velocity of the sphere over radius of the falling ball and the shear stress by

$2/9gR(\rho_s - \rho)$. Consider the acceleration regime of the settling sphere. Develop the friction factor and Reynolds number relationship during acceleration. Show how the falling ball viscometer can be used to obtain the viscosity and relaxation time information from experiments.

13. *Rotating Cylinder Viscometer*

Examine the rotating cylinder viscometer in transient and steady state conditions. The radii of the cylinders are 3.2 cm and 3 cm and the outer cylinder was suddenly rotated at 180 RPM. For a liquid filled in the annular space to a depth of 8 cm the torque produced on the inner cylinder is 10^4 Nm at steady state. Use the damped wave momentum transfer and relaxation equation and obtain the spatio-temporal velocity profile. Calculate the viscosity of the liquid. Develop a procedure to obtain the relaxation time of the liquid using the transient torque data.

14. *Transnational Oil Pipeline*

There is interest in a peace pipeline to bring gas from Iran to N.Delhi via different countries such as Afghanisthan, Pakisthan, possibly Turkeministhan and India. Prepare a)preliminary estimate of the pipe size required for a transcontinental pipeline between the gulf and N.Delhi. It has to handle 1000 std m^3/hr of natural gas at a average pressure drop of 3 atm abs at an average temperature of 25 $^{\circ}$C. What is the maximum force exerted by the pipe ? Is this during transience or during steady state ? What is the ideal friction factor relation to be used ?   What is the optimal number of pipes to minimize the total cost to achieve the objectives of the task ? Using the damped wave momentum transfer and relaxation time equation obtain the time taken to reach steady state and the conditions to avoid subcritical damped oscillations.

14. *Transient Laminar Flow in a Circular Pipe*

Consider a hot circular pipe through which the fluid is flowing in laminar flow.

Obtain the transient velocity profile using the damped wave momentum transfer and relaxation equation. Obtaining a average velocity use the governing equation for heat transfer and plug the derived expression for the velocity in the azimuthal direction.

$$\tau_r(\partial^2 T/\partial t^2 \; + <v_z>\partial^2 T/\partial t\partial z \; + \partial T/\partial t \qquad = \qquad \alpha(1/r \; \partial/\partial r \; (r\partial T/\partial r)$$

Obtain the transient temperature profile using the damped wave heat conduction and relaxation equation. Obtain the ratio of the boundary layer thickness hydrodynamic to the thermal boundary layer thickness. Discuss the implications of Prandlt number in transience.

15. *Generalized Substitution to Convert Hyperbolic PDE to Parabolic PDE*

In worked example 4.3 a general substitution was used to reduce the hyperbolic PDE in 1 space dimension into a parabolic PDE. Consider all the three space dimensions and seek a suitable general substitution to reduce the hyperbolic PDE in 3 space dimensions and time dimensions into a parabolic PDE.

16. *Hemispherical Cup*

Repeat worked exampled 4.1 and 4.2 for a hemispherical cup. Obtain the pressure and velocity distribution as a function of z and t in the cup using the

extended Euler equation in 1 dimension and 2 dimensions respectively. What is different in the predictions of the efflux time, velocity and pressure profile.

Volume of a partially filled sphere $= \pi/6h_1 \, (3r_2^2 + h_1^2)$

17. *Friction Factor for a Moving Bubble*

Develop the friction factor for a bubble moving through a liquid.   Obtain the transient and steady state relationships. What assumptions were necessary ?

18. *Elutriating Bed with Positive Coefficient of Permeability Coefficient*

Consider a elutriating bed of particles. Set up the governing equation for the v_z . Obtain for a circular pipe with the pressure drop vs flow rate with   a positive permeability coefficient the conditions where the velocity will   exhibit subcritical damped oscillations using the extended Euler equation   making allowance for the Darcy's law with a positive permittivity. In elutriating bed as the superficial velocity increases the pressure drop decreases. This is in contrast to the Darcy's law for packed beds when the   pressure drop increased for increased f low rates. Use the damped wave momentum transfer and relaxation equation.

5.0   Applications

Nomenclature

C_p	Heat Capacity (J/kg/K)
c	Concentration (mole/m^3)
C_T	Concentratin of the solute in the tissue space (mole/m^3)
D_{AB}	Binary Diffusivity (m^2/s)
D_A	Binary diffusivity of the drug (m^2/s)
D_e	Effective diffusivity (m^2/s)
D_n	Diffusivity of the neurotransmitter
D_{ip}	Diffusion coefficient of the polymer within the polymeric material
D	Diameter of Tube (m)
h	Heat Transfer Coefficient (w/m^2 /K)
h'	Clearance in the helical ribbon agitator (m)
<h>	Heat transfer coefficient defined wrt to space averaged temperature
q =	h(<T> - T_{fire})
H	Height of agitator (m)
k	Thermal Conductivity of suspension (w/m K)
k_+	Encounter rate constant (sec.lit/mol)
k_-	Encounter rate constant (sec.lit /mol)
k_b	Rate constant of backward reaction (1/s)
k_f	Rate constant of forward reaction (1/s)
I_0	Modified Bessel function of zeroth order and first kind
I_1	Modified Bessel function of the first order and first kind
I_2	Modified Bessel function of the second order and first kind
I_3	Modified Bessel function of the third order and first kind
J_0	Bessel functin of the zeroth order and first kind
J_1	Bessel function of the first order and first kind
J_2	Bessel function of the second order and first kind
J_3	Bessel function of the third order and first king
K_0	Modified Bessel function of the zeroth order and second kind
K_a	Equilibrium association constant (K_a = 1/K_d)
K_b	Binding constant (C_B/C_A)
K_d	Equilibrium dissociation constant (=1/K_a) = k_b/k_f
K_e	Equilibrium rate constant (K_e = C_{Be}/C_{Ae})
K_m	Michaelis constant
M_t	Total mass of drug released from the external surface at time t, (moles)
q	Heat flux (w/m^2)
r	Distance (m)
r_c	Radius of the blood capillary (m)
r_T	Radius of the tissue cylinder
R	Radius of cylinder (m)
S	Storage coefficient (W/m^3/K)
t	Time (sec)

t_m	Thickness of the capillary (m)
T	Temperature (K)
T_0	Bed Temperature
T_1	Hot Tube Wall Temperature
u	Dimensionless temperature $(T - T_{fire})/(T_0 - T_{fire})$
u^t	Transient temperature
u^{ss}	Steady state temperature
u	Dimensionless Temperature $(T - T_0)/(T_1 - T_0)$
$u_{\tau\tau}$	Second Partial Derivative of u with Respect to τ
$u_{\tau Y}$	Second Partial Derivative of u with Respect to Y and τ
u_Y	First Partial Derivative of u with Respect to Y
u_{YY}	Second Partial Derivative of u with Respect to Y
U_y	Velocity of Suspension (m/s)
V_h	Velocity of heat (m/s) sqrt(α/τ_r)
V_m	Velocity of mass (m/s) sqrt(D/τ_{mr})
V_{mom}	Velocity of momentum (m/s) sqrt(γ/τ_{mom})
x	Distance (m)
X	Dimensionless Distance $x/\sqrt{(\alpha\tau_r)}$
X_c	Critical Radius beyond which there is no mass transfer
X_{pen}	Penetration Distance, $X_{pen} = (4\pi^2 + \tau^2)^{1/2}$
Y	Dimensionless Distance $y/\sqrt{(\alpha\tau_r)}$ and

Greek

α	Thermal Diffusivity (m²/s)
β	First zero of $J_{1/2}(\beta) = 0$
ρ	Density (kg/m³)
τ	Dimensionless time (t/τ_r)
ψ	Fick number $(D_{gs}t/h^2)$
τ_r	Relaxation Time, heat (s)
τ_{mr}	Relaxation Time, mass (s)
τ_{mom}	Relaxation Time, momentum (s)
Ω	Effectiveness factor for drug delivery = $<C_A>/C_0$

Dimensionless Groups

Bi*	Modified Biot modulus (h/SR)
Bi	Biot Number (h/sqrt(Sk))

$1/X_L$	Reciprocal of dimensionless length $(D_{ip}\tau_{mr})^{1/2}/L$
θ_{th}	Thiele Modulus $h(k'''/D_{gs})^{1/2}$
k^*	Dimensionless rate constant $(k_b\tau_{mr})$
hy^*	Dimensionless Heat Transfer Coefficient $(h_y/\sqrt{(Sk)})$
k_b^*	Dimensionless rate constant $(k_b^* = k_b\ \tau_{mr})$
k_f^*	Dimensionless rate constant $(k_f^* = k_f\ \tau_{mr})$
K^*	Dimensionless equilibrium rate constant $K^* = (k_b^* + k_f^* + 1)$
M^*	Dimensionless release amount, $M_t/AL(c_{ext} + c_0)$
Nu	Nusselt Number $(h\ D/k)$
Pe	Peclect Number, (Heat Transfer), Convection/Conduction, $U_y/\sqrt{(\alpha/\tau_r)}$
Pe_m	Peclect Number (mass), $V_z/\sqrt{D_{gs}}/\tau_{mr}$
Rel	Relaxation Number, heat, $(\alpha\tau_r / D^2)$
Rel_m	Relaxation Number, mass, $(D_{ip}\tau_{mr}/L^2)$

5.1 Convective and Conductive Contributions in Circulating Fluidized Bed Heat Transfer

Circulating Fluidized Bed, CFB technology is an attractive choice and was specifically developed to address today's needs for fuel flexibility and low emissions. Foster Wheeler, Babcock & Wilcox, Pyropower, Keeler Dorr-Oliver, Tennessee Valley Authority, Westinghouse and Combustion Engineering have reported new product lines with this technology. Although a number of correlations and theories for heat transfer between fluidized beds and surfaces have been suggested [Saxena 1989] very little has been reported in convective heat transfer in CFB to a horizontal tube. During short time scales the system may be far from equilibrium and the high rate unsteady state transient processes cannot be fully described using Fourier's heat conduction law.

The Cattaneo and Vernote non-Fourier equation, for heat conduction which was shown to be the only mathematical modification [Boley, 1964] that can remove the singularity in surface flux is:

$$q = -k\, \partial T/\partial x - \tau_r\, \partial q/\partial t \qquad (5.1)$$

Combined with the energy equation this was solved for the semi-infinite case by the method of Laplace Transforms and method of relativistic coordinate transformation presented earlier. The velocity of heat is given by $\sqrt{(\alpha/\tau_r)}$. A generalized expression for heat flux may be written as;

$$q = -k\, \partial T/\partial x - \tau_r\, \partial q/\partial t - \tfrac{1}{2!}\, \tau_r^2\, \partial^2 q/\partial t^2 - 1/3!\quad \tau_r^3 \partial^3\, q/\partial t^3 \qquad (5.2)$$

In this paper the Cattaneo and Vernotte non-Fourier heat conduction equation is used in the energy balance in 1 dimension along with the convection term. Consider the flow of gas solid suspension past a horizontal tube in a CFB. Two cases can be identified;

Case 1: Flow of Gas – Solid Suspension Away from the Horizontal Tube

The governing equation from an energy balance may be written after obtaining the dimensionless variables as;

$$u_{\tau\tau} + Pe\,(u_{\tau Y} + u_Y) + u_\tau = u_{YY} \qquad (5.3)$$

Case 2:   Flow of Gas – Solid Suspension Toward the Horizontal Tube

The governing equation from an energy balance after obtaining the dimensionless variables may be written as;

$$u_{\tau\tau} - Pe\,(u_{\tau Y} + u_Y) + u_\tau = u_{YY} \qquad (5.4)$$

The minus sign indicates the relative velocity of the suspension with respect to the direction of heat conduction path. For the bottom of the tube the y axis is selected as bottom direction is positive. Where y is the direction of flow of gas and solid, T the temperature, t the time and q the flux., $u = (T - T_0)/(T_1 - T_0)$, $\tau = t/\tau_r$, $Y = y/\sqrt{(\alpha\tau_r)}$ and $Pe = U_y/\sqrt{(\alpha/\tau_r)}$.   The Peclet number, Pe represents the ratio of the gas velocity to the velocity of heat by conduction.   When Pe is 0,   Eqs. [5.3,5.4]   reduce to the hyperbolic heat wave propagative equation by conduction alone which was solved for and presented earlier for the semi-infinite medium case under CWT. Obtaining the Laplace transform of Eqs.[5.3,5,4] the second order ODEs result,

(top of tube) $\underline{u}\,(s^2 + s)$ $= d^2\underline{u}/dY^2 - d\underline{u}/dY\,Pe\,(1 + s)$ $\qquad$ (5.5)

(bottom of tube) $\underline{u}\,(s^2 + s)$ $= d^2\underline{u}/dY^2 + d\underline{u}/dY\,Pe\,(1 + s)$ $\qquad$ (5.6)

Solving for the ordinary differential equation given by Eqs. (5.5,5.6);

$$r_1, r_2 = \tfrac{1}{2}(\,Pe\,(1 + s)(1 \pm \sqrt{(1 + 4s/Pe^2(s+1))}\,) \qquad (5.7)$$

At $Y = \infty$, $u = 0$ and $Y = 0$, $u = 1$

So $\underline{u}_Y\,(Y = 0) = -\,1/2s\,(s+1)(Pe - \sqrt{(Pe^2 + 4s/(s+1)}\,))$ $\qquad$ (5.8)

The heat transfer coefficient obtained in the Laplace domain may be written as;

(top of tube) hy^* $= L^{-1}\,Pe/2s\,(\,-1 + \sqrt{(1 + 4s/(s+1)/Pe^2)}\,)$ $\qquad$ (5.9)

where hy^* $= h_y/\sqrt{(k\,\rho\,C_p\,/\tau_r)}$ $\qquad$ (5.10)

For the case of bottom of tube, the axes is drawn in a manner that y is positive in the bottom direction and;

(bottom of tube) $hy^* = L^{-1}\,Pe/2s\,(\,1 + \sqrt{(1 + 4s/(s+1)/Pe^2)})$ $\qquad$ (5.11)

Inverting the Laplace domain after expanding the square root using the binomial theorem we have;

$$h/\sqrt{(k\rho C_p/\tau_r)} = +\exp(-\tau)/Pe - (1-\tau)\exp(-\tau)/Pe^3 + 2\exp(-\tau)/Pe^5(1 - 2\tau + \tau^2/2) - \ldots$$

(top of tube)   (5.12)

$h/\sqrt{(k\rho C_p/\tau_r)}$ $= Pe + \exp(-\tau)/Pe - (1-\tau)\exp(-\tau)/Pe^3 + 2\exp(-\tau)/Pe^5(1 - 2\tau + \tau^2/2) - ...$
(bottom of tube)   (5.13)

This expansion is good for large Peclet numbers. The terms in the series were inverted by the residue theorem by Heaviside expansion as shown by Mickley, Sherwood and Reed [1957] and by expressing the term as $P(s)/Q(s)$ and finding the poles and writing the inversion .; The results are presented in Table 1. The results are shown in Figure 5.1 for three different Peclect numbers. It can be seen that at short contact times, the microscale conduction effects become a dominant contribution to the heat transfer coefficient. At large contact times, the heat transfer coefficient becomes independent of the contact time and is found to be function of the Peclect number.

The difference in values between top and bottom of the tube gives the twice the convective contribution. The driving force for heat transfer as the suspension moves towards the tube is away from the tube leading to a minus sign in the convective contribution in the governing equation. As the suspension moves away from the tube the heat transfer driving force and the velocity of the suspension are in the same direction. Hence the conduction and convection have to be added to obtain the total heat transfer. In the case of the bottom of the tube the convection contribution has to be subtracted from the conduction contribution to obtain the net heat transfer coefficient. Thus when the difference between the values from the top and bottom of the tube is obtained the contribution to convection alone will result and when the values from top and bottom are added the twice the conduction contribution will result. Hence Eq. [5.12-5.13] and normalizing for direction the convective contribution may be seen to be $Pe/2$. The conductive contribution is the addition of Eqs. [5.12-5.13] and dividing by 2 to give after converting heat transfer coefficient to a dimensionless Nusselt number,

$Nu = \sqrt{(D^2/\alpha\tau_r)}$ $(A_1(\tau)/Pe + A_2(\tau)/Pe^3 + A_3(\tau)//Pe^5 + ...)$
(top of tube)   (5.14)

$Nu = \sqrt{(D^2/\alpha\tau_r)}(Pe/2 + A_1(\tau)/Pe + A_2(\tau)/Pe^3 + A_3(\tau)//Pe^5 + ...)$
(bottom of tube)   (5.15)

As $\tau \to 0$, in the short time limit the transient solution for top of the tube becomes;

$Nu = \sqrt{(D^2/\alpha\tau_r)}$ $(1/Pe - 1/Pe^3 + 1//Pe^5 + ...)$ (5.16)

$= \sqrt{(1/Rel)}\,(1/Pe - 1/Pe^3 + 1//Pe^5 + ...)$ (5.17)

For very large Peclet numbers, the transient portion of the solution drops out and the portion that results is invariant with time. In the absence of convection, conduction alone is the mechanism of heat transfer. Here the open interval solution developed earlier can be used. $(Uy = 0)$ and is given for both the bottom and top of the tube by;

$hy^* = \exp(-\tau/2) I_0 (\tau/2)$ (5.18)

310

When the relaxation time becomes infinite as in the insulator of heat, the D'Alambert's wave solution results.

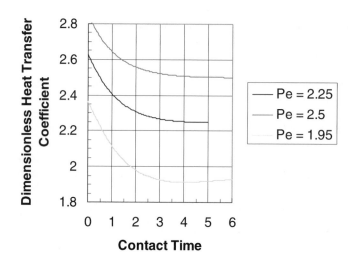

Figure 5.1 Dimensionless Heat Transfer Coefficient in CFB to and Away from a Tube

A second method of solution would be to invert Eq. [5.9] the case for bottom of tube, from the Laplace domain as follows;

$$hy^* \quad = L^{-1} Pe/2s (1 + \sqrt{(1 + 4s/(s+1)/Pe^2)} \quad (5.19)$$

$$= Pe/2s + A \sqrt{(s + B)}/2s/\sqrt{(s+1)} \quad (5.20)$$

where, $A = \sqrt{(4 + Pe^2)}; \quad B = Pe^2/(4 + Pe^2)$ \quad (5.21)

$$= Pe/2 + A/2 \exp(-\tau(1 + B)/2) I_0 (\tau(1 - B)/2) + AB/2 \int_0^\tau \exp(-p (1 + B)/2) I_0 (p(1 - B)/2)dp \quad (5.22)$$

$$hy^* \quad = Pe/2 + A/2 \exp(-\tau(1 + B)/2) I_0 (\tau(1 - B)/2) + AB/(1+ B)(\exp(-\tau(1 + B)/2) \\ (I_0 (\tau(1 - B)/2) + I_1 (\tau (1 - B)/2) - 1) \quad (5.23)$$

Table 5.1 Laplace Inverse of Each Term in the Infinite Series

No.	$A_m(\tau) = \exp(-\tau)(A_1 + A_2\tau + \ldots + A_m\,\tau^{m-1}/(m-1)!)$ $A_j = (-1)^{j+1}k!/(j-1)!^2/(k+1-j)!$	$L^{-1}(C\,s^k/(s+1)^{k+1})$
1.	$\mathrm{Exp}(-\tau)$	$1/(s+1)$
2.	$-\exp(-\tau)(1-\tau)$	$-s/(s+1)^2$
3.	$2\exp(-\tau)(1 - 2\tau + \tau^2/2!)$	$+2s^2/(s+1)^3$
4.	$-5\exp(-\tau)(1 - 3\tau + 3/2!\tau^2 - 1/3!\tau^3)$	$-5s^3/(s+1)^4$
5.	$14\exp(-\tau)(1 - 4\tau + 6/2!\tau^2 - 4/3!\tau^3 + 1/4!\tau^4)$	$14.s^4/(s+1)^5$
6.	$-42\exp(-\tau)(1 - 5\tau + 10/2!\tau^2 - 10/3!\tau^3 + 5/4!\tau^4 - 1/5!\tau^5)$	$-42s^5/(s+1)^6$
7.	$132\exp(-\tau)(1 - 6\tau + 15/2!\tau^2 - 20/3!\tau^3 + 15/4!\tau^4 - 6/5!\tau^5 + 1/6!\tau^6)$	$+132s^6/(s+1)^7$
8.	$-429\exp(-\tau)(1 - 7\tau + 21/2!\tau^2 - 35/3!\tau^3 + 35/4!\tau^4 - 21/5!\tau^5 + 7/6!\tau^6 - 1/7!\tau^7)$	$-429s^7/(s+1)^8$
9.	$1430\exp(-\tau)(1 - 8\tau + 28/2!\tau^2 - 56/3!\tau^3 + 70/4!\tau^4 - 56/5!\tau^5 + 28/6!\tau^6 - 8/7!\tau^7 + 1/8!\tau^8)$	$+1430s^8/(s+1)^9$
10.	$-4862\exp(-\tau)(1 - 9\tau + 36/2!\tau^2 - 84/3!\tau^3 + 126/4!\tau^4 - 126/5!\tau^5 + 84/6!\tau^6 - 36/7!\tau^7 + 9/8!\tau^8 - 1/9!\tau^9)$	$-4862s^9/(s+1)^{10}$
11.	$16796\exp(-\tau)(1 - 10\tau + 45/2!\tau^2 - 120/3!\tau^3 + 210/4!\tau^4 - 252/5!\tau^5 + 210/6!\tau^6 - 120/7!\tau^7 + 45/8!\tau^8 - 10/9!\tau^9 + 1/10!\tau^{10})$	$+16796s^{10}/(s+1)^{11}$

The inversions can be read from the tables provided in Mickley, Sherwood and Reed [22]. Eq. [5.23] is the exact solution and it includes modified Bessel function of the first kind and zeroth order. For the special case when Pe = 1,

$$hy^* = 1/2 + \sqrt{5}/2\,\exp(-\tau(1 + 1/5)/2)\,I_0(\tau(2/5) + \sqrt{5}/6(\exp(-\tau(3/5)(I_0(\tau(2/5) + I_1(\tau(2/5) - 1) \qquad (5.24)$$

Representative Experimental Data

The key parameter in using the model is the contact time of the suspension with the tube wall and the dimensionless Peclet. This may be a function of the operating parameters. In addition to this the physical properties of the suspension is required and the relaxation time of the suspension is an added intrinsic parameter of the system. Some experimental data was available for comparison from Sundaresan and Kolar [23]. They used silica at 3 mean sizes between 143 – 263 μm, air velocity 4 – 8 m/s, suspension density of 9-25 kg/m^3, 0.5 – 1.0 m high U horizontal tubes, 5.5 m CFB, solid circulation flux, 13-80 kg/m^2/s and reported heat transfer coefficients between 30 – 105 W/m^2/K in their core heat transfer studies in a CFB. A uniform distribution of the suspension packets with the wall tube is assumed as a first approximation and the Nusselt number calculated as (Nu = (h D/k) = 105 * .05/0.04 = 131). Where the diameter of the tube is taken as 2 inches and the effective thermal conductivity of the suspension calculated from a simple average with the voidage as the weight from the values of air and silica as 0.04 w/m K. Taking the effective thermal diffusivity of the suspension as 3 E-6 m^2/s. (Pe = U$_y$/ $\sqrt{(\alpha/\tau_r)}$. = 8/$\sqrt{(3 E-6/1E-3)}$ = 146; el = $(\alpha\tau_r / D^2)$ = 3E-6 * 1 E-3/.05^2; Short Time Limit : Nu = $\sqrt{(1/Rel)}$ (1/Pe + …) = 6.25). The relaxation time can be estimated from an analogy to the speed of molecules. The Maxwellian distribution for the speed of the molecules was used to get the root mean square velocity of the molecule and by analogy to heat transfer the velocity of heat transfer was estimated. The relaxation time was obtained from this estimate given the thermal diffusivity of the suspension. The convective contribution is thus Pe/2 = 146/2 = 73. The conductive contribution to the Nusselt number is then given a typical contact time of the suspension with the tube as 5 milliseconds (Nu = 73 + exp(-5)/73 - exp(-5)*(-4)/73^3 … = 73).

Experimental validation of the model will require contact time of the solids in the wall information, preferably the statistical distribution of their residence times at the surfaces. The relaxation time is another intrinsic property of matter and needs measurement prior to an elaborate study of experimental data. Thus, the Cattaneo and Vernotte non-Fourier heat conduction equation is used to model the mesoscale conduction effects in convection dominated heat transfer in CFB to a horizontal tube. Hence the exact solution is obtained for the dimensionless heat transfer coefficient by conduction and convection and found to be

$$\text{Pe/2} + \quad A/2 \exp(-\tau(1 + B)/2) I_0 (\tau(1 – B)/2) + AB/(1+ B)(\exp(-\tau(1 + B)/2) (I_0 (\tau(1 – B)/2) + I_1 (\tau (1 – B)/2) - 1)$$

for the case of bottom of tube. This can be used especially for Peclet numbers close to 1.

Two cases were identified, i.e., one for suspension flow toward the tube and the other for flow of suspension away from the tube. The Peclet number that gives the ratio of the convective fluid velocity to the velocity of heat conduction and the relaxation number that is the inverse of the Fourier number with the relaxation time substituted for time are important parameters for predicting the Nusselt number based upon the tube diameter. The convective contribution is given by Pe/2 = U$_y$/$\sqrt{(\alpha/\tau_r)}$/2.
A more usable form is obtained by expansion of the Laplace domain using binomial infinite series and upon obtaining the inversion by the method of residues. 13 terms in the binomial series for the case of bottom of tube is taken and results of the inversion presented in Table 1. This is useful at large Peclet numbers.

$$\text{Nu} \quad = \quad \sqrt{(D^2/\alpha\tau_r)} \ (Pe/2 + A_1(\tau)/Pe + A_2(\tau)/Pe^3 + A_3 (\tau)//Pe^5 +…)$$

The representative data available was used to show how the model can be applied to data from the experimental measurements from the CFB. In the short time limit the transient solution for top of the tube becomes;

$$Nu = \sqrt{(D^2/\alpha\tau_r)} \, (1/Pe \, - 1/Pe^3 + 1//Pe^5 + \ldots\ldots)$$

For very large Peclet numbers the transient portion of the solution drops out and the portion that results is invariant with time. In the absence of convection, conduction alone is the mechanism of heat transfer. Here the open interval solution developed earlier can be used. ($U_y = 0$). When the relaxation time becomes zero as in the insulator of heat, the D'Alambert's wave solution was developed as presented earlier.

5.2 CPU Overheating and the Storage Coefficient

The storage coefficient, S, with the units of W/m^3 /K, and given by $S = \rho \, C_p \, /\tau_r$ can be a critical parameter in the design of substrates of high speed processors in a similar vein to the thermal conductivity k (W /m/K) and heat transfer coefficient (W/m^2/K). The ratio of the thermal mass to the relaxation time of the material may be an indicator of the heat stored in the material. Especially during periodic phenomena this may become an important consideration. With the increase in speed of the microprocessors the periodic heating of the surface of a substrate becomes of interest. Given the time scales of the period of the microprocessor and some of the short time scale anomalies of Fourier heat conduction it is of interest to evaluate other expressions other than Fourier such the Cattaneo & Vernotte non-Fourier heat conduction and relaxation equation, Dual Phase Lag Model subject to a periodic boundary condition.

Heat sinks are used in electronic equipment designs to transfer heat from a heat source, such as an electronic component, to lower temperature surroundings. The objective of a heat sink is to lower the temperature of the heat generator to prevent performance degradation and prolong the life of the heat source. The effectiveness of a heat sink depends on its ability to transfer heat from the heat source to the lower temperature surroundings. Some factors influencing this ability include the heat transfer rate of the material from which the heat sink was constructed and the surface area of the heat sinks. The heat transferring ability of a heat sink may be increased using a material with a higher heat transfer rate to construct the heat sink. Heat sinks typically comprises of one solid piece of material that has a high conductivity with an adequate mechanical strength for secondary support functions. The materials that possess these qualities include metals or metallized plastics, such as aluminum and copper. Attempts have been made to increase the heat transferring ability of a heat sink may by increasing the surface area through which heat may be dissipated, e.g., lengthen the fins as described in a recent patent from Lucent technologies.

An integrated circuit operating at a constant fixed level of activity has an equilibrium temperature which is eventually reached. At equilibrium, the heat lost to the surroundings of the circuit on average equals the heat generated by the circuit. The temperature, therefore remains roughly constant. Unfortunately, the equilibrium temperature of an integrated circuit is often above the safe operating temperature and a cooling scheme is needed. Many high performance microprocessor systems experience excessive heat buildup in the central processing units (CPU) due to high power dissipation. Conventional cooling schemes for these systems use either a cooling fan or a large heat sink. Cooling fans move air past the integrated circuits and thereby increase the heat loss from the integrated circuits to the surroundings. This lowers the equilibrium temperature. However, cooling fans are large (compared to the scale of an integrated circuit), consume power, and can be noisy. Heat sinks are also large, in addition to being expensive and less effective. In portable computers where size and power consumption are critical, neither fans nor heat sinks are practical. The temperature of a CMOS integrated circuit depends on the power used by the integrated circuit which is proportional to the clock speed or frequency of operation of the integrated circuit. Prior art power control schemes can change clock speed and therefore change circuit temperature. However, existing power control schemes are aimed at preserving battery power in devices such as portable computers and not at monitoring integrated circuit temperature. Any effect of such power saving circuitry on temperature is coincidental and ineffective at limiting worse case heat buildup. Regulation of temperature is often a goal of the engineer. Power use regulator was described in a patent by Picopower [2]. For example, if the CPU is running at 33 MHz for 30 seconds, a certain temperature, such as 50.degree. C, will be reached. The actual values are experimentally determined. If the maximum temperature allowable for the CPU is set at 50 ^{0} C. then the regulation threshold must be set so that 30 seconds of 33 MHz operation causes the up/down counter to increment to the threshold value. The CPU clock is then forced to slow down for a "cooling time" during which time the counter will then decrement and the CPU will cool. The cooling time, for example, might be 100 milliseconds. While the circuit is cooling the temperature count on the up/down counter decreases since the regulator samples the clock speed as slow, or cool. After the cooling time is finished the CPU is then allowed to run at 33 MHz until the regulation threshold is again reached. Then the CPU is again forced to run at a slower clock rate. This cycle repeats throughout the operation of the circuitry. As is apparent from the above description, the heat buildup in an integrated circuit is monitored by sampling the clocking frequency of the integrated circuit at regular time intervals. When the sampling indicates a high frequency, this indicates that the integrated circuit is generating too much heat and its temperature is increasing. The temperature count is correspondingly increased. When the sampling indicates a low frequency, this indicates that the integrated circuit's temperature is either cool or cooling, such that the temperature count is thereby decreased. This method of monitoring and regulating temperature in an integrated circuit works because the heat generation in an integrated circuit is proportional to current flow, and current flow in a CMOS integrated circuit is directly proportional to the switching frequency of the circuit, commonly known as clock frequency. This indicates that heat generation is itself proportional to clock frequency. Often times the thermal conductivity and parabolic Fourier heat conduction law is used to design the regulation of the IC chip.

Although thermal conductivity of the heat sink material and the surface area through which the heat is removed have been considered as parameters in the design and operation of heat sinks the ratio of the thermal mass of the material to the characteristic time of the material the storage coefficient has not been considered. In this study, it will be shown that the effectiveness of the heat sink can be increased by consideration of the storage coefficient of the material. Especially during periodic phenomena the heat is stored in the material and with a thermal lag for transmission with the heterogeneous materials when the surface comes to its minimum temperature there exists a small region close to the surface at a higher temperature causing a reversal of heat flux. This can cause damage to sensitive electronic equipment. The damping of the fluctuations is an important consideration here. Some materials can dampen the fluctuations and here the reversal of heat flux would be minimal or even absent. Higher the thermal capacitance and lower the relaxation time, i.e., higher the storage coefficient the damping term will be less and it the reversal of heat more pronounced. Thus a mica substrate is a poor choice compared with the ceramic substrate as shown in Renganathan and Turton [1989]. A numerical study was conducted using the finite element method and the periodic boundary condition was studied using different parameters of the substrate material for a thin film sensor using the Fourier parabolic diffusion equations. The study could resolve an apparent discrepancy between reports of 100% fluctuation in temperature and 4 % fluctuations for what appeared to be similar conditions of the phenomena. It turns out the later used mica substrate and the former used the ceramic substrate. The parametric analysis revealed that the storage coefficient can be an important parameter. The storage coefficient for the ceramic substrate is lower compared with that of mica which also has a higher thermal conductivity. Less harm would be done by the reversal of heat flux in the ceramic compared with that of mica.

The analysis competed in section 2.2 can be applied for the design of substrate material for high speed microprocessors. The periodic boundary condition was evaluated in a semi-infinite medium for the hyperbolic PDE, parabolic PDE and the dual phase lag model.

5.3 Cosinous Concentration Profile during Scale-up of Grafting in Helical Ribbon Agitated CSTR

During the manufacture of engineering thermoplastic such as ABS, (Acrylonitrile Butadiene Styrene) and HIPS (High Impact Polystyrene) often times the CSTR configuration, Sharma [1997] is used in the continuous polymerization of the monomers and grafting onto polybutadiene. High viscous polymer syrups were found to be well agitated by the helical ribbon agitator. HRAs were found to exhibit good mixing characteristics. The circulatory pattern of down flow near the wall and up flow at the center can be observed. The initiated free radical reactions including grafting onto polybutadiene and matrix formation occurs during the flow of high viscous flow through close clearances. The problem of simultaneous diffusion and reaction during laminar flow has not been studied. This study explores the problem of diffusion and convective bulk flow effects on the scale-up of initiation and grafting steps. The initiation using peroxy initiators may be described by a simple first order rate expression. The grafting reaction can take the form of an autocatalytic effect. These are first examined using the parabolic equations of diffusion. The conditions were the concentration will exhibit pulsations in the space domain is determined. This can cause non-homogenous grafting or poor grafting sometimes leading to reactor build-up.

The mass balance on the clearance channel close to the wall of the HRA in transient condition including the bulk convective flow effects can be written down as;

$$\partial C_g/\partial t - V_z \partial C_g/\partial z = D_{gs}\partial^2 C_A/\partial z^2 + k''' C_g \qquad (5.25)$$

The bulk flow is assumed to be in the opposite direction of the diffusion. The grafting of growing SAN chains can have an autocatalytic effect on the grafting rate. This is denoted using a plus sign in the reaction rate term and represents the product rate. Let,

$$\psi = D_{gs} t/h^2 \; ; \; u = C_g/C_{gs}; \; X = z/h \qquad (5.26)$$

$$Pe_m = V_z h'/D_{gs} \; ; \; \theta_{th} = h'(k'''/D_{gs})^{1/2} \qquad (5.27)$$

Where, Pe_m is the Peclet number, mass and denotes the ratio of the convective flow to the diffusion normalized with the HRA clearance width, h', and the θ_{th} is the Thiele modulus that denoted the rate of grafting to the rate of diffusion, non-dimensionlaized using the flow width. In the dimensionless form, equation (1) becomes,

$$\partial u\partial \psi - Pe_m \partial u/\partial X - \theta_{th}^2 u = \partial^2 u/\partial X^2 \qquad (5.28)$$

The initial and boundary conditions can be written as;

$$\tau = 0, \; u = 0 \qquad (5.29)$$

$$\tau > 0, \; Z = 0, \; u = 1 \qquad (5.30)$$

$$Z = \infty, \; u = 0 \qquad (5.31)$$

Where H is the height of the agitator.

Suppose that the solution be expressed as a sum of transient and steady state parts;

$$u = u^t + u^s \qquad (5.32)$$

The non-homogeneity in the boundary condition can be lumped into the steady state part; Thus,

$$-Pe_m \, \partial \, u^s/\partial X \, - \theta_{th}{}^2 \, u^s \; = \; \partial^2 \, u^s/\partial X^2 \qquad (5.33)$$

$$Z = 0, \, u^s \; = 1 \qquad (5.34)$$

$$Z = H, \; u^s \; = 0 \qquad (5.35)$$

$$\partial \, u^t /\partial \psi \; - Pe_m \, \partial \, u^t /\partial X \, - \theta_{th}{}^2 \, u^t \; = \; \partial^2 u^t /\partial X^2 \qquad (5.36)$$

$$Z = 0, \, u^t \; = 0 \qquad (5.37)$$

$$Z = \infty, \; u^t \; = 0 \qquad (5.38)$$

The transient portion of the solution can be obtained by the method of separation of variables;

$$\text{Let } u^t = \; V(\tau) \, \phi \, (X) \qquad (5.39)$$

Then Eq. [5.36] becomes;

$$V' \, \phi \; - \; Pe_m \, V \, \phi' \; - \; \theta_{th}{}^2 \, V \, \phi \; = \; V \, \phi'' \qquad (5.40)$$

Dividing throughout the Eq. [5.40] by $V\phi$

$$V'/V \; - \; Pe_m \, \phi'/ \phi \, - \, \theta_{th}{}^2 \; = \; \phi''/\phi \qquad (5.41)$$

$$V'/V \; = \; \phi''/\phi \; + Pe_m \, \phi'/ \phi + \; \theta_{th}{}^2 \; = -\lambda_n{}^2 \qquad (5.42)$$

$$V \; = c_1 \, \exp(-\lambda_n{}^2\tau \,) \qquad (5.43)$$

The space domain portion of the solution is then;

$$\phi'' \qquad + Pe_m \, \phi' \; + \; (\theta_{th}{}^2 +\lambda_n{}^2)\phi \; = \; 0 \qquad (5.44)$$

$$\phi \; = \; c_2 \, \exp\text{-}X(Pe_m/2) \; \exp\text{-}X/2 sqrt((Pe_m{}^2 - 4(\theta_{th}{}^2 +\lambda_n{}^2)) \qquad (5.45)$$

From the boundary condition it can be seen that c_3 is zero. $-\lambda_n$ can be an arbitrary constant and can be set to 1 for purposes of further analysis. The 1 lets the transient solution be no different from most other transient diffusion problems using Fick's second law of diffusion. When $Pe_m < 2 \, sqrt(\theta_{th}{}^2 + 1)$ the character of the solution in Eq. [5.45] is bifurcated. Or when $V_z < 2 \, D_{gs} /h' \, sqrt(\, 1 + \, k''' \, h^2/D_{gs})$ the solution becomes ;

$$\phi \; = \; c_2 \, \exp\text{-}X(Pe_m/2) \; Cos(X/2 sqrt((\, -Pe_m{}^2 + 4\theta_{th}{}^2 + 4 \,) \,)) \qquad (5.46)$$

Hence the solution is bifurcated. From the boundary condition the c_2 is seen to be 1. The general solution can be written as;

For $Pe_m < 2\sqrt{\theta_{th}^2 + 1}$,

$$u = c_1 \exp(-\lambda_n^2\tau) + \exp\text{-}X(Pe_m/2)\ \exp\text{-}X/2\sqrt{((Pe_m^2 - 4(\theta_{th}^2 + 1)))} \tag{5.47}$$

From the initial condition, then,

$$u = (1 - \exp(-\tau))(\ \exp X(-Pe_m/2)\ (\exp\text{-}X/2\sqrt{((Pe_m^2 - 4(\theta_{th}^2 + 1))))} \tag{5.48}$$

$$Pe_m > 2\sqrt{\theta_{th}^2 + 1},$$

$$u = (1 - \exp(-\tau))(\ \exp\text{-}X(Pe_m/2)\ (\cos(X/2\sqrt{((4\theta_{th}^2 + 4 - Pe_m^2)))}) \tag{5.49}$$

For the case of the bulk flow along the direction of diffusion the solution remain except for a + sign on the $\exp(XPe_m/2)$ term.

5.4   Thermal Barrier Coating

A patent [Ellis, 1986] was obtained by Delphi Research, USA, to use Magneisum Oxychloride Sorrel cement coatings as a fire control method and a barrier system intended to prevent the ignition of and spread of flame along a combustible substrate. These can be applied to warheads in the Navy to prevent fire spread. The critical thickness of insulation has been discussed for the cylindrical geometry that denotes a radius below which the heat loss is more compared with the case without insulation. Beyond the critical thickness is the cost thickness tradeoff. No reports of transient heat conduction analysis to a cylinder coated with a high temperature barrier coating are known. Barrier coatings are used in applications with high fire potential. This prevents the temperature from reaching high levels in naval applications and keeps the material properties in tact. Ceramic coating by CVD process or spray process can be affected. The hyperbolic mass wave propagative equation was used to develop an expression for coat thickness. The thermal conductivity of the coating may be reduced by increasing the porosity of coat. Ignition and fire spread of structure is prevented by Sorrel cement coating. Magnesium oxychloride cement coating co-bonded with high alumina calcium aluminate cement colloidal silica can be effective. The UL 1709 test by ASTM standards can be used to measure the resistance of a given material to temperature rise fires. The time needed for the average temperature of a steel column insulated (steel protection factor) with the material being tested to exceed 1000 F or the amount of time needed for a single temperature of the steel coupon to exceed 1200 F. The super conductors of heat are materials that offer little or no resistance to heat conduction. Does it occur at low or high temperatures. The Copper type behavior is different from the Aluminum type behavior. In the Copper type the material thermal conductivity increases with decrease in temperature and in Aluminum type the conductivity values increase with increasing temperature.

The phenomenological relations of Fourier's heat conduction law, Fick's mass diffusion law, and Ohm's electrical conduction law is well established when the cross-effects in the equations between fluxes and forces can be neglected. When a system is close to equilibrium [Kondepudi and Prigogine, 1998] a general theory based upon relations between forces and flows can be formulated. For a small deviation in the forces from their equilibrium value of zero the flows can be expected to be linear functions of the forces. Cross effects such as Seebeck, Peltier, Soret and Dufour may result in one of driving force causing another flow. Onsager [1931] provided a well theoretically founded set of reciprocal relations for the phenomenological coefficients. Onsager's theory was based upon the principle of detailed balance or microscopic reversibility and is valid for systems at equilibrium. Experimental evidence for hyperbolic wave propagative heat equation provided by Mitra et al report relaxation time of the order of $20 - 30$ seconds. The work on Kaminski [1990] shows that the relaxation time can be as high as 1000 seconds as in falling rate drying [23]. The initial estimates of relaxation times by Luikov [1966] were in the order of 10^{-12} seconds. The range of relaxation time values reported vary from a few femto seconds to third of an hour. So the values of relaxation time s for useful materials are not clear. With the reports of higher relaxation times the use of higher order terms in the extended Fourier's law of heat conduction will be of consequence. In this paper the exact solution for the CWT problem in a cylinder in 1 dimension is studied by the method of separation of variables. The conditions were the pulsations in temperature occur is of special interest. The time taken to alarm in a Naval warhead upon application of a flame retardant coating is examined.

Consider a cylinder at initial temperature T_0. The surface of the cylinder is maintained at a constant temperature T_{fire} for times greater than zero. The entire naval warhead and coating is idealized as the coating. The estimates from such an estimation will be the ceiling or upper estimate on the real times. The heat propagative velocity is given as the square root of the ratio of thermal diffusivity and relaxation time.

$$V_h = sqrt(\alpha/\tau_r) \qquad (5.50)$$

The initial condition

$$t = 0, \ (all\ r), \ T = T_0 \qquad (5.51)$$

$$t > 0, \ r = 0, \ T = finite; \ \partial T/\partial r = 0 \qquad (5.52)$$

$$t > 0, \ r = R \ T = T_{fire} \qquad (5.53)$$

320

The governing equation can be obtained by eliminating q_r between the constitutive equation and the equation from energy balance of in − out − accumulation. This is achieved by differentiating equation (3) with respect to r and the energy equation with respect to t and eliminating the second cross derivative of q with respect to r and time. Thus,

$$\tau_r \, \partial^2 T/\partial t^2 \; + \; \partial T/\partial t \;\; = \; \alpha \, \partial^2 T/\partial r^2 \;\; + \alpha/r \; \partial T/\partial r \qquad (5.54)$$

Obtaining the dimensionless variables ;

$$u \; = (T - T_{fire})/(T_0 - T_{fire}); \; \tau \; = \; t/\tau_r; \;\; X \; = \;\; r/sqrt(\, \alpha \, \tau_r) \qquad (5.55)$$

The governing equation in the dimensionless form is then;

$$\partial u/\partial \tau \;\; + \;\; \partial^2 u/\partial \tau^2 \;\; = \; \partial^2 u/\partial X^2 \;\; + 1/X \;\; \partial u/\partial X \qquad (5.56)$$

5.4.1 Asymptotic Analysis of Governing Equation

In the asymptotic limit of infinite heat velocity the equation (9) reduces to the parabolic diffusion equation. The solution is obtained by the method of separation of variables. First the damping term is removed by the substitution, $u = exp(-n\tau)w \, (X, \tau)$. By choosing $n = 1/2$, the damping component of the equation is removed to yield;

$$-w/4 \;\; + \;\; \partial^2 w/\partial \tau^2 \;\; = \;\; \partial^2 w/\partial X^2 \;\;\;\; + \; 1/X \; \partial w/\partial X \qquad (5.59)$$

Rewriting Eq. [5.59] in terms of the dimensional variables showing the velocity of heat;

$$-w/4 \;\; + \; (\tau_r \, \partial^2 w/\partial t^2 \,) \; = \;\; (\alpha \tau_r \,) \;\; (\partial^2 w/\partial x^2 + \; 1/x \; \partial w/\partial x) \qquad (5.60)$$

$$or \; (1/v_h^2 \, \partial^2 w/\partial t^2 \,) \;\; - w/(4 \alpha \tau_r \,) \;\; = \;\; (\partial^2 w/\partial x^2 + \; 1/x \; \partial w/\partial x) \qquad (5.61)$$

In the asymptotic limit of infinite relaxation time or zero velocity of heat the Eq. [5.61] becomes ;

$$(1/v_h^2 \, \partial^2 w/\partial t^2 \,) \;\; = \; (\partial^2 w/\partial x^2 \;\;\;\; + \; 1/x \; \partial w/\partial x) \qquad (5.62)$$

$$\text{or } \partial^2 w/\partial\tau^2 \; = \; \partial^2 w/\partial X^2 \quad + \; 1/X \; \partial w/\partial X \tag{5.63}$$

Eq. [5.63] is the wave equation in cylindrical coordinates.

$$\text{Let } \eta \; = \; \tau^2 \; - \; X^2 \tag{5.64}$$

$$\partial w/\partial X \; = \; -\partial w/\partial\eta \; (2X) \tag{5.65}$$

$$\partial^2 w/\partial X^2 = \partial^2 w/\partial\eta^2 \; (4\,X^2\,) \; - \; 2 \; \partial w/\partial\eta \tag{5.66}$$

$$\partial^2 w/\partial\tau^2 \; = \partial^2 w/\partial\eta^2 \; (4\,\tau^2) \; + \; 2 \; \partial w/\partial\eta \tag{5.67}$$

Plugging Eq. [5.65-5.67] in Eq. [5.63]

$$\partial^2 w/\partial\eta^2 \; (4\,\eta\,) \quad = \quad -6\partial w/\partial\eta \tag{5.68}$$

$$\text{Let } s = \partial w/\partial\eta \tag{5.69}$$

Then,

$$\partial s/\partial\eta \; = \; -3s/2\eta \tag{5.70}$$

Integrating both sides,

$$\ln(s) \; = \; -3/2 \; \ln(\eta) + c \tag{5.71}$$

$$\text{Or } s = \partial w/\partial\eta \; = \; c'\eta^{-3/2} \tag{5.72}$$

Integrating Eq. [5.72] wrt η,

$$W \; = \; -1/2 \; c' \; \eta^{-1/2} \quad + c" \tag{5.73}$$

$$\text{Or } u \; = \exp(-\tau/2) \; (\; c" \; - \; c'/2/sqrt(\eta)) \tag{5.74}$$

The constants can be solved for the appropriate boundary conditions. In the asymptotic limit of zero thermal diffusivity which is another case of zero velocity of heat in addition to the relaxation time, Eq. [5.63] becomes,

$$-w/4 \quad + \; (\tau_r^2 \; \partial^2 w/\partial t^2\,) \; = \; 0 \tag{5.75}$$

$$\text{or } w \; = \quad A \exp(-t/2\tau_r\,) \; + \; B \exp(t/2\tau_r\,) \tag{5.76}$$

At large times, the system reaches steady state the value of u becomes 0. So B = 0 in Eq. [5.76]

$$\text{Or } u \; = A \exp(-\tau/2) \tag{5.77}$$

A can be seen as 1 from the initial condition , u = 1 at zero time.

5.4.2 Average Temperature (UL 1709 Test)

For purposes of estimation of average time taken for temperature rise as prescribed in the UL 1709 ASTM test the governing equation is subject to a convective boundary condition and the terms averaged in space domain as follows. A finite slab is considered with thickness 2R. After dividing the governing equation by $\exp(-\tau/2)$,

$$-W/4 \quad + \quad \partial^2 W/\partial\tau^2 \quad = \partial^2 W/\partial X^2 \tag{5.78}$$

The boundary condition at the surface can be written as;

$$\partial u/\partial X \ = -Bi <u> \tag{5.79}$$

$$\text{where } Bi = h/sqrt(Sk) \tag{5.80}$$

h is defined for convenience with respect to the space averaged temperature of the slab. where, S – storage coefficient and k – thermal conductivity

Integrating Eq. [5.78] wrt X from $-R/sqrt(\alpha \tau_r)$ to $+ R/ sqrt(\alpha \tau_r)$, and defining the space averaged temperature as;

$$<W> \ = \ sqrt(\alpha \tau_r)/2R\!\int_{-XR}^{+XR} WdX \tag{5.81}$$

Then,

$$<W>_{\tau\tau} - <W>/4 \ = -2Bi \, sqrt(\alpha \tau_r)/2R <W> \tag{5.82}$$

$$\text{or} <W>_{\tau\tau} - <W>/4 \ = -Bi^*<W> \tag{5.83}$$

$$Bi^* = h/SR \tag{5.84}$$

Where S is the storage coefficient. This second order ODE can be solved as;

$$<W> \ = A \exp(-\tau(sqrt(1/4 - Bi^*))) \ + B \exp(\tau(sqrt(1/4 - Bi^*))) \tag{5.85}$$

$$\text{or} <u> = A \exp(-\tau(1/2 + sqrt(1/4 - Bi^*))) \ + B \exp(\tau(-1/2 + sqrt(1/4 - Bi^*))) \tag{5.86}$$

For large times as the system reaches steady state the $<u>$ becomes zero. For certain values of Bi and times the contribution from the term with B will be non-zero which is physically unrealistic. A physically realistic solution can be obtained by setting B as zero. From the initial condition A can be seen as 1. So

$$<u> \ = \ \exp(-\tau(1/2 + sqrt(1/4 - Bi^*))) \tag{5.87}$$

Equation (38) can be used for estimates of the time taken to alarm as suggested by the UL 1709 test. Further when,

$$2Bi > \tfrac{1}{4} \tag{5.88}$$

$$<u> = \exp(-\tau/2)\exp(-i\tau/\mathrm{sqrt}(-1/4 + Bi^*))) \qquad (5.89)$$

By De Movrie's theorem and taking the real parts,

$$<u> = \exp(-\tau/2)\mathrm{Cos}(\tau\mathrm{sqrt}(-1/4 + Bi^*))) \qquad (5.90)$$

The temperature is damped oscillatory in the time domain! The dimensionless frequency of the pulsations are;

$$\varpi^* = 1/\tau_r \mathrm{sqrt}(h/RS - \tfrac{1}{4}) \qquad (5.91)$$

5.4.3 Exact Solution

The method of separation of variables can be used to obtain the solution of Eq. [5.78]

$$\text{Let } W = V(\tau)\,\phi\,(X) \qquad (5.92)$$

Plugging Eq. [5.92] into Eq. [5.78], and separating the variables that are a function of X only and τ only;

$$\phi'' + \phi'/X + \lambda^2\,\phi = 0 \qquad (5.92)$$

$$V'' = V(1/4 - \lambda^2) = 0 \qquad (5.93)$$

The solution for Eq. [5.92] is the Bessel function of 0th order and first kind;

$$\phi = c_1 J_0(\lambda X) + c_2 Y_0(\lambda X) \qquad (5.94)$$

It can be seen that $c_2 = 0$ as the concentration is finite at $X = 0$. Now from the BC at the surface Eq. [5.53],

$$\phi = c_1 J_0(\lambda R/\sqrt{\alpha\tau_r}) = 0 \qquad (5.95)$$

$$\text{or } \lambda_n R/\sqrt{D\tau_r} = 2.4048 + (n-1)\pi \qquad (5.96)$$

$$\text{for } n = 1,2.3... \qquad (5.97)$$

The solution for Eq. [5.93] is the sum of two exponentials. The term containing the positive exponential power exponent will drop out as with increasing time the system may be assumed to reach steady state. Thus,

$$V = c_4 \exp(-\tau \mathrm{sqrt}(1/4 - \lambda_n^2)) \qquad (5.98)$$

Or,

$$u = \sum_0^\infty c_n J_0(\lambda_n X) \exp(-\tau (1/2 - \text{sqrt}(1/4 - \lambda_n^2)))) \quad (5.99)$$

The c_n can be solved for from the initial condition by using the principle of orthogonality for Bessel functions. At time zero the LHS And RHS are multiplied by $J_{1/2}(\lambda_m X)$. Integration between the limits of 0 and R is performed. When n is not m the integral is zero from the principle of orthogonality. Thus when n = m,

Thus,

$$c_n = \int_0^R J_0(\lambda_n X) \ / \ \int_0^R J_0^2 (\lambda_n X) \quad (5.100)$$

It can be noted from Eq. [5.99] that when,

$$1/4 < \lambda_n^2 \quad (5.101)$$

the solution will be periodic with respect to time domain. This can be obtained by using De Movries theorem and obtaining the real part to $\exp(-i\tau \sqrt{(-1/4 + \lambda_n^2)})$. Also it can be shown that for terms in the infinite series after n or for

$$n > 1 + R /\pi \sqrt{\alpha \tau_r} - 2.4048/\pi \quad (5.102)$$

the contribution to the solution will be periodic. Thus a bifurcated solution is obtained. Also from Eq. [5.99] it can be seen that all terms in the infinite series will be periodic. i.e, even for n =1 when,

$$R < 4.8096\sqrt{\alpha \tau_r} \quad (5.103)$$

And

$$u = \sum_0^\infty c_n J_0 (\lambda_n X) \exp(-\tau/2 \ \text{Cos} (\text{sqrt}(-1/4 + \lambda_n^2))) \quad (5.104)$$

Thus the transient temperature profile in a cylinder is obtained for a step change in temperature at the surface of the sphere using the modified Fourier's heat conduction law. A lower limit on the radius is obtained to avoid cycling of temperature in the time domain during transience. The UL1709 test procedures can be followed and the time taken to alarm can be computed for a given thickness of coating of high temperature barrier coating. Thus, A lower limit on the thickness of the high temperature barrier coating at the surface exists and found to be $4.8096\sqrt{\alpha \tau_r}$. The exact solution for transient temperature profile using finite speed heat conduction is derived by the method of separation of variables. It is a bifurcated solution. For certain values of lamda the time portion of the solution is cosinous and damped and for others it is a infinite series of Bessel function of the first kind and 0th order and decaying exponential in time. For any dimension the term in the infinite series from which the contribution to the transient temperature profile is derived. The UL1709 test procedures can be followed to find the time taken to alarm in the navel warhead for example that is protected from flame spread by conduction using the high temperature Barrier coating. In the asymptotic limit of infinite heat velocity the governing equation becomes parabolic diffusion equation. In the limit of zero velocity of heat and infinite relaxation time the wave equation result and solution can be obtained by a relativistic coordinate transformation. In the asymptote of zero velocity of heat and zero thermal diffusivity the solution for the dimensionless temperature is a decaying exponential in time. The average temperature of the naval warhead as indicated by UL 1709 test was estimated by using a idealized finite slab, and Leibnitz rule and an analytical expression for the average temperature was obtained using convective boundary condition. The solution is;

For $\frac{1}{2} \geq$ Bi

$$<u> = \exp(-\tau(1/2 + \text{sqrt}(1/4 - Bi^*))) \qquad (5.105)$$

For Bi > ½, $\qquad (5.106)$

$$<u> = \exp(-\tau/2)\text{Cos}(\tau\text{sqrt}(-1/4 + Bi^*))) \qquad (5.107)$$

The average temperature is damped oscillatory in time domain.

The dimensionless frequency of the pulsations is;

$$\varpi^* = 1/\tau_r \text{ sqrt}(h/RS - ¼) \qquad (5.108)$$

5.5 Infinite Order PDE and Solution by Method of Laplace Transforms

The transient concentration profile in a semi-infinite medium at constant wall temperature is derived from an infinite order PDE. The solution is;

$$u = 1 - X^2/2! \exp(\tau) + X^4/4! \exp(\tau) (1 + \tau) - X^6/6! \exp(\tau) (1/2 + 2\tau + \tau^2/2!) + ...$$
$$(5.109)$$

The infinite order PDE was obtained by a recursive use of a two term Taylor series on the temporal derivative of heat flux. The method of Laplace transforms was used and the residue theorem was used for the inversion of the Taylor series expansion of the infinite terms. The surface heat transfer coefficient can be obtained by the method of Laplace transforms as. The heat transfer coefficient is given as ;

$$h = \text{sqrt}(k \rho C_p /\tau_r)/(\tau - ½)^{1/2}/ \text{gamma}(1/2) \qquad (5.110)$$

$$\tau > ½$$

Otherwise, h = 0.

A Taylor series expansion in two variables about the relaxation time of heat flux after correcting for the direction of heat transfer and supposing that the space functionality is only linear;

$$q = -k\, \partial T/\partial x \quad - \tau_r\, \partial q/\partial t \quad - \tfrac{1}{2}!\, \tau_r^2\, \partial^2 q/\partial t^2 - 1/3!\; \tau_r^3 \partial^3\, q/\partial t^3 -... \quad (5.111)$$

A recursive Taylor on $\partial J/\partial t$ when only the first two terms are included leads to the following infinite order PDE which is a reasonable approximation for small relaxation times:

$$q = -k\, \partial T/\partial x \quad - \tau_r\, \partial T/\partial t \quad -\tau_r^2\, \partial^2 T/\partial t^2 \quad - \tau_r^3 \partial^3\, T/\partial t^3 \quad -..... \quad (5.112)$$

Consider a semi-infinite medium at an initial temperature T_0 is subjected for times greater than zero at a surface temperature of T_s. The initial condition of the surrounding fluid;

$$t = 0, \;\; Vx \quad T = T_0 \qquad\qquad\qquad (5.113)$$

$$t > 0, \;\; x = 0, \;\; T = \;\; T_s \qquad\qquad\qquad (5.114)$$

$$t > 0, \;\; x = \infty, \;\; T = \;\; T_0 \qquad\qquad\qquad (5.115)$$

The governing energy balance (in – out = accumulation) can be obtained by eliminating q_x between Eq. [5.112] and the equation from energy balance of $(-\partial q/\partial x = (\rho\, C_p \partial T/\partial t)$ This is achieved by differentiating Eq. [5.112] with respect to x and the energy balance equation with respect to t several times and eliminating the cross derivatives of q with respect to x and time. Thus,

$$-\partial q/\partial x = \;\; (\rho\, C_p\,)\partial T/\partial t \qquad\qquad (5.116)$$

$$-\partial^2 q/\partial x \partial t \;\; = \partial^2\, T/\partial t^2 \qquad\qquad (5.117)$$

$$-\partial^3 q/\partial x \partial t^2 \;\; = \partial^3 T/\partial t^3 \qquad\qquad (5.118)$$

....
The governing equation is thus;

$$......\tau_r^3\, \tau_r\, \partial^4\, T/\partial t^4 + \tau_r^2\, \partial^3 T/\partial t^3 + \;\; \tau_r\, \partial^2\, T/\partial t^2 + \partial T/\partial t \;\; = \alpha\, \partial^2\, T/\partial x^2$$
$$(5.119)$$

Obtaining the dimensionless variables ;

$$u \;\; = (T - T_0)/(T_s - T_0); \tau \;\; = \;\; t/\tau_r \;\; ; X = \;\; x/\sqrt{(\alpha\, \tau_r)} \qquad (5.120)$$

The governing equation in the dimensionless form is then;

$$\ldots\ldots \partial^4 u/\partial\tau^4 \ + \ \partial^3 u/\partial\,\tau^3 \ + \ \partial^2 u/\partial\,\tau^2 \ + \ \partial u/\partial\,\tau \ \ = \partial^2 u/\partial\,X^2 \qquad (5.121)$$

The solution is obtained by the method of Laplace transforms.

$$d^2\underline{u}/dX^2 \ = \ \underline{u}\,(s \ + \ s^2 \ + \ s^3 \ + \ s^4 \ + \ s^5 \ +\ldots..) \qquad (5.122)$$

Adding and subtracting 1 to RHS of Eq. [5.122]

$$d^2\underline{u}/dX^2 \ = \ \underline{u}\,(-1 + 1/(\,1 - s\,)=\underline{u}\,(s/(\,1 - s\,)) \qquad (5.123)$$

$$\text{or } \underline{u} \ = \ 1/s \ \exp(-sqrt(s/(1-s)X) \qquad (5.124)$$

$$= \ 1/s \ \exp(-isqrt(s/s-1)X) \qquad (5.125)$$

This is after substituting the boundary conditions at X = infinity and $X = 0$. Expanding the terms in the Laplace domain by Taylor series and equating the real parts;

$$\underline{u} \ = \ 1/s(\ 1 - \ X^2/2! \ s/(s-1) + X^4/4! \ s^2\,/(s-1)^2 \ - \ \ldots\ldots) \qquad (5.126)$$

Inverting each term in the series from the Laplace domain;

$$u \ = \ 1 \ \ - X^2/2! \ \exp(\tau) \ + X^4/4! \ \exp(\tau) \ \ (1 + \tau) \ - \ X^6/6! \ \exp(\tau) \ \ (1/2 + 2\tau \ + \tau^2\,/2!\,) \ +\ldots.. \qquad (5.127)$$

The governing equation obtained from Eq. [5.111] is thus;

$$\partial u/\partial\tau \ \ \ + \ \partial^2 u/\partial\tau^2 \ \ + \ \ \tfrac{1}{2}!\partial^3 u/\partial\tau^3 + \ 1/3!\partial^4 u/\partial\tau^4 \ \cdots\cdots = \ \partial^2 u/\partial X^2 \qquad (5.128)$$

Obtaining the Laplace Transforms;

$$\underline{u}\,(\,s + \ s^2 \ + \ s^3/2! \ +\ldots..) \ \ \ \ = \ d^2\underline{u}/dX^2 \qquad (5.129)$$

$$\text{or } \underline{u} \ \ s\,(\ 1 + \ s + s^2/2! + \ldots) = \ \underline{u} \ \ s \ \exp(s) \qquad (5.130)$$

Solving the second order differential equation in Laplace domain;

$$\underline{u} \ = \ A \ \exp(-X \ sqrt(s)\exp(s/2) \qquad (5.131)$$

The BC is used to see that $A = 1/s$. The constant is seen to be 0 as at X tends to infinity the concentration is T_0. The flux is given by;

$$q^*\,(\exp(s)\,) \ = -d\underline{u}\,/dX \qquad (5.132)$$

$$q^* \ = \ \ \ 1/s^{1/2} \ \exp(-s/2)\exp(-X \ sqrt(s)\exp(s/2) \qquad (5.133)$$

$$\text{where } q^* = q/sqrt(k \ \rho \ C_p \ /\tau_r \ )/(\ T_s - T_0\,) \qquad (5.134)$$

At the surface, $X = 0$.

$$q^* = 1/s^{1/2} \exp(-s/2) \qquad (5.135)$$

The inverse of the Laplace transformed expression in Eq. [5.135] is within the tables [Mickley, Sherwood and Reed, 1957]

$$j = \tfrac{1}{2}; \quad k = \tfrac{1}{2} \qquad (5.136)$$

$$q^* = 0 \quad \text{for} \quad 0 < \tau < \tfrac{1}{2} \qquad (5.137)$$

$$= 1/(\tau - \tfrac{1}{2})^{1/2}/\,\text{gamma}(1/2) \quad \text{for} \quad \tau > \tfrac{1}{2} \qquad (5.138)$$

The heat transfer coefficient is given as ;

$$h = \text{sqrt}(k\,\rho\,C_p/\tau_r)/(\tau - \tfrac{1}{2})^{1/2}/\,\text{gamma}(1/2) \qquad (5.139)$$

for $\tau > \tfrac{1}{2}$

otherwise, $h = 0$

This can be used to model the heat transfer coefficient between gas-solid fluidized beds and immersed surfaces where the mechanism for heat transfer is inherently unsteady as a result of perpetual renewal of packets of solid particles by bubble motion at the surface.

5.6 Maxwellian Basis for Secondary Pressure Fluctuations in Gas-Solid Bubbling Fine Particle Fluidized Beds

Pressure fluctuations have been observed in gas-solid fluidized beds using single and double pressure pribes at various locations ranging from the plenum below the distributor plate to the freeboard region. The bubble phenomena observed have been largely attributed as the cause for these fluctuations. Various statistical studies of the data acquired using the pressure probes have been reported. Lirag and Litman [1971] used the PDF, probability density function analyses of the signals to confirm the periodic nature of the fluctuations as denoted by the saddle points. Fan, Ho, Hirakowa and Walawander [1981] showed that the power spectral density and other time series analyses can be effectively utilized in the study of hydrodynamics of fluidization and the change with temperature.

Sharma [1998] outlined the hierarchy of linear and other analyses that can be effectively utilized in the study of pressure fluctuations. A generalized normal distribution was suggested to capture the periodicity in the PDF analyses as the number of saddle points were reproducible. Pressure fluctuations can be used as a measure and in control of quality of fluidization as shown by Daw and Hawk [1995]. Roy, Davidson and Tuponogov [1990] dealt with the velocity of sound in fine particle gas-solid fluidized beds. They noted that dense bed will transmit small pressure fluctuations with considerable damping. Assuming that the gas and particle are in phase the velocity of sound in fluidzed beds would be 10 m/s compared with the 300 m/s in air. Depending on the frequency of fluctuations the fluidization regimes and types can be delineated (Geldart [1973] , Grace [1989], Kunii and Levenspiel [1991]). Baskakov, Tuponogov and Filippovsky [1986] recognized several types of fluidized bed pressure fluctuations. Systems with low distributor resistance and a large plenum chamber lead to self-sustained oscillations. The fluctuations were found to be random in character featuring a broad frequency range. A fundamental frequency was seen to dominate. Equations for fundamental frequency in the bed were provided by a few investigatos (Tamarin, [1964], Hilby [1967], Verloop and Heertjes [1974], Davidson [1989], Baskakov [1986]. Expressions for a natural frequency in the fluidzed bed were provided by Kang and Sutherland [1967]. Broadhurst and Becker [1975] and Sadasivan [1980]. Particle size effect on pressure fluctuations was reported by Svaboda et al [1984]. A theory from the fluid mechanics and stability to explain the pressure fluctuations was put forth first by Anderson and Jackson [1968]. A basis based upon the particle movement was given by Jones and Pyle [1971]. Sharma and Turton [1998] has found that pressure fluctuations can be used to correlate heat transfer events to surfaces. In this section, the origin of pressure fluctuations from the Maxwell [1873] expression for shear stress and velocity gradient is explored.

Consider the laminar flow in a circular pipe of narrow dimensions in transience. The bubbles in the fludized bed causes circulation motion that keeps the flow in channels formed by the interstitial voids in transience. Thus the flow of fluid in a cylindrical interstitial channel can be studied. The damped wave momentum transfer and relaxation equation is written as;

$$\Im_{xy} \ + \ \tau_{mom}\partial\Im\ /\partial t \ = \ - \ \mu \ \partial v/\partial r \qquad\qquad (5.140)$$

The governing equation for the axial velocity as a function of the radial direction can be written after combining the modified Newton's law of viscosity including the relaxation term with the momentum balance equation to yield:

$$\partial v/\partial t \ + \ \tau_{mom} \ \partial^2 v/\partial t^2 \qquad = \ \Delta P/L\rho \ + \ v/r \ \partial/\partial r(r \ \partial v/\partial r) \quad (5.141)$$

Defining the dimensionless variables;

$$u \ = \ v/v_{max}; \ X \ = \ r/sqrt(v\tau_{mom}); \ \tau \ = \ t/\tau_{mom}; \ P^* \ = \ \Delta P\tau_{mom}/l\rho v_{max} \ (5.142)$$

The dimensionless governing equation then becomes;

$$\partial u/\partial \tau \ + \ \partial^2 u/\partial \tau^2 \qquad = \partial^2 u/\partial X^2 \ + \ 1/X \ \partial u/\partial X \ + \ P^* \qquad\qquad (5.143)$$

The solution is assumed to be a sum of two parts, i.e., the steady state and transient part; Let $u = u^{ss} + u^t$. Then the governing equation becomes;

$$d^2 u^{ss}/dX^2 \ + \ 1/X \ du^{ss}/dX \ + P^* \ = 0 \qquad\qquad (5.144)$$

$$\partial u^t/\partial \tau \ + \ \partial^2 \ u^t \ /\partial \tau^2 \qquad = \partial^2 \ u^t \ /\partial X^2 \ + \ 1/X \ \partial \ u^t \ /\partial X \qquad\qquad (5.145)$$

Integrating the equation for the steady state component of the velocity wrt X;

$$X \ \partial u^{ss}/\partial X \ = \ -P^* \ X^2/2 \ + C' \qquad\qquad (5.146)$$

At X = 0, the gradient of the velocity is zero as a condition of extremama can be expected from symmetry considerations. So it can be seen that C' = 0. Integrating the resulting equation again wrt X,

$$u^{ss} \ = \ -P^* \ X^2/4 \ + D' \qquad\qquad (5.147)$$

From the boundary condition at $X = X_R$

$$D' = P*X_R^2/4 \text{ and} \tag{5.148}$$

$$u^{ss} = P* X_R^2/4(\nu\tau_{mom}) (1 - r^2/R^2) \tag{5.149}$$

The above relation is the Hagen-Poisuelle flow in laminar pipes at steady state where P* is given by Eq. [5.142]. The rest of the problem is obtaining the transient part. First the daming term is removed bu a substitution such as $u^t = W \exp(-n\tau)$. At $n= \frac{1}{2}$ the governing equation reduces to ;

$$-W/4 + \partial^2W/\partial\tau^2 = \partial^2W/\partial X^2 + 1/X \, \partial W/\partial X \tag{5.151}$$

The method of separation of variables can be used to obtain the exact solution to the above equation. The boundary and time conditions for the transient portion of the velocity are then;

$$\tau = 0, \, u^t = 1 \tag{5.152}$$

$$\tau = \text{infinity}, \, u^t = 0 \tag{5.153}$$

$$\tau > 0, \, X = X_R, \, u^t = 0 \tag{5.154}$$

$$X = 0, \text{ symmetry considerations}$$

$$\text{Let } W = V(\tau)\phi(X) \tag{5.155}$$

The wave equation becomes;

$$V''/V -1/4 = (\phi'' + \phi'/X)/\phi = -\lambda_n^2 \tag{5.156}$$

Thus,

$$X^2\phi'' + X\phi' + X^2\lambda_n^2\phi = 0 \tag{5.157}$$

This can be recognized as a Bessel equation of the first order [Mickley, Sherwood, Reed, 1955] and the solution can be written as;

$$\phi = c_1 J_0(\lambda_n X) \tag{5.158}$$

c_2 can be seen to be zero as ϕ is finite at $X = 0$. From the boundary condition at $X = R/sqrt(\nu\tau_{mom})$;

$$c_1 J_0(\lambda_n X_R) = 0 \tag{5.159}$$

$$\lambda_n = sqrt(\nu\tau_{mom})/R \; (2.4048 + (n-1)\pi) \tag{5.160}$$
$$n = 1, 2, 3....$$

Now the time domain part of the wave is obtained by solving the second order ODE.

$$V''/V - 1/4 = -\lambda_n^2 \tag{5.161}$$

Thus,

$$uexp(\tau/2) = W = (c_3 \exp(+\tau sqrt(1/4 - \lambda_n^2)) + c_3 exp(-\tau sqrt(1/4 - \lambda_n^2)))\phi \tag{5.162}$$

At infinite times the RHS becomes infinitely large. The LHS is zero multiplied by infinity and is zero. At steady state the velocity is bounded and hence the constant c_3 is set to zero. It can be seem that for small channel dimensions,

i.e., $R < 4.8096 \; sqrt(\nu\tau_{mom})$. The solution is periodic wrt to time. The general solution for such cases can be written as;

$$u = \Sigma_1^\infty C_n \exp(-\tau/2) \; Cos \; (\tau \; sqrt(\lambda_n^2 - 1/4)) \; J_0(\lambda_n X) \tag{5.163}$$

The initial condition can be taken to be maximum velocity in plug flow. In the fluidized state, the interstitial voids are formes into channels and renewed by bubbles and slugs. So the initial superficial velocity essentially at plug flow becomes a periodic profile as in Hagen-Poiseulle flow when channgels are formed. The transient portion is governed for small channels by the generalized Netwon's law of viscosity. So the initial condition prior to the onset of transience for one cycle is;

$$0 = \Sigma_1^\infty C_n J_0(\lambda_n X) \tag{5.164}$$

The constant canbe solved for by the orthogonality property.

$$C_n = -\int_0^{XR} J_0(\lambda_m X)dX / \int_0^{XR} J_0^2(\lambda_m X)dX \qquad (5.165)$$

The channel equivalent diameter in a fluidized bed Kunii and Levenspiel,[1991]) at minimum fludization can be written as;

$$R_c = \varphi_s d_p \varepsilon /3/(1-\varepsilon) \qquad (5.166)$$

Combining Eqs. [5.166] and [5.160], when,

$$d_p < 14.4288 \ sqrt(\nu\tau_{mom})(1 - \varepsilon)/(\varphi_s \varepsilon) \qquad (5.167)$$

The transient velocity profile in the channels will be subcritical damped oscillatory as shown in Eq. [5.163]. The frequencies of pulsations can be given by;

$$F_n = sqrt(\nu\tau_{mom} (4.8096 +(n-1)\pi)^2/R_c^2 -1/4) \qquad (5.168)$$

n = 1,2, 3…

These pulsations translate to pressure pulsations by the relation between velocity of flow to the pressure drop required. As the bubble events drives the transience these frequencies are secondary to the primary bubble frequency. The relaxation times that have been measures as 15 seconds for a few heterogeneous systems can provide a reasonable frequency estimate of the phenomena.

5.7 Delivery of Drugs and Finite Speed Diffusion

The better understanding of the process of drug ingestion and assimilation is a classical engineering endeavor where mathematical analysis and basic science can be used to achieve a desired objective. The effectiveness of drug therapy is the study of biophysics and physiology of drug movement through tissues. For example, consider the localized delivery of drug to a tissue by controlled release from an implanted, drug loaded polymer. In the period following implantation, drug molecules are slowly released from the polymer into the extracellular space of the tissue. The availability of drug to the tissue, i.e., the rate of release into the extracellular space into the implant by varying the chemical or physical properties of the polymer. The relationship between polymer chemistry, implant structure and the rate of drug release need be better understood.

Prescribed rates of drug delivery offer improved efficacy, safety and convenience. One example is the drug dosed by periodic pills and that by controlled release technology. When the drug is administered, the concentration of the drug rises sharply in the blood. (Figure 5.2) This rise can carry the drug concentration past the effective level and above the toxic level. The concentration then drops below the effective level. In contrast, when the drug delivered is by controlled release its concentration rises above the level required to be effective and stays there, without sudden excursions to toxic or ineffective levels. This steady controlled release is achieved by diffusion.

Figure 5.2 Drug Concentration with Time for Periodic Dose and Controlled Release Technology

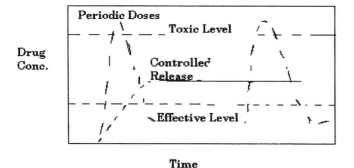

Time

The objective is to obtain the release rate of a solute like a drug to be constant with time. A zero order release is said to occur when the solute concentration varies linearly with time. Once the drug is released into the extracellular space, however, rates of drug migration throughj the tissue, uptake into the circulatory system, and elimination from the body depend on the characteristics of the drug and physiololgy of the delivery site. To fathom the barriers to drug delivery, various length scales are considered at a variety of length scales. The length scale of an organism of about 1.0 – 2.0m, length scale of an organ of about 1 – 100 mm, a tissue segment of length scale of 1 – 100 μm, or the length scale within the extracellular matrix of 1 – 100 nm. Implantable drug delivery system design requires good understanding of the molecular mechanisms for drug transport and elimination, particularly at the site of delivery.

Multiple forms of drug administration are available such as pills, injections, lotions and suppositories. Oral dosage is preferred as they are painless, uncomplicated, and self-administered. Many drugs are degraded within the gastrointestinal tract or nor absorbed in sufficient quantity to be effective. The route and method of drug administration will influence the kinetic of biodistribution and elimination and, therefore the effectiveness of therapy. Delivery of identical amounts of drug to tissue sites differing in local anatomy and physiology can result in measurable differences in pattern of delivery. For instance, when identical preparations of human immunoglobin G (IgG) were administered orally, subcutaneously, inramuscularly or intraveneously to human patients different patterns of IgG concentration in the plasma were observed over time.

5.7.1   Response to a Pulse in Mass Flux

The concentration of A as a function of time and distance from the site of initial injection can be predicted by solving the finite speed damped wave diffusion and relaxation equation in one dimension in cartesian coordainate system. Writing Eq. [C.21] from Appendix C , neglecting bulk flow and the y and z components, subject to the following initial and boundary conditions,

$$\tau_{mr}\partial^2 C_A/\partial t^2 + \quad \partial C_A/\partial t \quad = \quad D_{AB}\partial^2 C_A/\partial x^2 \tag{5.169}$$

$$-\partial C(x,t)/\partial x = \delta(x) M/A \tag{5.170}$$

where $J''' = M/A$ is a pulse of mass flux injected at time zero divided by the diffusivity and has units of (mol/m^4). M/A is the amount in moles injected divided by the cross-sectional area through which the mass diffusion occurs.

$$C(x,t) = 0, \ x = -\infty \tag{5.171}$$

$$C(x,t) = 0, \ x = \infty \tag{5.172}$$

Obtaining the Laplace tranform of Eq. [5.172], assuming that at $t = 0$, the concentration everywhere in the medium is zero,

$$\tau_{mr} s^2 \underline{C_A} + \quad s \underline{C_A} \quad = \quad D_{AB}d^2 C_A/dx^2 \tag{5.173}$$

$$\underline{C_A} (s^2\tau_{mr} + s) = D_{AB} d^2\underline{C_A}/dx^2 \tag{5.174}$$

$$\text{Or } \underline{C_A} = c_1 \exp((s^2\tau_{mr} + s)^{1/2}x/D_{AB}^{\ 1/2} + c_2\exp((s^2\tau_{mr} + s)^{1/2}-x/D_{AB}^{1/2} \tag{5.175}$$

The concentration can first be solved for in the domain $0 \le x \le \infty$, and then by symmetry in can be reflected for the domain $-\infty \le x \le 0$. At $x = \infty$, $C_A = 0$ and hence, c_1 need be set to zero.

$$\text{Or } \underline{C_A} = c_2\exp((s^2\tau_{mr} + s)^{1/2}-x/v_m \tag{5.176}$$

From the boundary condition given in Eq. [5.173],

$$J''' \quad = c_2 \ \text{sqrt}(s^2\tau_{mr} + s)/v_m \tag{5.177}$$

$$\text{Or } c_2 = J'''/\text{sqrt}(s^2\tau_{mr} + s)/v_m \tag{5.178}$$

Thus,

Or $\underline{C_A}$ = $[v_m \tau_{mr} J''']/\text{sqrt}(s^2 + 1/\tau_{mr} s) \exp(s(s+ 1/\tau_{mr}) - x/v_m$ (5.179)

where,

$$v_m = \text{sqrt}(D/\tau_{mr})$$ (5.180)

Inverting the Laplace transform as avialable from the tables in Mickley, Sherwood and Reid [1975].

$$C_A /[J''' v_m \tau_{mr}] = \exp(-t/2\tau_{mr}) I_0 [1/2\tau_{mr}(t^2 - x^2/v_m^2)^{1/2}]$$ (5.181)

5.7.2 Cylindrical Coordinates

Consider the movement of solute A from the surface of the cylinder of radius R, into a homogeneous tissue. For example, the cylinder might represent the external surface of a capillary that contains a high concentration of a drug. The concentration within the tissue, in the region r > R, can be determined by solving for the governing equation in polar coordinates from Appendix C from Eq. [C.22]:

$$\tau_{mr} \partial^2 C_A/\partial t^2 + \partial C_A/\partial t = D_{AB}(1/r \, \partial/\partial r \, (r \partial C_A/\partial r))$$ (5.182)

If the concentration of A in the tissue outside the cylinder is initially zero, and the concentration at the outer surface of the cylinder maintained at C_0, the initial and boundary conditions can be written as;

$$C(r, t) = 0; \text{ for } R \leq r; \ t = 0$$ (5.183)

$$C(r,t) = C_0; \text{ for } r = R; \ t > 0$$ (5.184)

$$C(r,t) = 0; \text{ for } r = \infty; \ t > 0 \tag{5.185}$$

Let $u = C/C_0$; $X = r/\text{sqrt}(D\tau_{mr})$; $\tau = t/\tau_{mr}$

$$\partial^2 u/\partial\tau^2 + \partial u/\partial\tau \quad = \quad (1/X \ \partial/\partial X \ (X\partial u/\partial X) \tag{5.186}$$

Let $u = w\exp(-\tau/2)$. As shown in the preceding sections the damping term is removed from the governing equation.

$$\partial^2 w/\partial\tau^2 - w/4 \quad = \quad (1/X \ \partial/\partial X \ (X\partial w/\partial X) \tag{5.187}$$

Let $\eta = \tau^2 - X^2$ for $\tau > X$

Then Eq. [5.189] is transformed into;

$$\eta^2 \ \partial^2 w/\partial\eta^2 + 3/2\eta \ \partial w/\partial\eta \ - \eta w/16 = 0 \tag{5.188}$$

Comparing Eq. [5.190] with the generalized Bessel equation as given in Eq. [2.105], the solution to the special relation in Eq. [2.128] is;

$$a = 3/2; \ b = 0; \ c = 0; \ d = -1/16; \ s = \tfrac{1}{2} \tag{5.189}$$

The order p of the solution is then $p = 2 \ \text{sqrt}(1/16) = \tfrac{1}{2}$ $\hspace{2cm}$ (5.190)

Or $\quad W = c_1 I_{1/2} \ (1/2 \ \text{sqrt}(\tau^2 - X^2)/(\tau^2 - X^2)^{1/4} + c_2 I_{-1/2} \ (1/2 \ \text{sqrt}(\tau^2 - X^2)/(\tau^2 - X^2)^{1/4}$
$$\tag{5.191}$$

c_2 can be seen to be zero as W is finite and not infinitely large at $\eta = 0$. An approximate solution can be obtaining by eliminating c_1 between Eq. [5.186] and the above equation. It can be noted that this is a mild function of time, however. As the general solution of PDE consists of n arbitrary functions when the order of the PDE is n compared with n arbitrary constants for ODE. From the boundary condition at, $X = X_R$,

$$1 = \exp(-\tau/2) \ c_1 I_{1/2} \ (1/2 \ \text{sqrt}(\tau^2 - X_R^2)/(\tau^2 - X_R^2)^{1/4} \tag{5.192}$$

or $\ u = [(\tau^2 - X_R^2)^{1/4}/(\tau^2 - X^2)^{1/4}][I_{1/2} \ (1/2 \ \text{sqrt}(\tau^2 - X^2) / I_{1/2} \ (1/2 \ \text{sqrt}(\tau^2 - X_R^2)]$
$$\tag{5.193}$$

In terms of elementary functions, Eq. [5.194] can be written as;

or $u = [(\tau^2 - X_R^2)^{1/2}/(\tau^2 - X^2)^{1/2}][\operatorname{Sinh}(1/2 \operatorname{sqrt}(\tau^2 - X^2) / \operatorname{Sinh}(1/2 \operatorname{sqrt}(\tau^2 - X_R^2)]$

$$(5.194)$$

In the limit of X_R going to zero, the expression becomes;

or $u = [\tau/(\tau^2 - X^2)^{1/2}][\operatorname{Sinh}(1/2 \operatorname{sqrt}(\tau^2 - X^2) / \operatorname{Sinh}(\tau/2)]$ (5.195)

This is for $\tau > X$

For $X > \tau$,

$u = [(X_R^2 - \tau^2)^{1/4}/(X^2 - \tau^2)^{1/4}][J_{1/2}(1/2 \operatorname{sqrt}(X^2 - \tau^2)/ I_{1/2}(1/2 \operatorname{sqrt}(\tau^2 - X^2)]$

$$(5.196)$$

Eq. [5.197] can be written in terms of trigonometric functions as;

$u = [(X_R^2 - \tau^2)^{1/2}/(X^2 - \tau^2)^{1/2}][\operatorname{Sin}(1/2 \operatorname{sqrt}(X^2 - \tau^2)/ \operatorname{Sinh}(1/2 \operatorname{sqrt}(\tau^2 - X^2)]$

$$(5.197)$$

In the limit of X_R going to zero, the expression becomes;

or $u = [\tau/(X^2 - \tau^2)^{1/2}][\operatorname{Sin}(1/2 \operatorname{sqrt}(X^2 - \tau^2)/ \operatorname{Sinh}(\tau/2)]$ (5.198)

The time lag at a point in the cylindrical medium to realize the disturbance can be calculated as;

$\tau_{lag}^2 = X^2 - 4\pi^2$ (5.199)

or $\tau_{lag} = \operatorname{sqrt}(X_p^2 - 4\pi^2)$ (5.200)

For a given instant in time, the penetration distance can be calculated from Eq. [5.198]:

$X_{pen} = (4\pi^2 + \tau^2)^{1/2}$ (5.201)

For times greater than the time lag and less than X_p the dimensionless concentration is given by Eq. [5.197]. For dimensionless times greater than X_p, the dimensionless concentration is given by Eq. [5.196]. For distances *closer to the surface compared with 2π the time lag will be zero*. For a solute with $D_A = 10^{-11}$ m²/s and a cylinder with radius R = 0.001 m, and a mass relaxation time of 30 seconds, the solute will penetrate in the first 6 hrs no further than a distance of 12.5 mm, in the first 24 hrs no further than a distance of 50 mm and in the first 100 hrs no further than a distance of 207.8 mm.

5.7.3 Spherical Coordinates

Biological systems sometimes exhibit spherical symmetry. For instance, certain cells are assumed to be spherical as are vesicles within cells and synthetic vesicles. The transport of solutes is often sought in the circular coordinate system. The governing equation from Appendix C, Eq. [C.23], neglecting convective effects, can be written as;

$$\tau_{mr}\partial^2 C_A/\partial t^2 + \partial C_A/\partial t = D_{AB}[1/r^2 \, \partial/\partial r \, (r^2 \, \partial C_A/\partial r) \tag{5.202}$$

The gradients in the θ and ϕ directions are neglected. Eq. [5.203] is rendered dimensionless by the following substitutions;

$$\text{Let } u = C/C_0; \quad X = r/\text{sqrt}(D\tau_{mr}); \quad \tau = t/\tau_{mr} \tag{5.203}$$

$$\partial^2 u/\partial \tau^2 + \partial u/\partial \tau = [1/X^2 \, \partial/\partial X \, (X^2 \, \partial u/\partial X) \tag{5.204}$$

Consider the flux of solute A towards a spherical cell in suspension. If the solute is consumed at the surface of the cell, and the rate of consumption is rapid, solute concentrations in the vicinity of the cell.

The time and space conditions are as follows;

$$u = 1, \quad \text{for } X \geq X_R, \quad \tau = 0 \tag{5.205}$$

$$u = 0, \quad X = X_R \tag{5.206}$$

$$u = 1, \quad X = \infty \tag{5.207}$$

At steady state,

$$u = -C_1/X + C_2 \tag{5.208}$$

From Eq. [5.208], $C_2 = 1$. From Eq. [5.207],

$C_1 \quad = X_R$ (5.209)

Thus,

$u^{ss} \quad = 1 - R/r$ (5.210)

Inorder to solve for the transient portion of the solution the initial value problem is changed to a boundary value problem by the following substitution;

Let $V = 1 - u$ (5.211)

The governing equation [5.205] can then be written as;

$\partial^2 V/\partial\tau^2 + \partial V/\partial\tau \quad = [1/X^2 \; \partial/\partial X \; (X^2 \; \partial V/\partial X)$ (5.212)

The damping term is removed from the governing equation. This is done realizing that the transient temperature decays with time in a exponential fashion. The other reason for this maneuver is to study the wave equation without the damping term. Let $u = w \exp(-n\tau)$.

$\exp(-n\tau) \, w \, (-n + n^2) \; + \; \exp(-n\tau) \, \partial w/\partial\tau \, (1 - 2n) \; + \exp(-n\tau) \, \partial^2 \, w/\partial\tau^2$
$= \text{RHS}$ (5.213)

By choosing $n = \frac{1}{2}$, the damping component of the equation is removed;

Thus for $n = \frac{1}{2}$

$\qquad -w/4 \; + \; \partial^2 \, w/\partial\tau^2 \; = \; \partial^2 \, w/\partial X^2 \qquad + 2/X \; \partial w/\partial X$ (5.214)

Eq. [5.215] can be solved by using the method of relativistic transformation of coordinates.

Consider the transformation variable η as;

$$\eta \; = \; \tau^2 - X^2$$ (5.215)

$$\text{for } \tau > X$$

$$\partial w/\partial\tau \; = \; (\partial w/\partial\eta) 2\tau$$ (5.216)

$$\partial^2 \, w/\partial\tau^2 \; = \; (\partial^2 \, w/\partial\eta^2) 4\tau^2 \; + \; 2(\partial w/\partial\eta)$$ (5.217)

In a similar fashion,

$$\partial w/\partial X = (\partial w/\partial \eta)2X \tag{5.218}$$

$$\partial^2 w/\partial X^2 = (\partial^2 w/\partial \eta^2)4X^2 + 2(\partial w/\partial \eta) \tag{5.219}$$

Plugging Eqs. [5.218- 5.220] into Eq. [5.215]

$$(\partial^2 w/\partial \eta^2)4(\tau^2 - X^2) + 8(\partial w/\partial \eta) - w/4 = 0 \tag{5.220}$$

$$4\eta^2 \partial^2 w/\partial \eta^2 + 8\eta \partial w/\partial \eta - \eta w /4 = 0 \tag{5.221}$$

or
$$\eta^2 \partial^2 w/\partial \eta^2 + 2\eta \partial w/\partial \eta - \eta w/16 = 0 \tag{5.222}$$

Comparing Eq. [5.223] with the generalized Bessel equation as given in Eq. [2.105], the solution to the special differential equation in Eq. [2.175] is;

$$a = 2; \ b = 0; \ c = 0; \ s = \tfrac{1}{2}; \ d = -1/16 \tag{5.223}$$

The order of the Bessel solution would be;

$$p = 2 \, sqrt(\tfrac{1}{4}) = 1 \qquad ; sqrt(|d|/s) = \tfrac{1}{2} \tag{5.224}$$

Hence the solution to Eq. [5.223] can be written as ;

$$w = c_1 I_1(1/2 \, sqrt(\tau^2 - X^2)/(\tau^2 - X^2)^{1/2} + c_2 K_1 (1/2 \, sqrt(\tau^2 - X^2)/(\tau^2 - X^2)^{1/2} \tag{5.225}$$

c_2 can be seen to be zero as W is finite and not infinitely large at $\eta = 0$. The solution is in terms of a composite modified Bessel function of the first order and first kind. Therefore the heat flux can be written as;

$$V = c_1 exp(-\tau/2)I_1 (1/2 \, sqrt(\tau^2 - X^2)/(\tau^2 - X^2)^{1/2} \tag{5.226}$$

From the boundary condition at the solid surface;

$$1 = c_1 exp(-\tau/2)I_1 (1/2 \, sqrt(\tau^2 - X_R^2)/(\tau^2 - X_R^2)^{1/2} \tag{5.227}$$

Dividing Eq. [5.227] by Eq. [5.228] the solution for u can be given in a more usable form as ;

$$V = [(\tau^2 - X_R^2)^{1/2}/(\tau^2 - X^2)^{1/2}] [I_1 (1/2 \, sqrt(\tau^2 - X^2)/ I_1 (1/2 \, sqrt(\tau^2 - X_R^2)] \tag{5.228}$$

This is valid for $\tau > X$. For $X > \tau$;

$$V = [(\tau^2 - X_R^2)^{1/2}/(X^2 - \tau^2)^{1/2}] \, [J_1 \, (1/2 \, sqrt(X^2 - \tau^2)/ \, I_1 \, (1/2 \, sqrt(\tau^2 - X_R^2)] \tag{5.229}$$

For $X = \tau$, the solution at the wave front result. This can be obtained by solving Eq. [5.233] at $\eta = 0$.

In the limit of X_R going to zero,

$$V = [\tau/(X^2 - \tau^2)^{1/2}] \, [I_1 \, (1/2 \, sqrt(\tau^2 - X^2)/ \, I_1 \, (\tau/2)] \tag{5.230}$$

This is valid for $\tau > X$. For $X > \tau$;

$$V = [\tau/(X^2 - \tau^2)^{1/2}] \, [J_1 \, (1/2 \, sqrt(X^2 - \tau^2)/ \, I_1(\tau/2)] \tag{5.231}$$

Three regimes can be identified. The first regime is that of the thermal lag and consist of no change from the initial concentration. The second regime is when

$$\tau_{lag}^2 = X^2 - (7.6634)^2 \tag{5.232}$$

or $\tau_{lag} = sqrt(X_p^2 - 7.6634^2)$ (5.233)

The first zero of $J_1(x)$ occurs at $x = 3.8317$. The 7.6634 is twice the first root of the Bessel function of the first order and first kind. For times greater than the time lag and less than X_p the dimensionless concentration is given by Eq. [5.232]. For dimensionless times greater than X_p, the dimensionless concentration is given by Eq. [5.229]. For distances closer to the surface compared with $7.6634sqrt(D\tau_r)$ the thermal lag time will be zero. The ballistic term manifests as a thermal lag at a given point in the medium.

Thus, In the limit of X_R going to zero,

$$u = 1 - [\tau/(\tau^2 - X^2)^{1/2}] \, [I_1 \, (1/2 \, sqrt(\tau^2 - X^2)/ \, I_1 \, (\tau/2)] \tag{5.234}$$

The above Eq. [5.235] is valid for $\tau > X$.

For $X > \tau$;

$$u = 1 - [\tau/(X^2 - \tau^2)^{1/2}] \, [J_1 \, (1/2 \, sqrt(X^2 - \tau^2)/ \, I_1 \, (\tau/2)] \tag{5.235}$$

344

When X = τ,

$$V = c'\exp(-\tau/2) = c'\exp(-X/2) \tag{5.236}$$

$$C' = \exp(X_R/2) \tag{5.237}$$

$$\text{Or } u = 1 - \exp(X_R - X)/2 \tag{5.238}$$

In the limit of X_R going to zero,

$$u = 1 - \exp(-X/2) \tag{5.239}$$

The dimensionless concentration profile is shown in Figure 5.2. The three regimes are piecewise continuous and the function is monotonic. For short times for distances greater than the said time, the dimensionless concentration is described by the Bessel composite function of the first kind and first order. At the occurrence of the first zero of the Bessel function the concentration will be undisturbed.

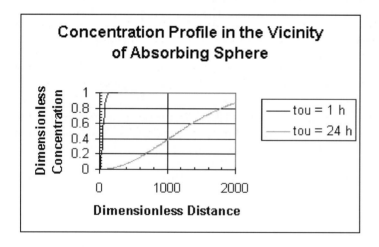

Figure 5.3 Dimensionless Concentration Profile Near the Absorbing Sphere

5.7.4 Solute Binding and Elimination

The concentration of solute/drug within a region of tissue will depend not only on damped wave diffusion and relaxation but also on physiological processes. For instance, a solute may be generated or consumed within a cell or tissue region by a chemical reaction, usually one mediated by enzymes. Alternatively, the solute could be eliminated from the diffusion process by immobilization to some fixed element, binding to the cytoskeleton or an organelle or by paritioning from the extracellular space into a capillary where it enters the circulatory system. Similarly, enzymatic conversion, can be used to eliminate the solute diffusing through the extracellular space. The elimination may be effected by immobilization by some fixed element and by internalization into cells or capillaries or generated by secretion from a cell. When a diffusing solute is generated or consumed homogeneously within some region of interest in a tissue, the differential mass balance in molar concentrations can be written from Eq. [C.21] neglecting convective effects and assuming constant diffusivity as;

$$\tau_{mr}\partial^2 C_A/\partial t^2 \;+\; \partial C_A/\partial t \qquad = \qquad D_{AB}(\partial^2 C_A/\partial x^2 \;+\; \partial^2 C_A/\partial y^2 + \partial^2 C_A/\partial z^2) \quad + R_A$$

$$(5.240)$$

where R_A is the molar rate of generation or consumption of the drug or solute per volume within the differential volume of the element. Consider a solute that diffuses within the extracellular space of the tissue, but also interacts with the tissue by reversibly binding to some fixed component of the tissue. For example, the diffusing solute might bind to a protein on the cell surface or to a protein in the extracellular matrix. When bound the solute may be considered to be completely immobilized and when released it is free to diffuse again within the extracellular space. If the conversion of the diffusible solute, A, to the bound form, B is rapid and characterized by an equilibrium constant $K_e = C_{Be}/C_{Ae}$.

$$A \xrightarrow[\;k_b\;]{\;k_f\;} B \qquad\qquad\qquad (5.241)$$

$$R_A \;=\; -(k_f + k_b)C_A \;+\; k_b C_{A0} \qquad\qquad (5.242)$$

At equilibrium the rate becomes zero,

and it can be seen that,

$$k_f/k_b \;=\; C_{Be}/C_{Ae} \;=\; K_e \qquad\qquad (5.243)$$

Thus,

346

$$-\partial J/\partial x \quad -(k_f + k_b)C_A \quad + k_b C_{A0} \quad = \quad \partial C_A/\partial t \qquad (5.244)$$

The damped wave diffusion and relaxation equation can be written as;

$$J = -D_A \, \partial C_A/\partial x \, - \, \tau_{mr} \, \partial J/\partial t \qquad (5.245)$$

Differentiating the mass balance equation with respect to time and the damped wave and relaxation equation with respect to space and eliminating the second cross-derivative derivative of mass flux with respect to time and space;

$$D_A \, \partial^2 C_A/\partial x^2 \; = \; \tau_{mr} \, \partial^2 C_A/\partial t^2 + (\, \tau_{mr}(\, k_f + k_b \,) + 1)\partial C_A/\partial t \quad + (k_f + k_b)C_A \, - \, k_b C_{A0}$$
$$(5.246)$$

$$\text{Let } X = x \text{sqrt}(D_A \, \tau_{mr}); \;\; \tau = t/\tau_{mr}; \;\; k^*_b = k_b \tau_{mr}; \;\; k^*_f = k_f \tau_{mr} \qquad (5.247)$$

$$\partial^2 C_A/\partial X^2 \; = \; \partial^2 C_A/\partial \tau^2 + (k_b{}^* + k_f{}^* + 1)\partial C_A/\partial \tau \quad + (k_b{}^* + k_f{}^*) \, C_A \, - \, k^*_b C_{A0}$$
$$(5.248)$$

$$\text{Let } u = (k_b{}^* + k_f{}^*) \, C_A \, - \, k^*_b C_{A0} \qquad (5.249)$$

Then,

$$1/((k_b{}^* + k_f{}^*)\partial^2 u/\partial X^2 \; = \; 1/(k_b{}^* + k_f{}^*)\partial^2 u/\partial \tau^2 + (k_b{}^* + k_f{}^* + 1)/(k_b{}^* + k_f{}^*)\partial u/\partial \tau \; + \; u$$
$$(5.250)$$

or,

$$\partial^2 u/\partial X^2 \; = \; \partial^2 u/\partial \tau^2 + (k_b{}^* + k_f{}^* + 1)\partial u/\partial \tau \; + \; u(k_b{}^* + k_f{}^*) \qquad (5.251)$$

Solutions to the above equation will be similar to the problem without any reaction except that the diffusion coefficient is reduced by a factor equal to the binding constant plus unity. These same equations can be used to evaluate penetration into tissues whenmore complicated equilibrium expressions are appropriate by substituting the non-linear equilibrium expression into the governing equation and solving the rest of the equation. Numerical values for K_e can be obtained from the information on the equilibrium association or dissociation constants.

$$\text{Let } K^* = \tau_{mr}(\, k_f + k_b + 1) \quad = (k_b{}^* + k_f{}^* + 1) \qquad (5.252)$$

Eq. [5.252] becomes;

$$\partial^2 u/\partial X^2 \; = \; \partial^2 u/\partial \tau^2 + \; K^* \, \partial u/\partial \tau \; + \; u(K^* - 1) \qquad (5.253)$$

$$\text{Let } u = W \exp(-n\tau) \qquad (5.254)$$

The damping term is first removed by a $u = w \exp(-n\tau)$ substitution.

$$\partial u/\partial \tau = -nw + w_\tau \qquad (5.255)$$

$$\partial^2 u/\partial \tau^2 = +n^2 w - 2nw_\tau + w_{\tau\tau} \qquad (5.256)$$

$$w_{XX} = (n^2 - nK^* + K^* - 1)w + w_\tau(K^* - 2n) + w_{\tau\tau} \qquad (5.257)$$

Choosing $n = K^*/2$ Eq. [5.253] becomes;

$$w_{XX} = -w/4(K^* - 2)^2 + w_{\tau\tau} \qquad (5.258)$$

Let the transformation be;

$$\eta = \tau^2 - X^2 \quad \text{for} \quad \tau > X \qquad (5.259)$$

$$4\eta \partial^2 w/\partial \eta^2 + 4\partial w/\partial \eta - w/4(|K^* - 2|^2 = 0 \qquad (5.260)$$

$$\text{or } \eta^2 \partial^2 w/\partial \eta^2 + \eta \partial w/\partial \eta - w\eta|K^* - 2|^2/16 = 0 \qquad (5.261)$$

Comparing Eq. [5.261] with the generalized Bessel equation [Mickley, Sherwwod, Reed, 1957];

$$a = 1; \ b = 0; \ c = 0; \ s = \tfrac{1}{2}; \ d = -1/16(K^* - 2)^2 \qquad (5.262)$$

The order $p = 0$. $\mathrm{sqrt}(d)/s = i\,|K^* - 2|^2$ and isimaginary. Hence the solution is;

$$w = c_1 I_0 \tfrac{1}{2} |K^* - 2|^2 \mathrm{sqrt}(\tau^2 - X^2) + c_2 K_0 \tfrac{1}{2} |K^* - 2|^2 \mathrm{sqrt}(\tau^2 - X^2) \qquad (5.263)$$

c_2 can be seen to be zero from the condition that at $\eta = 0$ W is finite.

From the boundary condition at the wavefront, at $\eta = 0$, $w = 1$ it can be seen that $c_1 = u_s$. The solution at the wavefront can be obtained by directly solving fopr Eq [5.263] at $\eta = 0$. The integration constant is solved for when $X = 0$, $u = u_s$. Thus.

$$u/u_s = \exp(-\tau K^*/2) I_0 \tfrac{1}{2} |K^* - 2| \mathrm{sqrt}(\tau^2 - X^2) \qquad (5.264)$$

This is valid for $K^* \neq 2$

For $K^* = 2$, Eq. [5.258] reverts to the wave equation.

It can be seen that the solution changes to a Bessel composite function of the first kind and first order when $X > \tau$,

$$u/u_s \;=\; \exp(-\tau K^*/2)\; J_0\; \tfrac{1}{2}\; |K^* - 2|\; \text{sqrt}(X^2 - \tau^2) \tag{5.265}$$

When the argument of the Bessel composite function of the first kind and zeroth order becomes 2.4048 the concentration u becomes zero.

Thus, for a given point in the medium, the time lag experienced prior to any appreciable change in the concentration can be estimated as;

$$\tau_{\text{lag}} \;=\; (Xp^2 \;\; - 23.13/|K^* - 2|^2)^{1/2} \tag{5.266}$$

For a given time instant, the penentration distance beyond which there the concentration is undisturbed can be estimated as;

$$X_{\text{pen}} \;=\; (\tau^2 \;+\; 23.13/|K^* - 2|^2)^{1/2} \tag{5.267}$$

Worked Example 5.7.1

Rewrite Eq. [5.247] in terms of a effective diffusivity and interpret the governing equation.

Eq. [5.245] can be written as;

$$-\partial J/\partial x \;\; + R_A \;=\; \partial C_A/\partial t \tag{5.268}$$

$$R_A \;=\; -(k_f + k_b)C_A \;\; + k_b C_{A0} \;=\; dC_A/dt = 1/\tau_{mr}\, dC_A/d\tau \tag{5.269}$$

$$R_B \;=\; k_f\, C_A \;-\; k_b\, C_B \;=\; 1/\tau_{mr}\, dC_B/d\tau \tag{5.270}$$

$$\text{Let } K_b = C_B/C_A \tag{5.271}$$

$$R_B \;=\; k_f\, C_A \;-\; k_b\, C_B \;=\; K_b/\tau_{mr}\, dC_A/d\tau \tag{5.272}$$

$$R_A \;=\; -R_B \tag{5.273}$$

Thus,

$$-\tau_{mr}\partial J/\partial x \;\; = \;(1 + K_b)\, \partial C_A/\partial \tau \tag{5.274}$$

Let X = x/sqrt($D\tau_{mr}$)   (5.275)

$-\partial J/\partial X = (1 + K_b)v_m \partial C_A/\partial \tau$ (5.276)

The damped wave diffusion and relaxation equation can be written as;

$J = -v_m \partial C_A/\partial X - \partial J/\partial \tau$ (5.277)

Differentiating Eq. [5.276] wrt τ and Eq. [5.277] wrt X,

$-\partial^2 J/\partial X \partial \tau = (1 + K_b)v_m \partial^2 C_A/\partial \tau^2$ (5.276)

The damped wave diffusion and relaxation equation can be written as;

$\partial J/\partial X = -v_m \partial^2 C_A/\partial X^2 - \partial^2 J/\partial \tau \partial X$ (5.277)

Eliminating the second cross-derivative wrt J between Eqs. [5.276] and [5.277],

$\partial^2 C_A/\partial X^2 = (1 + K_b) [\partial^2 C_A/\partial \tau^2 + \partial C_A/\partial \tau]$ (5.278)

Define a effective diffusivity,

$D_e = D/(1 + K_b)$ (5.279)

and a new ordinate, Y = X sqrt(D/D_e)

$\partial^2 C_A/\partial Y^2 = \partial^2 C_A/\partial \tau^2 + \partial C_A/\partial \tau$ (5.280)

The binding constant, K_b is thus lumped with the diffusivity to give a effective diffusivity. Solutions to Eq. [5.280] was presented in Chapter 3.0 for some boundary conditions.

The binding constant K_b can be obtained from information on the equilibrium association, K_a or dissociation K_d constants for receptor-ligand pairs. These constants may be defined as;

$$L + R \quad \overset{k_f}{\underset{k_r}{\Leftrightarrow}} \quad L\text{-}R \qquad (5.281)$$

$K_d = 1/K_a = [L][R]/[L\text{-}R] = k_r/k_f$ (5.282)

Where K is the ligand, R is the receptor and L-R is the ligand-receptor complex. Consider the case of a drug molecule, which can be assigned the binding ligand [L] = C_A interacting with a binding site, R which is present on the surface of the cells or stationary molecules in the extracellular space to form bound complexes, [L-R] = C_B. Assuming that the receptor concentration is constant in the tissue, the binding constant, K_b, is related to the dissociation constant:

$$K_b = C_B/C_A = [L-R]/[L] = [R]/K_d \qquad (5.283)$$

Some values of the dissociation constant is presented in Saltsman [2001].

Upon substitution of an appropriate kinetic expression for the rate of generation or consumption of solute within the tissue space, Eq. [5.247] can be solved to determine the spatio-temporal patterns exhibited by the concentration. For example, consider the one dimensional damped wave diffusion and relaxation, of solute fro the interface, where the concentration is maintained constant. If the diffusing solute is also eliminated from the tissue, such that the volumetric rate of elimination is first order with a characteristic rate constant, k, Eq. [5.247] can be reduced to;

$$D_A \, \partial^2 C_A/\partial x^2 = \tau_{mr} \, \partial^2 C_A/\partial t^2 + (k_f^* + 1)\partial C_A/\partial t + k_f \, C_A \qquad (5.284)$$

This equation can be solved subject to the initial and boundary conditions;

$$C_A(x,t) \qquad = 0, \text{ for } x \geq 0 \qquad (5.285)$$

$$C_A(x,t) \qquad = C_o; \text{ for } x = 0; \; t > 0 \qquad (5.286)$$

$$C_A(x,t) \qquad = 0; \text{ for } x = \infty, \quad t > 0 \qquad (5.287)$$

Eq. [5.284] can be solved as shown in chapter 3.0 and the solution written as;
Let ;

$$u = C_A/C_o; \; X = x/\sqrt{(D\tau_{mr})}; \; \tau = t/\tau_{mr} \qquad (5.288)$$

Eq. [5.284] becomes;

$$\partial^2 u/\partial X^2 = \partial^2 u/\partial \tau^2 + (k_f^* + 1)\partial u/\partial \tau + k_f^* \, u \qquad (5.289)$$

The Eq. [5.289] can be solved by the method of relativistic transformation of coordinates. The damping term is first removed by a $u = w \exp(-n\tau)$ substitution. Choosing $n = (1 + k^*)/2$ Eq. [5.289] becomes;

$$\partial^2 w/\partial X^2 = -w(1 - k_f^*)^2/4 + w_{\tau\tau} \qquad (5.290)$$

Let the transformation be;

$$\eta = \tau^2 - X^2 \text{ for } \tau > X$$

Eq. [5.290] becomes;

$$4\tau^2 \partial^2 w/\partial \eta^2 + 2\partial w/\partial \eta - w(1 - k^*)^2/4 = 4X^2 \partial^2 w/\partial \eta^2 - 2\partial w/\partial \eta \qquad (5.291)$$

or $\eta^2 \, \partial^2 w/\partial \eta^2 + \eta \partial w/\partial \eta \; - w\eta(1 - k^*)^2 \, /16 \; = 0$ \hfill (5.292)

Comparing Eq. [5.292] with the generalized Bessel equation [Mickley, Sherwwod, Reed, 1957];

$$a = \; 1; \; \; b = 0; \; \; c = 0; \; \; s = \tfrac{1}{2} \; ; \; d = -1/16(1 - k_f^*)^2 \hfill (5.293)$$

The order $p = 0$. $\mathrm{sqrt}(d)/s = \tfrac{1}{2} \, i(1 - k^*)$ and is imaginary. Hence the solution is;

$$w \; = \; c_1 \, I_0 \, \tfrac{1}{2} \, |1 - k_f^*| \; \mathrm{sqrt}(\tau^2 - X^2) \; + c_2 K_0 \, \tfrac{1}{2} \, |1 - k_f^*| \; \mathrm{sqrt}(\tau^2 - X^2) \hfill (5.294)$$

c_2 can be seen to be zero from the condition that at $\eta = 0$ W is finite.

$$w = \exp(-\tau(1 + k_f^*)/2) \, I_0 \, [\tfrac{1}{2} \, |1 - k^*| \; \mathrm{sqrt}(\tau^2 - X^2)] \hfill (5.295)$$

From the boundary condition at $X = 0$,

$$1 \; = \; \exp(-\tau(1 + k_f^*)/2) \, c_1 \, I_0 \, \tfrac{1}{2} \, [|1 - k_f^*\tau|] \hfill (5.296)$$

c_1 can be eliminated between Eqs. [5.296, 5.294] or calculated to yield;

$$u \; = \; I_0 \, [\tfrac{1}{2} \, |1 - k_f^*| \; \mathrm{sqrt}(\tau^2 - X^2)]/ \, I_0 \, \tfrac{1}{2} \, [|1 - k_f^*|\tau] \hfill (5.297)$$

This is valid for for $\tau > X$, $k_f^* \neq 1$

For $X > \tau$,

$$u \; = \; J_0 \, [\tfrac{1}{2} \, |1 - k_f^*| \; \mathrm{sqrt}(X^2 - \; \tau^2]/ \, I_0 \, \tfrac{1}{2} \, [\, |1 - k_f^*|\tau] \hfill (5.298)$$

At the wavefront , $\tau = X$, $u = \exp(-\tau(1 + k_f^*)/2)) = \exp(-X(1 + k_f^*)/2)$

The mass inertia can be calculated from the first zero of the Bessel function at 2.4048.

Thus,

$$\tau_{inertia} = \; \mathrm{sqrt}(X_p{}^2 - 23.1323/|1 - k_f^*|^2) \hfill (5.299)$$

The concentration at a interior point in the semi-infinite medium is shown in Figure 3.4. Three regimes can be identified. During the first regime of mass inertia there is no transfer of mass upto a certain tresholf time at the interior point $X_p = 10$. The second regime is given by Eq. [3.62] represented by a modified Bessel function of the first kind and zeroth order. The rise in dimensionless concentration proceeds from the dimensionless time 2.733 upto the wave front at $X_p = 10.0$. The third regime is given by

Eq. [3.61] and represents the decay in time of the dimensionless concentration.   It is given by the modified Bessel composite function of the first kind and zeroth order.   In Figure 3.5 is shown the three regimes of the concentration when $k_r^* = 2.0$. It can be seen from Figure 3.5 that the mass inertia time has increased to 8.767. The rise is nearly a jump in concentratin at the interior point $X_p = 10.0$. When $k^* = 0.25$ as shown in Figure 3.6 the inertia time is 7.673. In Figure 3.6 the three regimes for the case when $k_r^* = 0.0$ is plotted. In Table 3.1 the mass inertia time for various values of k_r^* for the interior point $X_p = 10.0$ is shown. k_r^* needs to be sufficiently far from 1 to keep the inertia time positive.

The corresponding steady-state solution of Eq. [5.284] can be obtained;

$$D_A \, \partial^2 C_A/\partial x^2 = \quad + k_r \, C_A \tag{5.300}$$

$$C_A = c_1 \exp(-x(k_r/D_A)^{1/2} \quad + c_2 \exp(+x(k_r/D_A)^{1/2} \tag{5.301}$$

At $x = \infty$, $C_A = 0$ and hence, $c_2 = 0$, and from the boundary condition given in Eq. [5.286], c_1 can be seen to be C_o.

$$C_A/C_o = \exp(-x(k_r/D_A)^{1/2} \tag{5.302}$$

5.7.4.1   Maternal Effect Genes

Solutions to the damped wave diffusion and relaxation equation in 1 dimension can be used to interpret a variety of biological phenomena.   Examples from developmental biology,   drug design and neuroscience.   Maternal efect genes are segregated to defined regions in the developing embryo.   One of these genes,   which encodes a protein called bicoid, is concentrated at the anterior end of Drosophilla embryos [Driever, 1988].   Bicoid produced at the anterior end diffuses toward the posterior pole. As the protein diffuses, is can be metabolized. Simultaneous diffusion and elimination produce a stable protein gradient.   The cells of the embryo respond to the local concentration of bicoid by expressing certain genes.   In this way, the bicoid gradient provides positional information to cells throughout the embryo. These events are critical for the formation of structures throughout the organism. They are achieved by a mechanism for gradient formation, diffusion with homogeneous elimination which can be expalined by the steady state model given in Eq. [5.302].

5.7.4.2 Drug Penetration in the Tissue

The damped wave diffusion and relaxation equation can be used to develop a simple, quantitative method for predicting the extent of drug penetration into a tissue following the introduction of a local source. Using Eq. [5.302] the drug penetration is quantitated. When a gradient of drug cocentration of present, the region of tissue neasrest the implant will be exposed to high, possibly toxic, drug levels. The region farthest from the implant will be untreated. A effectiveness factor for drug delivery can be defined as the ratio of the average cocentration at the implant/tissue interface, C_0,

$$\Omega = <C_A>/C_0 = \int_0^L C_A \, dx \, / \int_0^L dx \tag{5.303}$$

where L is the distance into the tissue that requiers treatment with the drug. For values of Ω near unity, the overall concentration profile is nearly flat, and solute delivery to this region of the tissue is effective. Substituting Eq. [5.302] into Eq. [5.303] yields;

$$\Omega = 1/\phi \, (1- \exp(-\phi)) \tag{5.304}$$

where the dimensionless parameter ϕ is defined after Thiele [1939] as;

$$\phi^2 = L^2 \, k_f/D \tag{5.305}$$

The effectiveness factor can be defined for the transient case also;

$$\partial^2 u/\partial X^2 = \partial^2 u/\partial \tau^2 + (k_f^* + 1)\partial u/\partial \tau + k_f^* u \tag{5.306}$$

Obtaining the Laplace transform of Eq. [5.305],

$$d^2\underline{u}/dX^2 = (s^2 + s(k_f^* + 1) + k_f^*) \, \underline{u} \tag{5.307}$$

$$\underline{u} = c_1 \exp(-X (s^2 + s(k_f^* + 1) + k_f^*) + c_2 \exp(X (s^2 + s(k_f^* + 1) + k_f^*) \tag{5.308}$$

From the boundary condition at $X = \infty$, c_2 can be seen to be zeo. From the boundary condition at $X = 0$,

$$1/s = c_1$$

$$\underline{u} = 1/s\exp(-X (s^2 + s(k_f^* + 1) + k_f^*)^{1/2} \tag{5.309}$$

Eq. [5.303] in the Laplace domain can be written as;

$$\underline{\Omega} \ = \ <C_A>/C_o \ = \ C_o \int_0^L \underline{u}\,dx / \int_0^L dx \tag{5.310}$$

Eq. [5.309] substituted in Eq. [5.309] yields;

$$\underline{\Omega} \ = \ C_o/L \ \ 1/s \ (s^2 + s(k_f^* + 1) \ + k_f^*)^{1/2} (1 - \exp(-L \ (s^2 + s(k_f^* + 1) \ + k_f^*)^{1/2}) \tag{5.311}$$

$$= \ C_o/L \int_0^\tau \exp(-p(k_f^* + 1)/2 \ ) \ I_0 \ p/2 \ |k_f^* - 1|dp$$

$$= \ C_o/L \int_0^\tau \exp(-p(k_f^* + 1)/2 \ ) \ I_0 \ |k_f^* - 1|/2(\tau^2 - L^2)^{1/2}dp \tag{5.312}$$

It can be seen that as ϕ increases the rate of drug diffusion decreases with respect to the rate of elimination. This results in steep concentration gradients or ineffective solute delivery to the tisue. The physico-chemical properties of the drug which determine ϕ on the extent of drug penetration in tissue [Saltzman and Radomsky, 1991].

5.7.4.3 Neurotransmission Across the Synaptic Cleft

These models can be used to predict the movement of molecules in more complex geometrical arrangements such as the synaptic cleft. Neurotransmitter molecules are released from vesciles in the presynaptic neurin into the synaptic cleft. The region of space between the presynaptic neuron sending a signal and the postsynaptic neuron receiving a signal can possess different characteristic geometries. The rectangular distance across the cleft, the spherical vesicle that releases the neurotransmitter and the cylindrical patch of receptors that sense the signal on the postsynaptic neuron. The concentration of neurotransmitter within the synaptic cleft, n can be solved for from the damped wave diffusion and relaxation equation. Written on a molar basis in three dimensions;

$$-\nabla J \ + f(x,y,z,t) \ = \ \partial n/\partial t \tag{5.313}$$

The damped wave diffusion and relaxation equation in three dimensions can be written as;

$$J = -\nabla C - \tau_{mr} \, \partial J / \partial t \qquad (5.314)$$

Differenting Eq. [5.313] wrt to t and operating Eq. [5.314] wrt ∇ and eliminating $\nabla \partial J / \partial t$,

$$D_n(\partial^2/\partial x^2 + \partial^2/\partial y^2 + \partial^2/\partial z^2)n + f(x,y,z,t) = (f + 1)\partial n/\partial t + \tau_{mr} \, \partial^2 n/\partial t^2$$
$$(5.315)$$

A continuous source function that is consistent with experimental data is;

$$f(x,y,z,t) = q. \exp(- \, 1/b(x^2 + y^2) - z^2/c) \, . \, t^\alpha \exp(-\beta t) \qquad (5.316)$$

where q is related to the number of neurotransmitter molecules initially in the vesicle, b and c are Gaussian variances in the lateral and z directions, limited by the size of the vesicle opening and cleft width d_{syn}, and α and β are parameters. Neurotransmitter molecules diffuse in the cleft and do not cross the membrane boundaries:

$$\partial n / \partial z = 0, \text{ at } z = 0 \text{ and } z = d_{syn} \qquad (5.317)$$

These equations were solved for the special case when $\tau_{mr} = 0$ by Klienle et al [1996].

5.7.4.4 Enzymatic Reactions

Biological systems are chemically and structurally complex with slow rates of diffusion. As the diffusing molecules become larger it is obstructed by physical structures. Growth and metabolism, the key processes of life, occur throough an orderly, regulated, coupled array of biochemical reactions. The reaction between any two biochemical substrates, A and B requires contact or collision between the two reactants. The rate of the reaction is often determined by the frequency of collisions. When the frequency of collisions is an important and significant consideration, the finite speed diffusion effects may become important. Most cellular functions depend on enzymatic reactions. Intracellular enzymes participate in several hundred reactions per second. A typical intracellular compartment with dimensions of the order of magnitude of 1 μm is mixed by the process of damped wave diffusion and relaxation. Every two

356

molecules collide at a frequency of 1 second in a intracellular compartment with a diffusion time of 10 ms and typical diffusivity 10^{-10} m²/s.

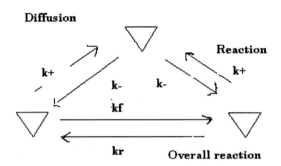

Figure 5.4 Mechanism for Diffusion Limited Reactions

       The enzyme-substrate is often analyzed by a mechanism as shown in Figure 5.3. The overall reaction of solutes A and B occurs in reversible steps. The first step involves collision in which the two solutes come in proximity with each other to allow for the reaction. The second step involves the intrinsic rate of reaction. Mass balance equations for the reaction kinetics are written for the species, A, B, A-B and A-B. The intermediate species A-B reaches steady state concentration rapidly compared to the other species.

$$dC_{AB}/dt = k_f\, C_A\, C_B - k_r\, C_{AB} \tag{5.318}$$

where $k_f = k_{+1}k_+/(k_- + k_1)$

$$k_r = k_-k_{-1}/(k_1 + k_-) \tag{5.319}$$

       The overall apparent rates of the forward and reverse reaction (k_f and k_r) can be written in terms of rates of encounter (k_+ and k_-) and the intrinsic forward and reverse reaction rates (k_+ and k_-). The encounter step involves the simultaneous diffusion of two separated solute species that collide during random motion. The kinetics of this event (k_+ and k_-) can be determined by analysis of the diffusion equation. The concept of binding constant was discussed in the previous section to describe the diffusion of solutes to the surface of the cell. If the diffusion is toward the solute molecule which is fixed in space and surrounded by solute B that can diffuse with diffusion coefficient D equal to D_A and D_B, the rate of collision is equal to the steady-state flux of B molecules to the surface of the A molecule:

$$k_+ \, C_{B,\infty} \; = \; -D \, (4\pi a^2) \, \partial C_B/\partial t \qquad\qquad (5.320)$$

where $C_{B,\infty}$ is the bulk concentration of solute B and a is the radius of the encounter complex A-B that is formed upon collision of B with A. Substituting the steady state concentration profile C_B ;

$$k_+ \; = 4\pi a D \qquad\qquad (5.321)$$

This reaction rate is for a single molecule of A. The reverse rate constant can be estimated in a similar fashion. The reverse reaction starts with a single B molecule bound in the encounter complex with volume 4 π $a^3/3$, which must diffuse subsequently into an unbounded fluid in which the concentration is negligible. The rate of dissociation of the complex is equated to the rate of diffusion of B molecules away from the complex surface:

$$k_- \, [1/4 \, \pi \, a^3/3] \; = -D \, (4\pi a^2)\partial C_B/\partial t \qquad\qquad (5.322)$$

Eq. [5.322] after substitution of C_B yields;

$$k_- \; = 3D/a^2 \qquad\qquad (5.323)$$

The encounter rate constants k_+ and k_- can usually be estimated from physical properties of the two reactants such as the diffusion coefficient and encounter radius. The relative importance of diffusion vs reaction in determining the overall rate of reaction can be estimated by using Eqs. [5.319].

The relative rates of diffusion and reaction are important in the regulation of biochemical pathways. Transcription factors are proteins that bind to specialized regions of DNA and thereby facilitate transcription by RNA polymerase. The binding of one class of transcription factors the basic leucine zipper, bZIP proteins has been found to be diffusion limited and the wave diffusion may be a important consideration. The bZIP factors have a C terminal leucine zipper domain, a basic region that binds to DNA, and a domain that is important for transcriptional regulation. Stable association of the bZIP dimer with DNA occur via a monomer of dimer pathway. Binding of both monomer and dimer was found to be diffusion limited. Analysis of the kinetics of both pathways suggests that the monomer pathway may have an overall kinetic advantage. The bZIP factors must dimerize for transcriptional activity and thererfore it is assumed that dimerization occurs before DNA binding.

The monomer and dimer binding, the reactions 3 and 2, were found to be rapid so that the dimerization reaction is not necessary for the function of bZIP factors.

358

The monomer pathway as shown in Figure 5.4, may provide kinetic advantages in transcriptional regulation. Monomeric bZIP can bind to DNA more rapidly than dimeric bZIP. Monomers can slide along the side of the DNA backbone. Silding represents an opportunity for bZIP to perform a one dimensional search of the DNA strand. Faster searching allows the transcription factor to find its target rapidly. Binding to the target is stabilized by dimerization of bZIP at the target site. The one dimensional diffusion in the function of the DNA binding proteins has been studied with the restriction endonuclease EcoRV, which cleaves the DNA polymer after recognizing a specific restriction site on the DNA strand. EcoRV recognizes and cleaves GAT↓ATC sites on DNA. The overall rate of cleavage was determined for PCR amplified DNA sequences of different length, but each containing one restriction site. The role of linear diffusion in the overall kinetics of cleavage can be modeled [Jeltsch and Pingoud, 1998]:

$$k_{on}/k_{+1} = \sum_{1}^{L} \exp(-(n-s)^2/P) \qquad (5.324)$$

where k_{on} is the overall rate of association with the target site at postiion s. k_{+1} is the rate of association of EcoRV with a site n, and P is the probability that the enzyme will diffuse one basepair along the DNA rather than dissociate from the DNA polymer. P is equal to the ratio of the rate constant for diffusion (k_{diff}) to the rate constant for dissociation (k_{off}).

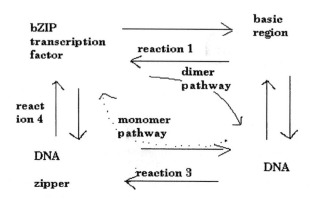

Figure 5.5 Mechanism for bZIP Transcription Factor

In most forms of drug delivery, spatial localization and duration of drug concentration are constrained by organ physiology and metabolism. Drugs administered orally will distribute to tissues based on principles of wave diffusion, relaxation, permeation and flow. Controlled delivery systems offer an alternative approach to regulating both the duration and spatial localization of therapeutic agents. In controlled delivery, the active agent is combined with other components to produce delivery system. Controlled drug delivery systems frequently involve, combinations of active agents with inert polymeric materials. Controlled delivery systems are distinguished from sustained release drug formulations. Sustained release is often achieved by mixing an active agent with binders that alter the agent's rate of dissolution in the intestinal tract or adsorption from a local injection site. Controlled delivery systems include a component that can be engineered to regulate an essential characteristic, e.g., duration of release, rate of release or targeting and have a duration of action longer than a day.

5.7.5 Transdermal Delivery Systems

Many classes of polymeric materials are now available for use as biomaterials. Silicone, polyethylene, polyurethanes, PMMA and EVAc account for the majority of polymeric materials currently used in clinical applications. Reservoir drug delivery devices in which a liquid reservoir of drug is enclosed in a silicone elastomer tube, were demonstrated to provide controlled release of small molecules several decades ago [Folkman and Long, 1964]. Lipophilic drugs to the eye or skin such as Ocusert, Alza corp., Estraderm and Trasderm Nitro, Ciba Giegy, offers a number of advantages. In most common reservoir and transdermal systems release from the reservoir into the external solution occurs in three steps;

1) dissolution of the drug in the polymer
2) diffusion of drug across the polymer membrane
3) dissolution of the drug into the external phase

Assuming that the rate of diffusion across the membrane is much slower than the rate of dissolution partiopning at either interface so that partioning can be assumed to be at equilibrium, the kinetics of release can be modeled by performing a mass balance on the drug within the polymer membrane.

For a slice control volume in the membrane with thickness Δx a mass balance on the diffusing drug molecule yields;

$$\tau_{mr}\partial^2 c_p/\partial t^2 \quad + \quad \partial c_p/\partial t \quad = \quad D_{i:p} \; \partial^2 c_p/\partial x^2 \tag{5.325}$$

360

where D_{ip} is the diffusion coefficient for the drug within the polymer material and c_p is the concentration of the drug (mg/mL) within the polymer. Eq. [5.325] implies that there is no bulk flow or generation/consumptionof drug within the polymer. For a drug diffusing within a polymeric material, the density does not change with time, and the drug is always present at low concentrations within the material. The concentration within the membrane canbe solved for subject to the following conditions;

Membrane

Drug
Reser.

Membrane

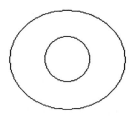

Figure 5.6 Reservoir Delivery Systems Based on Rate Limiting Polymer Membranes

$$t = 0, \quad 0 < x < L, \quad c_p = c_{p,0} \tag{5.326}$$

$$t > 0, \quad x = 0, \quad c_p = c_{p,1} \tag{5.327}$$

$$t > 0, \quad x = L, \quad c_p = c_{p,2} \tag{5.328}$$

The initial condition is given by Eq. [5.326] and the boundary conditions are given by Eqs. [5.327, 5.328]. The concentration may be assumed to be the sum of the transient and steady state part;

$$\text{Let } c_p = c_p^{ss} + c_p^{t} \tag{5.329}$$

$$0 = D_{i:p} \; \partial^2 c_p{}^{ss}/\partial x^2 \tag{5.330}$$

$$c_p{}^{ss} = c_1 x + c_2$$

From Eq. [5.327], c_2 can be seen to be $c_{p,1}$. From Eq. [5.328],

$$c_1 = (c_{p,2} - c_{p,1})/L \tag{5.331}$$

Thus,

$$(c_p{}^{ss} - c_{p,1})/(c_{p,2} - c_{p,1}) = x/L \tag{5.332}$$

The rest of the problem is solving for the transient portion of the solution;

$$\tau_{mr} \partial^2 c_p{}^t/\partial t^2 + \partial c_p{}^t/\partial t = D_{i:p} \; \partial^2 c_p{}^t/\partial x^2 \tag{5.333}$$

The boundary conditions will now be homogeneous;

$$c_p{}^t = 0, \; x = 0 \tag{5.334}$$

$$c_p{}^t = 0, \; x = L \tag{5.335}$$

$$\text{Let } \tau = t/\tau_{mr}; \; X = x/(D_{ip}\tau_{mr})^{1/2} \tag{5.336}$$

$$\partial^2 c_p{}^t/\partial \tau^2 + \partial c_p{}^t/\partial \tau = \partial^2 c_p{}^t/\partial X^2 \tag{5.337}$$

Let $c_p = w \exp(-\tau/2)$ and Eq. [5.337] less the damping term can be written in terms of w as shown in the earlier sections as;

$$\partial^2 w/\partial \tau^2 - w/4 = \partial^2 w/\partial X^2 \tag{5.338}$$

Eq. [5.338] can be solved by the separation of variables as shown in Chapter 3.0.

$$\text{Let } w = g(X) \; V(\tau) \tag{5.339}$$

$$V''/V - \tfrac{1}{4} = g''/g = -\lambda_n{}^2 \tag{5.340}$$

$$g = c_1 \operatorname{Sin}(\lambda_n X) + c_2 \operatorname{Cos}(\lambda_n X) \tag{5.341}$$

At $X = 0$, $g = 0$ so c_2 can be seen to be zero. At $X = L/(D_{ip}\tau_{mr})^{1/2}$, $g = 0$

$$\lambda_n L/(D_{ip}\tau_{mr})^{1/2} = n\pi, \; n = 1,2,3..... \tag{5.342}$$

$$V = c_3 \exp(+\tau \text{sqrt}(1/4 - \lambda_n^2)) + c_4 \exp(-\tau \text{sqrt}(1/4 - \lambda_n^2)) \qquad (5.343)$$

When $\tau = \infty$, $\quad c_p{}^t = \quad w \exp(-\tau/2) = 0 \qquad\qquad (5.343)$

Thus it can be seen that c_3 can be set to zero in Eq. [5.343].

Thus,

$$c_p = \sum_1^\infty c_n \sin(\lambda_n X) \exp(-\tau \text{sqrt}(1/4 - \lambda_n^2)) \qquad (5.344)$$

From the initial condition,

$$c_{p,0} = \sum_1^\infty c_n \sin(\lambda_n X) \qquad\qquad (5.345)$$

Multiplying both sides by $\sin(\lambda_m X)$ and integrating between $X = 0$, X_L. It can be seen that all the terms other than the one when $\lambda_n = \lambda_m$ will become zero. Thus,

$$2c_{p,0}(1 - (-1)^n)/n\pi = c_n \qquad\qquad (5.346)$$

Further it can be seen that when $L < \pi(D_{ip}\tau_{mr})^{1/2}$ subcritical damped oscillations in concentration can be seen. The solution for the concentration can be seen as;

$$c_p = c_{p,1} + (c_{p,2} - c_{p,1})x/L$$
$$+ 2 c_{p,0} \sum_1^\infty (1 - (-1)^n)/n\pi \, \sin(\lambda_n X) \exp(-\tau/2)\exp(-\tau \text{sqrt}(1/4 - \lambda_n^2))$$
$$(5.347)$$

For small membranes, $L < \pi(D_{ip}\tau_{mr})^{1/2}$, subcritical damped oscillations in concentration can be found.

$$c_p = c_{p,1} + (c_{p,2} - c_{p,1})x/L$$
$$+ 2 c_{p,0} \sum_1^\infty (1 - (-1)^n)/n\pi \, \exp(-\tau/2)\sin(\lambda_n X) \cos(\tau \text{sqrt}(\lambda_n^2 - 1/4))$$

$$(5.347)$$

or

$$c_p = c_{p,1} + (c_{p,2} - c_{p,1})x/L$$

$$+ c_{p,0} \sum_1^{\infty} (1 - (-1)^n)/n\pi \, \exp(-\tau/2)[\text{Sin}(\lambda_n X + \tau(\lambda_n^2 - 1/4)^{1/2})$$

$$+ \text{Sin}(\lambda_n X - \tau(\lambda_n^2 - 1/4)^{1/2}] \qquad (5.348)$$

The gradient of concentration at the external surface of the membrane for small systems are;

$$-\partial c_p/\partial X|_{XL} = (c_{p,2} - c_{p,1})/L$$

$$+ (c_{p,0} (D_{ip}\tau_{mr})^{1/2}/L)\sum_1^{\infty}(1 - (-1)^n)\exp(-\tau/2)[\text{Cos}(n\pi + \tau[D_{ip}\tau_{mr}(n\pi/L)^2 - 1/4)]^{1/2})$$

$$+ \text{Cos}(n\pi + \tau[D_{ip}\tau_{mr}(n\pi/L)^2 - 1/4)]^{1/2}] \qquad (5.349)$$

The mass flux is then given by;

$$J = \exp(-t/\tau_{mr}) + (c_{p,2} - c_{p,1})/L$$

$$+ (D_{ip}c_{p,0} (D_{ip}\tau_{mr})^{1/2}/L)\sum_1^{\infty}(1 + (-1)^n)\exp(-\tau/2)[\text{Sin}(\tau[D_{ip}\tau_{mr}(n\pi/L)^2 - 1/4)]^{1/2})$$

$$+ \text{Sin}(\tau[D_{ip}\tau_{mr}(n\pi/L)^2 - 1/4)]^{1/2}] \qquad (5.349)$$

The total amount of material leaving the membrane, M_t, can be found by integrating the mass flux through the external surface with respect to time:

$$M_t = A \int_0^t J_{x|x=L}dt = (At/L)D_{i,p}(c_{p,2} - c_{p,1}) - \tau_{mr}\exp(-t/\tau_{mr})$$

$$+ (D_{ip}c_{p,0} (D_{ip}\tau_{mr})^{1/2}/L)\sum_1^{\infty}(-1 + (-1)^{n+1})[D_{ip}\tau_{mr}(n\pi/L)^2 - 1/4)]^{1/2}\exp(-\tau/2)[\text{Cos}(\tau[1$$

$$+ D_{ip}\tau_{mr}(n\pi/L)^2 - 1/4)]^{1/2}) - 2$$

$$+ \text{Cos}(\tau[D_{ip}\tau_{mr}(n\pi/L)^2 - 1/4)]^{1/2}] \qquad (5.350)$$

For small systems, such as when $L < \pi (D_{ip}\tau_{mr})^{1/2}$ as shown in Eq. [5.350] the total amount of material leaving the membrane is subcritical damped oscillatory. A is the cross-sectional area of the external surface of the device which is equal to the cross-sectional area for the transdermal system and twice the area of the transdermal system and twice the cross-sectional area for the planar reservoir system.

5.7.6   Diffusion Through Cylindrical Membranes

For implantable reservoir devices, a cylindrical geometry is more practical than a planar arrangement. Consider a cylindrical reservoir surrounded by a polymeric membrane. The cylinder has a length, L, cross-sectional radius b, and a wall thickness b-a. The rate of drug release from this cylinder can be modified by changing the geometry of the device by changing b, b/a or L or by changing the drug/polymer combination. The governing equation;

$$\tau_{mr} \, \partial^2 c/\partial t^2 \; + \; \partial c/\partial t \; = \; D_{ip} \; 1/r \; (\partial/\partial r \; r \partial c/\partial r) \qquad (5.351)$$

If the inside of the cylinder is maintained at constant concentration of drug, $c = c_1$ at $r = a$, and the outside of the cylinder is free of drug, $c = 0$ at $r = b$ and the cylinder wall is initially saturated with drug, $c = c_1$ at $a < r < b$, then Eq. [5.351] can be solved analytically to obtain c as a function of position in the cylinder wall.

$$\text{Let } \tau = t/\tau_{mr}; \quad u = c/c_1; \quad X = r/\text{sqrt}(D_{ip}\,\tau_{mr}) \qquad (5.352)$$

$$\partial^2 u/\partial \tau^2 \; + \; \partial u/\partial \tau \; = \; 1/X \; \partial/\partial X(X \partial u/\partial X) \qquad (5.353)$$

The dimensionless concentration may be assumed to consist of the sum of the steady state and transient parts;

$$\text{Let } u \; = \; u^t \; + u^{ss} \qquad (5.354)$$

$$u^{ss} \quad = c\ln X + c" \qquad (5.355)$$

From the boundary conditions,

$$X = X_a, \; u = 1 \qquad (5.356)$$

$$X = X_b \, , \; u = 0 \qquad (5.357)$$

$$1 \; = \; c' \ln X_a \; + c" \qquad (5.358)$$
$$0 \; = \; c\ln X_b \; + c" \qquad (5.359)$$

$$-1 \; = \; c\ln(b/a) \qquad (5.360)$$

$$c' \quad = \; 1/\ln(a/b) \qquad (5.361)$$

$$c" \; = \; \ln X_b/\ln(b/a) \qquad (5.362)$$

$$u^{ss} \quad = \; \ln(X/X_b \,)/\ln(a/b) \qquad = \ln(r/b)/\ln(a/b) \qquad (5.363)$$

The steady state concentration profile can be seen to be independent of the diffusivity. The total mass of the drug released from the external surface of the cylinder released at time t , M_t is found by first calculating the mass flux at the external surface r = b, and then integrating with respect to time.

$$M_t = 2\pi bL \int_0^t J_{r=b}dt = 2\pi bLtv_mc_1/\ln(b/a) \qquad (5.364)$$

where v_m is the velocity of mass diffusion. Eq. [5.364] is assuming that the mass flux is at steady state during the entire time of consideration. This may be an idealization. For short times the mass flux can expected to be transient. The transient portion of the concentration profile can be obtained as follows;

$$\partial^2u^t/\partial\tau^2 + \partial u^t/\partial\tau = 1/X\,\partial/\partial X(X\partial u^t/\partial X) \qquad (5.365)$$

The boundary conditions are now homogeneous;

$$X = X_a,\ u^t = 0 \qquad (5.366)$$

$$X = X_b,\ u^t = 0 \qquad (5.367)$$

$$\tau = 0,\ u^t = 1 \qquad (5.368)$$

The damping term in Eq. [5.365] can be removed by a u = w exp(-τ/2) to yield:

$$\partial^2w/\partial\tau^2 - w/4 = 1/X\,\partial/\partial X(X\partial w/\partial X) \qquad (5.369)$$

Eq. [5.369] can be solved for by the method of separation of variables;

Let w = V(τ)g(X) $\qquad (5.370)$

$$V''/V - \tfrac{1}{4} = (g'/X + g'')/g = -\lambda_n^2 \qquad (5.371)$$

$$X^2g'' + Xg' + X^2g\lambda_n^2 = 0 \qquad (5.372)$$

$$g = c_1 J_0(\lambda_nX) + c_2Y_0(\lambda_nX) \qquad (5.373)$$

Eq. [5.373] will become infinity at X = 0. The dimensionless concentration is always finite and hence c_2 can be set to zero.

From the boundary condition,

$$X = X_a,\ u^t = 0 \qquad (5.374)$$

$$\lambda_n X_a = 2.4048 + (n-1)\pi \tag{5.375}$$

$$n = 1,2,3\ldots\ldots$$

$$V = c_3 \exp(+\tau \text{ sqrt}(1/4 - \lambda_n^2)) + c_4 \exp(-\tau\text{sqrt}(1/4 - \lambda_n^2)) \tag{5.376}$$

As time becomes infinitely large the transient portion of the solution becomes zero and hence c_3 needs to be set to zero.

Thus,

$$u^t = \sum_1^\infty c_n J_0(\lambda_n X) \exp(-\tau/2)\exp(-\tau\text{sqrt}(1/4 - \lambda_n^2)) \tag{5.377}$$

when λ_n subcritical damped oscillations can be seen;

$$a < 4.8096(D_{ip}\tau_{mr})^{1/2} \tag{5.378}$$

Then using De Movrie's theorem and taking the real parts;

$$u^t = \sum_1^\infty c_n J_0(\lambda_n X) \exp(-\tau/2)\text{Cos}(\tau\text{sqrt}(\lambda_n^2 - 1/4)) \tag{5.379}$$

The c_n can be solved for from the initial condition and using the principle of orthogonality and multiplying the infinite series with $J_0(\lambda_m X)$ and integrating between the limits of X_a and X_b.

$$c_n = \int_{X_a}^{X_b} [[1 - \ln(X/X_b)]/\ln(a/b)]J_0(\lambda_n X)dX / \int_{X_a}^{X_b} J_0^2(\lambda_n X)dX \tag{5.380}$$

It can be seen that the λ_n was solved for from the boundary condition at $X = X_a$. When $\lambda_n X$ is increased from X_a for large b prior to reaching X_b the zero of the Bessel function may arise and the concentration will become zero from that point onwards. This is the critical region beyond which there is no transfer of concentration. In such cases, the total amount of drug released from the external surface will be less than predicted from calculations. As can be seen from the above analysis, the integrand when solving for the total drug released passes from zero time through a transient stage prior to reaching steady state. For common situations, the time t is often times less than within a constant multiple of the time taken to reach steady state. It may contain a zero contribution during the transient stage. Thus for large membrane systems there can be expected a time lag prior to the origin of the external release of the drug. Thus $(b-a)/v_m$ is an estimate of the time taken for the external surface to see the drug from the reservior prior to its release (Figure 5.5) . This time lag can be calculated. At the region where the concentration ceases to move, the $\partial c/\partial r = 0$,

$$c_1 /(D_{ip}\tau_{mr})^{1/2}\partial u^t/\partial X \;=\; \Sigma_1^\infty \; c_n \; \lambda_n \; J_1(\lambda_n X) \; \exp(-\tau/2)Cos(\tau sqrt(\lambda_n^2 - 1/4){=}0 \tag{5.381}$$

$$\text{when } (2.4048 + (n\text{-}1)\pi) \; X_c \;=\; 3.8317 + (n\text{-}1)\pi \tag{5.382}$$

$$\text{or } X_c \;=\; (3.8317 + (n\text{-}1)\pi)/(2.4048 + (n\text{-}1)\pi) \tag{5.383}$$

For n = 1,

When, $X_c > 1.5934$, the first term in the infinite series has to be set to zero,

For n = 2, when the zero of the Bessel occurs at

$X_c > 1.258$ the second term in the infinite series has to be set to zero inorder to keep the predictions for concentration from becoming negative. Thus for $X_c > 1.5934$ or

$$a < \; r_c/(D_{ip}\tau_{mr})^{1/2} < \; b \tag{5.384}$$

where $r_c > 1.5934(D_{ip}\tau_{mr})^{1/2}$ and less than b the concentration will be zero.

5.7.7   Matrix Delivery System

Proteins for instance, do not diffuse readily through any of the hydrophobic biocompatible polymers that are usually used for implantable reservoir systems or they diffuse slowly. Membrane materials cannot be found for some therapeutic agents to provide adequate permeability to permit release from a reservoir device. Matrix system for delivery consists of dissolution and dispersion of drug molecules out a solid polymer phase. Biodegradable materials disappear after implantation. The design of the material and device, can control the rate of polymer degradation and dissolution and the rate of drug delivery. In matrix drug delivery system, molecules of drug are dissolved in a biocompatible polymer, producing a homogeneous device with drug molecules uniformly dispersed throughout the material. The drug molecules are released by diffusing through the polymer to the surface of the device from which they are released into the external environment.

The release of a drug that is initially dissolved within a polymer matrix can be predicted by solving for the damped wave diffusion and relaxation equation within the polymer slab;

$$\partial^2 u/\partial\tau^2 + \partial u/\partial\tau = \partial^2 u/\partial X^2 \tag{5.385}$$

where $u = (c - c_{ext})/(c_0 - c_{ext})$; $\tau = t/\tau_{mr}$; $X = x/(D_{ip}\tau_{mr})^{1/2}$ (5.386)

and c_{ext} is the concentration of the drug in the external reservoir. The damping term can be removed from Eq. [5.385] by a $u = w\exp(-\tau/2)$,

$$\partial^2 w/\partial\tau^2 - w/4 = \partial^2 w/\partial X^2 \tag{5.387}$$

By the method of separation of variables the solution for w can be obtained for the following time and space conditions as shown in chapter 2.0 for transient temperature in a finite slab.

$$u = 1, \quad \tau = 0 \tag{5.388}$$

$$u = 0, \quad X = 0, \ L/(D_{ip}\tau_{mr})^{1/2} \tag{5.389}$$

$$u = 0, \quad \tau = \infty \tag{5.390}$$

$$\partial u/\partial X = 0, \ X = 0 \tag{5.391}$$

A bifurcated solution results. For small width of the slab, $L < 2\pi \ \text{sqrt}(D_{ip}\tau_{mr})$, the transient concentration is subcritical damped oscillatory. An exact well bounded solution that is bifurcated depending on the width of the slab is provided. For $L \geq 2\pi \ \text{sqrt}(D_{ip}\tau_{mr})$,

$$u = \Sigma_1^\infty c_n\exp(-\tau/2) \exp(-\text{sqrt}(1/4 - \lambda_n^2) \ \tau) \ \text{Cos}(\lambda_n X) \tag{5.392}$$

Where $c_n = 4(-1)^{n+1}/(2n-1)\pi$ and $\lambda_n = (2n-1)\pi \ \text{sqrt}(D_{ip}\tau_{mr})/L$ (5.393)

For $L < 2\pi \ \text{sqrt}(D_{ip}\tau_{mr})$,

$$u = \Sigma_1^\infty c_n\exp(-\tau/2) \ \text{Cos}(\text{sqrt}(\lambda_n^2 - 1/4) \ \tau) \ \text{Cos}(\lambda_n X) \tag{5.394}$$

The total amount of drug released from the matrix can be determined by integration;

$$M_t = c_0 AL - \int_{-L/2}^{L/2} c(x,t) \ A \ dx \tag{5.395}$$

$M_t/AL(c_{ext} + c_0) = M^*$

$M^* = 1 - 8C^*/(\pi^2 \, Rel_m)\sum_1^\infty \exp(-\tau/2)/(2n-1)^2 Cos(\tau sqrt(\lambda_n^2 - 1/4))$

(5.396)

Eq. [5.396] is valid for small membranes, when $L < 2\pi \, sqrt(D_{ip}\tau_{mr})$. For large membrane systems, when $L \geq 2\pi \, sqrt(D_{ip}\tau_{mr})$,

$M^* = 1 - 8C^* / (Rel_m\pi^2)\sum_1^\infty \exp(-\tau/2)/(2n-1)^2 \exp(-\tau sqrt(1/4 - \lambda_n^2))$

(5.397)

Where $C^* = (c_0 - c_{ext})/(c_{ext} + c_0)$

Relaxation number (mass), $Rel_m = D_{ip}\tau_{mr}/L^2$ (5.398)

Eq. [5.396] is shown in Figure 5.6. The time taken to steady state or the value at infinite time can be read off from the graph. In this case at $\tau = 0.35$ the M^* reaches 1.

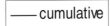

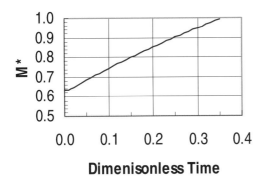

Dimenisonless Time

Figure 5.7 Dimensionless Total Drug Released from Matrix System

This approach to model the release of dispersed or dissolved drugs from a matrix can be easily extended to cylindrical or spherical shapes by solving the gioverning equation in the cylindrical and spherical coordinates. The macroscopic geometry of a matrix can influence the rate and pattern of protein release.

5.8 Krogh Tissue Cylinder

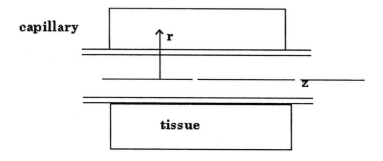

Figure 5.8 Geometry of Krogh Tissue Cylinder

A microscopic view of capillaries in tissue indicates a repetitive arrangement of capillaries surrounded by a cylindrical layer of tissue. A idealized sketch of the capillary bed and the corresponding layer of tissue idealized into a cylinder is shown in Figure 5.7. Let the radius of the tissue layer be r_T. The residence time of the blood in the capillary is in the order of 1 second. The wave diffusion and relaxation time is comparable in magnitude to the residence time in the blood. Krogh [1919] showed this cylindrical capilary tissue model to study the supply of oxygen to muscle. The tissue space surrounding the capillary is considered a continuous phase albeit it consists of discrete cells. An effective diffusivity D_T can be used to represent the diffusion process in the tissue. The driving force for the diffusion is driven by the consumption of the solute by the cells within the tissue space.

The Michaelis-Menten equation can be used to describe the metabolic consumption of the solute in the tissue space. The equation may be written as;

$$R = V_m C_T /(K_m + C_T) \tag{5.399}$$

Where C_T is the concentration of the solute in the tissue space. For consumption of the solute R will have a positive value and for solute production it will have a negative value. V_m represents the maximum reaction rate. The maximum reaction rate occurs when $<C> >> K_m$. The reaction rate is then in zero order in solute concentration. The blood flows through the capillary with an average velocity of V. A steady state shell balance on the solute in the blood from z to $z + \Delta z$ can be written as;

$$-V \, dC/dz = 2/r_c \, K_0 \, (C - C_T|_{rc + tm}) \tag{5.400}$$

where K_o is represented by a overall mass transfer coefficient. The overall mass transfer coefficient represents combined resistance of fluid flowing through the capillary k_m and the permeability of the solute in the capillary wall P_m. A steady state shell balance at a given value of z from r to $r + \Delta r$ may also be written for the solute concentration in the tissue space;

$$D_T/r \, d/dr \, (r \, dC_T/dr) - R = 0 \tag{5.401}$$

The boundary conditions for Eqs. [5.400, 5.401], are;

$$z = 0, \quad C = C_o \tag{5.402}$$

$$r = r_c + t_m, \quad C_T = C_T|_{rc + tm} \tag{5.403}$$

$$r = r_T, \quad dC_T/dr = 0 \tag{5.404}$$

The axial diffusion is neglected in the tissue space in comparison with the radial diffusion. From the zero order rate of reaction $R = R_o$ a constant. Solving for Eq. [5.401] And the boundary conditions given in Eqs. [5.403,5.404],

$$C_T - C_T|_{rc + tm} = (r^2 - (r_c + t_m)^2)R_o/4D_T - r_T{}^2 R_o/2D_T \ln(r/(r_c + t_m)) \tag{5.405}$$

The variation of concentration as a function of z can be calculated by equating the change in solute concentration within the blood to the consumption of solute in the tissue space;

$$C = C_o - R_o/V r_c{}^2 (r_T{}^2 - (r_c + t_m)^2)z \tag{5.406}$$

Eq. [5.406] is combined with Eq. [5.400];

$$C_T|_{rc+tm} - C_0 = \quad - R_o/Vr_c^2 \, (r_T^2 - (r_c + t_m)^2)z \quad - R_o/2r_cK_o \, (r_T^2 - (r_c + t_m)^2)$$

$$(5.407)$$

Combining Eqs.[5.405-5.407],

$$C_T - C_0 = (r^2 - (r_c + t_m)^2)R_o/4D_T - r_T^2R_o/2D_T \, \ln(r/(r_c + t_m))$$
$$- R_o/Vr_c^2 \, (r_T^2 - (r_c + t_m)^2)z \quad - R_o/2r_cK_o \, (r_T^2 - (r_c + t_m)^2)$$

$$(5.408)$$

It can be deduced that under certain conditions some regions may not receive any solute. A critical radius of tissue can be idenfied, $r_{critical}$ and defined as the distance beyond which no solute is present in the tissue.

$$\text{At } r = r_{critical}, \quad dC_T/dr = 0 \text{ and } C_T = 0 \qquad (5.409)$$

This can be solved for from Eq. [5.408] after replacing r_T with $r_{critical}$. The equation is nonlinear.

Worked Example 5.7.2

Idealize Figure 5.7 in the cartesian coordinates and obtain the solution for the concentration of the solute in the tissue space.

The governing equations for the concentration of the solute in the capillary and in the tissue can be written after taking the r in Figure 5.7 as x,

$$-V \, dC/dz = 2/r_c \, K_o \, (C - C_T|_{rc+tm}) \qquad (5.410)$$

Considering the effects of diffusion in x direction only in the tissue and assuming a zeroth order reaction rate

$$D_{AB}\partial^2 C_T/\partial x^2 = R_o \qquad (5.411)$$

Integrating, and substituting for the boundary conditions ;

$$x = x_c + t_m , \quad C_T = C_T|_{xc+tm} \qquad (5.413)$$

$$x = x_T, \qquad dC_T/dx = 0 \qquad (5.414)$$

$$-R_o/D_{AB}x_T = c_1 \qquad (5.415)$$

$C_T - C_T|_{x_c + tm} = (R_o/2D_{AB})(x^2 - (x_c + t_m)^2) - R_o x_T /D_{AB} (x - (x_c + t_m)))$ (5.416)

The variation of concentration as a function of z can be calculated by equating the change in solute concentration within the blood to the consumption of solute in the tissue space;

$$V A C_0 - V A C = R_o z A_T$$ (5.417)

$$C = C_0 - R_o z A_T/VA$$ (5.418)

Eq. [5.418] is combined with Eq. [5.410];

$$R_o A_T/A = 2/x_c K_0 (C - C_T|_{rc + tm})$$ (5.419)

$$C_T|_{rc + tm} = C - K_0 x_c R_o A_T/2A$$ (5.420)

Therefore,

$C_T - C_0 = R_o z A_T/VA + K_0 x_c R_o A_T/2A + (R_o/2D_{AB})(x^2 - (x_c + t_m)^2) - R_o x_T /D_{AB} (x - (x_c + t_m))$
(5.421)

At a critical distance from the capillary wall the concentration in the solute will become zero. This can be solved for from the above equation.

At and beyond the critical distance,

$$dC_T/dx = 0 = C_T$$ (5.422)

Replacing x_T with $x_{critical}$,

$0 = C_0 + R_o z A_T/VA + K_0 x_c R_o A_T/2A + (R_o/2D_{AB})(x^2 - (x_c + t_m)^2) - R_o x_{critical} /D_{AB} (x - (x_c + t_m))$
(5.423)

$x_{critical}^2 (-R_o/2D_{AB}) = C_0 + R_o z A_T/VA + K_0 x_c R_o A_T/2A - (R_o/2D_{AB}) (x_c + t_m)^2 - R_o x_{critical} /D_{AB} (x - (x_c + t_m))$
(5.424)

The quadratric equation in $x_{critical}$ is then,

$$A x_{critical}^2 + B x_{critical} + C = 0$$ (5.425)

Where,

$$A = -(R_o/2D_{AB})$$ (5.426)

$$B = + (x_c + t_m) R_o/D_{AB}$$ (5.427)

$$C = C_0 + R_ozA_T/VA + K_{0}x_c\ R_oA_T/2A - (R_o/2D_{AB})x_c + t_m)^2 + R_o(x_c + t_m)/D_{AB}$$

$$(5.428)$$

When the solution of the quadratic expression for the critical distance in the tissue are real, and found to be less than the thickness of the tissue then the onset of zero concentration will occur prior to the periphery of the tissue. This zone can be seen as the anaroxic or oxygen depleted regions in the tissue.

Summary

The heat transfer to a suspension moving away from a hot tube and moving toward a hot tube was studied using the damped wave heat conduction and relaxation equation. The relative velocity of the suspension and direction compred with the heat transfer path leads to a positive contribution from the bulk flow convective term for the flow of suspension away from and a negative contribution when the flow direction is opposite to the heat tranfer path as is the case when the cold suspension moves toward the hot tube. The method of Laplace transforms were used and the binomial expansion used and then the term by term inversion using residue theorem was obtained. The results are provided in a Table for 11 terms. The exact solution is also obtained. The two cases are added with each other and divided by two to provide the conduction and relaxation contributions to the heat transfer coefficient at the surface to the bed. When the two cases are subtracted and divided by two the convective contribution resuls. Some experimental data was used to support the theory from the literature.

With the increased speed of microprocessors the CPU overheating may be a increasing problem as the power consumed reached 100 w and speeds over 2 GHz. The thermal management of the CPU involves use of efficient heat sink devices. A study of the transient heat conduction under a periodic boundary condition was studied using the Fourier parabolic heat equation, damped wave conduction and relaxation hyperbolic PDE and a dual phase lag equation. The storage coefficient $S = (\rho C_p/\tau_r)$, with units W/m^3 K in addition to the thermal conductivity is a important parameter in the design of the substrate. The thermal relaxation time, τ_r, participates in the attenuation and thermal lag of the process. Under certain conditions, the heat flux may reverse in direction and backflow to the device causing harm.

Helical ribbon agitators are used in the pilot plant to scale-up continuous polymerization process to manufacture ABS from the bench unit to the manufacturing. The bulk flow and the autocatalytic free radical grafting polymerization reactions were studied and a cosinuous concentration profile was derived under certain conditions.

Magnesium oxychloride cement coatings are used to impart flame retardancy. Naval warheads can be coated with the sorrel cement to delay fire spread during a accident. The transient temperature in the film coating using the damped wave conduction and relaxation equation was studied. The asymptotic limits at infinite relaxation time and the infinite velocity of heat and zero thermal diffusivity were derived. The method of separation of variables was used to obtain the exact solution of the transient temperature in the film. A critical thickness below which subcritical damped oscillations can be seen was derived. A thickness above which the payoff is small can also be seen. Hence there exists a optimal thickness of coating. ASTM test UL 1709 procedures were followed and the time taken to alarm was found.

An infinite order PDE was solved for in VLSI, silicon doping application to obtain the transient concentration profiles. Two approaches were taken. One is a infinite series sum of Laplace transformed terms. The other is a expansion of terms and term by term inversion from the tables obtained.

The pressure fluctuations in gas-solid fluidized beds was attempted to be interpreted using the damped wave momentum transfer and relaxation equation. The bubble events drives the major frequency of pressure fluctuations. When the superficial gas velocity in the cylindrical passages during one bubble cycle goes from plug/uniform to prabolic. As the dimensions of the void path are small, from the method of separation of variables, subcritical damped oscillations in the velocity can be seen when d_p < 14.4288 sqrt($\nu\tau_{mom}$)(1 - ε)/($\varphi_s\varepsilon$). The frequencies of pulsations can be given by; F_n = sqrt($\nu\tau_{mom}$) (4.8096 +(n-1)π)^2/R_c^2 -1/4) , n = 1,2, 3... . These pulsations translate to pressure pulsations by the relation between velocity of flow to the pressure drop required. As the bubble events drives the transience these frequencies are secondary to the primary bubble frequency.

Controlled drug delivery system offers a better alternative compared with the periodic does. In the periodic dosage the toxic levels are exceeded and the minimal level required is not attained some of the time. The response to a pulse of mass flux was derived in cartesian coordinates. The transient concentration profile in cylindirical and spherical coordinates were obtained using the damped wave diffusion and relaxation equation. The solute binding and elimination was studied using a reversible rate of reaction and a effective diffusivity. The mass inertial lag time associated with the propagation of concentration disturbance was calculated in cylindrical and spherical coordiantes. Protein profile in maternal effect genes, drug penetration in the tissue and neurotransmission across the synaptic cleft using the damped wave diffusion and relaxation equation was studied. Enzymatic reactions are diffusion limited and their schemes were studied. The overall mechanism is represented by reversible reactions. Reservoir delivery systems were modeled. The cumulative mass transported across the membrane was calculated. The diffusion in cylindrical membranes were considered. Transdermal systems were mentioned. The drug released from a matrix delivery system was derived and shown as a Figure.

The Krogh tissue cylinder was modeled. At steady state for the cylindrical geometry a critical radius beyond which zero mass diffusion can be calculated was derived. The governing equation requires numerical solution. In cartesian coordinates, the quadratic equation to represent the zone of zero concentration or anaroxic or oxygen depleted region was derived. A $x_{critical}^2$ + B$x_{critical}$ + C = 0 where A, B and C are given in terms of the thickness of the capillary, capillary diameter, tissue diameter, binary diffusivity, entrance concentration and area of the tissue.

Exercises

1. *Wilting of the Lettuce*

   Lettuce leaves in a salad wilt. The process of wilting is accelerated if the letuce is salted. The water droplets on the surface of the leaves comes from the interior of the lettuce plant cells. Consequently the turgor pressure and internal rigidity of the leaves is lowered and they wilt. The process of water transport out of the cells caused by increase in external salt concentration is an example of osmosis. Dutrochet made systematic observations of osmotic pressure in the 1800s. He observed that (Weiss, 1996) small animal bladders filled with dense solution and completely closed and plunged in water became turgid and swollen excessively. Water flowed into the bladder so as to dilute the solution inside. Van't Hoff noted that the osmotic pressure was proportional to the product of the solute concentration and the absolute temperature with a constant of proportinality that equalled the molar gas constant R. The Darcy's law provied the solvent flux as a function of the pressure gradient and the constant of proportaionality called hydraulic permeability. Darcy's law can be generalized to include both hydraulic and osmotic pressures in a porous medium and the accumulation effects can also be included to give an expression for the solvent flux as;

   $$J_{solv} = -\kappa\, \partial(p - \pi)/\partial x - \tau_{pr}\, \partial J_{solv}/\partial t$$

   Where, J_{solv} is the solvent flux, τ_{pr} is the relaxation time of pressure, κ is the hydraulic permeability and π is the osmotic pressure that can be written as RTC_{sol}, where C_{sol} is the solute concentration. For the wilting of the lettuce show that the governing equations can be written asuming the salt can permeate through the lettuce as neglecting the hydraulic pressure gradient as,

   $$\tau_{mr}\partial^2 C_{solv}/\partial t^2 + \partial C_{solv}/\partial t = \kappa RT\partial^2 C_{solv}/\partial x^2$$

   with the following space and time conditions,

   $$x = \delta,\ C_{solv} = 0$$

   $$x = 0,\ C_{solv} = C_{solv0}$$

   $$t = 0,\ C_{solv} = 0$$

Show that if the lettuce is semi-permeable membrane, at steady state, the solvent transport can be given by;

$$p - p_0 - (\pi - \pi_0) = -J_{solv}/\kappa (x - x_0)$$

where x_0 is the reference location at which the hydraulic and osmotic pressures are known.

2. *Restriction Mapping*

Endonucleases or restirction enzymes cut the unmethylated DNA at several sites and restrict their activity. About 300 restriction enzymes are known and they act upon 100 distinct restriction sites that are palindromes. Sme cut leaves blunt ends and others leave them sticky. The restriction fragment lengths can be measured by using the technique of gel electrophoresis. The solid matrix is the gel usually agarose or polyacrylamide which is permeated with liquid buffer. As DNA is a negatively charged molecule when placed in a electric field the DNA Migrates toward the positive pole. DNA migration is a function of its size. Calibration is used to relate the migration distance as a function of size. migration distance of DNA under a field for a set time is measured. The DNA molecule is made to fluoresce and made visible under ultraviolet light by staining the gel with ethidium bromide. A second method is to tag the DNA with a radio active label and then to expose the X-ray film to the gel. The migratrion under gel electrophoresis is by a process of damped wave diffusion and relaxation. The migration rate under electrophoresis can be given by;

$$J_{frag} = -(z_A u_A F)^* \partial E/\partial x - D_{frag} \partial C_A/\partial x - \tau_{mr} \partial J_{frag}/\partial t$$

Show that the governing equation can be written in 1 dimension as;

$$\tau_{mr} \partial^2 C_A/\partial t + \partial C_A/\partial t = D_{frag} \partial^2 C_A/\partial x^2 + -(z_A u_A F)^* \partial^2 E/\partial x^2$$

3. *Relativistic Transformation of Coordinates*

The governing equation in damped wave transport and relaxation can be written in 1 dimension as a hyperbolic PDE in two variables as;

$$\partial^2 u/\partial \tau^2 + \partial u/\partial \tau = \partial^2 u/\partial X^2$$

Rewrite the governing equation as;

$$\partial u/\partial \tau = \partial^2 u/\partial X^2 - \partial^2 u/\partial \tau^2$$

Letting $\eta = X^2 - \tau^2$, for $X > \tau$, transform only the RHS of the above equation.

$$\partial u/\partial \tau = 4\eta \partial^2 u/\partial \eta^2 + 4\partial u/\partial \eta$$

Use the method of separation of variables to solve the above equation for non-zero values of X. Show the solution can be written as;

$$u = c'\exp(-c^2\tau)J_0c(X^2 - \tau^2)^{1/2}$$

Choose $c^2 = \frac{1}{2}$ and solve for the c' from the condition at the wave front, $u = \exp(-\tau/2)$ and show that this is equal to 1. Confirm that using this analysis that the u $= w\exp(-\tau/2)$ substitution. Further for $\tau > X$, show that the solution can be written as;

$$u = \exp(-\tau/2)I_0c(X^2 - \tau^2)^{1/2}$$

4.   *Krebs Cycle and Reactions in Circle*

The kinetics of reactions of a class such as those involved in Krebs cycle metabolic pathways and reversible terpolymerization was proposed to be treated as a system of reactions in circle. In Krebs cycle the oxalacetic acid (A) is converted, to citric acid (B), to isocitric acid (C), to α - ketogluaric acid (D), to succyl coA (E), succinic acid (F), to fumaric acid (G), to maleic acid (H) and back to A, oxalaacetic acid. Wherever, the reactant is consumed as well as generated with several intermediate steps the kinetic of the system of reactions need be tested for a Brusselator. The thermodynamic validity of such kinetics is also of concern as the system entropy of real processes always increases. Damped oscillatory kinetics was shown to obey the Clausius inequality.

The kinetics of a system of 3 reactants in circle canbe written, assuming to obey first order kinetics could be written as;

$$\begin{aligned}
d\,C_A/dt &= - k_1\,C_A + k_3\,C_C \\
d\,C_B/dt &= - k_2\,C_B + k_1\,C_A \\
d\,C_C/dt &= - k_3\,C_C + k_2\,C_B
\end{aligned}$$

where C_A, C_B, C_C are the concentrations of reactants A, B and C at a given instant t. If the initial concentration of A is given by A_0 and that of B and C is zero, and obtaining the Laplace transforms of above Eqs. It was shown by Sharma [2003] that,

or $\underline{C_A}$ $= A_0\,(s + k_3\,)(s + k_2)\, /\,s(\,s^2 + s\,(k_3 + k_2 + k_1) + k_3\,k_2 + k_1\,k_3 + k_2\,k_1\,)$

The above Eq. can be inverted using the residue theorem. When the poles of the denominator of above Eq. is complex damped oscillatory kinetics can be expected. Thus when $b^2 - 4\ ac < 0$ or when,

or $k_3 < (sqrt(k_2) + sqrt(k_1))^2$

This expression is symmetrical with respect to reactants 1, 2 and 3 . When the relation holds good for 1 reaction rate constant as less than the square of the sum of the square root of the rate constants of the other two reactions the damped oscillatory kinetics can be expected.

In order to obtain the concentration profiles as a function of time of the reactants   A, B and C the kinetic equations of A, B and C are decoupled by differentiating it again with respect to time. Thus;

$$d^2C_A/dt^2 + \alpha\ dC_A/dt + \gamma\ C_A = A_0\ k_2\ k_3$$

$$\alpha = k_1 + k_2 + k_3;\ \gamma = k_1k_2 + k_3\ k_1 + k_2k_3$$

In a similar manner the equations for B and C can be written. The auxiliary quadratic equation for   the above ODE   with constant coefficients for the homogeneous part of the solution is solved for the particular integral combined to give the general solution for the three reactants with time as;

$$C_A = C_1 \exp(-\alpha t/2) \exp(tsqrt(\alpha^2 - 4\gamma)) + C_2 \exp(-\alpha t/2) \exp(-tsqrt(\alpha^2 - 4\gamma)) + C'$$

$$C_B = C_3 \exp(-\alpha t/2) \exp(tsqrt(\alpha^2 - 4\gamma)) + C_4 \exp(-\alpha t/2) \exp(-tsqrt(\alpha^2 - 4\gamma)) + C''$$

$$C_C = C_5 \exp(-\alpha t/2) \exp(tsqrt(\alpha^2 - 4\gamma)) + C_6 \exp(-\alpha t/2) \exp(-tsqrt(\alpha^2 - 4\gamma)) + C'''$$

Differentiating the above Eqs. with respect to time and multiplying both sides by $\exp(\alpha t/2)$ it can be seen that at infinite time the LHS becomes zero as the reaction rate drops to zero. In order for RHS also to be zero, C_1 , C_3, C_5 need be set to zero. Then at infinite time the concentrations of A, B and C adds to A_0. As the reactions are symmetrical and simple first order a good approximation is that;

$$C' = C'' = C''' = A_0/3$$

The remaining three constants are solved for from the initial concentrations for A, B and C as A_0, 0 and 0. Thus for $\alpha^2 > 4\gamma$,

$$C_A = A_0/3 (1 + 2 \exp(-\alpha t/2) \exp(-t/2sqrt(\alpha^2 - 4\gamma)))$$

$$C_B = A_0/3 (1 - \exp(-\alpha t/2) \exp(-t/2sqrt(\alpha^2 - 4\gamma))) = C_C$$

The solution is bifurcated and for values of $\alpha^2 < 4\gamma$, the argument within the square root sign in above Eqs. become imaginary and using the De Movrie's theorem and equating the real parts;

$$C_A = A_0/3 \left(1 + 2 \exp(-\alpha t/2) \cos\left(t/2\sqrt{4\gamma - \alpha^2}\right) \right)$$

$$C_B = A_0/3 \left(1 - \exp(-\alpha t/2) \cos\left(t/2\sqrt{4\gamma - \alpha^2}\right) \right) = C_C$$

Thus for certain values of the rate constants the concentration becomes subcritical damped oscillatory. With a oscillatory reacting species show that the governing equation for the simulataneous damped wave diffusion and relaxation and reaction can be written as;

$$\tau_{mr} \partial^2 C_A/\partial t^2 + \partial C_A/\partial t - k'''\tau_{mr}\gamma C_0/3 \sin\gamma t + k'''C_0/3 \cos\gamma t = D\partial^2 C/\partial x^2$$

5. In section 3.1, 3.2, 3.3 the method of relativistic transformation of coordinates was used to derive the spatio-temporal concentration in a semi-infinite medium, infinite cylinder and infinite sphere. Use the transformation $\eta = X^2 - \tau^2$. Show that the Bessel differential equation obtained for $X > \tau$, results in a Bessel function solution. Combined with the boundary condition relation in terms of the modified Bessel function show that the results shown in the sections can be derived for $X > \tau$.

6. In the manufacture of HIPS, high impact polystyrene, ABS acrylonitrile butadiene and styrene using continuous polymerization technology the FSD, falling strand devolatilizer is used to remove the unreacted monomers and diluent from the reactor syrups prior to being recyled back to the reactors. During the residence time of 1 hour in the two stage FSD at 240 ^{0} C, calculate the mass flux from the falling strands of polymer and from this obtain the devolatilization efficiency. The vaccum in the second stage DV is 5 torr. The unreacted monomers are styrene, acrylonitrile and diluent is ethylbenzene. Use the damped wave diffusion and relaxation equation to obtain the spatio-temporal concentration of the monomers and diluent in the vapor phase. The interface concentration can be taken as the saturated case and the bulk vapor composition can be calculated from the vaccum pressure. The damped wave diffusion and relaxation equation may be used.

A second contribution can be calculated from the surface of the syrups residing in the second stage of the DV, devolatilzer. Compare the results with that obtained using the parabolic diffusion equations.

7. The simultaneous grafting reaction with an autocatalytic effect along with diffusion and bulk viscous flow effects in a close clearance Helical Ribbon Agitator may be

modeled using parabolic diffusion equations.   Show that the exact solution is bifurcated. Show that when the bulk flow velocity, V_z < 2 D_{gs} /h sqrt(1 + k''' h^2/D_{gs}), the concentration profile in the spatial domain changes in character from decaying exponential to cosinous.   The non-homogeneity in the boundary condition can be accounted for assuming a steady state and a transient part to the solution and superposing them.

During the manufacture of engineering thermoplastic such as ABS, (Acrylonitrile Butadiene Styrene) and HIPS (High Impact Polystyrene) often times the CSTR configuration is used in the continuous polymerization of the monomers and grafting onto polybutadiene. High viscous polymer syrups were found to be well agitated by the helical ribbon agitator. HRAs were found to exhibit good mixing characteristics. The circulatory pattern of down flow near the wall and up flow at the center can be observed. The initiated free radical reactions including grafting onto polybutadiene and matrix formation occurs during the flow of high viscous flow through close clearances. The problem of simultaneous diffusion and reaction during laminar flow canm be studied. Explore the problem of diffusion and convective bulk flow effects on the scale-up of initiation and grafting steps. The initiation using peroxy initiators may be described by a simple first order rate expression. The grafting reaction can take the form of an autocatalytic effect. Examined these reactions using the parabolic equations of diffusion. Derive the conditions were the concentration will exhibit pulsations in the space domain. This can cause non-homogenous grafting or poor grafting sometimes leading to reactor build-up.

The mass balance on the clearance channel close to the wall of the HRA in transient condition including the bulk convective flow effects can be written down as;

$$\partial C_g/\partial t - V_z \partial C_g /\partial z = D_{gs}\partial^2 C_A/\partial z^2 + k''' C_g$$

The bulk flow is assumed to be in the opposite direction of the diffusion. The grafting of growing SAN chains can have an autocatalytic effect on the grafting rate. This is denoted using a plus sign in the reaction rate term and represents the product rate.

Let, $\tau = D_{gs} t /h^2$; u = C_g/C_{gs};   X = z/h

$Pe_m = V_z h / D_{gs}$; $\theta_{th} = k''' h/ D_{gs}$

Where, Pe_m is the Peclet number, mass and denotes the ratio of the convective flow to the diffusion normalized with the HRA clearance width and the θ_{th} is the Thiele modulus that denoted the rate of grafting to the rate of diffusion, non-dimensionlaized using the flow width. In the dimensionless form, the governing equation becomes,

$$\partial u\partial \tau - Pe_m \partial u/\partial X - \theta_{th}^2 u = \partial^2 u/\partial X^2$$

The initial and boundary conditions can be written as;

$\tau = 0$, $u = 0$

$\tau > 0$, $Z = 0$, $u = 1$

$\qquad Z = \infty$, $u = 0$

Where H is the height of the agitator.

Suppose that the solution be expressed as a sum of transient and steady state parts;

$u = u^t + u^s$

The non-homogeneity in the boundary condition can be lumped into the steady state part. Thus,

$-Pe_m \, \partial \, u^s / \partial X - \theta_{th}^2 \, u^s \qquad = \partial^2 u^s / \partial X^2$

$Z = 0$, $u^s = 1$

$Z = H$, $u^s = 0$

$\partial \, u^t / \partial \tau - Pe_m \, \partial \, u^t / \partial X - \theta_{th}^2 \, u^t = \partial^2 u^t / \partial X^2$

$Z = 0$, $u^t = 0$

$Z = \infty$, $u^t = 0$

The transient portion of the solution can be obtained by the method of separation of variables;

Let $u^t = V(\tau) \, \phi \, (X)$

Then Eq.[12] becomes;

$V' \, \phi - Pe_m \, V \, \phi' - \theta_{th}^2 \, V \, \phi = V \, \phi''$

Dividing throughout the above equation by $V\phi$

$V'/V - Pe_m \, \phi'/\phi - \theta_{th}^2 = \phi''/\phi$

$V'/V = \phi''/\phi \qquad + Pe_m \, \phi'/\phi + \theta_{th}^2 \qquad = -\lambda_n^2$

$V = c_1 \exp(-\lambda_n^2 \tau)$

The space domain portion of the solution is then;

$$\phi" \quad + Pe_m \, \phi' + (\theta_{th}^2 + \lambda_n^2)\phi = 0$$

$$\phi = c_2 \, exp\text{-}X(Pe_m/2) \, exp\text{-}X/2sqrt((Pe_m^2 - 4(\theta_{th}^2 + \lambda_n^2)))$$

From the boundary condition it can be seen that c_3 is zero. $-\lambda_n$ can be an arbitrary constant and can be set to 1 for purposes of further analysis. Th e1 lets the transient solution be no different from most other transient diffusion problems using Fick's second law of diffusion. When

$Pe_m < 2 \, sqrt(\theta_{th}^2 + 1)$ the character of the solution in the governing equation is bifurcated;

Or when $V_z < 2 \, D_{gs} /h \, sqrt(1 + k''' \, h^2/D_{gs})$ the solution becomes ;

$$\phi = c_2 \, exp\text{-}X(Pe_m/2) \, Cos(X/2sqrt((\text{-}Pe_m^2 + 4\theta_{th}^2 + 4)))$$

Hence the solution is bifurcated. From the boundary condition the c_2 is seen to be 1. The general solution can be written for $Pe_m < 2 \, sqrt(\theta_{th}^2 + 1)$ as;

$$u = c_1 \, exp(-\lambda_n^2 \tau) + exp\text{-}X(Pe_m/2) \, exp\text{-}X/2sqrt((Pe_m^2 - 4(\theta_{th}^2 + 1)))$$

From the initial condition then,

$$u = (1 - exp(-\tau))(\, expX(-Pe_m/2) \, (exp\text{-}X/2sqrt((Pe_m^2 - 4(\theta_{th}^2 + 1)))))$$

$Pe_m > 2 \, sqrt(\theta_{th}^2 + 1)$,

$$u = (1 - exp(-\tau))(\, exp\text{-}X(Pe_m/2) \, (Cos(X/2sqrt((\, 4\theta_{th}^2 + 4\text{-}Pe_m^2)))))$$

For the case of the bulk flow along the direction of diffusion the solution remain except for a + sign on the $exp(XPe_m/2)$ term.

8. Repeat the analysis of the above problem of simultaneous autocatalytic reaction, convective flow in Helical Ribbon Agitators, HRA along with damped wave diffusion and relaxation equation.

9. Since the first introduction of permselective membranes by Monsanto in 1977 to separate hydrogen from a stream of gases there is interest in this area. Separationf of hydrogen, carbon dioxide, hydrogen sulfide and nitrogen can be effected. Liquid separations of amines, organic acids from esterification batches are possible. Membrane separations offer advantages of lower energy compared with the distillation, and is flexible and offers lower equipment cost and easier to handle. Permselectivity is the preferred diffusion or permeation of one molecule

in the membrane often times made of polymer, compared with other molecules over a said length the thickness of the membrane in a given time. Membranes are judged by the permselectivity parameter. It offers information on selectivity during separation. The permeability across a membrane of thickness, a can be written as;

$$J = D\,H/aRTc\,(\,p_{10} - p_{11})$$

Where H is the partition coefficient, p_{10} and p_{11} refers to the feed and permeate Sides respectively. Obtain a expression for the transient permeabilty and permselectivity using the damped ave diffusion and relaxation equation. Obtain the transient concentration profile for a finite length membrane for the two species of interest. The surface flux expressions can be obtained. The ratio of the surface flux can be defined as the permselectivity. Discuss the thickness of the film as a parameter. Show that for membrane thickness less than 2π sqrt($D\tau_r$) the concentration will exhibit subcritical damped oscillations. When the concentration undergoes subcritcal damped oscillations, what happens to the permeability and permselectivity. What happens when the flux is an exponential decay ? Can the mebrane separation unit be operated when the membrane is in transient mode. Does the membrane have to be made greater than a certain thickness ?

10. Consider the diffusion of a species 1 that moves not only because of the driving force of its concentration gradient but also because of the concentration gradient of another species. As a first approximation the term describing the electrophoretic mobility is not included in the governing equation. The field is a function of space and independent of the concentration of the species. Thus for two species the coupled diffusion equations for two species are as follows using the Fick's law of diffusion;

$$J_1 = -D_{11}\,\partial\,C_1/\partial x - D_{12}\,\partial\,C_2/\partial x$$

$$J_2 = -D_{22}\partial\,C_2/\partial x - D_{21}\partial C_1/\partial x$$

For 1 dimensional diffusion, the accumulation term can be equated as follows by a mass balance;

$$-\partial\,J_1/\partial x = \partial\,C_1/\partial t$$

$$-\partial\,J_2/\partial x = \partial\,C_2/\partial t$$

Eliminating J_1, J_2 between the mass balance and constitutive equations,

$$\partial\,C_1/\partial t = D_{11}\,\partial^2\,C_1/\partial x^2 + D_{12}\,\partial^2\,C_2/\partial x^2$$

$$\partial\,C_2/\partial t = D_{22}\,\partial^2\,C_2/\partial x^2 + D_{21}\,\partial^2\,C_1/\partial x^2$$

C_2 can be eliminated between equations (5) and (6) by double differentiation of Eqs. [5-6] wrt x,

$$\partial^3 C_1/\partial t\, \partial x^2 \quad = D_{11}\, \partial^4 C_1/\partial x^4 + D_{12}\, \partial^4 C_2/\partial x^4$$

$$\partial^3 C_2/\partial t\, \partial x^2 \quad = D_{22}\, \partial^4 C_2/\partial x^4 + D_{21}\, \partial^4 C_1/\partial x^4$$

$$\partial^3 C_1/\partial t\, \partial x^2 = D_{11}\, \partial^4 C_1/\partial x^4 + D_{12}/D_{22}\ (\partial^3 C_2/\partial t\, \partial x^2 - D_{21}/\, \partial^4 C_1/\partial x^4)$$

Now

$$\partial^3 C_2/\partial t\, \partial x^2 \quad = 1/D_{12}\, \partial^2 C_1/\partial t^2 - D_{11}/D_{12}\, \partial^3 C_1/\partial x^2 \partial t$$

Thus;

$$\partial^3 C_1/\partial t\, \partial x^2 = D_{11}\, \partial^4 C_1/\partial x^4 + D_{12}/D_{22}\ (1/D_{12}\, \partial^2 C_1/\partial t^2 - D_{11}/D_{12}\, \partial^3 C_1/\partial x^2 \partial t$$
$$- D_{21}/\, \partial^4 C_1/\partial x^4)$$

$$(1+ D_{11}/D_{22})\, \partial^3 C_1/\partial t\, \partial x^2 = (D_{11} - D_{21})\partial^4 C_1/\partial x^4 + 1/D_{22}\, \partial^2 C_1/\partial t^2$$

For the special case when $D_{11} = D_{21}$, the solution can be shown to be a error function solution with a composite variable of space and time. For other cases, the governing equation becomes;

$$\alpha\, \partial^3 C_1/\partial t\, \partial x^2 = \beta\ \partial^4 C_1/\partial x^4 + \gamma\, \partial^2 C_1/\partial t^2$$

where

$$(1+ D_{11}/D_{22}) = \alpha\ ; \quad = (D_{11} - D_{21}) = \beta\ ; \quad 1/D_{22} = \gamma$$

Solve the governing equation by the method of separation of variables. The governing equation is a fourth order partial differential equation in two variables. Six boundary conditions and initial conditions are needed to fully describe the problem. The two species values at the two boundaries make four of these conditions and the initial and final condition makes the 5th and 6th conditions. The conditions were the concentration exhibits subcritical damped oscillatory behavior is proposed to be identified. Obtain the exact solutions in various regimes for the governing equation.

11. Repeat the analysis in Problem 10, using the damped wave diffusion and relaxation equation as the constitutive law.

Sources, References and Further Reading

[1] Abramowitz, M, and Stegun, I. A., 1965, *Handbook of Mathematical Functions,* Dover, New York, NY, USA

[2] Alberts, B., Bray, D., Lewis, J., Raff, M, Roberts, K., and Watson, J. D., 1989, *The Cell,* Garland Publishing, New York, NY, USA

[3] Alberts, B., Bray, D., Lewis, J., Raff, M, Roberts, K., and Watson, J. D., 1989, *Molecular Biology of the Cell,* Garland Publishing, New York, NY, USA

[4] Alder, B. J., and Wainright, T., 1969, *Molecular Dynamics by Electronic Computers, in Transport Processes in Statistical Mechanics,* Interscience, New York, NY, USA

[5] Andrews, L. C., 1985, *Special Functions for Engineers and Applied Mathematicians,* Macmillan, New York, NY, USA

[6] Archimides, 287 BC – 212 BC, Buoyancy

[7] Aris, R., 1962, *Vectors, Tensors, and the Basic Equations of Fluid Mechanics,* Prentice Hall, Englewood Cliffs, NJ, USA

[8] Aris, R., 1978, *Mathematical Modeling Techniques,* Pitman, London, UK

[9] Aris, R., 1975, *The Mathematical Theory of Diffusion and Reaction in Permeable Catalysis,* 1, Clarendon Press, Oxford, UK

[10] Arpaci, V. S., *Conduction Heat Transfer,* Addison Wesley, Reading, MA, USA

[11] Ans Magna, 1501-1576, Renaissance Mathematics

[12] Antaki, P. J., 1998, Int. J. Heat and Mass Transfer, **41**, 14, 2253

[13] Baker, R. W., 1987, *Controlled Release of Biologically Active Agents,* Wiley, New York, NY, USA

[14] Barbee J. H., and Cokelet, G. R., 1971, *Prediction of Blood Flow in Tubes with Diameters as Small as 29 μm,* Microvasc. Res., **3,** 17

[15] Barbee J. H., and Cokelet, G. R., 1971b, *The Fahraeus Effect,* Microvasc. Res., **3,** 16

[16] Bateman, H., 1964, *Partial Differential Equations of Mathematical Physics,* Cambridge University Press, Cambridge, UK

[17] Bai, C. and Lavine, A. S., 1991, *Hyperbolic Heat Conduction in a Superconducting Film,* ASME/JSME Thermal Engineering Joint Conference, Reno, NV, USA

[18] Bai, C. and Lavine, A. S., 1995, <u>ASME, J of Heat Transfer,</u> **117,** 256-263

[19] Barletta, A. and Zanchini, E., 1997, <u>Int. J. Heat and Mass Transfer,</u> **40,** 5, 1007-1016

[20] Barnes, and Walters, 1985, <u>Rheological Acta,</u> 323

[21] Baskakov A. P., and Suprun, V. M., 1972, <u>Ind. Eng. Chem.,</u> 12, 324

[22] Baskakov A. P, Tuponogov, V. G. and Filippovsky, N. P., 1986, <u>Powder Technology,</u> **45,** 113-117

[23] Batchelor, G. K., 1967, *An Introduction to Fluid Mechanics,* Cambridge University Press, UK

[24] Baumeister, K. J., and Hamill, T. D., 1971, <u>ASME J of Heat Transfer,</u> **93,** 126-128

[25] Baumeister, K. J., and Hamill, T. D., 1969, <u>ASME J of Heat Transfer,</u> **91,** 543-548

[26] Bejan, A., 1988, *Advanced Engineering Thermodynamics,* John Wiley, New York, NY, USA

[27] Bellman, R. E., and Roth R. S., 1984, *The Laplace Transform,* World Scientific Publishing, Singapore

[28] Benard, H., 1901, <u>Annales de Chimie et appliquees,</u> **23,** 62-144

[29] Berg, H. C., 1983, *Random Walks in Biology,* Princeton University Press, Princeton, NJ, USA

[30] Berger, C., et al., 1998, <u>FEBS Letters,</u> **425,** 14-18

[31] Bernoulli, D., 1738, *Hydrodynamica,* Dulsecker, Strousbourg, Germany

[32] Biot, A., 1970, *Variational Principles in Heat Transfer,* Oxford University Press, UK

[33] Bird, R. B., Da, G. C., and Yaruso, B. J., 1983, *Reviews in Chemical Engineering,* 1,1

[34] Bird, R. B., Stewart, W. E. and Lightfoot E. N., 1960, *Transport Phenomena,*

John Wiley, New York, NY, USA

[35] Bird, R. B., Armstrong, R. C., and Hassager, O., 1987, *Dynamics of Polymeric Liquids, Vol I, Fluid Mechanics,* Wiley Interscience, New York, NY, USA

[36] Blausius, H., 1908, Math Phys., 56, 1

[37] Botterill, J. S.M and Williams, J. R., 1975, *Fluid Bed Heat Transfer,* Academic Press, UK

[38] Boley, B. A.., 1964, *Heat Transfer Structures and Materials*, Pergamon, NY, USA

[39] Boltzmann, L., 1878, Wied Ann, 5, 430

[40] Brazhnikov, A. M., Karpychev, V. A. and Luikova, A. V., 1975, Inzhenerno Fizicheskij Zhurnal, **28**, 4, 677-680

[41] Brinkman, H. C., 1947, *A Calculation of the Viscous Force Exerted by a Flowing Field on a Dense Swarm of Particles*, Appl. Sci. Res., **A1**, 27

[42] Broadhurst T. E. and Becker, H. A., 1975, AIChE J., **21**, 238

[43] Brown and Churchill, S. W., 1989, *Experimental Measurement of Second Sound*, AIChE Annual Meeting, San Francisco , CA, USA

[44] Buladi, P. M., Chang, C. C., Belovich, J. M. and Gatica, J. E., 1996, *Transport Phenomena and Kinetics in an Extravascular Bioartificial Pancreas*, AIChE J, **42**, 2668

[45] Bungay, P. M., and Brenner, H., 1973, *The Motion of a Closely Fitting Sphere in a Fluid Filled Tube*, Int. J. Multiph. Flow, **1**, 25

[46] Callen, H. B., 1985, *Thermodynamics*, Wiley, New York, NY, USA

[47] Campbell G. A. and Foster, R. M., 1948, *Fourier Integrals for Practical Applications*, D. Van Nostrand Company, Princeton, NJ, USA.

[48] Carey, G. F. and Tsai, M., 1982, J of Numerical Heat Transfer, 5, 309-314

[49] Casimir, H. B. G., 1938, Physica, **5,** 495-500

[50] Carslaw, H. S. and Jaeger, J. C. 1948, *Conduction of Heat in Solids,* Clarendon Press, Oxford, UK

[51] Carslaw, H. S. and Jaeger, J. C., 1963, *Operational Methods in Applied Mathematics*, Dover, New York, NY, USA

[52] Cattaneo, C, 1958, <u>Comptes Rendus</u>, **247**, 431

[53] Chandrasekhar, S., 1960, *Radiative Transfer*, Dover, New York, NY, USA

[54] Chandrasekhar, S., 1943, <u>Review Modern Physics</u>, **15**, 1-89

[55] Chandrasekhar, S., 1961, *Hydrodynamic and Hydromagnetic Instability,* Oxford University Press, UK

[56] Charm, S. E. and Kurland, G. S., 1974, *Blood Flow and Microcirculation*, Wiley, New York, NY, USA

[57] Chen, P. J. and Amos, D. E., 1970, <u>ASME J of Heat Transfer</u>, **37**, 1145-1146

[58] Chen, G., 2001, <u>Phys. Rev. Lett.</u>, **86,** 11, 2297 – 2300

[59] Cheng, K. J., 1989, <u>ASME J of Heat Transfer</u>, **111**, 225 – 231

[60] Chester, M., 1963, <u>Physical Review</u>, **131**, 5, 2013

[61] Choi, S. H. and Wilhelm, H. E., 1976, <u>Phy. Review</u>, **14**, 1825-1834

[62] Churchill, R. V., and Brown, J. W., 1987, *Fourier Series and Boundary Value Problems*, McGraw Hill, New York, NY, USA

[63] Churchill, R. V., 1972, *Operational Mathematics*, McGraw Hill, New York, NY, USA

[64] Clausius, R., 1867, *Mechanical Theory of Heat*, John van Voorst, UK

[65] Cima, L. G., Vacanti, J. P., Vacanti, C., Ingber, D., Mooney, D. and Langer, R., 1991, *Tissue Engineering by Cell Transplantation using Degradable Polymer Substrates,* <u>J. of Biomechancial Eng.</u>, **113**, 143

[66] Cooper, J. R., Bloom, F. E.., and Roth, R. H., 1996, *Biochemical Basis of Neuropharmacology*, Oxford University Press, UK

[67] Courant, R., and Friedrichs, K. O., 1948, *Supersonic Flow and Shock Waves*, Wiley Interscience, New York, NY, USA

[68] Courant, R. and Hilbert, D., 1962, *Methods of Mathematical Physics : Partial Differential Equations*, Interscience, New York, NY, USA

[69] Clift, R., Grace J. R., and Weber, M. E., 1978, *Bubbles, Drops and Particles,* Academic Press, New York, NY, USA

390

[70] Coleman, B. D., Fabrizio, M. and Owen, D. R., 1982, <u>Arch. of Rational Mechanical Analysis</u>, **80,** 135-158

[71] Colton, C. K, 1995, *Implantable Biohybrid Artificial Organs*, <u>Cell Transplant</u>, **4,** 415

[72] Conant, J. B., 1957, *Harvard Case Histories in Experimental Science, Vol I,* Harvard University Press, Cambridge, MA, USA

[73] Constantinides, A., 1988, *Applied Numerical Methods in Engineering,* McGraw Hill, New York, NY, USA

[74] Courant, R., and Hilbert, D., 1966, *Methods of Mathematical Physics,* Interscience, New York, NY, USA

[75]   Crank, J., 1975, *The Mathematics of Diffusion*, Oxford University Press, Oxford, UK

[76] Curry, F. E. and Michel, C. C., 1980, *A Fiber Matrix Model of Capillary Permeability,* <u>Microvasc. Res.</u>, **20,** 96

[77] Cussler, E. L. 1997, Cambridge University Press, Cambridge, UK

[78] Cussler, E. L., 1982, <u>AIChE J</u>, **28,** 500

[79]   Cussler, E. L. and Featherstone, J. D., 1981, <u>Science,</u> **213,** 1018

[80] Daw, C. S. and Hawk, J. A., 1995, *Fluidization Quality Analyzer for Fluidized Beds,* US Patent 5, 435, 972

[81] de Donder, T., and van Rysselberghe, P., 1936, *Affinity,* Stanford University Press, Menlo Park, CA, USA

[82] de Groot, S. R., and Mazur, P., 1969, *Non-Equilibrium Thermodynamics,* Amsterdam, Netherlands

[83] Decker N. A., and Glicksman, L. R., 1983, <u>Int. J Heat Mass Transfer</u> , 26, 1307

[84] Deen, W. M., 1987, *Hindered Transport of Large Molecules in Liquid-Filled Pores,* <u>AIChE J</u>, **33,** 1409

[85]   De Groot, S. R. and Mazur, P., 1962, *Non-Equilibrium Thermodynamics,* North Holland Publishing, Amsterdam, Netherlands

[86]   Driever, W., and Nusslein-Volhard, 1988, <u>Cell</u>, **54,** 83-93

[87] Eckert, E. R.G, 1950, *Introduction of the Transfer of Heat and Mass*, McGraw Hill, New York, NY, USA

[88] Einstein, A. 1906, Ann Physik, **19**, 286

[89] Ellis, 1986, *Fire barrier coating composition containin Magnesium oxychlorides and high* alumina calcium aluminate cements or magnesium oxysulphate, US Patent 4,572,862

[90] Erdelyi, A., 1956, *Asymptotic Expansions*, Dover, New York, NY, USA

[91] Euler, L., 1755, *Memoires de Academie de Science,* **11**, 217

[92] Fan, L. T., Ho T. C., Hirakoa and Walawander, W. P., 1981, AIChE J, **27**, 38

[93] Fauske, H. K., 1973, J. of Nuclear Science and Engineering, **51**, 95-101

[94] Fermi, E., 1937, *Thermodynamics*, Prentice Hall, New York, NY, USA

[95] Ferry, J. D., 1980, *Viscoelastic Properties of Polymers*, Wiley, New York, NY, USA

[96] Fick, A., 1855, Ann Physik, Leipzig, **170**, 59

[97] Field, R. J., and burger, M., 1985, *Oscillations and Traveling Waves in Chemical Systems,* Wiley, New york, NY, USA

[98] Folkman, J., and Long, D., 1964, J. of Surgical Research, **4**, 139-142

[99] Forland, K. S., 1988, *Irreversible Thermodynamics : Theory and Applications*, New York, Wiley, New York, NY,USA

[100] Fourier, J. B., 1822, *Theorie analytique de la chaleur*, English Translation by A. Freeman, 1955, Dover Publications, New York, NY, USA

[101] Frankel, J. I., Vick, B., and Ozisik, M. N., 1987, Int. J. Heat and Mass Transfer, **30**, 1293-1305

[102] Frankel, N. A., and Acrivos, A., 1967, Chem. Eng. Sci., 847-853

[103] Gaehtgens, P., 1980, *Flow of Blood Through Narrow Capillaries: Rheological Mechanisms Determining Capillary Hematocrit and Apparent Viscosity,* Biorheology, **17**, 183

[104] Geldart, D., 1973, Powder Technology, **7**, 285

[104] Garabedian, P. R., 1964, *Partial Differential Equations*, Wiley, New York, NY, USA

[105] Gibbs, J. W., 1878, *On the Equilibrium of Heterogeneous Substances,* Trans. Conn. Acad. Sci., 343

[106] Glandsdorff, P., and Prigogine, I., 1971, *Thermodynamics of Structure Stability and Fluctuations,* Wiley, New York, NY, USA

[107] Grace, J. R., 1986, Can. J. of Chem. Eng., **64**, 353

[108] Gray, D. N., 1981, *The Status of Olefin-SO2 Copolymers as Biomaterials*, in *Biomedical and Dental Applications of Polymers*, ed. Gebelein C. G., Koblitz F. F., 21, Plenum, New York, NY, USA

[109] Greenberg, M. D., 1971, *Applications of Green's Function in Science and Engineering*, Prentice Hall, Englewood Cliffs, NJ, USA

[110] Grober, H., Erk, S., and Grigull, U., 1961, *Fundamentals of Heat Transfer*, McGraw Hill, New york, NY, USA

[111] Gurtin, M. E., and Pipkin, A. C., 1968, Arch. of Rational Mechanical Analysis, **31**, 113-126

[112] Guggenheim, E. A., 1960, *Elements of the Kinetic Theory of Gases,* McGraw Hill, New York, NY,USA

[113] Gyftopoulos, E. P. and Beretta, G. P., 1991, *Thermodynamics: Foundations and Applications*, MacMillan, New York, NY, USA

[114] Haase, R., 1990, *Thermodynamics of Irreversible Processes,* Dover, New York, NY, USA

[115] Haberman, R., 1987, *Elementary Applied Partial Differential Equations,* Prentice Hall, Englewood Cliffs, NJ, USA

[116] Hagen, G., 1839, Ann-Phys. Chem., 46, 423

[117] Halliday, D., and Resnick, R., 1988, *Fundamentals of Physics,* Wiley, NY, USA

[118] Hartnett, J., P., and Hu, 1989, J of Rheology, 53, 671

[119] Haynes, R. F., 1960, *Physical Basis of the Dependence of Blood Viscosity on Tube Radius,* Am. J. Physiol., **198**, 1193

[120] Heisler, M. P., 1947, *Temperature Charts for Induction and Constant*

Temperature Heating, Trans. ASME, **69**, 227

[121]   Heleshaw, H., J.,. S., 1898, *Investigation of the Nature of the Surface Resistance of Water and of Streamline Motion under certain Experimental Conditions*, Trans. Inst. Nav. Arch.

[122]   Hess, B., and Mikhailov, A., 1994, Science, **264**, 223-224

[123]   Hilby, J. W., 1967, Proceedings of the Int. Symp. on Fluidization, Eindoven, **99**, Netherlands University Press

[124]   Holman, J.P, 1976, *Heat Transfer*, McGraw Hill, New York, NY, USA

[125]   Hougen, O. A., and Watson, K. M., 1947, *Chemical Process Principles*, Wiley, New York, NY, USA

[126]   Ince, E. L., 1956, *Ordinary Differential Equations*, Dover, New York, NY, USA

[127]   Incropera, F. P., and D. P. De Witt, 1981, *Fundamentals of Heat Transfer*, Wiley, New York, NY, USA

[128]   Jackson, R. and Anderson, T. B., 1968, Ind. Eng. Chem. Fund., **7**, 1, 12-21

[129]   Jackson, H. E. and Walker, C. T., 1971, Phys. Review B, **3**, 1428-1439

[130]   Jahnke, E., and Emde, F., 1943, *Tables of Functions*, Dover Publications, New York, NY, USA

[131]   Jeffreys, H., 1924, *The Earth*, Cambridge University Press, UK

[132]   Jeffrey, A., 1976, *Quasilinear Hyperbolic Systems and Waves*, Pitman, London, UK

[133]   John, F., 1978, *Partial Differential Equations*, Springer Verlag, New York, NY, USA

[134]   Jeltsch, A., and Pingoud A., 1998, Biochemistry, **37**, 2160-2169

[135]   Jeltsch, A., et al, 1996, EMBO J, **15**, 18, 5104-5111

[136]   Jones, B. R. E. and Pyle D. L., 1971, *On Stability Dynamics and Bubbling in Fluidized Beds*, AIChE Symp. Ser., 116, **67**, 1-10

[137]   Joseph, D. D. and Preziosi L., 1989, Reviews of Modern Physics, **61**, 41

[138]   Joseph, D. D. and Preziosi L., 1990, Reviews of Modern Physics, **62**, 375

[139] Jost, J., 1960, *Diffusion in Solids, Liquids and Gases,* Academic Press, New York, NY, USA

[140] Jou. D., Casas-Vazquez, J. and Lebon, G., 1992, <u>Phys., Rev. A.</u>, **45**, 8371-8373

[141] Jou. D., Casas-Vazquez, J. and Lebon, G., 1988, <u>Reports on Prog. in Physics</u>, **51**, 1105-1179

[142] Jou, D., 1996, *Extended Irreversible Thermodynamics*, Springer Verlag, New york, NY, USA

[143] Kaminski, W., 1990, <u>ASME J. of Heat Transfer</u>, **112**, 555-560

[144] Kang, W. K., Sutherland, J. P., and Osberg G. L., 1967, <u>Ind. Eng. Chem, Fund.</u>, **6,** 4, 449-504

[145] Kao, T. T., 1977, <u>ASME J. of Heat Transfer</u>, **99**, 343-345

[146] Kapila, A. K., 1983, *Asymptotic Treatment of Chemically Reacting Systems,* Pitman Advanced Publishing Program, Boston, MA, USA

[147] Kaplan, D., 1991, <u>Science</u>, **252**, 554-557

[148] Katchalsky, A., 1965, *Non-Equilibrium Thermodynamics and Biophysics,* Harvard University Press, Boston, MA, USA

[149] Kato, H. and Nara, K., 1998, *Method and Apparatus for Measurement of Thermal Diffusivity,* US Patent 5,713,665

[150] Kelly, D. C., 1968, <u>American J. Physics</u>, **36**, 585-591

[151] Kevorkian, J., and Cole, J. D., 1981, *Perturbation Methods in Applied Mathematics*, Springer Verlag, New York, NY, USA

[152] Kittel, C., 1986, *Introduction to Solid State Physics*, John Wiley, New York, NY, USA

[153] Kirkaldy, J. S., and Brown, L. C., 1963, <u>Can. J. Metallurgical Quarterly</u>, **2**, 89

[154] Kleinle, J. et al. 1996, <u>Biophysical Journal</u>, **71**, 2413-2426

[155] Kost, J., 1990, *Pulse and Self-Regulated Drug Delivery,* CRC Press, Boca Raton, FL

[156] Kraegen, E. W., Young, J. D., George, E. P. and Lazarus, L., 1972, *Oscillations in Blood Glucose and Insuling after Oral Glucose*, <u>Horm. Met. Res.</u>, **4**, 409

[157] Kreith, F., 1973, *Principles of Heat Transfer*, Intext, New York, NY, USA

[158] Kreith, F., and Black, W. Z., 1979, *Basic Heat Transfer*, Harper and Row, New York, NY, USA

[159] Kreyszig, E., 1993, *Advances Engineering Mathematics*, Wiley, New York, NY, USA

[160] Krogh, A., 1919, <u>J Physiol.</u>, **52**, 409 - 415

[161] Kuhn, T., 1977, *The Essential Tension*, University of Chicago Press, Chicago, IL, USA

[162] Kuiken, G.D.C., 1994, *Thermodynamics of Irreversible Processes*, Wiley, New York, NY, USA

[163] Kunii, D. and Levenpiel, O., 1991, *Fluidization Engineering*, Butterworth Heineman, NY, USA

[164] Lacy, P. E., 1995, *Treating Diabetes with Transplanted Cells*, <u>Sci. Amer.</u>, **273**, 501

[165] Lagerlund, T. D., and Low, P. A., 1993, *Mathematical Modeling of Time Dependent Oxygen Transport in Rat Peripheral Nerve*, <u>Comput. Biol. Med.</u>, **23**, 29

[166] Landau, L., and Liftshitz, E. M., 1987, *Fluid Mechanics*, Pergamon, UK

[167] Langer, R. S., and Wise, D. L., 1984, *Medical Applications of Controlled Release*, CRC Press, Baco Raton FL, USA

[168] Lauffenburger, D. A., and Linderman, J. J., 1993, *Receptors: Models for Binding, Trafficking, and Signalling*, Oxford University Press, UK

[169] Lax, P. D., 1971, *Shock Waves in Entropy in Contributions to Nonlinear Functional Analysis*, Zarantonello, H., Academic Press, New York, NY, USA

[170] Lebedev, N. N., 1972, *Special Functions and Their Applications*, Dover, New York, NY, USA

[171] Lebrat, J. P., and Varma A., 1993, <u>AIChE J.</u> **39**, 1732-1734

[172] Lennard-Jones, J. E., 1924, Proc. Royal Society, **A106**, 441-462

[173] Lienhard, J., H., 1981, *A Heat Transfer Textbook*, Prentice Hall, Englewood Cliffs, New Jersey, USA

[174] Levenspiel, O, 1998, *Engineeing Flow and Heat Exchange,* Kluwer, Amsterdam, Netherlands

[175] Levenspiel, O., 1962, *Chemical Reaction Engineering*, Wiley, NY, USA

[176] Levenspiel, O., 1989, *Chemical Reactor Omnibook*, Oregon State University Press, Corwallis, OR, USA

[177] Lide, D. R., 1994, *CRC Handbook of Chemistry and Physics,* CRC Press, Ann Arbor, MI, USA

[178] Lightfoot, E. N., 1995, in *Biomedical Engineering Handbook*, ed. Bronzino, J. D., CRC Press, Boca Raton, FL, USA

[179] Lirag and Littman, 1971, *Statistical Study of Pressure Fluctuations in a Fluidzed Bed,* AIChE Symposium Series, **67**, Fluidization, 11-22

[180] Lauffenburger, D., Rothman, C., and Zigmond, S., 1983, J. of Immunology, **131**, 940-947

[181] Leicester, H. M., 1971, *The Historical Background of Chemistry*, Dover, New York, NY, USA

[182] Letcher, J. S., 1969, ASME J. of Heat Transfer, **91**, 585-587

[183] Loney, 2002, *Laplace Transform methods in Class*, Chemical Engineering Education, 33, 4

[184] Luikov A. V., Bubnov, V. A. and Soloviev, I. A., 1976, Int. J Heat and Mass Transder, **19,** 245-248

[185] Luikov, A. V., 1966, Int. J. Heat and Mass Transfer, 9, 139-152

[186] Mach. E., 1986, *Principles of the Theory of Heat*, D. Reidel, Boston, MA, USA

[187] MacRobert, T. M., 1948, *Spherical Harmonics : An Elementary Treatise on Harmonic Functions with Applications,* Dover, New York, NY, USA

[188] Mason, S. F., 1962, *History of the Sciences*, Collier Books, New york, NY, USA

[189] Maxwell, J. C., 1867, <u>On the Dynamical Theory of Gases</u>, Phil. Trans. Roy. Soc., **157**, 49

[190] Maxwell, J. C., 1873, *A Treatise on Electricity and Magnetism*, **Vol. I**, 365. Oxford University Press, UK.

[191] McCabe, W. L., Smith, J., C., and Hariott, P., 1993, *Unit Operations in Chemical Engineering*, McGraw Hill, NY, USA

[192] Miller, R., 1967, *Transient Heat Conduction in Finite Slab with Position Dependant Heat Generation*, NASA, Washington, DC, USA

[193] Mendoza, E., 1977, *Reflections on the Motive Force of Fire by Sadi Carnot and other Papers on the Second Law of Thermodynamics by E. Clapeyron and R. Clausius,* Peter Smith, Glouster, MA, USA

[194] Mikhailov, M. D., and Ozisik, M. N., 1984, *Unified Analysis and Solutions of Heat and Mass Diffusion,* Wiley, New York, NY, USA

[195] Michalowski, S., Mitura, E. and Kaminski, W., 1982, <u>Hungarian J of Ind. Chem.</u>, **10,** 387-394

[196] Mickley, H. S., Sherwood, T. S. and Reed C. E., 1957, *Applied Mathematics in Chemical Engineering*, McGraw Hill, USA

[197] Minkowycz, W. J., Pletcher, R. H., Schneider, G. E. and Sparrow, E. M., 1988, *Handbook of Numerical Heat Transfer,* Wiley, New York, USA

[198] Mitra, K, Kumar, S., Vedavarz, A., and Moallemi, M. K., 1995, <u>ASME J Of Heat Transfer</u>, **117**, 568

[199] Mitura, E., Michalowski, S. and Kaminski, W., 1988, <u>Drying Technology</u>, **6**, 1, 113-137

[200] Majumdar, A., 1993, <u>ASME J of Heat Transfer</u>, **115**, 7-16.

[201] Murray, J. D., 1984, *Asymptotic Analysis*, Springer Verlag, New York, NY, USA

[202] Mickley, H. S. and Fairbanks, L. R., 1955, <u>AICHE J</u>, **3**, 374

[203] Morse, P. M. and Feshbach, H., 1953, *Methods of Theoretical Physics*, McGraw Hill, New York, NY, USA

[204] Myers, G. M., 1971, *Analytical Methods in Conduction Heat Transfer,* McGraw Hill, New York, NY, USA

398

[205] Myint-U, T., and Debnath, L., 1987, *Partial Differential Equations for Scientists and Engineers,* North Holland, New York, NY, USA

[206] National Bureau of Standards and Technology, 1971, News Bull, 55, 3

[207] National Bureau of Standards and Technology, 1982, *Tables of Chemical Thermodynamics Properties,* NBS, Washington, DA, USA

[208] Narayanamurti, V. and Dynes R. C., 1972, Phys. Rev. Letters, 1428-1439

[209] Navier, C. L. H. M., 1821, Ann Chimie, **19**, 244 - 260

[210] Nernst, W., 1969, *A New Heat Theorem,* Dover, New York, NY, USA

[211] Sir Isaac Newton (1643 – 1727), *Philosophiae Naturalis Principia Mathematica*

[212] Nixon, F. E., 1965, *Handbook of Laplace Transform Tables and Theorems,* Holt, Rinehart and Winston, New York, NY, USA

[213] Nugent, L. J. and Jain, R. K., 1984a, *Extravascular Diffusion in Normal and Neoplatic Tissues,* Cancer Res., **44**, 238

[214] Nugent, L. J. and Jain, R. K., 1984b, *Pore and Fiber-Matrix Models for Diffusive Transport in Normal and Neoplastic Tissues,* Microvas. Res., **28**, 270

[215] Oldroyd, J. G., 1953, Proc. of Royal Society, **A218**, 122-132

[216] Oleinik, O. A., 1963, *Uniqueness and Stability of the Generalized Solution to the Cauchy Problem for a Quasi-Linear Equation*, Americal Mathematical Society Translational Services, 2, **33**, 285 – 290

[217] Opong, W. S. and Zydney, A. L., 1991, *Diffusive and Convective Protein Transport Trough Asymmetric Membranes,* AIChE J, **37**, 1497

[218] Onsager, L, 1931, Phys. Rev., **37,** 405-426

[219] Ozisik, M. N., 1985, *Heat Transfer - A Basic Approach*, McGraw Hill, New York, NY, USA

[220] Ozisik, M. N., 1989, *Boundary Value Problems of Heat Conduction*, Dover, New York, NY, USA

[221] Ozisik, M. N., 1980, *Heat Conduction*, Wiley, New York, NY, USA

[222] Ozisik, M. N. and Tzou, D. Y., 1992, *On the Wave Theory in Heat Conduction,* ASME Winter Annlual Meeting, Anaheim, CA, USA.

[223] Ozisik, M. N. and Tzou, D. Y., 1994, ASME J of Heat Transfer, **116**, 526-535

[224] Ozisik, M. N. and Vick, B., 1984, Int. J. Heat and Mass Transfer, **27**, 1845-1854

[225] Pahence, J. M., and Kohn, J., 1997, *Biodegradable Polymers for Tissue Engineering, in Principles of Tissue Engineering*, ed., Lanza, R. P., Langer, R., and Chick, W. L., Landes, R. G., 273, Boulder, CO, USA

[226] Pathak, C. P., Sawhney, A. S. and Hubell, J. A., 1992, *Rapid Photopolymerization of Immunoprotective Gels in Contact with Cells and Tissues,* J. Am. Chem. Soc., **114**, 8311

[227] Perez, E. P., Merril E. W., Miller, D., Cima, L. G., 1995, *Corneal Epithelial Wound Healing on Bilayer Composite Hydrogels*, Tissue Engineering, **1**, 263

[228] Perry, R. H. and Green D. W., 1997, *Chemical Engineers Handbook*, McGraw Hill, NY, USA

[229] Peshkov, V, 1944, J of Physics, USSR, **8**, 381-386

[230] Planck, M., 1945, *Treatise on Themodyamics,* Dover, New York, NY, USA

[231] Poiseuille, J. L., 1841, Comptes Rendes

[232] Popov, S. amd Poo, M. M., 1992, J. of Neuroscience, **12**, 1, 77-85

[233] Powell, R. W., Ho, C. Y. and Liley, P. E., 1966, *Thermal Conductivityof Selected Materials,* National Bureau of Standards and Technology, Washington, DC, USA

[234] Prandtl, L., 1952, *Essentials of Fluid Dynamics*, Harper, NY, USA, 105

[235] Prigogine, I., and Defay, R., 1967, *Chemical Thermodynamics*, Longman, UK

[236] Prigogine, I., and Petrosky, T., 1996, Chaos, Solitons and Fractals, **7**, 441.

[237] Prigogine, I and Kondepudi, D, 1998, *Modern Thermodynamics, From Heat Engines to Dissipative Structures,* John Wiley, New York, NY, USA

[238] Prigogine, I., and Stengers, I., 1984, *Order Out of Chaos*, Bantam, New York, NY, USA

[239] Prigogine, I., 1967, *Introduction to Thermodynamics of Irreversible Processes*, Wiley, New York, NY, USA

[240] Prigogine, I., 1980, *From Being to Becoming*, W. H. Freeman, San Francisco, CA, USA

[241] Qiu, T. Q. and Tien, C. L., 1992, Int. J. Heat and Mass Transfer, **35**, 719-726.

[242] Radomsky, M. L., et al, 1990, Biomaterials, **11**, 619-624

[243] Rainville, E. D., 1963, *The Laplace Transform*, MacMillan, New York, NY, USA

[244] Ramachandran, P. A. and Mashelkar, R. A., 1980, *Lumped Parameter Model for Hemodialyzer with Application to Simulation of Patient -Artificial Kidney System,* Med. Biol. Eng. Comput., **18**, 179

[245] Ramkrishna, D., and Amundson, N. R., 1985, *Linear Operator Methods in Chemical Engineering*, Prentice Hall, Englewood Cliffs, NJ, USA

[246] Rayleigh, L., 1873, Proc. Math. Soc. London, **4**, 357

[247] Lord Rayleigh (J. W. Strutt), 1916, Phil. Mag., 6, **32**, 529-546

[248] Reid, R. C., Prausnitz, J. M., and Poling, B. E., 1987, *The Properties of Gases amd Liquids*, McGraw Hill, New York, NY, USA

[249] Renganathan, K., 1990, *Correlation of Heat Transfer Using Pressure Fluctuations in Gas-Solid Fluidized Beds*, Ph.D Dissertation, West Virginia University, Morgantown, WV, USA

[250] Renganathan, K., and Turton,R., 1989, *Thermal Inertia Effects in Fluidized Bed to Surface Heat Transfer,* 10th International Conference on Fluidized Bed Combustion, May

[251] Renganathan, K., and R. Turton, 1989, *Mesoscopic Correlation of Heat Transfer in Fluidized Beds to Immersed Surfaces*, 81st AIChE Annual Meeting, San Francisco, CA, USA, November

[252] Renganathan, K. and Turton R., 1989, *Approximate Solution of Hyperbolic Heat Equation to Predict Heat Transfer in Fluidized Beds to Immersed Surfaces*, 81st AIChE Annual Meeting, San Francisco, CA, USA, November

[253] Renganathan, K. and Turton R., 1989, *Data Reduction from Thin-Film Heat Gauges in Fluidized Beds*, Powder Technology, 249-254.

[254] Renkin, E. M., 1954, *Filtration, Diffusion, and Molecular Sieving Through Porous Cellulose Membranes*, J. Gen. Phys., **38**, 225

[255] Rhee, H. K., Aris, R., and Amundson, N. R., 1986, *First Order Partial Differential Equations*, 1,2, Prentice Hall, Englewood Cliffs, NJ, USA

[256] Reynolds, O., 1883, *An Experimental Investigation of the Circumstances which Determine whether the Motion of Water shall be direct or Sinous and of the Laws of Resistance in Parallel Channels,* Trans. Royal Society, **174,** London, UK

[257] Roach, G. F., 1982, *Green's Functions*, Cambridge University Press, Cambridge, UK

[258] Roetman, E. L., 1975, Int. J. of Eng. Sci., **13**, 699-701

[259] Rohsenow, W. M., Hartnett, J. P., and Cho, Y. I., 1996, *Handbook of Heat Transfer*, McGraw Hill, New York, NY, USA

[260] Roy, R., Davidson, J. F., Tupanogov, V. G., 1990, Chemical Eng. Sci., **45**, 11, 3233-3245

[261] Ruschak, K. L., and Miller, C. A., 1972, IEC Fund., **11**, 534

[262] Sadasivan, N., Barrereau, D. and Laqueric, 1980, Powder Technology, **26**, 67

[263] Saltzman, W. M., 2001, *Drug Delivery, Engineering Principles for Drug Therapy,* Oxford University Press, UK

[264] Saltzman, W. M., and Radomsky, M. L., 1991, Chem. Eng. Sci., **46**, 2429-2444

[265] Saxena, S. C., 1989, *Advances in Heat Transfer*, 19, 97

[266] Scales, J. A. and Snieder R., 1999, Nature, **401,** 739-740

[267] Schlichting, 1979, *Boundary Layer Theory*, McGraw Hill, NY, USA

[268] Schneider, P. J., 1955, *Conduction Heat Transfer,* Addison-Wesley, Reading, MA, USA

[269] Segre, E., 1984, *From Falling Bodies to Radio Waves*, W. H. Freeman, New York, NY, USA

[270] Sharma, K. R., 2003aa, *Mesoscopic Heat Conduction and Onset of Periodicity in the Solution ofHyperbolic Damped Wave Conduction and Relaxation Equation*, CHEMCON, 56th Annual Meeting of the Indian Institute of

402

Chemical Engineers, Bhubaneshwar, Orissa, India, December

[271] Sharma, K. R., 1997, *A Statistical Design to Define Process Capability of Continuous Mass Polymerization of ABS at the Pilot Plant,* **124th ACS National Meeting,** Las Vegas, NV, USA, September

[272] Sharma, K. R. and Turton , R., 1998a, Powder Technology, **99,** 109-118

[273] Sharma,K. R., 1998b, *Periodicity in Pressure Fluctuations in Gas-Solid Fluidized Beds,* Fluidization IX, Durango, CO, USA, May

[274] Sharma, K. R., 1998c, *Role of Turbulence, Time Dependence on Heat Transfer in Circulating Fluidized Beds,* 90th AIChE Annual Meeting, Miami, FL, USA, November

[275] Sharma, K. R., 1999a, *Mesoscopic Estimation of Intraocular Pressure Measurements in the Diagnosis of Glaucoma,* 91st AIChE Annual Meeting, Dallas, TX, USA, October/November

[276] Sharma, K. R., 1999b, *Amplitude Damping of Signals from Thin-film Heat Gauges in Circulating Pressurized Fluidized Beds,* 91st AIChE Annual Meeting, Dallas, TX, USA, October/November

[277] Sharma, K. R., 1999c, *Data Interpretation from Thin Film Heat Gauges in Bubbling Fluidized Beds,* 91st AIChE Annual Meeting, Dallas, TX, USA, October/November

[278] Sharma, K. R., 1999d, *Mesoscopic Correlation for Heat Transfer Coefficients with Pressure Fluctuations in Pressurized Fluidized Bed Combustors,* 33rd National Heat Transfer Conference, NHTC, Albuquerque, NM, USA, August

[279] Sharma, K. R., 1999e, *Heat Transfer Coefficient at Minimum Fluidization Velocity for Gas-Solid Fluidized Beds,* 33rd National Heat Transfer Conference, NHTC, Albuquerque, NM, USA, August

[280] Sharma, K. R., 1999f, *Thermal Time Constant of a Bubble and Interpretation of Heat Transfer Signals,* 33rd National Heat Transfer Conference, NHTC, Albuquerque, NM, USA, August

[281] Sharma, K. R,. 1999g, *Rectilinear Migration of a Drop in a Nonlinear Temperature Gradient,* 33rd National Heat Transfer Conference, NHTC, Albuquerque, NM, USA, August

[282] Sharma, K. R., 1999h, *Development of a Mechanistic Heat Transfer Model that Accounts for Turbulence, Time Dependence and Non-Fourier Heat Propagation in Circulation Pressurized Fluidized Beds,* International Mechanical Engineering Congress and Exposition, IMECE 99, Nashville, TN,

USA, November

[283] Sharma, K. R., 1999i, *Gas-Solid Mass Transfer in Fluidized Bed Degasifier*, International Mechanical Engineering Congress and Exposition, IMECE 99, Nashville, TN, USA, November

[284] Sharma, K. R, 2001a, *Mechanism of Heat Transfer to a U Tube in a CFB*, 17th National Convention of Chemical Engineers, Kochi, India, October

[285] Sharma, K. R., 2001b, *Convective and Relaxation Effects in Heat Transfer in CFB to a Horizontal Tube,* CHEMCON 2001, Chemical Engineering Congress, Chennai, India, December

[286] Sharma, K. R., 2001c, Bounded *Solution for HHC Wave Propagative Equation,* CHEMCON 2001, Chemical Engineering Congress, Chennai, India, December

[287] Sharma, K. R. 2003, *Lecture Notes in Computational Molecular Biology*, SASTRA University Press, Thanjavur, India.

[288] Sharma, K. R., 2003a, *Pulse injection and Decay in Infinite Media*, 225th ACS National Meeting, New Orleans, LA, USA, March 23rd – March 27th

[289] Sharma, K. R., 2003b, *Critical Radii of Zirconium Nuclear Fuel Rod Neither Greater than the Shape Limit Nor Less than Cycling Limit*, AIChE Spring National Meeting, Presentation Record, New Orleans, LA, USA, March 30th – April 3rd

[290] Sharma, K. R., 2003c, *Surface Renewal Theory for Mass Transfer Coefficient using Infinite Order PDE Inorder to Capture non-Fickian Diffusion*, 225th ACS National Meeting, New Orleans, LA, USA, March 23rd – March 27th

[291] Sharma, K. R., 2003d, *Temperature Solution in Semi-Infinite Medium under CWT using Cattaneo and Vernotte for non-Fourier Heat Conduction*, 225th ACS National Meeting, New Orleans, LA, March 23rd – March 27th

[292] Sharma, K. R., 2003e, *Storage Coefficient of Substrate in a 2 GHz Microprocessor,* 225th ACS National Meeting, New Orleans, LA, USA, March 23rd - March 28th

[293] Sharma, K. R., 2003f, *Temperature Dependent Heat Source Effect in Cooling Nuclear Reactors,* 225th ACS National Meeting, New Orleans, LA, USA, March 23rd – March 28th

[294] Sharma, K. R., 2003g, *Transient Temperature Solution under CWT for a Sphere using the Cattaneo and Vernotte non-Fourier Equation,* 225th ACS National Meeting, New Orleans, LA, USA, March 23rd - March 28th

[295] Sharma, K. R., 2003h, *Time Taken to Alarm in High Temperature Barrier Coating*, AIChE Spring National Meeting, New Orleans, LA, USA, March 30th - April 3rd

[296] Sharma K. R., 2003i, *Unsteady Viscous Flow - Flow Near a Wall Suddenly Set in Motion using the Hyperbolic Momentum Wave Propagative Equation*, AIChE Spring National Meeting, New Orleans, LA, USA, March 30th - April 3rd

[297] Sharma, K. R., 2003j, *Molecular Sieve Design for Reuse of Army Vehicle Washeries under Mild Conditions*, AIChE Spring National Meeting, New Orleans, LA, USA, March 30th - April 3rd

[298] Sharma, K. R., 2003k, *Adsorber Design for Removal of Acrylonitrile from Water*, AIChE Spring National Meeting, New Orleans, LA, USA, March 30th - April 3rd

[299] Sharma, K. R. 2003l, *Finite Speed Diffusion in Restriction Fragment Mapping*, 40th Indian Chemical Society Meeting, Jhansi, U.P, India, December

[300] Sharma, K. R. 2003m, *Fluid Mechanics and Machinery*, Anuradha Publishers, Kumbakonam, India

[301] Sharma, K. R, 2004a, *Plane of Zero Concentration in Finite Speed Diffusion*, 41st Annual Convention of Chemists, Meeting, Delhi University, N.Delhi India, December

[302] Sharma, K. R., 2004b, *Some Issues in Thermal Barrier Coating*, CHEMCON 2004, Mumbai, India, December

[303] Sharma, K. R., 2004c, *Minimum Width of a Cylindrical Pellet to Avoid Subcritical Damped Oscillations*, CHEMCON 2004, Mumbai, India, December

[304] Sharma, K. R., 2004d, *A Solution by Wave Coordinate Transformation of Hyperbolic Diffusion & Relaxation Equation with Surface Reaction*, CHEMCON 2004, Mumbai, India, December

[305] Sharma, K. R., 2000t1, *Fluidized Bed Heat Transfer in Combustors - Hyperbolic Heat Wave Partial Differential Equation Solution by the Method of Laplace Transform and Series Solution by the Method of Frobenius*, ACS Symp. Arch., 52nd Southeast Regional Meeting of the ACS, SERMACS, New Orleans, LA, USA, October

[306] Sharma, K. R., 2000t2, *Transient Fixed Bed Adsorption Examination using Hyperbolic Partial Differential Equations*, ACS Symp. Arch., 52nd Southeast Regional Meeting of the ACS, SERMACS, New Orleans, LA, USA, October

[307] Sharma, K. R., 2001t1, *Hyperbolic Wave Propagative Partial Differential Equation Solution in Elution Chromatography,* ACS Symp. Arch., 221st ACS National Meeting, San Diego, CA, USA, April

[308] Sharma, K. R, 2001t2, *Thermal Time Constant of a Sphere using Hyperbolic Heat Wave Equation*, ACS Symp. Arch., 34th Middle Atlantic Regional Meeting of the ACS, Baltimore, MD, USA, June

[309] Sharma, K. R., 2001t3, *Unsteady Evaporation by Hyperbolic Mass Propagative Equation*, ACS Symp. Arch., 33rd ACS Central Regional Meeting/Great Lakes Meeting, Grand Rapids, MI, USA, June

[310] Sharma, K. R., 2001t4, *Unsteady Diffusion with First Order Reaction*, ACS Symp. Arch., 33rd ACS Central Regional Meeting/Great Lakes Meeting, Grand Rapids, MI, USA, June.

[311] Sharma, K. R., 2001t5, *Gas Absorption with Rapid Chemical Reaction*, ACS Symp. Arch., 33rd ACS Central Regional Meeting/Great Lakes Meeting, Grand Rapids, MI, USA, June

[312] Sharma, K. R., 2001t6, *Diffusion and Chemical Reaction in Isothermal Laminar Flow along a Soluble Flat Plate*, ACS Symp. Arch., 33rd ACS Central Regional Meeting/Great Lakes Meeting, Grand Rapids, MI, USA, June

[313] Sharma, K. R., 2001t7, *Quenching of a Steel Billet*, ACS Symp. Arch., 222 ACS National Meeting, Chicago, IL, USA, August

[314] Sharma, K. R., 2001t8, *Modified Bessel Composite Function of the First Kind Solution of Hyperbolic Heat Wave Equation*, ACS Symp. Arch., 222 ACS National Meeting, Chicago, IL, USA, August

[315] Sharma, K. R., 2001t9, *Thermal Conductivity Change with Temperature for Metallic Solids*, ACS Symp. Arch., 53rd Southeast Regional Meeting of the ACS, SERMACS, Savannah, GA, USA, September

[316] Sharma, K. R, 2001t10, *Generalized Normal Distribution and Periodicity*, AIChE Symp. Arch., 93rd AIChE Annual Meeting, Reno, NV, USA, November

[317] Sharma, K. R., 2001t11, *Compound Semiconductor Dopant Diffusion in VLSI/SLSI*, AIChE Symp. Arch., 93rd AIChE Annual Meeting, Reno, NV, USA, November

[318] Sharma, K. R., 2001t12, *Calculation of Thermal Diffusvity from Periodic Heating of Sample,* AIChE Symp. Arch., 93rd AIChE Annual Meeting, Reno, NV, USA, November

[319] Sharma, K. R., 2002t1, *CWF Problem with Hyperbolic Heat Wave Propagative*

Equation, ACS Symp. Arch., 223rd ACS National Meeting, Orlando, FL, USA, April

[320] Sharma, K. R., 2002t2, *Optimization of Heat Sinks in SLSI*, ACS Symp. Arch., 223rd ACS National Meeting, Orlando, FL, USA, April

[321] Sharma, K. R., 2002t3, *On the Inverse Heat Transfer Problem*, ACS Symp. Arch., 223rd ACS National Meeting, Orlando, FL, USA, April

[322] Sharma, K. R., 2002t4, *Modified Heisler/Grober Charts using Hyperbolic Partial Differential Equation*, ACS Symp. Arch., 223rd ACS National Meeting, Orlando, FL, USA

[323] Sharma, K. R., 2002t8, *Hyperbolic Wave Propagative Equation Solution to Chemisorption of Organic Sulfides onto Palladium,*, ACS Symp. Arch., Rocky Meeting Regional Meeting of the ACS, RMRM 02, Albuquerue, NM, USA, October

[324] Sharma, K. R., 2002t9, *Generalized Fick's Diffusion Law and Dopant Concentration Profile*, AIChE Symp. Arch., 94th AIChE Annual Meeting, Indianapolis, IN, USA, November

[325] Sharma, K. R., 2002t10, *Concentration Polarization Layer in Reverse Osmosis*, AIChE Symp. Arch., 94th AIChE Annual Meeting, Indianapolis, IN, USA, November

[326] Sharma, K. R., 2003t1, *Finite Speed Heat Conduction - I Relativistic Coordinate Transformation in Cartesian Coordinates*, 36th Middle Atlantic Regional Meeting of the ACS, Princeton, NJ, USA, June

[327] Sharma, K. R., 2003t2, *Finite Speed Heat Conduction - II Relativistic Transformation in Cylindrical Coordinates,* 36th Middle Atlantic Regional Meeting of the ACS, Princeton, NJ, USA, June

[328] Sharma, K. R., 2003t3, *Finite Speed Heat Conduction - III Relativistic Transformation in Spherical Coordinates,* 36th Middle Atlantic Regional Meeting of the ACS, Princeton, NJ, USA, June

[329] Sharma, K. R., 2003t4, *Finite Speed Heat Conduction - IV Pulse Decay in Infinite Media*, 36th Middle Atlantic Regional Meeting of the ACS, Princeton, NJ, USA, June

[330] Sharma, K. R.,, 2003t5, *Finite Speed Heat Conduction - V Critical Coating Thickness of Barrier Coating*, 36th Middle Atlantic Regional Meeting of the ACS, Princeton, NJ, USA, June

[331] Sharma, K. R., 2003t6, *Finite Speed Heat Conduction - VI Critical Thickness*

of Sphere to Avoid Pulsations by Method of Separation of Variables, 36th Middle Atlantic Regional Meeting of the ACS, Princeton, NJ, USA, June

[332] Sharma, K. R., 2003t7, *Finite Speed Heat Conduction - VII Critical Thicknes of Finite Slab to Avoid Pulsations,* 36th Middle Atlantic Regional Meeting of the ACS, Princeton, NJ, USA, June

[333] Sharma, K. R., 2003t8, *Finite Speed Mass Diffusion - VIII Fast Simultaneous Reaction and Diffusion in a Finite Slab,* 36th Middle Atlantic Regional Meeting of the ACS, Princeton, NJ, USA, June

[334] Sharma, K. R., 2003t9, *Finite Speed Mass Diffusion - IX Fast Simultaneous Diffusion and Reaction in a Cylinder*, 36th Middle Atlantic Regional Meeting of the ACS, Princeton, NJ, USA, June

[335] Sharma, K. R., 2003t10, *Finite Speed Heat Conduction - X Fast Simultaneous Diffusion and Reaction in a Sphere*, 36th Middle Atlantic Regional Meeting of the ACS, Princeton, NJ, USA, June

[336] Sharma, K. R., 2003t11, *Finite Speed Heat Conduction -XI Nulcear Heat Source in a Cylinder,* 36th Middle Atlantic Regional Meeting of the ACS, Princeton, NJ, USA, June

[337] Sharma, K. R., 2003t12, *Finite Speed Heat Conduction - XII Storage Coefficient of Substrate in Microprocessor*, 36th Middle Atlantic Regional Meeting of the ACS, Princeton, NJ, USA, June

[338] Sharma, K. R., 2003t13, *Finite Speed Mass Diffusion - XIII Infinite Relaxation Number Limit & D'Alambert's Solution,* 36th Middle Atlantic Regional Meeting of the ACS, Princeton, NJ, USA, June

[339] Sharma, K. R., 2003t14, *Finite Speed Heat Conduction - XIV Infinite Order PDE for Transient Temperature Representation,* 36th Middle Atlantic Regional Meeting of the ACS, Princeton, NJ, USA, June

[340] Sharma, K. R., 2003t15, *Finite Speed Heat Conduction - XV CWF in Semi-Infinite Medium*, 36th Middle Atlantic Regional Meeting of the ACS, Princeton, NJ, USA, June

[341] Sharma, K. R., 2003t16, *Finite Speed Heat Conduction - XVI Convection & Conduction to and fro a Horizontal Tube in a CFB,* 36th Middle Atlantic Regional Meeting of the ACS, Princeton, NJ, USA, June

[342] Sharma, K. R., 2003t17, *Method of Complex Temperature Solution to Cattaneo and Vernotte Non-Fourier Finite Speed Heat Conduction Equation*, 35th Great Lakes Regional Meeting of the ACS, GLRM 03, Chicago, IL,USA, June

408

[343] Sharma, K. R., 2003t18, *Transient Analysis of Laminar Flow in Circular Pipe – Occurence of Pulsations below Critical Value*, 58th Northwest Regional Meeting of the ACS, Bozeman, Montana, USA, June

[344] Sharma, K. R., 2003t19, *Melting of Ice Cube including the Finite Speed Heat Conduction Effect*, 58th Northwest Regional Meeting of the ACS, Bozeman, Montana, USA, June

[345] Sharma, K. R., 2003t20, *Diffusion Issues in DNA Sequencing using Sanger's Method*, 58th Northwest Regional Meeting of the ACS, Bozeman, Montana, USA, June

[346] Sharma, K. R., 2003t21, *Dopant Transient Concentration Profile Representation using Third Order PDE*, Northeast Regional Meeting of the ACS, NERM 03, Saratoga Springs, New York, USA, June

[347] Sharma, K. R., 2003t22, *Where is the Second Law Violation in Hyperbolic Damped Wave Finite Speed Diffusion Equation Solution ?* 226th ACS National Meeting, New York, NY, USA, September

[348] Sharma, K. R., 2003t24, *Secondary Pressure Fluctuations in Gas Solid Geldart A Fluidized Beds and a Maxwellian Explaination*, 226th ACS National Meeting, New York, NY, USA, September

[349] Sharma, K. R., 2003t25, *Dissolution of Spherical Drop by Finite Speed Diffusion*, 95th AIChE Annual Meeting, San Francisco, CA, USA, November

[350] Sharma, K. R., 2003t26, *Diffusion and Convection during Chemisorption*, 95th AIChE Annual Meeting, San Francisco, CA, USA, November

[351] Sharma, K. R., 2003t27, *Cosinous Temperature Profile during Reaction and Heat Transfer in CFB*, 95th AIChE Annual Meeting, San Francisco, CA, USA, November

[352] Sharma, K. R., 2003t28, *Radial Flow Chromatography and Transient Concentration Profile using Finite Speed Diffusion*, 95th AIChE Annual Meeting, San Francisco, CA, USA, November

[353] Sharma, K. R., 2003t29, *Ratio of Covection to Storage and Pulsations in Finite Speed Heat Conduction*, Central Regional Meeting of the ACS, Pittsburgh, PA, October

[354] Sharma, K. R., 2004t1, *Minimum width of a Cylindrical Pellet to Avoid Subcritical Damped Oscillations*, 228th ACS National Meeting, Philadelphia, PA, USA, August

[355] Sharma, K. R., 2004t2, *Cosinous Concentration Profile during Simultaneous*

Grafting, Diffusion and Bulk Flow in Helical Ribbon Agitator using Parabolic Diffusion Equation, 228th ACS National Meeting, Philadelphia, PA, USA, August

[356] Sharma, K. R., 2004t3, *Plane of Zero Concentration in a Slab with Reacting Interface,* 228th ACS National Meeting, Philadelphia, PA, USA, August

[357] Sharma, K. R., 2004t4, *Confounding Effect of Charge on Gel Electrophoresis Measurements,* 59th Northwest Regional Meeting of the American Chemical Society, NORM/RMRM, Utah State University, UT, USA, June

[358] Sharma, K. R., 2004t5, *On a Explicit Expression for Plasma Layer Thickness,* 60th Soutwest Regional Meeting of the American Chemical Society, Fort Worth, TX, USA, September/October

[359] Sharma, K. R., 2004t6, *Critical Region of Zero Concentration in Finite Speed Diffusion and Relaxation,* 60th Soutwest Regional Meeting of the American Chemical Society, Fort Worth, TX, USA, September/October

[360] Sieniutycz, S., 1977, Int. J. Heat Mass Transfer., **20**, 1221-1231

[361] Sieniutycz, S., 1981, Int. J. of Eng. Sci., **36**, 621-624

[362] Schneider, P. J., 1963, *Temperature Response Charts*, Wiley, New York, NY, USA

[363] Schneider, P. J., 1955, *Conduction Heat Transfer*, Addison-Wesley, Reading, MA, USA

[364] Schwedoff, T., 1890, J. Physique, 2, **9**, 34

[365] Scriven, L. E., 1960, Chem. Eng. Sci., **12**, 98-108

[366] Scriven, L. E. and Sternling, C. V., 1960, Nature, 187, 186-188

[367] Smith, K. L. and Herbig, S. M., 1992, *Controlled Release in Membrane Handbook,* W. S. Ho and K. K. Sirkar, van Nostrand Reinhold, NewYork, NY, USA

[368] Smoluchowski, M., 1917, Z. Physok. Chem. , **92**, 9, 129-168

[369] Sneddon, I. N., 1957, *Elements of Partial Differential Equations*, McGraw Hill, New York, NY, USA

[370] Sparrow, E. M., and Cess, R. D., 1976, *Radiation Heat Transfer*, Cole Brooks, New York, NY, USA

410

[371] Spera, F. J., and Trial, A., 1993, *Verification of the Onsager Reciprocal Relations in Molten Silicate Solution*, Science, **259**, 204

[372] Spiegel, M. R., 1968, *Mathematical Handbook for Scientists and Engineers*, McGraw Hill, New York, NY, USA

[373] Steffens, H. J., 1979, *James Prescott Joule and the Concept of Energy*, Science History Publications, New York, NY, USA

[374] Stokes, G. G., 1845, Trans. Camb. Phil. Soc., **8**, 287 - 305

[375] Streeter, V. L., and Wylie, E. B., 1983, *Fluid Mechanics*, McGraw Hill, NY, USA

[376] Streeter, V. L., 1948, *Fluid Dynamics*, McGraw Hill, NY, USA

[377] Streeter, R. L., 1973, *The Analysis and Solution of Partial Differential Equations*, Brooks Cole, Montery, CA, USA

[378] Sundaresan R., and Kolar, 2002, Core Heat Transfer Studies in a CFB, Powder Technology, 124, 138-151

[379] Sunderland, J. E., and Johnson, K. R., 1964, *Shape Factors for Heat Conduction Through Bodies with Isothermal or Convective Boundary Conditions*, Trans. ASHRAE, **70**, 237

[380] Svaboda, K., Cermak, J., Hartman, M., Drakos, J. and Selunchy, K., 1984, AICHE J, **30**, 3, 513-517

[381] Swartz, E. T., and Pohl R. O., 1989, Reviews of Modern Physics, **61**, 605-668

[382] Taitel, Y., 1972, Int. J. Heat Mass Transfer, 15, 369–371

[383] Tarcha. P. J., 1990, *Polymers for Controlled Drug Delivery*, CRC Press, Boca Raton, FL, USA

[384] Taylor, G. I., 1936, Proc. Of Royal Society, **A157**, 546-564

[385] Tamarin, A. I., 1964, Int. Chem. Eng., **4**, 1, 50-54

[386] Tang D. W., and Araki N., 1996, Int. J. Heat Mass Transfer, **39**, 15, 3305

[387] Thiele, E. W., 1939, Ind. Eng. Chem., **31**, 7, 916-920

[388] Toor, H. L., 1971, AIChE J, **17**, 5

[389] Touloukian, Y. S., Liley, P. E. and Saxena, S. C., 1970, *Thermophysical Properties of Matter, Vol 3, Thermal Conductivity – Nonmetallic Liquids and Gases,* Plenum Data Corp., New York, NY, USA

[390] Tschoegl, N. W., 1989, *The Phenomenological Theory of Linear Viscoelatic Behavior,* Springer Verlag, Berlin, Germany

[391] Titchmarsh, E. C., 1946, *Eigenfunction Expansions,* Oxford University Press, Oxford, UK

[392] Tzou, D. Y., 1989, Int. J. Heat and Mass Transfer, **32**, 1979-1987

[393] Tzou, D. Y., 1990a, ASME. J. Heat Transfer, **112**, 21-27

[394] Tzou, D. Y., 1990b, Int. J. Heat and Mass Transfer, **33**, 877- 885

[395] Tzou, D. Y., 1990c, Int. J. Eng. Sci., **28**, 1003-1017

[396] Tzou, D. Y., 1991a, ASME J Heat Transfer, **113,** 242-244

[397] Tzou, D. Y., 1993, Int. J. Heat Mass Transfer, **36**, 7, 1845-1850

[398] Tzou, D. Y., 1992, Int. J. of Heat Mass Transfer, 35, 2437-2356.

[399] Tzou, D. Y., 1995, ASME J of Heat Transfer, **117**, 8-16.

[400] Tzou, D. Y., 1997, *Macro-to Micro Scale Heat Transfer: The Lagging Behavior,* Taylor and Francis, Washington, DC, USA

[401] Varma, A. and Morbidelli, M., 1997, *Mathematical Methods in Chemical Engineering,* Oxford University Press, UK

[402] Varma, A., 1982, *Some Historical Notes on the Use of Mathematics in Chemical Engineering, in A Century of Chemical Engineering,* Furter, W. F., Plenum Press, New York, NY, USA

[403] Verloop, J. and Heertjes, P. M., 1974, Chem. Eng. Sci., **29**, 1035

[404] Vernotte, P., 1958, Les *paradoxes de la theorie continue de l'equation de la chaleur,* C. R. Hebd. Seanc. Acad. Sci. Paris, 246, **22**, 3154

[405] Vreedenberg, H. A., 1958, Chem. Eng. Sci., 9, 52

[406] Wakeham, Nagashima, A., and Sengers, J. V., 1991, *Merasurement of the Transport Properties of Fluids,* CRC Press, Boca Raton, FL, USA

[407] Wang, L., 2000, Int. J of Heat and Mass Transfer, **43**, 365-373

[408] Watson, G. N., 1966, *A Treatise on the Theory of Bessel Functions,* Cambridge University Press, Cambridge, UK

[409] Weinberg, S., 1980, *The First Three Minutes,* Bantam, New York, NY, USA

[410] Weinberger, H. F., 1965, *A First Course in Partial Differential Equations,* Blaisdell, Waltham, MA, USA

[411] Weiss, T. F., 1996, *Cellular Biophysics,* MIT Press, Cambrdge, MA, USA

[412] Welty, J. R., Wicks, C. E., and Wilson, R. E., 1984, *Fundamentals of Momentum, Heat and Mass Transfer,* Wiley, New York, NY, USA

[413] Weymann, H. D., 1967, American J of Physics, **35**, 488-496

[414] Whitman, G. B., 1974, *Linear and Nonlinear Waves,* Wiley, New York, NY, USA

[415] Wiggert, D. C., 1977, ASME J. of Heat Transfer, **19**, 245-248

[416] Widder, 1941, *The Laplace Transform,* Princeton University Press, Princeton, NJ, USA

[417] Wilhelm, H. E. and Choi, S. H., 1975, J of Chemical Physics, **63**, 2119-2123

[418] Wolf. H. F., 1983, *Heat Transfer,* Harper and Row, New York, NY, USA

[419] Xu, M., and Wang, L, 2002, Int. J of Heat and Mass Transfer, **45**, 1055-1061

[420] Zachmanoglou, E. C., and Thoe D. W., 1986, *Introduction to Partial Differential Equations with Applications,* Dover, New York, NY, USA

Appendix A

Bessel's Differential Equation

The linear second order differential equation shown in Eq. [A.1], is referred to as Bessel's equation and the solutions termed as "Bessel" functions. The solutions of this equation is available in Watson [1966]. Certain types of differential equations are amenable to a solution expressed as a power series. Such a series is said to converge if it approaches a finite value as n approaches infinity. The simplest test for convergence is the ratio test within the interval of convergence. The method of Frobenius is a convenient method to obtain power series solution to linear homogeneous second order differential equation with variable convergent coefficients.

$$x^2 \, d^2y/dx^2 + xdy/dx + (x^2 - p^2)y = 0 \qquad (A.1)$$

Eq. [A.1] when expressed in the standard form can be written as;

$$d^2y/dx^2 + (1/x)dy/dx + (x^2 - p^2)/(x^2)y = 0 \qquad (A.2)$$

The second order homogeneous general second order ordinary differential equation can be expressed in the standard form as [Varma and Morbidelli, 1997],

$$R(x) \, d^2y/dx^2 + (P(x)/x)dy/dx + V(x)/x^2(y) = 0 \qquad (A.3)$$

Comparing Eq. [A.2] with the standard form [which is;

$$R(x) = 1; \quad P(x) = 1; \quad V(x) = x^2 - p^2 \qquad (A.4)$$

The functions $R(x)$, $P(x)$ and $V(x)$ need be expanded as power series and the coefficients of the power series calculated as;

$$R_0 = 1; \quad R_1 = R_2 = R_3 \; \dots\dots R_n = 0 \qquad (A.5)$$

$$P_0 = 1; \quad P_1 = P_2 = P_3 \; \dots\dots P_n = 0 \qquad (A.6)$$

$$V_0 = -p^2; \quad V_2 = 1; \qquad (A.7)$$

$$V_1 = V_3 = V_4 = \dots\dots = V_n \qquad (A.8)$$

The solution to Eq. [A.3] by the method of Frobenius [Mickley, Sherwood and Reed, 1975] states that there is atleast one solution of the following form ;

$$y = x^s \sum_{0}^{\infty} A_n x^n \qquad (A.9)$$

Substituting Eq. [A.2] into Eq. [A.4], the indicial equation obtained can be written as;

$$s^2 + (P_0 - 1)s + V_0 = 0 \quad \text{or} \quad s^2 - p^2 = 0 \tag{A.10}$$

$$s_1 = p; \quad s_2 = -p \tag{A.11}$$

The recurrence relation for A_n in Eq. [A.9] can be seen to be ;

$$A_n = - \sum_1^n g_k(s + n)A_{n-k} /f(s+n) \tag{A.12}$$

The two solutions for the two roots in Eq. [A.11] are;

$$y_1(x) = A_0 x^p \left(1 + \sum_1^\infty (-1)^k x^{2k} /[(1+p)(2+p).......(k+p)2^{2k}k!] \right) \tag{A.13}$$

$$y_2(x) = B_0 x^{-p} \left(1 + \sum_1^\infty (-1)^k x^{2k} /[(1-p)(2-p).......(k-p)2^{2k}k!] \right) \tag{A.14}$$

Eqs. [A.13, A.14] can be expressed as a more useful form by making use of the gamma function. The gamma function can be defined in the Euler form as;

$$\Gamma(p) = \int_0^\infty \exp(-x)x^{p-1}dx, \qquad p > 0 \tag{A.15}$$

Some mentionable properties of the gamma function are;

$$\Gamma(p+1) = p\Gamma(p) = , \qquad p > 0$$

$$\Gamma(p + k) = (p+ k-1)(p + k - 2)......(p+1)(p)\Gamma(p)$$

If p is a positive integer, $\qquad \Gamma(n+1) = n! \tag{A.16}$

Gamma function generalizes the use of factorial to noninteger positive values of p. Thus,

$$\Gamma(9/2) = (7/2)(5/2)(3/2)(1/2)\,\Gamma(1/2)$$

$$\Gamma(1/2) = \int_0^\infty \exp(-p)p^{-1/2}dp, = (\pi)^{1/2} \tag{A.17}$$

The defenition can be extented to negative non-integer values but not for zero and negative numbers. For large values of the argument the Stirling approximation may be used;

$$\Gamma(p) \sim (2\pi/p)^{1/2} (p/e)^p \tag{A.18}$$

With the use of gamma function Eq [A.13] becomes;

$$y_1(x) = A_0 \sum_0^\infty (-1)^k (x/2)^{2k+p} /(k!\,\Gamma(k+ p +1)) \tag{A.19}$$

Using the notation for the Bessel function of the first kind and p^{th} order,

$$\infty$$

$$J_p(x) \quad = \sum_0 (-1)^k (x/2)^{2k+p} /(k! \; k+p)!)$$
(A.20)

Eq. [A.19] becomes;

$$y_1(x) = C_1 \; J_p(x)$$

In a similar vein, Eq. [A.14] when p is neither zero nor a positrive integer can be written as;

$$y_2(x) = C_2 \; J_{-p}(x)$$
where,

$$J_{-p}(x) \quad = \sum_0^\infty (-1)^k (x/2)^{2k-p} /(k! \; k-p)!)$$
(A.21)

When p is zero or positive integer it can be shown that,

$$y_2(x) = \quad C_2 \; Y_p(x)$$

where $Y_p(x)$ is the Bessel function of the second kind anf p^{th} order. The Weber form can be written as;

$$Y_p(x) \quad = 2/\pi\{(\gamma + \ln(x/2))J_p(x) \quad 1/2\sum_0^\infty (-1)^{k+1} (\phi(k) + \phi(k+p)) (x/2)^{2k+p}/k!(k+p)!\}$$
(A.22)

where, γ is Euler's constant,

$$\gamma = 0.577215...$$

In this case, the roots of the indicial equation are both equal to zero and the second linearly independent form can be written as Bessel function multiplied with a logarithm function and a second infinite power series. Thus the complete solution of the Bessel equation when,
p is a positive integer or zero can be written as;

$$y \quad = C_1 \; J_p(x) + C_2 \; Y_p(x)$$
(A.23)

When p is neither a integer nor zero,

$$y = C_1 \; J_p(x) + C_2 \; J_{-p}(x)$$
(A.24)

The linear second order ordinary differential equation givenin Eq. [A.25] can be transformed into the Bessel equation given in Eq. [A.1] by a substitution,
z = ix

$$d^2y/dx^2 + xdy/dx - (x^2 + p^2)y = 0$$
(A.25)

$$z^2d^2y/dz^2 + zdy/dz + (z^2 - p^2)y = 0$$

The solution can then be written as;

p is a positive integer or zero can be written as;

$$y = C_1 J_p(ix) + C_2 Y_p(ix)$$

or $\qquad y = C_1 I_p(x) + C_2 K_p(x)$ \hfill (A.26)

When p is neither a integer nor zero,

$$y = C_1 J_p(ix) + C_2 J_{-p}(ix)$$

or $\qquad y = C_1 I_p(x) + C_2 I_{-p}(x)$ \hfill (A.27)

$I_p(x)$ is referred to as the modified Bessel function of the first kind and p^{th} order and is defined by the expression;

$$I_p(x) = \sum_{0}^{\infty} (x/2)^{2k+p}/(k!\,k+p)!)$$ \hfill (A.28)

$K_p(x)$ is referred to as the modified Bessel function of the second kind and p^{th} order and is defined by the expression;

$$K_p(x) = \pi/2\; i^{p+1} (J_p(ix) + iY_p(ix))$$ \hfill (A.29)

The generalized form of Bessel's equation can be written as;

$$x^2\, d^2y/dx^2 + x(a + 2bx^r)\, dy/dx + (c + dx^{2s} - b(1 - a - r)x^r + b^2x^{2r})y = 0$$ \hfill (A.30)

Eq. [A.30] can be reduced to the Eq. [A.1] after suitable transformations {Mickley, Sherwood and Reed, 1975]. The generalized solution for Eq. [A.30] may be written as;

$$Y = x^{(1-a)/2}\exp(-bx^r/r)\{ c_1 Z_p(|d|^{1/2}/s\; x^s) + c_2 Z_{-p}(|d|^{1/2}/s\; x^s)$$ \hfill (A.31)

Where $p = \mathrm{sqrt}[(1-a)^2 - 4c]/(2s)$ \hfill (A.32)

Table A.1 Forms of Bessel Solution

S.No	$|d|^{1/2}/s$	p	Z_p	Z_{-p}
1.	Real	Neither zero nor integer	J_p	J_{-p}
2.	Real	Either zero or integer	J_p	Y_p
3.	Imaginary	Neither zero nor integer	I_p	I_{-p}
4.	Imaginary	Either zero or integer	I_p	K_p

p is the order of the Bessel equation. The different forms of the Bessel solution assumes depending on the nature of $\mathrm{sqrt}(|d|)/s$ and p is given in Table A.1. For small values of x, the following approximations can be made for the Bessel functions;

$$J_p(x) \sim 1/2^p p! \, x^p \qquad \text{(A.33)}$$

$$J_{-p}(x) \sim 2^p/(-p)! x^{-p} \qquad \text{(A.34)}$$

$$Y_n(x) \sim -x^{-n} (2^n (n-1)!)/\pi \qquad \text{(A.35)}$$

$$Y_0(x) \sim 2/\pi \ln(x) \qquad \text{(A.36)}$$

$$I_p(x) \sim 2^p/(p)! \, x^p \qquad \text{(A.37)}$$

$$I_{-p}(x) \sim 2^p/(-p)! \, x^{-p} \qquad \text{(A.38)}$$

$$K_n(x) \sim 2^{n-1}(n-1)! \, x^{-n} \qquad \text{(A.39)}$$

$$K_0(x) \sim -\ln(x) \qquad \text{(A.40)}$$

For large values, the general character may be obtained by the following substitution;

$$y = x^{-1/2} u \qquad \text{(A.41)}$$

Eq. [A.1] then becomes;

$$x^2 d^2y/dx^2 = +3y/4 \ -x^{1/2}du/dx \ -x^{1/2}du/dx + x^{3/2}d^2u/dx^2$$

$$xdy/dx = -y/2 \ + x^{1/2}du/dx$$

$$+ d^2u/dx^2 + u(1 \ -(p^2-1/4)/x^2 \ ) = 0 \qquad \text{(A.42)}$$

For large values of x, approximately it can be shown that,

$$J_p(x) \sim (2/\pi x)^{1/2} \, \mathrm{Cos}(x - \pi p/2 \ - \pi/4) \qquad \text{(A.43)}$$

$$Y_p(x) \sim (2/\pi x)^{1/2} \, \mathrm{Sin}(x - \pi p/2 \ - \pi/4) \qquad \text{(A.44)}$$

In a similar vein, the modified Bessel function can be approximated as;

$$I_p(x) \sim \exp(x)/(2\pi x)^{1/2} \qquad \text{(A.45)}$$

$$K_p(x) \sim \exp(-x) (\pi/2x)\exp(-x) \qquad \text{(A.46)}$$

The first zero of the Bessel function of the first kind occures for the zeroth order $J_0(x)$ at 2.4048, for the first order, $J_1(x)$ at 3.8317, for the second order, $J_2(x)$ at 5.1356, for the third order at 6.3802, and $J_4(x)$ for the fourth order at 7.5883. The zeros of the Bessel function of the second kind occurs for the zeroth order $Y_0(x)$ at 0.8936, for the first order, $Y_1(x)$ at 2.1971, for the second order, $Y_2(x)$ at 3.3842, for the third order, $Y_3(x)$

at 4.5270, and $Y_4(x)$ for the fourth order at 5.6451. Both $J_p(x)$ and $Y_n(x)$ oscillate like damped sinusoidal functions and approach zero as x tends to infinity. The amplitude of the oscillations about zero decreases as x increases, and the distance between successive zeros of both functions decreases toward a limit of π as x increases. The zeros of $J_{p+1}(x)$ separate the zeros of $J_p(x)$. $I_p(x)$ in contrast increases continuously witrh x, and K_n decreases continuously. Bessel functions of order equal to half an odd integer can be represented in terms of the elementary functions;

$$J_{1/2}(x) = (2/\pi x)^{1/2}Sin(x) \tag{A.47}$$

$$J_{-1/2}(x) = (2/\pi x)^{1/2}Cos(x) \tag{A.48}$$

$$I_{1/2}(x) = (2/\pi x)^{1/2}Sinh(x) \tag{A.49}$$

$$I_{-1/2}(x) = (2/\pi x)^{1/2}Cosh(x) \tag{A.50}$$

The recurrence relations among Bessel functions can be given by;

$$J_{n+1/2}(x) = (2n-1)/x\ J_{n-1/2}(x) - J_{n-3/2}(x) \tag{A.51}$$

$$I_{n+1/2}(x) = -(2n-1)/x\ I_{n-1/2}(x) + I_{n-3/2}(x) \tag{A.52}$$

The following relations can be proved by using Eqs. [A.21, A.28].

$$d/dx\ (x^p Z_p(\alpha x)) = \alpha x^p Z_{p-1}(\alpha x), \qquad Z = J, Y, I \tag{A.53}$$

$$d/dx(x^p K_p(\alpha x)) = -\alpha x^p K_{p-1}(\alpha x) \tag{A.54}$$

$$d/dx\ (x^{-p}Z_p(\alpha x)) = -\alpha x^p Z_{p+1}(\alpha x), \qquad Z = J, Y, K \tag{A.55}$$

$$d/dx\ (x^{-p}I_p(\alpha x)) = \alpha x^{-p}I_{p+1}(\alpha x), \tag{A.56}$$

$$d/dx\ (Z_p(\alpha x)) = \alpha Z_{p-1}(\alpha x) - p/x Z_p(\alpha x), \qquad Z = J, Y, I \tag{A.57}$$

$$d/dx(K_p(\alpha x)) = -\alpha x^p K_{p-1}(\alpha x) - p/x\ K_p(\alpha x) \tag{A.58}$$

$$d/dx\ (Z_p(\alpha x)) = -\alpha Z_{p+1}(\alpha x) + p/x Z_p(\alpha x), \qquad Z = J, Y, K \tag{A.59}$$

$$d/dx(I_p(\alpha x)) = \alpha x^p I_{p+1}(\alpha x) + p/x\ I_p(\alpha x) \tag{A.60}$$

$$2d/dx\ I_p(\alpha x) = \alpha(I_{p-1}(\alpha x) + I_{p+1}(\alpha x)) \tag{A.61}$$

$$2d/dx\ K_n(\alpha x) = -\alpha(K_{n-1}(\alpha x) + K_{n+1}(\alpha x)) \tag{A.62}$$

$$Z_p(\alpha x) = \alpha x/2p\ (Z_{p+1}(\alpha x) + Z_{p-1}(\alpha x)), \qquad Z = J, Y \tag{A.63}$$

$$I_p(\alpha x) \quad = -\alpha x/2p \ (I_{p+1}(\alpha x) - I_{p-1}(\alpha x)) \qquad (A.64)$$

$$K_n(\alpha x) \quad = \alpha x/2p \ (K_{n+1}(\alpha x) - K_{n-1}(\alpha x)) \qquad (A.65)$$

When n is zero or a integer,

$$J_{-n}(\alpha x) \quad = (-1)^n J_n(\alpha x) \qquad (A.66)$$

$$I_{-n}(\alpha x) \quad = I_n(\alpha x) \qquad (A.67)$$

$$K_{-n}(\alpha x) \quad = K_n(\alpha x) \qquad (A.68)$$

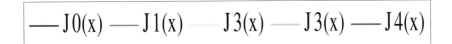

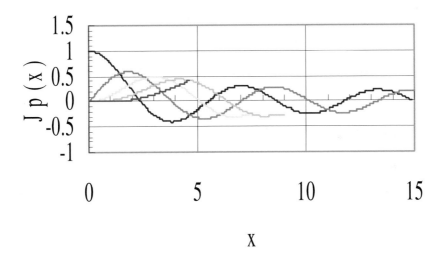

Figure A.1 Bessel function of the First Kind and pth order (p = 0,1,2,3,4..)

Appendix B

Inverse of Laplace Transforms

S.No.	Transform	Function f(t)
1.	$\Gamma(k)/s^k$, $k > 0$	t^{k-1}
2.	$1/s^{n+1/2}$, $n = 0,1,2,3....$	$2^n t^{n-1/2}/[1.3.5....(2n-1)]/\pi$
3.	$1/(s-a)^n$, $n = 1,2,3.....$	$\exp(at)t^{n-1}/(n-1)!$
4.	$1/(s^2 + a^2)$	$1/a \, \mathrm{Sin}(at)$
5.	$s/(s^2 + a^2)$	$\mathrm{Cos}(at)$
6.	$1/(s^2 - a^2)$	$1/a \mathrm{Sinh}(t)$
7.	$s/(s^2 - a^2)$	$1/a \mathrm{Cosh}(t)$
8.	$s/(s^4 + 4a^4)$	$\mathrm{Sin}(at)\mathrm{Sinh}(at)/(2a^2)$
9.	$1/s \, (s-1)^n/s^n$	$\mathrm{Exp}(t)/n! \; d^n/dt^n(t^n\exp(-t))$ Laquerre polynomial of degree n
10.	$(s-a)^{1/2} - (s-b)^{1/2}$	$\frac{1}{2}(\exp(bt) - \exp(at))/(\pi t^3)$
11.	$1/(s^{1/2} + a)$	$(\pi t)^{-1}$ - $a\exp(a^2 t)\mathrm{erfc}(at^{1/2})$
12.	$1/(s+a)/(s+b)^{1/2}$	$1(b-a)^{1/2} \exp(-at)\mathrm{erf}(t(b-a))^{1/2}$
13.	$(1-s)^n/s^{n+1/2}$	$(\pi t)^{-1/2} \, n!/(2n)! \, H_{2n}(t)$ Hermite polynomial $H_n(t) = \exp(-t^2)d^n/dt^n \exp(-t^2)$
14.	$(s+a)^{-1/2}(s+b)^{-1/2}$	$\exp(-(a+b)t/2)I_0((a-b)t/2)$
15.	$(s^2 + a^2)^{-1/2}$	$J_0(at)$
16.	$(s^2 - a^2)^{-k}$, $k > 0$	$(\pi)^{1/2}/\Gamma(k) \, (t/2a)^{k-1/2} I_{k-1/2}(at)$
17.	$\exp(-sk)/s$	$S_k(t) = 0$, $0 < t < k$ $= 1$, $t > k$, Heaviside unit step function
18.	$\exp(-sk)/s^j$, $j > 0$	0, $O < t < k$ $(t - k)^{j-1}/\Gamma(j)$, $t > k$
19.	$(1 - \exp(-sk))/s$	1 when $0 < t < k$ 0 when $t > k$
20.	$1/2s + \coth(sk/2)/2s$	$S(k,t) = n$, when $(n-1)k < t < nk$ $n = 1,2.3...$
21.	$1/[s \, (\exp(sk) - a)]$	0, when $0 < t < k$ $1 + a + a^2 +... + a^{n-1}$, when $nk < t < (n+1)k$, $n = 1,2,3...$
22.	$1/s \tanh(sk)$	$M(2k,t) = (-1)^{n-1}$ when $2k(n-1) < t < 2kn$

S.No.	Transform	Function f(t)
23.	$1/[s(1 + \exp(-sk))]$	$\frac{1}{2}M(k,t) + \frac{1}{2} = (1 - (-1)^n)/2$ when $(n-1)k < t < nk$
24.	$1/s^2 \tanh(sk)$	$H(2k,t) = t$ when $0 < t < 2k$ $= 4k - t$ when $2k < t < 4k$
25.	$1/(s\mathrm{Sinh}(sk))$	$2S(2k, t+k) - 2 = 2(n-1)$ when $(2n-3)k < t < (2n-1)k, \ t > 0$
26.	$1/(s\mathrm{Cosh}(sk))$	$M(2k, t+3k) + 1 = 1 + (-1)^n$ When $(2n-3)k < t < (2n-1)k, \ t > 0$
27.	$\mathrm{Coth}(sk)/s$	$2S(2k,t) - 1 = 2n-1$ when $2k(n-1) < t < 2kn$
28.	$k/(s^2 + k^2) \mathrm{Coth}(\pi s)/2k$	$\lvert \sin(kt) \rvert$
29.	$1/[(s^2 + 1)(1 - \exp(-\pi s))]$	Sint when $(2n=2)\pi < t < (2n-1)\pi$ 0, when $(2n-1)\pi < t < 2n\pi$
30.	$1/s \exp(-k/s)$	$J_0(2(kt)^{1/2})$
31.	$\exp(-k/s)/s^{1/2}$	$\mathrm{Cos}(2(kt)^{1/2})/(\pi t)^{1/2}$
32.	$\exp(k/s)/s^{1/2}$	$\mathrm{Cosh}(2(kt)^{1/2})/(\pi t)^{1/2}$
33.	$\exp(-k/s)/s^{3/2}$	$\mathrm{Sin}(2(kt)^{1/2})/(\pi k)^{1/2}$
34.	$\exp(k/s)/s^{3/2}$	$\mathrm{Sinh}(2(kt)^{1/2})/(\pi k)^{1/2}$
35.	$1/s^j \exp(-k/s), \ j > 0$	$(t/k)^{(j-1)/2} J_{i-1}(2(kt)^{1/2})$
36.	$1/s^j \exp(k/s), \ j > 0$	$(t/k)^{(j-1)/2} I_{i-1}(2(kt)^{1/2})$
37.	$\exp(-k(s)^{1/2}), \ k > 0$	$k/[2(\pi t^3)^{1/2}] \exp(-k^2/4t)$
38.	$1/s \exp(-ks^{1/2}), \ k > 0$	$\mathrm{erfc}(k/2t^{1/2})$
39.	$1/s^{1/2} \exp(-ks^{1/2}), \ k > 0$	$1/(\pi t)^{1/2}\exp(-k^2/4t)$
40.	$1/s^{3/2} \exp(-ks^{1/2}), \ k > 0$	$2(t/\pi)^{1/2} \exp(-k^2/4t) - k\mathrm{erfc}(k/2t^{1/2})$
41.	$a\exp(-s^{1/2}k)/(s(a + s^{1/2})), \ k > 0$	$(-\exp(ak)\exp(a^2t)\mathrm{erfc}(at^{1/2} + k/2t^{1/2})$ $+ \mathrm{erfc}(k/2t^{1/2})$
42.	$\exp(-s^{1/2}k)/s^{1/2}/(a + s^{1/2})$	$\mathrm{Exp}(ak)\exp(a^2t)\mathrm{erfc}(at^{1/2} + k/2t^{1/2})$
43.	$\exp(-k(s(s+a))^{1/2})/(s(s+1))^{1/2}$	$0, \ 0 < t < k$ $\exp(-at/2)I_0(a/2(t^2 - k^2)^{1/2})$
44.	$\exp(-k(s^2 + a^2)^{1/2})/(s^2 + a^2)^{1/2}$	0, when $0 < t < k$ $J_0(a(t^2 - k^2)^{1/2}$, when $t > k$
45.	$\exp(-k(s^2 - a^2)^{1/2})/(s^2 - a^2)^{1/2}$	0, when $0 < t < k$ $I_0(a(t^2 - k^2)^{1/2}$, when $t > k$

46.	$\exp(-k(s^2 + a^2)^{1/2} - s)/(s^2 + a^2)^{1/2}$	$J_0(a (t^2 + 2kt)^{1/2})$
47.	$\exp(-sk) - \exp(-k(s^2 + a^2)^{1/2})$	0, when $0 < t < k$ $ak/(t^2 - k^2) J_1(a (t^2 - k^2)^{1/2})$
48.	$\exp(-k(s^2 - a^2)) - \exp(-sk)$	0, when $0 < t < k$ $ak/(t^2 - k^2)^{1/2} I_1(a (t^2 - k^2)^{1/2})$, $t > k$
49.	$a^j \exp(-k(s^2 + a^2)^{1/2})/(s^2 + a^2)^{1/2}/(s^2 + a^2)^{1/2} + s)^j$, $j > -1$	0, when $0 < t < k$ $[(t - k)/(t + k)]^{1/2 j} J_j(a(t^2 - k^2)^{1/2})$, $t > k$
50.	$1/s \ln s$	$\lambda - \ln t$, $\lambda = -.5772$
51.	$1/s^k \ln s$	$t^{k-1} (\lambda/\Gamma(k)^2 - \ln t/\Gamma(k))$
52.	$\ln s/(s-a)$	$\exp(at) (\ln a - E_i(-at))$
53.	$\ln s/(s^2 + 1)$	$\text{Cos}t \, \text{Si}(t) - \text{sin}t \, \text{Ci}(t)$
54.	$s \ln s/(s^2 + 1)$	$-\text{sin}t \, \text{Si}(t) - \text{cos}t \, \text{Ci}(t)$
55.	$1/s \ln(1 + sk)$	$-\text{Ei}(-t/k)$
56.	$\ln[(s-a)/s-b)]$	$1/t (\exp(bt) - \exp(at))$
57.	$1/s \ln(1 + k^2s^2)$	$-2\text{Ci}(t/k)$
58.	$1/s \ln(s^2 + a^2)$, $a > 0$	$2\ln a - 2 \, \text{Ci}(at)$
59.	$1/s^2 \ln(s^2 + a^2)$, $a > 0$	$2/a(at \ln a + \text{Sin}at - at\text{Ci}(at))$
60.	$\ln(s^2 + a^2)/s^2$	$2/t(1 - \text{Cos}(at))$
61.	$\ln(s^2 - a^2)/s^2$	$2/t(1 - \text{Cosh}(at))$
62.	$\tan^{-1}(k/s)$	$1/t \, \text{Sin}kt$
63.	$1/s \tan^{-1}k/s$	$\text{Si}(kt)$
64.	$\exp(s^2k^2)\text{erfc}(sk)$, $k > 0$	$1/(k\pi^{1/2}) \exp(-t^2/4k^2)$
65.	$1/s \exp(s^2k^2)\text{erfc}(sk)$, $k > 0$	$\text{erf}(t/2k)$
66.	$\exp(sk) \text{erfc}(sk)^{1/2}$, $k > 0$	$k^{1/2}/[\pi(t)^{1/2}(t + k)]$
67.	$1/s^{1/2} \text{erfc}(sk)^{1/2}$	0, $0 < t < k$ $(\pi t)^{-1/2}$, $t > k$
68.	$1/s^{1/2} \exp(sk) \text{erfc}(sk)^{1/2}$, $k > 0$	$(\pi(t + k))^{-1/2}$
69.	$\text{erf}(k/s^{1/2})$	$1/\pi t \, \text{Sin}(kt^{1/2})$
70.	$1/s^{1/2} \exp(k^2/s) \text{erfc}(k/s^{1/2})$	$1/(\pi t)^{1/2} \exp(-2kt^{1/2})$
71.	$K_0(sk)$	0, $0 < t < k$ $(t^2 - k^2)^{-1/2}$, when $t > k$
72.	$K_0(ks^{1/2})$	$1/2t \exp(-k^2/4t)$
73.	$1/s \exp(sk) K_1(sk)$	$1/k (t(t + 2k))^{1/2}$
74.	$1/s^{1/2} K_1 (ks^{1/2})$	$1/k \exp(-k^2/4t)$
75.	$1/s^{1/2} \exp(k/s) K_0(k/s)$	$2/(\pi t)^{1/2} K_0 (2kt)^{1/2}$
76.	$\pi \exp(-sk) I_0(sk)$	$(t (2k - t))^{-1/2}$, $0 < t < 2k$ 0 , $t > 2k$
78.	$-(\gamma + \ln s)/s$, $\gamma = $ Euler's constant $= 0.5772156$	$\ln t$

79.	$1/as^2$ tanh(as/2)	Triangular wave function
80.	$1/s$ tanh(as/2)	Square wave function
81.	$\pi a/(a^2s^2 + \pi^2)$ coth(as/2)	Rectified sine wave function
82.	$\pi a/[(a^2s^2 + \pi^2)(1 - \exp(-as)]$	Half rectified sine wave function
83.	$1/as^2 - \exp(-as)/s(1 - \exp(-as))$	Saw tooth wave function
84.	Sinhsx/(sSinhsa)	$x/a + 2/\pi \sum_1^\infty (-1)^n/n$ sin(nπx)/a cos(nπt/a)
85.	Sinhsx/(sCoshsa)	$4/\pi \sum_1^\infty (-1)^n/(2n-1)$ sin(2n-1πx)/2a Sin(2n-1)πt/2a
86.	Coshsx/sSinhsa	$t/a + 2/\pi \sum_1^\infty (-1)^n/n$ Cos(nπx)/a Sin(nπt/a)
87.	Coshsx/sCoshsa	$1 + 4/\pi \sum_1^\infty (-1)^n/(2n-1)$ Cos(2n-1)πx)/2a Cos(2n-1)πt/2a
88.	Sinhsx/s^2 Coshsa	$x + 8a/\pi^2 \sum_1^\infty (-1)^n/(2n-1)^2$Sin(2n-1)$\pi$x)/2a Cos(2n-1)$\pi$t/2a
89.	Coshsx/s^2Sinhsa	$t^2/2a + 2a/\pi^2 \sum_1^\infty (-1)^n/(n^2)$ Cos(nπx)/a $(1 - $Cos(n$\pi$t/a)
90	Sinhxs$^{1/2}$/Sinhas$^{1/2}$	$2\pi/a^2 \sum_1^\infty (-1)^n n \exp(-n^2\pi^2 t)/a^2$ Sin(nπx/a)
91.	Coshxs$^{1/2}$/s$^{1/2}$Sinhas$^{1/2}$	$1/a + 2/a \sum_1^\infty (-1)^n n \exp(-n^2\pi^2 t)/a^2$ Cos(nπx/a)

Appendix C

Equation of Continuity and Equation of Motion and Equation for Temperature and Equation for Concentration in Cartesian, Cylindrical and Spherical Coordinates

The Navier-Stokes equation of motion is extended to include the fluid inertial and transient effects using the damped wave momentum transport and relaxation equation to give;

$$\tau_{mom}\rho D\partial V/\partial t \,/Dt \;+\; \rho DV/Dt \quad +\; \tau_{mom}\nabla(\partial p/\partial t) \;=\; \mu\,\nabla^2 V \,-\nabla p \;+\; \rho g \tag{C.1}$$

Table C.1 Equation of Continuity

Cartesian Coordinates
$$\partial\rho/\partial t \;+\; \partial(\rho v_x)/\partial x \;+\; \partial(\rho v_y)/\partial y \;+\; \partial(\rho v_z)/\partial z \;=\; 0 \tag{C.2}$$
Polar Coordinates
$$\partial\rho/\partial t \;+\; 1/r\;\partial(\rho r v_r)/\partial r \;+\; 1/r\partial(\rho v_\theta)/\partial\theta \;+\; \partial(\rho v_z)/\partial z \;=\; 0 \tag{C.3}$$
Spherical Coordinates
$$\partial\rho/\partial t \;+\; 1/r^2\;\partial(\rho r^2 v_r)/\partial r \;+\; 1/r\mathrm{Sin}\theta\;\partial(\rho v_\theta\,\mathrm{Sin}\theta)/\partial\theta \;+\; 1/r\mathrm{Sin}\theta\partial(\rho v_\phi)/\partial\phi \;=\; 0 \tag{C.4}$$

Equation of Motion in Terms of Velocity for a Newtonian Fluid with Constant Density and Viscosity

Table C.2 Extended Navier Stokes Equation including the Ballistic Accumulation Term in Cartesian Coordinates

---

x component

$$\rho\tau_{mom}\,[\partial^2 v_x/\partial t^2 + v_x\partial^2 v_x/\partial t\partial x + v_y\partial^2 v_x/\partial t\partial y + v_z\partial^2 v_x/\partial t\partial z\,] +$$

$$\rho[(\tau_{mom}\,\partial v_x/\partial t + v_x)(\,\partial v_x/\partial x\,) + (\tau_{mom}\,\partial v_y/\partial t + v_y)(\,\partial v_x/\partial y\,) + (\tau_{mom}\,\partial v_z/\partial t + v_z)($$

$$\partial v_x/\partial z)] \quad + \tau_{mom}\partial^2 p/\partial t\,\partial x + \rho\partial v_x/\partial t$$

$$= -\partial p/\partial x + \mu(\partial^2 v_x/\partial x^2 + \partial^2 v_x/\partial y^2 + (\partial^2 v_x/\partial z^2) + \rho g_x$$

(C.5)

---

y component

$$\rho\tau_{mom}\,[\partial^2 v_y/\partial t^2 + v_x\partial^2 v_y/\partial t\partial x + v_y\partial^2 v_y/\partial t\partial y + v_z\partial^2 v_y/\partial t\partial z\,] +$$

$$\rho[(\tau_{mom}\,\partial v_x/\partial t + v_x)(\,\partial v_y/\partial x\,) + (\tau_{mom}\,\partial v_y/\partial t + v_y)(\,\partial v_y/\partial y\,) +$$

$$(\tau_{mom}\,\partial v_z/\partial t + v_z)(\,\partial v_y/\partial z)] \quad + \tau_{mom}\partial^2 p/\partial t\,\partial y + \rho\partial v_y/\partial t$$

$$= -\partial p/\partial y + \mu\,(\partial^2 v_x/\partial x^2 + \partial^2 v_x/\partial y^2 + (\partial^2 v_x/\partial z^2) + \rho g_y$$

(C.6)

---

z component

$$\rho\tau_{mom}\,[\partial^2 v_z/\partial t^2 + v_x\partial^2 v_z/\partial t\partial x + v_y\partial^2 v_z/\partial t\partial y + v_z\partial^2 v_z/\partial t\partial z +$$

$$\rho[(\tau_{mom}\,\partial v_x/\partial t + v_x)(\,\partial v_z/\partial x\,) + (\tau_{mom}\,\partial v_y/\partial t + v_y)(\,\partial v_z/\partial y\,) +$$

$$(\tau_{mom}\,\partial v_z/\partial t + v_z)(\,\partial v_z/\partial z)] + \rho\partial v_z/\partial t + \tau_{mom}\partial^2 p/\partial t\,\partial z$$

$$= -\partial p/\partial z + \mu\,(\partial^2 v_x/\partial x^2 + \partial^2 v_x/\partial y^2 + (\partial^2 v_x/\partial z^2) + \rho g_z$$

(C.7)

Table C.3 Extended Navier Stokes Equation including the Ballistic Accumulation Term in Polar Coordinates

---

r component

$\rho\tau_{mom}$ [$1/r\partial^2(rv_r)/\partial t^2$ + $(v_r/r)\partial^2(rv_r)/\partial t\partial r$ + $v_\theta/r^2\,\partial^2(rv_r)/\partial t\partial\theta$ + $v_z\partial^2v_r/\partial t\partial z$ - $2/rv_\theta$ $\partial v_\theta/\partial t$] + ρ[($\tau_{mom}/r\partial(rv_r)/\partial t$ + v_r)($1/r\partial(rv_r)/\partial r$)

+ ($\tau_{mom}\,\partial v_\theta/\partial t$ + v_θ)($1/r^2\partial(rv_r)/\partial\theta$) + ($\tau_{mom}\,\partial v_z/\partial t$ + v_z)($\partial v_r/\partial z$)]

+ $\tau_{mom}\partial^2p/\partial t\,\partial r$ + $\rho/r\partial(rv_r)/\partial t$ - $\rho v_\theta^2/r$

= -$\partial p/\partial r$ + $\mu[\partial/\partial r(1/r\,\partial/\partial r(rv_r)$ + $1/r^2\partial^2v_r/\partial\theta^2$ -$2/r^2\partial v_\theta/\partial\theta$ + $(\partial^2v_r/\partial z^2)]$ + ρg_r

$$(C.8)$$

θ component

$\rho\tau_{mom}$ [$\partial^2v_\theta/\partial t^2$ + $v_r\partial^2v_\theta/\partial t\partial r$ + $v_\theta/r\,\partial^2v_\theta/\partial t\partial\theta$ + $v_z\partial^2v_\theta/\partial t\partial z$

+ $v_\theta/r\,\partial v_r/\partial t$ + $v_r/r\,\partial v_\theta/\partial t$] +

ρ[($\tau_{mom}/r\partial(rv_r)/\partial t$ + v_r)($\partial v_\theta/\partial r$) + ($\tau_{mom}\,\partial v_\theta/\partial t$ + v_θ)($1/r\partial v_\theta/\partial\theta$)

+ ($\tau_{mom}\,\partial v_z/\partial t$ + v_z)($\partial v_\theta/\partial z$)]

+ $\tau_{mom}\partial^2p/\partial t\partial\theta$ + $\rho\partial v_\theta/\partial t$ + $\rho\,v_r v_\theta/r$

= -$1/r\partial p/\partial\theta$ + $\mu[\partial/\partial r(1/r\,\partial/\partial r(rv_\theta)$ + $1/r^2\partial^2v_\theta/\partial\theta^2$ +$2/r^2\partial v_r/\partial\theta$ + $(\partial^2v_\theta/\partial z^2)]$ + ρg_θ

$$(C.9)$$

---

z component

$\rho\tau_{mom}$ [$\partial^2v_z/\partial t^2$ + $v_r\partial^2v_z/\partial t\partial r$ + $v_\theta/r\partial^2v_z/\partial t\partial\theta$ + $v_z\partial^2v_z/\partial t\partial z$]

+ ρ[($\tau_{mom}/r\partial(rv_r)/\partial t$ + v_r)($\partial v_z)/\partial r$)

+ ($\tau_{mom}\,\partial v_\theta/\partial t$ + v_θ)($1/r\partial v_z/\partial\theta$) + ($\tau_{mom}\,\partial v_z/\partial t$ + v_z)($\partial v_z/\partial z$)]

+ $\tau_{mom}\partial^2p/\partial t\,\partial z$ + $\rho\partial v_z/\partial t$

= -$\partial p/\partial z$ + $\mu[\partial/\partial r(r\partial v_z/\partial r)$ + $1/r^2\partial^2v_z/\partial\theta^2$ -$2/r^2\partial v_\theta/\partial\theta$ + $(\partial^2v_z/\partial z^2)]$ + ρg_z

$$(C.10)$$

Table C.4 Extended Navier Stokes Equation including the Ballistic Accumulation Term in Circular Coordinates

---

r component

$\rho\tau_{mom}$ [$1/r^2 \partial^2(r^2v_r)/\partial t^2$ + $v_r/r^2 \partial^2(r^2v_r)/\partial t\partial r$ + $v_\theta/r^3 \partial^2(r^2v_r)/\partial t\partial\theta$ + $v_\phi/r^3 Sin\theta$ $\partial^2(r^2v_r)/\partial t\partial\phi$ - $2v_\theta \partial v_\theta/\partial t$ - $2v_\phi \partial v_\phi/\partial t$] + ρ[($\tau_{mom}/r^2 \partial(r^2v_r)/\partial t$ + v_r)($1/r^2 \partial(r^2v_r)/\partial r$)

+ ($\tau_{mom} \partial v_\theta/\partial t$ + v_θ)($1/r^2 \partial(r^2v_r)/\partial\theta$) + ($\tau_{mom} \partial v_\phi/\partial t$ + v_ϕ)$1/rSin\theta(\partial v_r/\partial\phi$)]

+ $\tau_{mom}\partial^2 p/\partial t \partial r$ + $\rho/r^2 \partial(r^2 v_r)/\partial t$ - $\rho(v_\phi^2 + v_\theta^2)/r$

= -$\partial p/\partial r$ + μ[$1/r^2 \partial^2(r^2v_r)/\partial r^2$ + $1/r^2Sin\theta \partial/\partial\theta(Sin\theta\partial v_r/\partial\theta)$ + $1/r^2Sin^2\theta$ $(\partial^2v_r/\partial\phi^2)$] + ρg_r (C.11)

θ component

$\rho\tau_{mom}$ [$\partial^2 v_\theta/\partial t^2$ + $v_r\partial^2 v_\theta/\partial t\partial r$ + $v_\theta/r \partial^2 v_\theta /\partial t\partial\theta$ + $v_\phi/rSin\theta\partial^2 v_\theta/\partial t\partial\phi$

+ $v_\theta/r \partial v_r/\partial t$ + $v_r/r \partial v_\theta/\partial t$ - $v_\phi^2 Cot\theta/r$] +

ρ[($\tau_{mom}/r\partial(rv_r)/\partial t$ + v_r)($\partial v_\theta/\partial r$) + ($\tau_{mom} \partial v_\theta/\partial t$ + v_θ)($1/r\partial v_\theta /\partial\theta$)

+ ($\tau_{mom} \partial v_\phi/\partial t$ + v_ϕ)($1/rSin\theta\partial v_\theta/\partial\phi$)]

+ $\tau_{mom}\partial^2 p/\partial t\partial\theta$ + $\rho\partial v_\theta/\partial t$ + $\rho v_r v_\theta/r$ - $\rho v_\phi^2 Cot\theta/r$

= -$1/r\partial p/\partial\theta$ +
μ[$1/r^2\partial/\partial r(r^2 \partial v_\theta/\partial r$) + $1/r^2\partial/\partial\theta 1/Sin\theta \partial (v_\theta Sin\theta)/\partial\theta$ + $1/r^2 Sin^2\theta \partial^2v_\theta/\partial\phi^2$
+ $2/r^2 \partial v_r/\partial\theta$ - $2/r^2 Cos\theta/r^2Sin^2\theta \partial v_\phi/\partial\phi$] + ρg_θ (C.12)

ϕ Component

$\rho\tau_{mom}$ [$\partial^2 v_\phi /\partial t^2$ + $v_r\partial^2 v_\phi/\partial t\partial r$ + $v_\theta/r \partial^2 v_\phi/\partial t\partial\theta$ + $v_\phi/rSin\theta\partial^2 v_\phi /\partial t\partial\phi$

+ $v_\phi/r \partial v_r/\partial t$ + (v_r/r + $v_\theta Cot\theta/r)\partial v_\phi /\partial t$ - $\partial v_\theta/\partial t Cot\theta/r$] +

ρ[($\tau_{mom}/r\partial(rv_r)/\partial t$ + v_r)($\partial v_\phi/\partial r$) + ($\tau_{mom} \partial v_\theta/\partial t$ + v_θ)($1/r\partial v_\phi /\partial\theta$)

+ ($\tau_{mom} \partial v_\phi/\partial t$ + v_ϕ)($1/rSin\theta\partial v_\phi/\partial\phi$)]

+ $\tau_{mom}/rSin\theta\partial^2 p/\partial t\partial\phi$ + $\rho\partial v_\phi/\partial t$ + $\rho v_r v_\phi/r$ - $\rho v_\phi v_\theta Cot\theta/r$

= -$1/r\partial p/\partial\theta$ +
μ[$1/r^2\partial/\partial r(r^2 \partial v_\theta/\partial r$) + $1/r^2\partial/\partial\theta 1/Sin\theta \partial (v_\theta Sin\theta)/\partial\theta$ + $1/r^2 Sin^2\theta \partial^2v_\theta/\partial\phi^2$
+ $2/r^2 \partial v_r/\partial\theta$ - $2/r^2 Cos\theta/r^2Sin^2\theta \partial v_\phi/\partial\phi$] + ρg_θ

(C.13)

Equation for Temperature for a Newtonian Fluid with Constant Density and Viscosity

Table C.5 Damped Wave Conduction and Relaxation Equation for Temperature including the Convective Effects for Fluids of Constant Density, Thermal Conductivity and Zero Viscosity

Cartesian Components
$\tau_r[\partial^2 T/\partial t^2 + v_x \partial^2 T/\partial t\partial x + v_y \partial^2 T/\partial t\partial y + v_z \partial^2 T/\partial t\partial z] + \partial T/\partial t$ $(\tau_r \partial v_x/\partial t + v_x)\partial T/\partial x + (\tau_r \partial v_y/\partial t + v_y)\partial T/\partial y + (\tau_r \partial v_z/\partial t + v_z)\partial T/\partial z +$ $= \quad \alpha(\partial^2 T/\partial x^2 + \partial^2 T/\partial y^2 + \partial^2 T/\partial z^2)$ $\hfill\text{(C.14)}$

Polar Coordinates
$\tau_r(\partial^2 T/\partial t^2 + v_r/r\partial^2(rT)/\partial t\partial r + v_\theta/r\partial^2 T/\partial t\partial\theta + v_z\partial^2 T/\partial t\partial z +$ $\partial T/\partial r(\tau_r/r\,\partial(rv_r)/\partial t + v_r) + 1/r\partial T/\partial\theta(\tau_r\partial v_\theta/\partial t + v_\theta) + (\tau_r\partial T/\partial t + v_z)(\partial T/\partial z))$ $+ \partial T/\partial t \quad = \quad \alpha(1/r\,\partial/\partial r\,(r\partial T/\partial r) + 1/r^2\,\partial^2 T/\partial\theta^2 + \partial^2 T/\partial z^2$ $\hfill\text{(C.15)}$

Spherical Coordinates
$\tau_r\,(\partial^2 T/\partial t^2 + v_r/r^2\partial^2(r^2\,T)/\partial t\partial r + v_\theta/(r\mathrm{Sin}\theta)\partial^2\,(\mathrm{Sin}\theta\,T/\partial t\partial\theta + v_\phi 1/r\mathrm{Sin}\theta\,\partial^2 T/\partial t\partial\phi$ $+ \partial T/\partial r(\tau_{mr}/r^2\partial\,(r^2 v_r)/\partial t + v_r) + 1/r\partial T/\partial\theta(\tau_{mr}/\mathrm{Sin}\theta\partial(\mathrm{Sin}\theta\,v_\theta)/\partial t + v_\theta) +$ $1/r\mathrm{Sin}\theta(\tau_{mr}\partial v_\phi/\partial t + v_\phi)(\partial T/\partial\phi)) + \partial T/\partial t$ $= \alpha[1/r^2\,\partial/\partial r\,(r^2\,\partial T/\partial r) + 1/r^2\mathrm{Sin}\theta\,\partial/\partial\theta\,(\mathrm{Sin}\theta\,\partial T/\partial\theta) + 1/r^2\,\mathrm{Sin}^2\theta\,\partial^2 T/\partial\phi^2)$ $\hfill\text{(C.16)}$

The equation of energy including the viscous dissipation effects can be written as;

$$\rho C_v \, DT/Dt = -(\nabla.\mathbf{q}) - p \, (\nabla.\mathbf{v}) - (\tau : (\nabla \mathbf{v})) \qquad (C.17)$$

The ballistic accumulation term can be included as follows;
    Differentiating Eq. [C.11] with respect to t and the damped wave heat conduction and relaxation equation with respect to space dimensions,

$$\rho C_v \, D\partial T/\partial t/Dt = -(\nabla. \partial\mathbf{q}/\partial t) - p \, (\nabla. \partial\mathbf{v}/\partial t) - \partial/\partial t(\tau : (\nabla \mathbf{v})) \qquad (C.18)$$

$$\nabla.\mathbf{q} = -k \, \nabla^2 T - \tau_r \partial/\partial t \, \nabla q \qquad (C.19)$$

Substituting Eq. [C.13] into Eq. [C.12]

$$\tau_r \, \rho C_v \, D\partial T/\partial t/Dt + \rho C_v \, DT//Dt = \nabla^2 T - p \, \tau_r \, (\nabla. \partial\mathbf{v}/\partial t) - \tau_r \, \partial/\partial t(\tau : (\nabla \mathbf{v}))$$
$$- p \, (\nabla.\mathbf{v}) - (\tau : (\nabla \mathbf{v})) \qquad (C.20)$$

The shear term in Eq. [C.14] need be eliminated to obtain the Equation for temperature that included the viscous heat dissipation and the reversible rate of internal energy increase per unit volume by compression. The techniques described in Chapter 4 can be used to achieve this outcome.

Equation for Concentration for a Constant Density Diffusivity

Table C.6 Damped Wave Conduction and Relaxation Equation for Concentration including the Convective Effects

Cartesian Coordinates

$$\tau_{mr}[\partial^2 C_A/\partial t^2 + v_x \partial^2 C_A/\partial t\partial x + v_y \partial^2 C_A/\partial t\partial y + v_z \partial^2 C_A/\partial t\partial z] + \partial C_A/\partial t$$
$$(\tau_{mr}\partial v_x/\partial t + v_x)\partial C_A/\partial x + (\tau_{mr}\partial v_y/\partial t + v_y)\partial C_A/\partial y + (\tau_{mr}\partial v_z/\partial t + v_z)\partial C_A/\partial z +$$
$$= \quad D_{AB}(\partial^2 C_A/\partial x^2 + \partial^2 C_A/\partial y^2 + \partial^2 C_A/\partial z^2) \quad + R_A$$

$$(C.21)$$

Polar Coordinates

$$\tau_{mr}(\partial^2 C_A/\partial t^2 + v_r \partial^2 C_A/\partial t\partial r + v_\theta/r \partial^2 C_A/\partial t\partial\theta + v_z \partial^2 C_A/\partial t\partial z +$$
$$\partial C_A/\partial r(\tau_{mr}\partial v_r/\partial t + v_r) + 1/r\partial C_A/\partial\theta(\tau_{mr}\partial v_\theta/\partial t + v_\theta) + (\tau_{mr}\partial v_z/\partial t + v_z)(\partial C_A/\partial z))$$
$$+ \partial C_A/\partial t \quad = \quad D_{AB}(1/r\,\partial/\partial r\,(r\partial C_A/\partial r) + 1/r^2\,\partial^2 C_A/\partial\theta^2 + \partial^2 C_A/\partial z^2 + R_A$$

$$(C.22)$$

Spherical Coordinates

$$\tau_{mr}(\partial^2 C_A/\partial t^2 + v_r \partial^2 C_A/\partial t\partial r + v_\theta/r \partial^2 C_A/\partial t\partial\theta + v_\phi\,1/r\mathrm{Sin}\theta\,\partial^2 C_A/\partial t\partial\phi +$$
$$\partial C_A/\partial r(\tau_{mr}\partial v_r/\partial t + v_r) + 1/r\partial C_A/\partial\theta(\tau_{mr}\partial v_\theta/\partial t + v_\theta) + 1/r\mathrm{Sin}\theta(\tau_{mr}\partial v_\phi/\partial t + v_\phi)(\partial C_A/\partial\phi))$$
$$+ \partial C_A/\partial t$$
$$= D_{AB}[1/r^2\,\partial/\partial r\,(r^2\,\partial C_A/\partial r) + \quad 1/r^2\mathrm{Sin}\theta\,\partial/\partial\theta\,(\mathrm{Sin}\theta\,\partial C_A/\partial\theta) + 1/r^2\,\mathrm{Sin}^2\theta\,\partial^2 C_A/\partial\phi^2)$$
$$+ R_A$$

$$(C.23)$$

Answers

Chapter 1.0

1. the velocity distribution function varies with time. The temperature and velocity relation can be used to derive an expression for heat flux.

2. $q = -k\partial T/\partial y - \beta_1(\partial T/\partial y)^2 - \beta_2(\partial T/\partial y)^3 - \ldots$

3. $\partial u/\partial t$, the acceleration of the molecules needs to be related to the accumulation of temperature by the Boltzmann relation at equilibrium.

4. $(U'''/\rho C_p)T + \alpha\partial^2 T/\partial x^2 = \tau_r\partial^2 T/\partial t^2 + (1 - U'''/S)\partial T/\partial t$

5. when $\partial T/\partial t \gg \exp(t/\tau_r)$, $\alpha(\partial^2 T/\partial x^2 + \partial^2 T/\partial y^2 + \partial^2 T/\partial x^2) = \tau_r\partial^2 T/\partial t^2$

6. cross over takes place when $(k\rho C_p/\pi t)^{1/2} = t\exp(t/\tau_r)$. solution may be obtained by graphical means, or by a Taylor series expansion on $\exp(t/\tau_r)$.

7. ratio test. For the surface flux the ratio and rabii test show divergence.

8. $(T - T_0) = Q/\mathrm{sqrt}(kC_p \, \rho/\tau_r) \; 1/\mathrm{sqrt}(\pi\tau) \exp(-X^2/4\tau)$.

9. $Z = -1 - Pe^2/4$. Hyperbolic.

10. $\alpha\partial^2 T/\partial x^2 = \partial T/\partial t + \tau_r\partial^2 T/\partial t^2 + \tau_r^2/2! \; \partial^3 T/\partial t^3 + \tau_r^2/3! \; \partial^4 T/\partial t^4$.

11. $\alpha\partial^2 T/\partial x^2 = \partial T/\partial t + \tau_r\partial^2 T/\partial t^2 + B\tau_r(C_{po}/C_p)(\partial T/\partial t)^2$
    wave dominance when, $\partial T/\partial t \gg \exp(t/\tau_r)/(1 - \gamma\exp(t/\tau_r))$, where $\gamma = B\tau_r(C_{po}/C_p)$.

12. For regions close to the boundary, $\partial u/\partial\gamma + 1/\delta\partial^2 u/\partial\gamma^2 + Pe \, \partial^2 u/\partial Y\partial\gamma = 0$
    For regions far from the boundary, $\partial u/\partial\gamma + Pe \, \partial^2 u/\partial Y\partial\gamma = \partial^2 u/\partial Y^2$

Chapter 2.0

4. $u(X,\tau)/u(X_p,\tau) = I_0 1/2(\tau^2 - X^2)^{1/2}/ I_0 1/2(\tau^2 - X_p^2)^{1/2}$

5. 211.2 sec. 5.52 cm. a) 247.3 sec. b) 213.4 sec. c) $r = 8.7$ cm d) $r = 9.15$ cm.

6. $\partial^2 q*/\partial\tau^2 + \partial q*/\partial\tau = (1/X \, \partial/\partial X \, (X\partial q*/\partial X) + 1/X^2 \, \partial^2 q*/\partial\theta^2 + \partial^2 q*/\partial Z^2$

   $q* = \exp(-\tau/2) \, I_{3/2} \, \tfrac{1}{2} \, (\tau^2 - X^2 - X^2\theta^2 - Z^2)^{1/2}/((\tau^2 - X^2 - X^2\theta^2 - Z^2)^{3/4}$

   $u = \int \partial q*/\partial X \, d\tau + C(X)$

7. $u = (Q*/\kappa)[\tan(\omega\tau)\mathrm{Sin}(\kappa X - \omega\tau) - \mathrm{Cos}(\kappa X - \omega\tau)]/[\,\mathrm{Cos}(\kappa X_a - \omega\tau) - (\,\mathrm{Sin}(\kappa X_a - \omega\tau)]$

8. $u = [I_{5/2} \, \tfrac{1}{2} \, (\tau^2 - X^2 - Y^2)^{1/2}/ \, I_{5/2} \, (\tau/2)][\tau^{5/2}/((\tau^2 - X^2 - Y^2)^{5/4}]$.

9. $u = \sum_1^\infty c_n \exp(-\tau/2) \exp(-\mathrm{sqrt}(1/4 - \lambda_n^2) \, \tau) \, \mathrm{Sin}(\lambda_n X) + 1 + (x/L)(T_2 - T_0)/(T_0 - T_1)$
   where $\lambda_n = n\pi(\alpha\tau_r)^{1/2}/L$. $c_n = 2(1 - (-1)^n)n\pi$. Bifurcated result. For $L < 2\pi$
   $\mathrm{sqrt}(\alpha\tau_r)$, even for $n = 1$, $\lambda_n > \tfrac{1}{2}$,
   $\quad u = \sum_1^\infty c_n \exp(-\tau/2)\mathrm{Cos}(\mathrm{sqrt}(\lambda_n^2 - 1/4) \, \tau) \, \mathrm{Sin}(\lambda_n X) + 1 + (x/L)(T_2 - T_0)/(T_0 - T_1)$

10. $u = \sum_0^\infty c_n J_0(\lambda_n X) \exp(-\tau/2 - \tau(1/4 - \lambda_n^2)^{1/2}) + u_{ss}$
    $c_n = -\int_0^R J_0(\lambda_n X) / \int_0^R J_0^2(c\lambda_n X)$
    when $R_i < 2\pi(\alpha\tau_r)^{1/2}$
    $u = \sum_2^\infty c_n J_{1/2}(\lambda_n X) \exp(-\tau/2 \, \mathrm{Cos}(\tau(\lambda_n^2 - 1/4)^{1/2}) + u_{ss}$
    where, $u_{ss} = c_1 J_0(\lambda_n X) + c_2 Y_0(\lambda_n X)$
    $c_1 = J_0(\lambda_n X) /(\, J_0(\lambda_n X_i) \, Y_0(\lambda_n X_0) - J_0(\lambda_n X_0)Y_0(\lambda_n X_i))$

11. 136.1 sec. 8.98 cm.

12. 185 °C. 2.3 mm x 2.3 mm. 125 °C.

13. $\langle u \rangle = 1/2\exp(-\tau/2)\exp(-\tau/2(1 - 4Bi_0^*)^{1/2} + 1/2\mathrm{Cos}w*\tau +$
    for $Bi_0^* > \tfrac{1}{4}$, $\langle u \rangle = 1/2\exp(-\tau/2)\mathrm{Cos}(\tau/2(4Bi_0^* - 1)^{1/2}) + 1/2\mathrm{Cos}w*\tau$

15. Time to steady state from Figure 2.13 is 4.5 sec.

16. 6.2 secs.

17. 6.4secs; 5.12 cm; 6.56 secs.

18. $u = \sum_0^\infty c_n \exp(-\tau/2)\mathrm{Cos}(\mathrm{sqrt}(\lambda_n^2 - 1/4)\tau) \, \mathrm{Sin}(\lambda_n X) + u_1(1 - BiX/(1 + BiX_L)$.
    $\lambda_n = -Bi\tan(\lambda_n X_L)$. $c_n = 2(\lambda_n^2 + Bi^2)/(X_L(\lambda_n^2 + Bi^2) + Bi) \int \mathrm{Sin}(\lambda_n X_L)$

23. 7.84 secs. 10.5 secs. 3.9 secs.

24. 18 msec. 12.1 msec. 9 msec.

25. 7.2 cm.

26. 162 sec. 2.2 cm. 233 w/m^2.

27. 53 secs. 79 secs.

Chapter 3.0

1. $\partial C/\partial t = D\left(\partial^2 C_1/\partial z^2 + a_0/2\, \partial^2 C/\partial t \partial z\right)$

2. $u = (1 - \exp(-\tau/2)/(I_0(\tau/2) - 1)I_0(1/2(\tau^2 - X^2)^{1/2} + \exp(-\tau/2)$
   $u = (1 - \exp(-\tau/2)[\tau/\mathrm{sqrt}(\tau^2 - X^2 - Y^2 - Z^2)]\, I_1\,(\mathrm{sqrt}(\tau^2 - X^2 - Y^2 - Z^2)/2)/\, I_1(\tau/2) + \exp(-\tau/2)$

3. $\underline{u} = J_0(ix(s(s+1))^{1/2})/(sJ_0(ix_R\,(s(s+1))^{1/2}))$

4. $C^{ss} = C_0\, Pe_m\, \mathrm{Cosh}(k^{*1/2}X)/(\mathrm{Cosh}(k^{*1/2}X_a) - \mathrm{Cosh}(k^{*1/2}X_a))$
   $C^t = \sum_0^\infty C_n\exp(-(1 + k^*)\tau/2)\exp(-\mathrm{sqrt}((\,|1 - k^*|\,)^2/4 - \lambda_n^2)\tau)\mathrm{Cos}(\lambda_n X)$
   where $\lambda_n = \tan^{-1}(Pe_m/X_a) + 2n\pi,\ n = 0,1,2....$

5. for $\tau > X$, $\quad u = [\tau/(\tau^2 - X^2)^{1/2}]\,[I_1\,(1/2\,\mathrm{sqrt}(\tau^2 - X^2)/\,I_1\,(\tau/2)]$
   for $X > \tau$, $\quad u = [\tau/(X^2 - \tau^2)^{1/2}]\,[J_1\,(1/2\,\mathrm{sqrt}(X^2 - \tau^2)/\,I_1\,(\tau/2)]$
   for $X = \tau$, $\quad u = \exp(-X/2)$

6. $u^* = \exp(-\tau/2)\, I_0\,(1/2\,\mathrm{sqrt}(\tau^2 - X^2))$
   $u^* = \exp(-X^2/4\tau)$

7. $u = (1 - \mathrm{erf}\,(x/(4Dt)^{1/2})$
   $u = I_0(1/2(\tau^2 - X^2)^{1/2}/I_0(\tau/2)$

8. $C^t = \sum_1^\infty A_n\,\exp\,(-(1 + k^*)/2\tau)\,\exp\,(-1/2 - \mathrm{sqrt}((1 - k^*)^2/4 - \lambda_n^2)\tau)\,J_0\,(\lambda_n X)$
   Where $A_n = C_0\int_0^{XR} J_0\,(\lambda_n X)dX\ /\int_0^{XR} J_0\,^2(\lambda_n X)\,dX$
   For values of $R < \mathrm{sqrt}(D\,\tau_r)(4.8096)/|1 - k^*|$
   $C^t = C_0\sum_1^\infty A_n\,\exp\,(-\tau(1 + k^*)/2)\,\mathrm{Cos}\tau(\mathrm{sqrt}(\lambda_n^2 - (1 - k^*|^2/4)\,J_0(\lambda_n X)$
   Where $\lambda_n = \mathrm{sqrt}(D\,\tau_r)/R(2.4048 + (n-1)\pi),\ $ where $n = 1,2.3....$

9. $C^t = C_0\sum_1^\infty A_n\,\exp\,(-\tau/2)\,\mathrm{Cos}\tau(\mathrm{sqrt}(\lambda_n^2 - 1/4)\,\mathrm{Sin}(\lambda_n\,Z)\exp(-X/Pe_m)$
   $A_n = 2(1 - (-1)^n)/n\pi$

10. $C = A_n\,\exp(Pe_m Z/2)\,\exp(-Z/2\,(Pe_m^2 - 4c^2)^{1/2})\exp(-c^2\tau)$
    $C = A_n\,\exp(Pe_m Z/2)\,\exp(-Z/2\,(Pe_m^2 - 4c^2)^{1/2})\exp(-\tau/2\tau)\exp(-\tau/2(1 - 4\lambda_n^2)^{1/2})$

11. $h(t) = H\,(\,(1 + (Ct/H^2))^{1/2} -1)$, H is the initial height of the gas column and
    $C = cD_{AB}\,(x_{A1} - x_{A2})/(\rho^A/M_A)\,(x_B)_{ln}$

12. $u = (\,1 + C^*\,(J^* - 1))\,I_0(1/2\,(\,\tau^2 - X^2)^{1/2})/I_0(\tau/2)$
    where $C^* = C_s/(C_{sio} - C_s);\ J^* = J_w\,\tau_r\tau/\,\delta$

13. $L_{int} = (D_{12}\,\tau_{r1})^{1/2}\,(39.4784 + t^2/\tau_{r1}^2)^{1/2} + (D_{21}\,\tau_{r2})^{1/2}(39.4784 + t^2/\tau_{r2}^2)^{1/2}$

Chapter 4.0

4. $t_{efflux} = (\rho_{eff}/\rho_1) (2H/g)^{1/2}(A_{tube}/A_{orifice})$, $\rho_{eff} = (\rho_1 h_1 + \rho_2 h_2 + ...)/(H)$

6. $\delta^c < \pi (\gamma^c \tau_{mom}{}^c)^{1/2}$

7. $u^{ss} = Mom/2 (X_a^2 - X^2)$

8. $u^{ss} = P^*/4 X_R^2 (\kappa^2 -1) + \ln(r/\kappa R) (\kappa^2 - 1)/4\ln(\kappa)$
    for small R, $\lambda_n = (\gamma \tau_{mom}/\kappa^2 R^2)^{1/2} (2.4048 + (n-1)\pi)$
    subcritical oscillations when $R < 4.8096(\gamma\tau_{mom}/\kappa^2)^{1/2}$
    $f = 16/[Re (\kappa^4 - 1)/(\kappa^2 - 1) + (\kappa^2 -1)/\ln(\kappa)]$

10. $f = 16/Re (m^2 -1)/(2m\ln(m) - m^3 + 1)$

11. $T = \pi\mu\omega R^4/(2+ \operatorname{Sin}\theta)$; $P = \pi\mu\omega R^4/(2l\operatorname{Sin}\theta)$

12. $f = 4/Re$

13. $V = V^{ss} + V^t$; $V^t = \sum_1^\infty A_n \exp (-\tau/2)\exp(-\tau(1/4 - \lambda_n^2)^{1/2}) J_1(\lambda_n X)$
    where $\lambda_n = sqrt(\gamma \tau_{mom})/R(3.8317 + (n-1)\pi)$, where n = 1,2,3....
    $V = u/X$; $u^{ss} = u_i^{ss} + (X - X_i)(u_0^{ss}X_0 - u_iX_i)/[2(X_0^2 - X_i^2)]$

14. $L = 2.26816\ 10^5$ km. $w_p = 1.1$ Mw (work done by pump from friction losses only)
    12" pipe; Re: = 145,915. optimal number of pipes = 1.92. Cost of pipe = \$113.4 million. Utility cost = \$523.1 million/per year.

15. $\delta_T/\delta_h = (39.4784 + t^2/\tau_r^2)^{1/2}/(39.4784 + t^2/\tau_{mom}^2)^{1/2}$

16. $\theta = Z/\sqrt{3} + Y/\sqrt{3} + X/\sqrt{3} + \tau(1 - \beta)^{1/2}$

17. $t_{efflux} = \pi/5A_{orf}(2g)^{1/2}D^{5/2}$

18. $f = 4/Re$

19. $\tau_{mom} \partial^2 v_z/\partial t^2 + \partial v_z/\partial t + \gamma/\kappa v_z = 0$
    when, $Pb > \frac{1}{2}$, $u = (2gH)^{1/2} \exp(-\tau/2)\operatorname{Cos}(\tau/2 \ sqrt(4Pb^2 - 1)$
    $Pb = (\gamma\tau_{mom} /\kappa)$

Chapter 5.0

6. $\partial u/\partial \tau + \partial^2 u/\partial \tau^2 = \partial^2 u/\partial X^2$
    $\tau = 0$, u = 1; $\tau = \infty$, u = 0; X = 0, $-\partial u/\partial X = Pe_m u$; $Pe_m = k_{la}/v_m$; $v_m = (D_{AB}/\tau_{mr})^{1/2}$
    parabolic; $\partial u/\partial \tau = \partial^2 u/\partial X^2$. combine with solutions of finite slag, cylinder or sphere

RETURN TO: PHYSICS LIBRARY

351 LeConte Hall   510-642-3122

LOAN PERIOD 1 **1-MONTH**	2	3
4	5	6

ALL BOOKS MAY BE RECALLED AFTER 7 DAYS.

Renewable by telephone.

DUE AS STAMPED BELOW.

in PHYSICS LIBRARY until **MAR 1 2 2006**	
APR 2 2 2006	